Lecture Notes in Electrical Engineering

Volume 516

The book series *Lecture Notes in Electrical Engineering* (LNEE) publishes the latest developments in Electrical Engineering - quickly, informally and in high quality. While original research reported in proceedings and monographs has traditionally formed the core of LNEE, we also encourage authors to submit books devoted to supporting student education and professional training in the various fields and applications areas of electrical engineering. The series cover classical and emerging topics concerning:

- Communication Engineering, Information Theory and Networks
- Electronics Engineering and Microelectronics
- Signal, Image and Speech Processing
- Wireless and Mobile Communication
- Circuits and Systems
- Energy Systems, Power Electronics and Electrical Machines
- Electro-optical Engineering
- Instrumentation Engineering
- Avionics Engineering
- Control Systems
- Internet-of-Things and Cybersecurity
- Biomedical Devices, MEMS and NEMS

For general information about this book series, comments or suggestions, please contact leontina.dicecco@springer.com.

To submit a proposal or request further information, please contact the Publishing Editor in your country:

China
Jasmine Dou, Associate Editor (jasmine.dou@springer.com)

India
Swati Meherishi, Executive Editor (swati.meherishi@springer.com)
Aninda Bose, Senior Editor (aninda.bose@springer.com)

Japan
Takeyuki Yonezawa, Editorial Director (takeyuki.yonezawa@springer.com)

South Korea
Smith (Ahram) Chae, Editor (smith.chae@springer.com)

Southeast Asia
Ramesh Nath Premnath, Editor (ramesh.premnath@springer.com)

USA, Canada:
Michael Luby, Senior Editor (michael.luby@springer.com)

All other Countries:
Leontina Di Cecco, Senior Editor (leontina.dicecco@springer.com)
Christoph Baumann, Executive Editor (christoph.baumann@springer.com)

** Indexing: The books of this series are submitted to ISI Proceedings, EI-Compendex, SCOPUS, MetaPress, Web of Science and Springerlink **

More information about this series at http://www.springer.com/series/7818

Qilian Liang · Xin Liu ·
Zhenyu Na · Wei Wang ·
Jiasong Mu · Baoju Zhang
Editors

Communications, Signal Processing, and Systems

Proceedings of the 2018 CSPS Volume II:
Signal Processing

Volume 2

 Springer

Editors
Qilian Liang
Department of Electrical Engineering
University of Texas at Arlington
Arlington, TX, USA

Zhenyu Na
School of Information Science
and Technology
Dalian Maritime University
Dalian, China

Jiasong Mu
College of Electronic
and Communication Engineering
Tianjin Normal University
Tianjin, China

Xin Liu
School of Information
and Communication Engineering
Dalian University of Technology
Dalian, China

Wei Wang
College of Electronic
and Communication Engineering
Tianjin Normal University
Tianjin, China

Baoju Zhang
College of Electronic
and Communication Engineering
Tianjin Normal University
Tianjin, China

ISSN 1876-1100   ISSN 1876-1119   (electronic)
Lecture Notes in Electrical Engineering
ISBN 978-981-13-6503-4   ISBN 978-981-13-6504-1   (eBook)
https://doi.org/10.1007/978-981-13-6504-1

Library of Congress Control Number: 2019930976

This Springer imprint is published by the registered company Springer Nature Singapore Pte Ltd.
The registered company address is: 152 Beach Road, #21-01/04 Gateway East, Singapore 189721, Singapore

Contents

Part IV
Feature Selection

Dual-Feature Spectrum Sensing Exploiting Eigenvalue and Eigenvector of the Sampled Covariance Matrix

Yanping Chen[1] and Yulong Gao[2(✉)]

[1] Harbin University of Commerce, Harbin 150028, Heilongjiang, China
yanping1009@163.com
[2] Harbin Institute of Technology, Harbin 150080, Heilongjiang, China
ylgao@hit.edu.cn

Abstract. The signal can be charactered by both eigenvalues and eigenvectors of covariance matrix. However, the existing detection methods only exploit the eigenvalue or eigenvector. In this paper, we utilize both eigenvalues and eigenvectors of the sampled covariance matrix to perform spectrum sensing for improving the detection performance. The features of eigenvalues and eigenvectors are considered integratedly, and the relationship between the false-alarm probability and the decision threshold is offered. To testify this method, some simulations are carried out. The results demonstrate that the method shows some advantages in the detection performance over the conventional method only adapting eigenvalues or eigenvectors.

Keywords: Dual-feature · Spectrum sensing · Cognitive radio · Eigenvalue and eigenvector

1 Introduction

At present, wireless spectrum resources are assigned statically, and the given spectrum bands are authorized to some communication systems. Simultaneously, other systems cannot occupy these spectrum bands. For the limited wireless resources, the static style of working results in a precipitous decline in available spectrum resources. However, some of these licensed frequency bands are underutilized [1–3]. To solve the unbalanced utilization of spectrum resources, cognitive radio is presented. Many institutions and scholars put more attentions on key issues of cognitive radio [4–7], such as spectrum sensing, spectrum sharing.

Nowadays, conventional spectrum sensing methods are composed of matched filter detection, energy detection, likelihood ratio test detection, and cyclostationary detection [8–11]. When the signal is corrupted by the Gaussian white

© Springer Nature Singapore Pte Ltd. 2020
Q. Liang et al. (eds.), *Communications, Signal Processing,*
and Systems, Lecture Notes in Electrical Engineering 516,
https://doi.org/10.1007/978-981-13-6504-1_103

noise, matched filter detection has optimal detection performance. Unfortunately, some prior information about licensed users must be known. More difficultly, precise synchronization between licensed users and secondary users must also be achieved. It is mostly impossible for secondary users to acquire this prior knowledge. Energy detection is most commonly used in practice due to its simplicity and low computational complexity. No additional prior information about licensed users is required for energy detection. However, energy detection suffers from noise uncertainty and SNR wall, which degrade the detection probability and the false-alarm probability severely. Likelihood ratio test-based detection method has optimal detection performance under the Neyman–Pearson criterion. This method detects the signal by virtue of the difference of probability density function between the licensed user and noise. Obviously, the algorithm requires the corresponding prior knowledge. For cyclostationary detection method, the inherent features of the licensed user, which arise from modulation style, signal rate or other parameters, are exploited. This method can overcome the effect of fading and shadow and has a better performance than other detection methods in the low SNR region. But, the computational complexity is unaffordable for the practical application in most cases.

From the previous analysis, we can see that the above-mentioned methods have advantages and disadvantages. So the detection method based on random matrix was first introduced by Cardoso to deal with these problems. This method can cope with the noise uncertainty of energy detection and require no prior information about the licensed user. So this method attracts more attentions. In [12], the ratio of the maximum eigenvalue and the minimum eigenvalue is adopted as the test statistic to perform spectrum sensing, and the corresponding threshold is calculated according to MP law of random matrix. But the requirement of too many samples is unsatisfactory. Aiming to solve this problem, the authors in [13] exploit the Tracy–Widom distribution of maximum eigenvalue to modify the threshold and propose the maximum–minimum eigenvalue (MME) method. In light of these work, many modifications were made to improve the sensing performance.

In fact, it is well-known that eigenvectors also contain related information about the presence of signal. Compared to the method exploiting the eigenvalue, only a few works about the utilization of eigenvector are reported. In terms of principal component analysis, eigenvector-based feature template matching (FTM) method was proposed [14]. This method utilizes the eigenvector of sampled covariance matrix as the test statistic to carry out spectrum sensing. Additionally, the literature [15] combines kernel function of machine learning with FTM to propose the kernel feature template matching method. It is pointed that feature template matching (FTM) method is a special case of kernel feature template matching method. The authors of the literature also derived the false-alarm probability and decision threshold and analyzed the factors of affecting the detection performance.

Stimulated by the eigenvalue-based method and FTM algorithm, we combine the eigenvalue and eigenvector to carry out spectrum sensing.

In the proposed method, the product of main eigenvector and the ratio of maximum and minimum eigenvalue is utilized as the test statistic. In terms of the concept of random matrix, the false-alarm probability is derived by approximating some results. Finally, the decision threshold is calculated.

2   System Model and Algorithm Description

We assume that there are K sensing nodes. For the ith node, the binary hypothesis test can be expressed as

$$y_i(n) = \begin{cases} \omega_i(n), & H_0 \\ h_i(n)x_i(n) + \omega_i(n), & H_1 \end{cases} \tag{1}$$

where $\omega_i(n)$, $x_i(n)$, and $h_i(n)$ denote the additive noise, the licensed user signal, and channel gain, respectively. Two different segments signals are exploited to construct the received signal matrix $\mathbf{Y}_1$ and $\mathbf{Y}_2$

$$\mathbf{Y}_1 = \begin{bmatrix} y_1(1)\, y_1(2) \dots y_1(N) \\ y_2(1)\, y_2(2) \dots y_2(N) \\ \vdots \quad \dots \quad \vdots \\ y_K(1)\, y_K(2) \dots y_K(N) \end{bmatrix} \tag{2}$$

$$\mathbf{Y}_2 = \begin{bmatrix} y_1(N+1)\, y_1(N+2) \dots y_1(2N) \\ y_2(N+1)\, y_2(N+2) \dots y_2(2N) \\ \vdots \quad \dots \quad \vdots \\ y_K(N+1)\, y_K(N+2) \dots y_K(2N) \end{bmatrix} \tag{3}$$

where N is the number of signal samples. Their corresponding covariance matrix $\mathbf{R}_1$ and $\mathbf{R}_2$ are expressed as

$$\mathbf{R_1} = \frac{1}{N}\mathbf{Y}_1\mathbf{Y}_1{}^{\mathrm{T}}, \quad \mathbf{R_2} = \frac{1}{N}\mathbf{Y}_2\mathbf{Y}_2{}^{\mathrm{T}} \tag{4}$$

If only the noise exists, the maximum eigenvalue $\lambda_{\max}$ and minimum eigenvalue $\lambda_{\min}$ of the sampled covariance matrix have $\lambda_{\max} = \lambda_{\min} = \sigma^2$. But, when the signal is present, the maximum eigenvalue and minimum eigenvalue satisfy $\lambda_{\max} > \lambda_{\min} = \sigma^2$. Therefore, we have

$$\left(\frac{\lambda_{\max}}{\lambda_{\min}}\right)_{H_0} < \left(\frac{\lambda_{\max}}{\lambda_{\min}}\right)_{H_1} \tag{5}$$

Analogously, for the main eigenvector of H_0 and H_1 case, the following relation holds

$$(|\langle \mathbf{a_1}, \mathbf{b_1}\rangle|)_{H_0} < (|\langle \mathbf{a_1}, \mathbf{b_1}\rangle|)_{H_1} \tag{6}$$

where $\mathbf{a}_1$ and $\mathbf{b}_1$ are the corresponding main eigenvectors of the sampled covariance matrix $\mathbf{R}_1$ and $\mathbf{R}_2$. Jointly considering (8) and (9), we can obtain

$$\left(\frac{\lambda_{\max}}{\lambda_{\min}}\right)_{H_0}(|\langle \mathbf{a_1}, \mathbf{b_1}\rangle|)_{H_0} < \left(\frac{\lambda_{\max}}{\lambda_{\min}}\right)_{H_1}(|\langle \mathbf{a_1}, \mathbf{b_1}\rangle|)_{H_1} \tag{7}$$

We can observe that the product of the ratio of the maximum eigenvalue and minimum eigenvalue and the correlation of main eigenvectors for the H_0 and H_1 case differ obviously. Compared to the method only employing eigenvalue or eigenvector, more obvious difference can be obtained for the product. Thus, we select the product as the test statistic of spectrum sensing. The proposed method is summarized as follows.

3   Solving the False-Alarm Probability and the Decision Threshold

From the definition of false-alarm probability, we have

$$P_{fa} = P(D_1|H_0) = P(T > \gamma|H_0) = P\left(\rho_1\rho_2\left|\langle \mathbf{a}_1, \mathbf{b}_1 \rangle\right| > \gamma|H_0\right) \tag{8}$$

When only the noise exists, the covariance matrix of the received signal is Wishart matrix. From the MP law, the maximum eigenvalue and minimum eigenvalue in the H_0 case are

$$\lambda_{\max} = \sigma^2\left(1 + \sqrt{c}\right)^2, \quad \lambda_{\min} = \sigma^2\left(1 - \sqrt{c}\right)^2 \tag{9}$$

where c is the ratio of the number of nodes and the number of samples. The ratio of the maximum eigenvalue and the minimum eigenvalue is

$$\eta = \frac{\lambda_{\max}}{\lambda_{\min}} = \frac{\left(1 + \sqrt{c}\right)^2}{\left(1 - \sqrt{c}\right)^2} \tag{10}$$

For the matrix $\mathbf{R}_1$ and $\mathbf{R}_2$, we can calculate the corresponding η_1 and η_2.

It is pointed by the literature [16] that the eigenvalue and eigenvector of Wishart matrix are independent to each other. Let $\gamma = \varepsilon\eta^2$. Coupled with (10), the false-alarm probability can be expressed in the form

$$P_{fa} = P\left(\rho_1\rho_2\left|\langle \mathbf{a}_1, \mathbf{b}_1 \rangle\right| > \gamma|H_0\right) \approx P\left(\left|\langle \mathbf{a}_1, \mathbf{b}_1 \rangle\right| > \varepsilon|H_0\right) \tag{11}$$

To calculate the false-alarm probability, we derive the probability density function of $\left|\langle \mathbf{a}_1, \mathbf{b}_1 \rangle\right|$. In terms of the concept of eigenvalue decomposition, we can get

$$\mathbf{R}_1 = \mathbf{A}\boldsymbol{\Lambda}_1\mathbf{A}^T, \quad \mathbf{R}_2 = \mathbf{B}\boldsymbol{\Lambda}_2\mathbf{B}^T \tag{12}$$

where $\boldsymbol{\Lambda}_1$ and $\boldsymbol{\Lambda}_2$ are the diagonal matrix containing eigenvalues of $\mathbf{R}_1$ and $\mathbf{R}_2$. The columun vectors of and correspond to eigenvectors. And the main eigenvectors $\mathbf{a}_1$ and $\mathbf{b}_1$ of the sampled covariance matrix $\mathbf{R}_1$ and $\mathbf{R}_2$ are the first column of matrix $\mathbf{A}$ and $\mathbf{B}$. Because $\mathbf{A}$ and $\mathbf{B}$ are unitary matrix, we have $f\left(\mathbf{A}^T\mathbf{B}\right) = f\left(\mathbf{B}\right)$. We can say that the elements of $\mathbf{A}^T\mathbf{B}$ and $\mathbf{B}$ in the same position follow the same distribution, i.e., $f\left(\langle \mathbf{a}_1, \mathbf{b}_1 \rangle\right) = f\left(b_{11}\right)$.

From the properties of unitary matrix, b_{11} obeys the beta distribution with parameters $\alpha = \frac{1}{2}$ and $\beta = \frac{N-1}{2}$

$$f\left(x\right) = \frac{\left(1 - x^2\right)^{(n-1)/2-1}}{B\left(\frac{1}{2}, \frac{N-1}{2}\right)} \tag{13}$$

Because the distribution are computationally complex, we approximate it with the Gaussian distribution for the large N [15]. According to the result derived in the literature [15], $f(x)$ can be replaced by $h(x)$

$$h(x) = \frac{\sqrt{N}}{\sqrt{2\pi}} e^{-\frac{N x^2}{2}} \tag{14}$$

Substituting (14) into (11) yields

$$P_{fa} \approx P(|\langle \mathbf{a}_1, \mathbf{b}_1 \rangle| > \varepsilon | H_0) = 2Q\left(\sqrt{N}\varepsilon\right) \tag{15}$$

When the false-alarm probability is preset, we get the parameter ε. Therefore, the corresponding threshold can be calculated as

$$\gamma = \varepsilon\eta^2 = \frac{(1 + \sqrt{c})^4}{(1 - \sqrt{c})^4} Q^{-1}\left(\frac{P_{fa}}{2}\right) \frac{1}{\sqrt{N}} \tag{16}$$

4 Numerical Simulation and Analysis

We first verify the rationality of the decision threshold. We offer the relationship among the test statistic of H_0 case, test statistic of H1 case, and the decision threshold in Fig. 1 when $K=10$ and $SNR = -10$ dB. It is demonstrated that the threshold varies dynamically with the number of samples. Additionally, we can also observe that the difference of the test statistic for H0 and H1 case is obvious for all the samples and the decision threshold is between the test statistics for

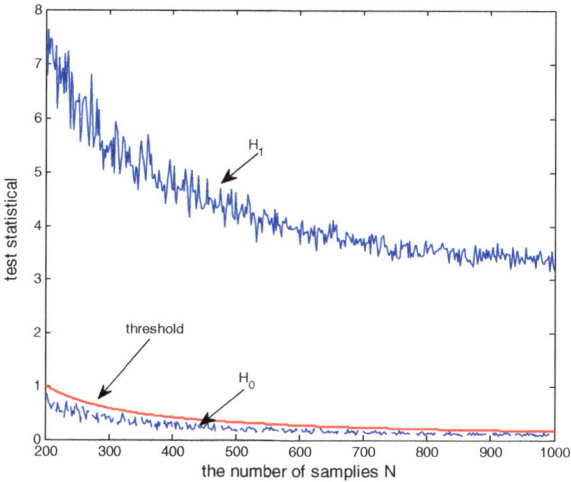

Fig. 1. Estimated pdf and theoretical result of the main eigenvector for H_0

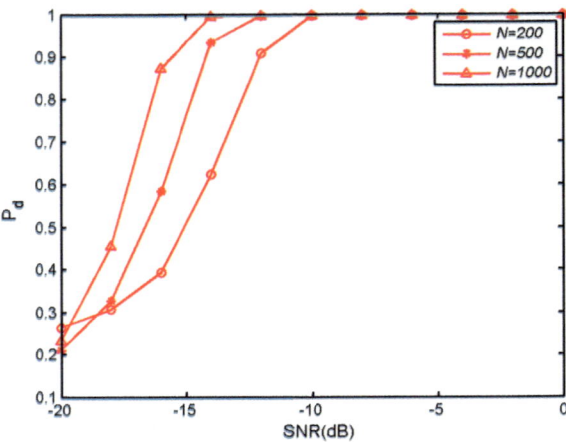

Fig. 2. Detection probability for the different samples and SNR

H_0 and H_1 case. All these results guarantee the proper threshold for obtaining satisfactory performance.

We now verify the detection performance of the proposed method. The simulated parameters are set as follows. $K = 10$, $N = 200$, 500, 1000. The number of Monte Carlo simulation is 2000. Figure 2 plots the detection probability for the different SNR. We can draw three conclusions. (1) The detection probability is improved with the increasing of samples for the low SNR region. For example, the detection probability is 0.5 and 1 for $N = 200$ and 1000 when $SNR = -15$ dB. (2) When SNR is more than -10 dB, the detection probability remains a constant 1 for the different samples. (3) when the samples remain unchanged, the detection probability increases proportionately with SNR.

To testify the superiority of the proposed method over other algorithms, we show the detection probability of FFM, MME, and the proposed method in Fig. 3. The simulated parameters are set as follows. $K = 10$, $N = 500$. The number of Monte Carlo simulation is 2000. It is noted that the samples are divided into two segments for FTM and the proposed method, and each segment has $N/2$ samples. We can observe that the proposed method has good detection performance over other two algorithms.

5   Conclusion

Inspired by the FTM and eigenvalue-based spectrum sensing, we introduced the spectrum sensing algorithm based on dual-feature consisting of eigenvalues and eigenvector. We defined the product of the ratio of the maximum eigenvalue and the minimum eigenvalue and main eigenvector as the test statistic. This test statistic considers the eigenvalue and eigenvector comprehensively. Based on the test statistic, we derived the false-alarm probability and offer the closed

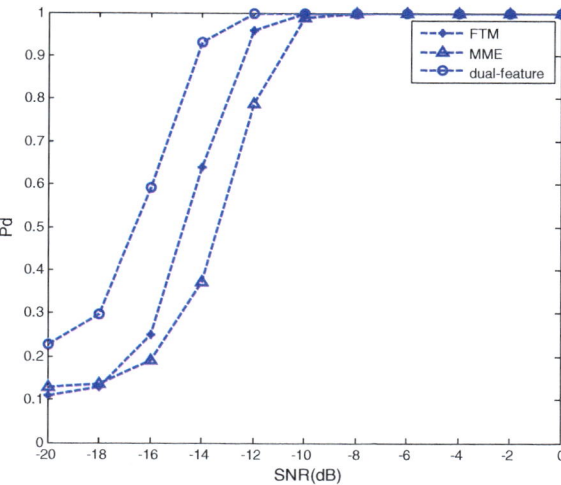

Fig. 3. Detection probability for the different samples and SNR

form of threshold under Neyman–Pearson criterion. In simulations, we testified the correctness of Gaussian approximation of probability density function and compared the proposed method with FTM and MME algorithms. The simulation results showed that the proposed method outperforms the FTM and MME algorithms with lower computational complexity.

Acknowledgments. This work is supported by National Natural Science Foundation of China (NSFC) (Grant No. 61671176).

References

1. Abdelmohsen A, Hamouda W. Advances on spectrum sensing for cognitive radio networks: theory and applications. IEEE Commun Surv Tutorials. 2017;19(2):1277–304.
2. Guo H, Jiang W, Luo W. Linear soft combination for cooperative spectrum sensing in cognitive radio networks. IEEE Commun Lett. 2017;21(7):1573–6.
3. Wang B, Liu KJR. Advances in cognitive radio networks: a survey. IEEE J Sel Top Sig Process. 2011;5(1):5–23.
4. Mitola J, Maguire GQ. Cognitive radio: making software radios more personal. IEEE Pers Commun. 1999;6(4):13–8.
5. Mchenry M, Livsics E, Nguyen T, Majumdar N. XG dynamic spectrum access field test results [Topics in radio communications]. IEEE Commun Mag. 2007;45(6):51–7.
6. Liu Chang, Li Ming, Jin Ming-Lu. Blind energy-based detection for spatial spectrum sensing. IEEE Wirel Commun Lett. 2015;4(1):91–8 Feb.
7. Cabric D, Brodersen RW. Physical layer design issues unique to cognitive radio systems. Proc IEEE Pimrc. 2005;2:759–63.

8. Cabric D, Tkachenko A, Brodersen R. Spectrum sensing measurements of pilot, energy, and collaborative detection. In: IEEE Conference on Military Communications; 2006. p. 1–7

9. Sonnenschein A, Fishman PM. Radiometric detection of spread-spectrum signals in noise of uncertain power. IEEE Trans Aerosp Electron Syst. 1992;28(3):654–60.

10. Yucek T, Arslan H. A survey of spectrum sensing algorithms for cognitive radio applications. IEEE Commun Surv Tutorials. 2009;11(1):116–30.

11. An T, Kim D, Song I, et al. Cooperative spectrum sensing based on generalized likelihood ratio test under impulsive noise circumstances. In: 2012 IEEE military communications conference; 2012. p. 1–6

12. Cardoso LS, Debbah M, Bianchi P, et al. Cooperative spectrum sensing using random matrix theory. In: International symposium on wireless pervasive computing; 2008. p. 334–8

13. Zeng Y, Liang YC. Maximum-minimum eigenvalue detection for cognitive radio. In: IEEE international symposium on personal, indoor and mobile radio communications; 2007. p. 1–5

14. Zhang P, Qiu R, Guo N. Demonstration of spectrum sensing with blindly learned features. IEEE Commun Lett. 2011;15(5):548–50.

15. Hou S, Qiu RC. Kernel feature template matching for spectrum sensing. IEEE Trans Veh Technol. 2014;63(5):2258–71.

16. Eldar YC, Chan AM. On the asymptotic performance of the decorrelator. IEEE Trans Inf Theor. 2003;49(9):2309–13.

Adaptive Scale Mean-Shift Tracking with Gradient Histogram

Changqing Xie, Wenjing Kang, and Gongliang Liu[(✉)]

School of Information and Electrical Engineering, Harbin Institute of Technology (Weihai), Weihai, China
15650131973@yeah.net, {kwjqq,liugl}@hit.edu.cn

Abstract. The mean-shift (MS) tracking is fast, is easy to implement, and performs well in many conditions especially for object with rotation and deformation. But the existing MS-like algorithms always have inferior performance for two reasons: the loss of pixel's neighborhood information and lack of template update and scale estimation. We present a new adaptive scale MS algorithm with gradient histogram to settle those problems. The gradient histogram is constructed by gradient features concatenated with color features which are quantized into the $16 \times 16 \times 16 \times 16$ bins. To deal with scale change, a scale robust algorithm is adopted which is called background ratio weighting (BRW) algorithm. In order to cope with appearance variation, when the Bhattacharyya coefficient is greater than a threshold the object template is updated and the threshold is set to avoid incorrect updates. The proposed tracker is compared with lots of tracking algorithms, and the experimental results show its effectiveness in both distance precision and overlap precision.

Keywords: Object tracking · Mean-shift · Scale estimation · Gradient

1 Introductions

Visual object tracking has always been a challenging work especially in the sequences with the deformation and rotation. The MS tracking has an outstanding performance and is easy to implement. It tracks by minimizing the Bhattacharyya distance between two probability density functions represented by a target and target candidate histograms. The histogram is a statistical feature that does not depend on the spatial structure within the search window. This makes it more robust than other algorithms. But it lacks the essential template update and pixel's neighborhood information leading to a worse accuracy.

The mean-shift algorithm is a nonparametric mode-seeking method for density functions proposed by Fukunaga and Hostetler [1]. Comaniciu et al. [2, 3] use it to track object. And Comaniciu et al. change the window size over multiple runs by a constant factor but produces little effect because the smaller windows usually have higher similarity. The image pyramids and an additional mean-shift algorithm for scale selection had been used after estimating the position to confirm the window size in Collins [4]. But its speed is lower than the conventional MS algorithm. A new histogram that exploits the object neighborhood has been proposed to help discriminate

Q. Liang et al. (eds.), *Communications, Signal Processing, and Systems*, Lecture Notes in Electrical Engineering 516,
https://doi.org/10.1007/978-981-13-6504-1_104

the target which is called background ratio weighting (BRW) in Vojir et al. [5]. This approach is faster than others and has a superior effect in sequences with scale change but performs poorly for grayscale sequences.

Gradient information is crucial for appearance representation since it contains pixel's neighborhood information and is insensitive to illumination variation but it always be ignored. Based on this observation, we present a novel adaptive scale MS algorithm with gradient histogram. The gradient information is calculated by Canny edge detector which was developed by Canny [6].

Moreover, the BRW algorithm has been used to improve the performance of videos with scale change. The template of target will be updated by liner interpolation only if conditions are met to avoid addition of incorrect information. The template update also can cope with appearance variation of target. The proposed tracker is compared with lots of algorithms, and the experiment results show that it is more robust and accurate.

2   Canny Edge Detector

The Canny edge detector is an edge detection operator that uses a multi-stage algorithm to detect a wide range of edges in images. To remove the noise, a Gaussian filter is applied to the image; the Gaussian filter kernel of size $(2k + 1) \times (2k + 1)$ is given by:

$$H_{ij} = \frac{1}{2\pi\sigma^2} \exp\left(-\frac{(i - (k+1))^2 + (j - (k+1))^2}{2\sigma^2} \right); \quad 1 \leq i, j \leq (2k+1) \tag{1}$$

Let I_o denote the original image and H is a classic Gaussian filter matrix; we get the image without noise $I = H * I_o$. Then we extract intensity gradient of the image with the Sobel operator proposed by Sobel [7]. The gradient information in horizontal and vertical directions is G_x and G_y. From this, the edge gradient can be determined by

$$G = \sqrt{G_x^2 + G_x^2} \tag{2}$$

where

$$G_x = \begin{bmatrix} -1 & 0 & 1 \\ -2 & 0 & 2 \\ -1 & 0 & 2 \end{bmatrix} * I \text{ and } G_y = \begin{bmatrix} -1 & -2 & -1 \\ 0 & 0 & 0 \\ 1 & 2 & 1 \end{bmatrix} * I. \tag{3}$$

The edge extracted from the gradient value is still quite blurred after processing the G with Gauss filter and Sobel operator. The non-maximum suppression and double-threshold joint should be applied to the processed image to improve effect of gradient. Then we can get the gradient image I_g to track the target.

CSK ━━━ ASMS ━━━ LOT ━━━ MS ━━━ OAB ━━━ OURS

Fig. 1. Result of six trackers in six sequences. (The sequence names from top to bottom are dog1, freeman1, freeman3, jogging, mountainbike, and singer1)

3 The Tracking Algorithm

Different from the conventional MS tracking algorithm, we append the I_g to original image to get I_e. After the combination of the images, we extract the histogram $\hat{q}$ from I_e. To cope with the problem which caused by the size of target changes, we use the BRW-MS instead of conventional MS. We can get $\hat{q}$ from:

$$\hat{q}_u = C \sum_{i=1}^{N} k\left(\frac{\left(x_i^{*1}\right)^2}{a^2} + \frac{\left(x_i^{*2}\right)^2}{b^2}\right) \delta\left[b\left(x_i^*\right) - u\right] \qquad (4)$$

and an ellipsoidal region is used to represent target $\frac{\left(x_i^{*1}\right)^2}{a^2} + \frac{\left(x_i^{*2}\right)^2}{b^2} < 1$ in current frame. The target candidate is given by

$$\hat{p}_u(y, h) = C_h \sum_{i=1}^{N} k\left(\frac{\left(y^1 - x_i^1\right)^2}{a^2 h^2} + \frac{\left(y^2 - x_i^2\right)^2}{b^2 h^2}\right) \delta[b(x_i - u)] \qquad (5)$$

where the h is the scale factor. The location of the target is obtained by

$$\hat{y}_1^1 = \frac{1}{a^2} m_k^1(\hat{y}_0, h_0) + \hat{y}_0^1, \ \hat{y}_1^2 = \frac{1}{b^2} m_k^2(\hat{y}_0, h_0) + \hat{y}_0^2 \tag{6}$$

$$h_1 = \left[1 - \frac{A}{K}\right] h_0 + \frac{1}{h_0} \frac{B}{K} \tag{7}$$

where A is

$$\sum_{i=1}^{N} w_i k \left(\frac{(y_0^1 - x_i^1)^2}{a^2 h_0^2} + \frac{(y_0^2 - x_i^2)^2}{b^2 h_0^2} \right), \tag{8}$$

B is

$$\sum_{i=1}^{N} w_i \left(\frac{(y_0^1 - x_i^1)^2}{a^2} + \frac{(y_0^2 - x_i^2)^2}{b^2} \right) g \left(\frac{(y_0^1 - x_i^1)^2}{a^2 h_0^2} + \frac{(y_0^2 - x_i^2)^2}{b^2 h_0^2} \right). \tag{9}$$

And $g(x) = -k'(x)$ is the derivative of $k(x)$. w_i can be obtained by

$$w_i = \max(0, W) \tag{10}$$

where W is

$$W = \left(\sum_{u=1}^{m} \frac{1}{\hat{\rho}[\hat{\boldsymbol{p}}(\hat{\boldsymbol{y}}_0, h_0), \hat{\boldsymbol{q}}]} \sqrt{\frac{\hat{q}_u}{\hat{p}_u(\hat{\boldsymbol{y}}_0, h_0)}} - \frac{1}{\rho\left[\hat{\boldsymbol{p}}(\hat{\boldsymbol{y}}_0, h_0), 3\widehat{\boldsymbol{bg}}\right]} \sqrt{\frac{\widehat{bg_u}}{\hat{p}_u(\hat{y}, h_0)}} \right) \delta[b(\boldsymbol{x}_i) - u] \tag{11}$$

The $\widehat{\boldsymbol{bg}}$ is the histogram of background computed over the neighborhood of the target in the first frame. Let us denote

$$K = \sum_{i=1}^{N} w_i g \left(\frac{(y_0^1 - x_i^1)^2}{a^2 h_0^2} + \frac{(y_0^2 - x_i^2)^2}{b^2 h_0^2} \right) \tag{12}$$

and

$$\mathbf{m_k}(\hat{\boldsymbol{y}}, h_0) = \frac{\sum_{i=1}^{N} x_i w_i g \left(\frac{(y_0^1 - x_i^1)^2}{a^2 h_0^2} + \frac{(y_0^2 - x_i^2)^2}{b^2 h_0^2} \right)}{K} - \hat{\boldsymbol{y}}_0 \tag{13}$$

When the location was determined, the $\hat{q}$ is updated by

$$\hat{\mathbf{q}}_{\mathbf{new}} = \begin{cases} \hat{\boldsymbol{q}}_{\mathbf{old}} & \text{if } \rho[\hat{\mathbf{p}}(\mathbf{y}), \hat{\mathbf{q}}_{\mathbf{old}}] \leq \alpha \\ (1 - \lambda)\hat{\boldsymbol{q}}_{\mathbf{old}} + \lambda \hat{\boldsymbol{p}}_u(\boldsymbol{y}) & \text{if } \rho[\hat{\mathbf{p}}(\mathbf{y}), \hat{\mathbf{q}}_{\mathbf{old}}] > \alpha \end{cases} \tag{14}$$

4   Experiment

Experiments are conducted on sequences from Object Tracking Benchmark2013 (OTB2013) dataset [8]. The sequences in OTB2013 not only suffer the deformation but also have other change such as fast motion, background clutter, motion blur, and so on. So, we selected six sequences from OTB2013 dataset to show the results. We compared the proposed algorithm with conventional and state-of-the-art algorithms which are available as source code. They are conventional mean-shit algorithm [2], ASMS [5], OAB [9], LOT [10], and CSK [11]. The parameters for those algorithms are set default.

Figure 1 shows the result of six trackers in sequences. The score of distance precision (DP) rate, overlap success (OS) rate, and center location error (CLE) can be obtained from Table 1.

In general, the gradient histogram improves performance of MS algorithm for the gradient histogram. The data in Table 1 show that proposed algorithm has a higher score than others in DP, OS, and CLE which means our tracker is better than others. The sequence dog1 suffers the scale change, and results show standard MS failed to track target because it only uses the gray levels to calculate histogram and it lacks template update. But our tracker can deal well with this condition for the adopting of gradient histogram and RBW algorithm. The proposed tracker has a better performance than other trackers in sequence singer1 which suffers illumination variation since the addition of gradient histogram which does not dependent on the current pixel value but the difference of adjacent pixel. What is more we find that our tracker has a great improvement in gray image compared to the colorful image than the ASMS algorithm since the ASMS algorithm only can acquire information in the gray channel which is scanty.

Table 1. Scores of six trackers in six sequences

Sequence name		Dog1	Freeman1	Freeman3	Jogging1	Mountainbike	Singer1
Our tracker	OS rate (%)	1	0.975	0.998	0.974	0.996	0.806
	DP rate (%)	1	0.837	0.304	0.958	0.965	0.299
	CLE (pixel)	3.18	6.23	6.57	8.1	8.36	33.39
MS	OS rate (%)	0.224	0.423	0.598	0.15	0.75	0.14
	DP rate (%)	0.054	0.012	0.02	0.078	0.706	0.259
	CLE (pixel)	48.46	35.02	30.6	83.93	22.33	33.95
ASMS	OS rate (%)	0.842	0.696	0.696	0.228	0.132	0.792
	DP rate (%)	0.841	0.387	0.304	0.218	0.132	0.299
	CLE (pixel)	9.98	19.93	37.5	93.69	238.2	33.91
CSK	OS rate (%)	0.926	0.371	0.813	0.231	0.996	0.521
	DP rate (%)	0.59	0.224	0.165	0.208	0.925	0.296
	CLE (pixel)	8.91	182.64	33.94	112.91	8.49	17.45
LOT	OS rate (%)	0.507	0.11	0.593	0.192	0.715	0.165
	DP rate (%)	0.506	0.08	0.085	0.023	0.671	0.208
	CLE (pixel)	24.16	137.09	42.09	89.68	25.52	145.06
OAB	OS rate (%)	0.999	0.552	0.741	0.893	0.724	0.786
	DP rate (%)	0.583	0.236	0.122	0.827	0.684	0.248
	CLE (pixel)	7.36	18.09	51.7	13.9	13.85	15.46

(The best results are in italics)

5   Conclusion

In this paper, an adaptive scale mean-shift algorithm with gradient histogram has been proposed to improve the tracking performance of MS-like algorithms. The gradient histogram is constructed by color histogram and gradient feature calculated by Canny edge detector. To deal with the scale change, the RBW algorithm is adopted. Template update is used to cope with appearance variation when the Bhattacharyya coefficient between the current frame and the template is greater than the threshold. The setting of the threshold makes tracker more robust for incorrect information. The proposed tracker is compared with a lot of algorithms in OTB2013 dataset. The experiment results show the tracker's effectiveness in deformation, rotation, scale change, and illumination variation. Moreover, our tracker has a better preference in gray sequences than conventional MS-like algorithms.

Acknowledgments. This work was supported by the National Natural Science Foundation of China (Grant No. 61501139) and the Natural Scientific Research Innovation Foundation in Harbin Institute of Technology (HIT.NSRIF.2013136).

References

1. Fukunaga K, Hostetler L. The estimation of the gradient of a density function, with applications in pattern recognition. IEEE Trans Inf Theor. 1975;21(1):32–40.
2. Comaniciu D, Ramesh V, Meer P. Real-time tracking of non-rigid objects using mean shift. In: Proceedings IEEE conference on computer vision and pattern recognition, 2000. IEEE; 2000, vol. 2. p. 142–9.
3. Comaniciu D, Ramesh V, Meer P. Kernel-based object tracking. IEEE Trans Pattern Anal Mach Intell. 2003;25(5):564–77.
4. Collins RT. Mean-shift blob tracking through scale space. In: Proceedings IEEE computer society conference on computer vision and pattern recognition, 2003. IEEE; 2003, vol. 2. p. II-234.
5. Vojir T, Noskova J, Matas J. Robust scale-adaptive mean-shift for tracking. Pattern Recogn Lett. 2014;49:250–8.
6. Canny J. A computational approach to edge detection. In: Readings in computer vision; 1987. p. 184–203.
7. Sobel I. An isotropic 3×3 image gradient operator. In: Machine vision for three-dimensional scenes; 1990. p. 376–9.
8. Wu Y, Lim J, Yang MH. Online object tracking: A benchmark. In: 2013 IEEE conference on computer vision and pattern recognition (CVPR). IEEE; 2013. p. 2411–8.
9. Grabner H, Grabner M, Bischof H. Real-time tracking via on-line boosting. Bmvc; 2006, vol. 1, no. 5. p. 6.
10. Oron S, Bar-Hillel A, Levi D, et al. Locally orderless tracking. Int J Comput Vision. 2015;111(2):213–28.
11. Henriques JF, Caseiro R, Martins P, et al. Exploiting the circulant structure of tracking-by-detection with kernels. In: European conference on computer vision. Springer, Berlin, Heidelberg; 2012. p. 702–15.

Improved Performance of CDL Algorithm Using DDELM-AE and AK-SVD

Xiulan Yu$^{(\boxtimes)}$, Junwei Mao, Chenquan Gan, and Zufan Zhang

Chongqing Key Labs of Mobile Communications Technology, Chongqing University of Posts and Telecommunications, Chongqing 400065, China
yuxl@cqupt.edu.cn, mjwlmt@163.com

Abstract. Due to the poor robustness and high complexity of the concentrated dictionary learning (CDL) algorithm, this paper addresses these issues using denoising deep extreme learning machine based on autoencoder (DDELM-AE) and approximate k singular value decomposition (AK-SVD). Different from the CDL algorithm, on input, DDELM-AE is added for enhancing denoising ability and AK-SVD replaces K-SVD for improving running speed. Additionally, experimental results show that the improved algorithm is more efficient than the original CDL algorithm in terms of running time, denoising ability, and stability.

Keywords: Signal compression · CDL · Deep learning · DDELM-AE · AK-SVD

1 Introduction

With the advent of the era of big data, increasing the storage space or transmission path will bring a great burden on communication system. To meet these challenges, signal compression algorithm plays an important role in the field of signal processing.

Signal sparsity has gained considerable interests, especially since Aharon et al. [1,2] introduced the K-SVD algorithm (e.g., see Refs. [3–6]). In [3], an online dictionary learning algorithm was proposed and widely used in hyperspectral images. In [4], the inner product of signals in sparse domain was preserved by linear mapping so that the compression of original signal could be completed. The important conclusion drawn from [5] is that the features of low-dimensional space preserve the main features of original signal. The emergence of compressed sensing (CS) made that the original signal can be recovered from the low-frequency sampled signal [6]. However, most algorithms based on CS

Junwei Mao, M.S., Chongqing University of Posts and Telecommunications. His current main research interest includes: signal processing and deep learning.

© Springer Nature Singapore Pte Ltd. 2020
Q. Liang et al. (eds.), *Communications, Signal Processing,
and Systems*, Lecture Notes in Electrical Engineering 516,
https://doi.org/10.1007/978-981-13-6504-1_105

must obey the restricted isometry property (RIP) constraint, which poses a huge challenge for signal compression.

In 2016, Silong et al. [7] proposed a nonlinear dimension reduction algorithm based on K-SVD, namely the CDL algorithm, which preserves the geometric characteristics of the original signal and overcomes the limitations of RIP. However, when the data size is too large, the SVD in the iterative process will bring a great burden on the running speed of the algorithm itself. Thus, in this paper, the DDELM-AE neural network [8] and AK-SVD algorithm [9] are considered to reduce computational complexity and improve denoising ability of the CDL algorithm.

First, the original signal enters the DDELM-AE neural network, which can remove useless information and get a better denoising representation. Second, put the denoising representation as an input of the AK-SVD algorithm, then the denoising dictionary can be obtained. Finally, simulation results reveal that the improved algorithm not only improves the running speed, but also enhances the denoising ability and stability of the system.

The organization of the rest of this paper is as follows: Sect. 2 presents the improved CDL algorithm in detail. In Sect. 3, experimental results are explained. Finally, Sect. 4 summarizes our work.

2   Improved CDL Algorithm

The main idea of CDL algorithm is based on the conclusion drawn in [10, 11] that for original signal X and compressed signal Y, the high-dimensional dictionary D and the low-dimensional dictionary P share a sparse representation coefficient α. Then, α becomes a bridge between X and Y:

$$\begin{cases} X = D\alpha + \varepsilon_1, \\ Y = P\alpha + \varepsilon_2, \end{cases} \tag{1}$$

where $\varepsilon_1 \in R^m, \varepsilon_2 \in R^d$ are noise. Therefore, the derivation of P becomes an important goal of this paper.

2.1   Improving the Denoising Ability of the CDL Algorithm.

Similar to [12], the number of each output layer in DDELM-AE network (see Fig. 1) equals to its input layer number. In this way, a higher level feature representation X' of the signal can be obtained. Furthermore, take the X' as AK-SVD input to generate a denoising dictionary, then the performance of AK-SVD algorithm can be improved.

In particular, when the hidden layer number of adjacent two nodes in the model is equal, orthogonal procrustes can be applied to solve the weight. The closed form solution can be obtained by finding the nearest orthogonal matrix

$$M = H^T X, \tag{2}$$

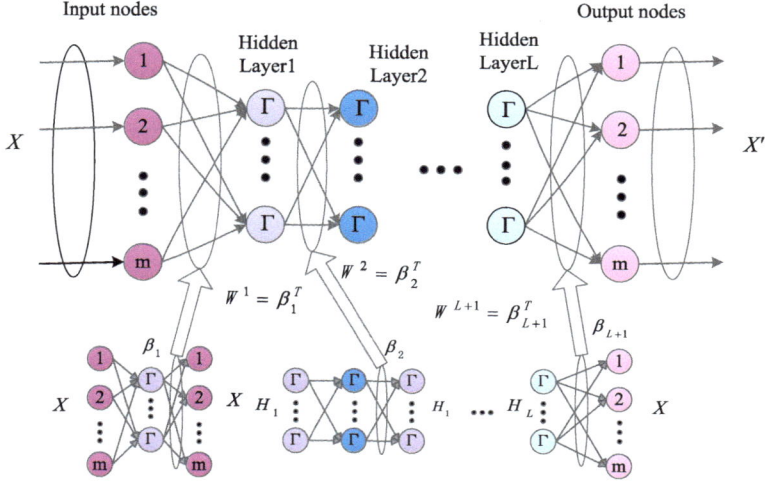

Fig. 1. A denoising deep ELM-AE model for the samples of training set X

where H is hidden layer output matrix. To find the orthogonal weight matrix $\hat{\beta}$, the SVD of M:

$$M = USV^T \tag{3}$$

can be calculated. Thus, $\hat{\beta}$ can be computed by

$$\hat{\beta} = UV^T. \tag{4}$$

2.2 Updating Dictionary with Denoising Deep ELM-AE

Put the output signal X' of the DDELM-AE as an input of AK-SVD algorithm and note that the sparse representation:

$$X' = DC, \tag{5}$$

and

$$D = X'C^T, \tag{6}$$

where $D = \{d_1, d_2, \ldots, d_k\}$ and $C = \{c_1, c_2, \ldots, c_n\}$ are sparse coefficient matrices. Additionally, in the K-SVD algorithm, the objective function:

$$\|X' - DC\|_F^2 = \left\| X' - \sum_{i=1}^{n} d_{(:,i)} c_{(i,:)} \right\|_F^2 \tag{7}$$

$$= \left\| R^{(l)} - d_{(:,l)} c_{(l,:)} \right\|_F^2$$

does not need an exact solution. Thus, the SVD process can be replaced by the formulas (8)–(10) in K-SVD algorithm [9], which greatly reduces the computational complexity and improves the running speed of the CDL algorithm.

$$d_l \leftarrow X'c_i^T - \left(\sum_{i \neq l} d_l c_i\right) c_i^T, \tag{8}$$

$$d_l \leftarrow d_l / \|d_l\|_2, \tag{9}$$

$$c_i \leftarrow X'^T d_l - \left(\sum_{i \neq l} d_l c_i\right)^T d_l. \tag{10}$$

The specific steps of the improved CDL algorithm are shown in Algorithm 1.

Algorithm 1. The improved CDL algorithm

Input: Signal $X = \{x_1, x_2, \ldots, x_n\}$, number of iterations m
Output: Trained dictionary D and P
1: Init: Set $D = D_0$, X as input to DDELM-AE, is trained to get X'
2: for $n = 1, 2, \ldots, m$ do
3:    $\forall i$: $\alpha_i \leftarrow \min \alpha_i \frac{1}{2} \|x'_i - D\alpha_i\|_2^2 + \lambda \|\alpha_i\|_1$
4:    $D^T D$ singular value decomposition: $D^T D = u \Lambda v^T$
5:    Updating dictionary D: $D' = \Gamma(D)$
6:    for $j = 1, 2, \ldots, L$ do
7:       $D_j = 0$
8:       $I =$ indices of the signals in X' whose representations use d_j
9:       $g = \alpha_{j,I}^T$
10:      $d = X'_{Ig} - D\alpha_{Ig}$
11:      $d = d / \|d\|_2$
12:      $g = X'^T_I d - (D\alpha_I)^T d$
13:      $D_j = d$
14:      $\alpha_{j,I} = g^T$
15:   end if
16: end for
17: Dictionary D singular value decomposition: $D = U\Theta V^T$
18: low-dimensional dictionary $P = U^T D$

---

3   Experimental Results and Analysis

3.1   Complexity Analysis

Under the same noise condition, taking Lena and Boat as the experimental objects, the difference between the original algorithm and the improved algorithm is test at the compression ratio (CR) of 16. The denoising ability of the

(a)   **(b)**   **(c)**   **(d)**

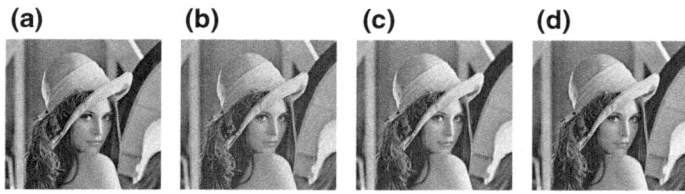

Fig. 2. Reconstructed images of Lena: **a** original image, **b** noisy image, **c** K-SVD, **d** DAK-SVD ($\sigma = 10, CR = 16$)

(a)   **(b)**   **(c)**   **(d)**

Fig. 3. Reconstructed images of Boat: **a** original image, **b** noisy image, **c** K-SVD, **d** DAK-SVD ($\sigma = 10, CR = 16$)

AK-SVD algorithm and K-SVD are used to compress and reconstruct the noisy image. The denoising effect is shown in Figs. 2 and 3.

Table 1 shows that the performance of the DAK-SVD algorithm is better than the K-SVD algorithm, according to the PSNR and TIME standard.

Table 1. Comparison of reconstructed images

Name of the image	Parameters	K-SVD	Proposed system
Lena	PSNR	31.5358	32.3233
	TIME	96.3186	73.0783
Boat	PSNR	29.7238	30.6521
	TIME	135.1286	103.3922

3.2 Stability Analysis

In the experiment, the original algorithm and the improved algorithm are tested under no noise condition and adding noise with standard deviations of 10, 20, 40 (Figs. 4, 5, 6 and 7) condition, respectively. Using PSNR to judge the quality of the decompressed image.

Table 2 reveals that in no noise condition, CDL compression performance is better than the proposed algorithm under different CR. However, after adding different noise, as shown from Figs. 5, 6, 7 and Table 3, the proposed algorithm

Fig. 4. Reconstructed images of Boat: **a** original image, **b** CDL, **c** proposed algorithm (no noise, $CR = 16$)

is more effective than the CDL algorithm in the same compression ratio. Due to the denoising processing of the deep learning model, the denoising ability of the original algorithm and the anti-interference ability are improved, which increases the system robustness.

Table 2. Comparison of reconstructed Boat (no noise)

CR		4	16	32	64
PSNR	CDL	35.22	34.26	30.56	28.23
	Proposed system	34.46	33.85	28.63	25.98

Fig. 5. Reconstructed images of Boat: **a** original image, **b** noisy image, **c** CDL, **d** proposed algorithm ($\sigma = 10, CR = 16$)

Table 3 and Fig. 8 indicate that, in the same compression ratio, the improved algorithm is more obvious advantage in the noise environment.

4   Conclusions

In this paper, the performance of CDL algorithm has been improved. The denoising ability has been enhanced by adding DDELM-AE on input, and the running speed has been improved by using AK-SVD. Additionally, experimental results have indicated that the improved algorithm is faster and more stable than the original algorithm under noisy environment.

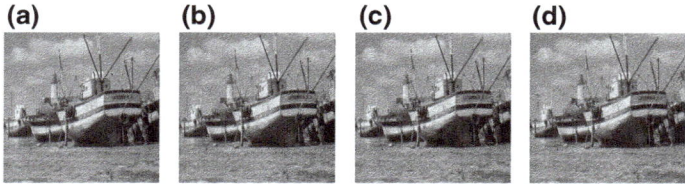

Fig. 6. Reconstructed images of Boat: **a** original image, **b** noisy image, **c** CDL, **d** proposed algorithm ($\sigma = 20, CR = 16$)

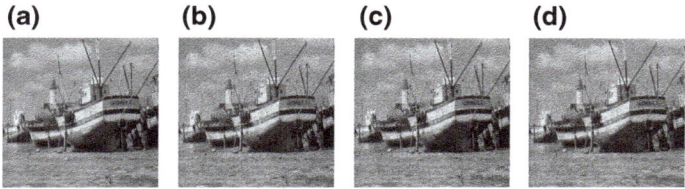

Fig. 7. Reconstructed images of Boat: **a** original image, **b** noisy image, **c** CDL, **d** proposed algorithm ($\sigma = 40, CR = 16$)

Table 3. Comparison of reconstructed Boat ($CR = 16$)

Noise		10	20	40
PSNR	CDL	31.53	29.52	20.50
	Proposed system	32.75	31.44	25.88

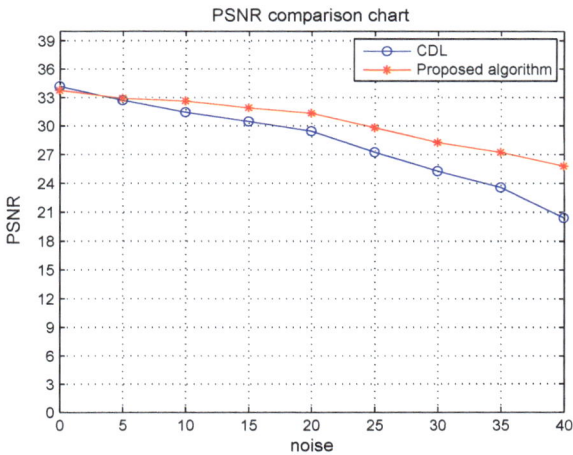

Fig. 8. Comparison of PSNR for CDL and proposed model

Acknowledgments. This work is supported by Natural Science Foundation of China (Grant No. 61702066), Scientific and Technological Research Program of Chongqing Municipal Education Commission (Grant No. KJ1704080), and Chongqing Research Program of Basic Research and Frontier Technology (Grant No. cstc2017jcyjAX0256).

References

1. Aharon M, Elad M, Bruckstein A. K-SVD: design of dictionaries for sparse representation. In: Proceedings of the workshop on signal processing with adaptive sparse structured representations (SPARS05); 2005. p. 9–12.
2. Aharon M, Elad M, Bruckstein A. K-SVD: an algorithm for designing overcomplete dictionaries for sparse representation. IEEE Trans Image Process. 2006;54(11):4311–22.
3. Wei F, Shutao L, Leyuan F, Benediktsson JA. Contextual online dictionary learning for hyperspectral image classification. IEEE Trans Geosci Remote Sens. 2017;1–12.
4. Gkioulekas IA, Zickler T. Dimensionality reduction using the sparse linear model. In: Proceedings of the 2011 advances in neural information processing systems; 2011. p. 271–9.
5. Calderbank R, Jafarpour S, Schapire R. Compressed learning: universal sparse dimensionality reduction and learning in the measurement domain. Princeton University, USA; 2009. Technical Report.
6. Donoho DL. Compressed sensing. IEEE Trans Inf Theor. 2006;52(4):1289–306.
7. SiLong Z, YuanXiang L, Xian W, XiShuai P. Nonlinear dimensionality reduction based on dictionary learning. In: ACTA AUTOMATICA SINICA; 2016. p. 1065–76.
8. Xiangyi C, Huaping L, Xinying X, Fuchun S. Denoising deep extreme learning machine for sparse representation. Memetic Comp. 2017;9(3):199–212.
9. Shuting C, Shaojia W, Binling L, Daolin H, Simin Y, Shuqiong X. A dictionary-learning algorithm based on method of optimal directions and approximate K-SVD. In: 2016 35th Chinese control conference (CCC); 2016. p. 6957–61.
10. Zeyde R, Elad M, Protter M. On single image scale-upusingsparse-representations. In: Proceedings of the 7th international conference on curves and surfaces; 2012. p. 711–30.
11. Jianchao Y, Wright J, Thoams H, Yi M. Image super resolution via sparse representation. IEEE Trans Image Process. 2010;19(11):2861–73.
12. Vincent P, Larochelle H, Lajoie I, Bengio Y. Stacked denoising autoencoders: learning useful representations in a deep network with a local denoising criterion. J Mach Learn Res. 2010;11(6):3371–408.

Body Gestures Recognition Based on CNN-ELM Using Wi-Fi Long Preamble

Xuan Xie, We Guo, and Ting Jiang$^{(\boxtimes)}$

Key Laboratory of Universal Wireless Communication,
Beijing University of Posts and Telecommunications, Beijing, China
{xiexvan,tjiang}@bupt.edu.cn

Abstract. Recently, researchers around the world have been striving to develop human–computer interaction systems. Especially, neither special devices nor vision-based activity monitoring in home environment has become increasingly important,and has had the potential to support a broad array of applications. This paper presents a novel human dynamic gesture recognition system using Wi-Fi signals. Our system leverages wireless signals to enable activity identification at home. In this paper, we present a novel Wi-Fi-based body gestures recognition model by leveraging the fluctuation trends in the channel of Wi-Fi signals caused by human motions. We extract these effects by analyzing the long training symbols in communication system. USRP-N210s are leveraged to set up our test platform, and 802.11a protocol is adopted to implement body gestures recognition system. Besides, we design a novel and agile segmentation algorithm to reveal the specific pattern and detect the duration of the body motions. Considering the superiority of feature extraction, convolutional neutral networks (CNN) is adopted to extract gesture features, and extreme learning machine (ELM) is selected as classifier. This system is implemented and tested in ordinary home scenario. The result shows that our system can differentiate gestures with high accuracy.

Keywords: Body gesture recognition · CNN · ELM · Long training sequence · USRP · 802.11a · Wi-Fi

1 Introduction

Wearable devices basis of the sensor have gradually integrated into our daily life. Activity monitor in home environment like elder care and child safety most are implemented in video surveillance. These applications have been accepted by the masses as a part of life. While it is not pleasant to wear some sensors or be kept on an electronic eye each moment even though the surveillance is well

This work was supported by the National Natural Sciences Foundation of China (NSFC) (No. 61671075) and Major Program of National Natural Science Foundation of China (No. 61633003).

Q. Liang et al. (eds.), *Communications, Signal Processing, and Systems*, Lecture Notes in Electrical Engineering 516,
https://doi.org/10.1007/978-981-13-6504-1_106

intentioned. Naturally, how to implement above function and make less adverse difference are worthy for research. Since wireless signals do not require line-of-sight and can traverse through walls, and some wireless signals such as Wi-Fi are widely used in home environment, suppose we can realize activity monitor or gesture recognition by wireless signals, user experience could be boosted greatly. Some wireless signals-based human activity recognition and localization systems have been proposed. WiSee [1] recognizes body gesture by extracting the micro-doppler shift from Wi-Fi signals. As we know, people makes an action, which results in a pattern of Doppler shifts at the wireless receiver, they proposed algorithms to extract Doppler shift from OFDM symbols. WiTrack [2] tracks the 3D motion of a user through the wall from the radio signals reflected off the body. They measures time it takes for its signal to travel from its transmit antenna to the reflecting body, and then back to each of its receive antennas. They leverage the geometric placement of its antennas to localize the moving body in 3D. WiHear [3] detects and analyzes fine-grained radio reflections from mouth movements by MIMO technology to recognize spoken words, they use PHY layer channel state information (CSI) on WLAN device to analyze micro-motion, then use multipath effect and discrete wavelet packet transformation to achieve lip reading with Wi-Fi. Besides, as we know, different gestures between transmitter and receiver have different effects on received signals, [4] extract the long preamble of Wi-Fi signals, and adopt the traditional improved classification arithmetic to classify and recognize different hands gestures.

This paper fully analyze the frame structure of IEEE 802.11a WLAN protocol. Without effecting communication quality, we recognize the human gestures by extracting and analyzing the long training sequence of Wi-Fi frame. In the process of communication, data is a kind of stochastic series, specific transmitting content have a decisive impact on waveforms. While, preamble sequence, short and long included, which are responsible for frame detection, frequency offset correction and symbol alignment, is fixed in different data frame. Inside, short training sequence, only two to three are available in signal detection [5], thus, the long training sequence is implemented to recognize the human motion. Long training sequence is used as frequency offset correction and symbol alignment in communication, which is composed of a fixed length pattern [6].

Our gesture recognition is omplemented in following steps. Firstly, long training sequence is extracted from wireless signals (IEEE 802.11a), the 802.11a system that [7] provided are adopted to extract long training symbols. Secondly, continuous periodic activities product, which make periodic influence on long preamble in communication system. A method is proposed to make sample extraction. Thirdly, features are extracted from the data by convolutional neutral networks (CNN). Finally, Extreme Learning Machine (ELM) is leveraged to classify and recognize different body gestures.

This paper was organized as follows. In Sect. 2, we introduce the features extraction algorithm and classifier. In Sect. 3, a brief introduction on experiment platform and scenario is provided, and the method that extracts long preamble is introduced. Then, we describe how to test fluctuation caused by body gestures.

In Sect. 4, we introduce our experiment and analyze the result. Finally, a conclusion is drawn in Sect. 5.

2   Recognition Approach

2.1   Convolutional Neutral Networks

Convolutional neutral networks (CNN) were first presented by Kunihiko Fushima; in recent years, CNNs fall in the category of deep learning techniques, which have been employed in many other fields such as speech recognition and images recognition [8,9]. In this paper, instead of learning an ad-hoc CNN for the body gestures recognition problem, CNN is used to choose the most informative features and the best combination of features. In our experiment, we can find that features extracted by CNN do better in our body gestures recognition than normal features selected manually.

CNN takes an input data, processes it with a series of different transformation layers, finally produce a prediction of the class. The transformation layers mainly contain three classes: input layer, hidden layer, and output layer, and convolutional layers are the main type of the hidden layers in CNNs. These are designed to perform a convolutional operation on the data. Besides, pooling layers are used between adjacent layers, which is average or maximum filters, they can reduce the impact of small variations and descend the dimension of previous layer output data. Dropout layers are used to fix overfitting, which could occur when the number of training data and CNN parameters are mismatching. When the CNNs are set, training data are used to train the net iteratively and the weight and bias of every neuron are adjusted until the loss function is fitted. The CNN network structure of this paper is Fig. 1.

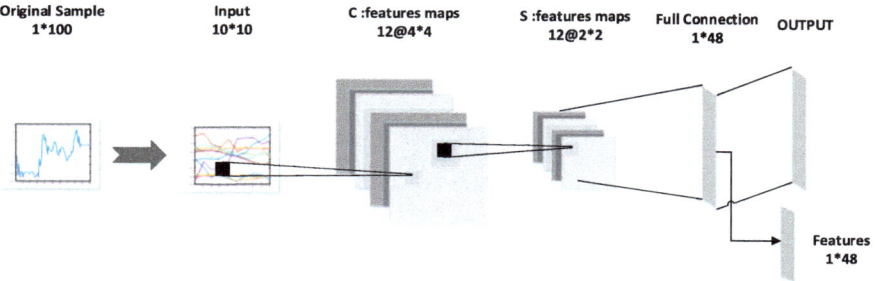

Fig. 1. CNN network structure.

Extreme Learning Machine Extreme learning machine (ELM) [10] is a kind of learning algorithm for the single-hidden-layer feed-forward neural networks. Comparing with the deep learning algorithm, it overcomes the slow training speed and overfitting problems. And ELM also has milder optimization constraints than support vector machine (SVM). ELM is based on empirical risk minimization theory and its learning process needs only a single iteration. The algorithm avoids multiple iterations and local minimization. Especially in small dataset, due to better generalization ability and fast learning rate, it is used in various fields. We adopt the ELM in our recognition.

ELM is initially developed for a single-hidden-layer feed-forward neural networks (SLFNs). Given an input sample (X_i, y_i), where $X_i = [x_{i1}, x_{i2}, x_{i3}, \ldots, x_{in}]^T \in R^n$, $y_i = [y_{i1}, y_{i2}, y_{i3}, \ldots, y_{im}]^T \in R^m$. The output of ELM with L hidden nodes is designed as

$$\sum_{i=1}^{L} \beta_i g(\omega_i \cdot X_j + b_j) = o_j, \quad j = 1, \ldots, N \tag{1}$$

where β_i is the hidden-layer bias, ω_i is the input weight of the ith neuron, g(x) is the activation function, and o_j is the output of hidden layer. $\omega_i \cdot X_j$ means the inner product of ω_i and X_j. In case of ELM approximates the data perfectly for all samples, we have

$$\sum_{j=1}^{L} \|o_j - y_j\| = 0 \tag{2}$$

that means existing β_i, ω_i, and b_j such that

$$\sum_{i=1}^{L} \beta_i g(\omega_i \cdot X_j + b_j) = y_j, j = 1, \ldots, N \tag{3}$$

Denoted by matrix

$$H\beta = Y \tag{4}$$

where

$$Y = (y_1, y_2, y_3, \ldots, y_N)^T$$
$$\beta = (\beta_1, \beta_2, \beta_3, \ldots, \beta_L)^T$$

and H is the hidden-layer output matrix:

$$A = \begin{bmatrix} g(\omega_1 \cdot X_1 + b_1) & \cdots & g(\omega_L \cdot X_1 + b_L) \\ \vdots & \cdots & \vdots \\ g(\omega_1 \cdot X_N + b_1) & \cdots & g(\omega_L \cdot X_N + b_L) \end{bmatrix} \tag{5}$$

In the process of training, ELM aims at $\widehat{\omega}_i$, $\widehat{b}_i$, and $\widehat{\beta}_i$, which minimum the training error:

$$\|H(\widehat{\omega}_i, \widehat{b}_i)\widehat{\beta}_i\| = \max_{\omega, \beta, b} \|H(\omega_i, b_i) - Y\| \tag{6}$$

Which is equivalent to minimizing the loss function

$$E = \sum_{j=1}^{L} \left(\sum_{i=1}^{L} \beta_i g(\omega_i \cdot X_j + b_j) - y_j \right)^2 \tag{7}$$

According to ELM theory, input weights ω and b can be randomly assigned, and $\beta = H^*Y$ where H^* is the Moore–Penrose of H.

3   Data Processing

Experiment Scenario We adopt a complete 802.11a communication system [7], which is implemented based on GNU Radio and fitted for operation on an Ettus USRP N210. USRP N210 is a kind of generic software defined radios (SDRs). We set the packet transmission rate to 50 packets per second to obtain sufficient information from body gestures.

We set up our system in a meeting room, as depicted in Fig. 2. The testing area is 6 m × 6 m. And the volunteer performs different gestures in room.

The distance between transmitter and receiver is 4.0 m. The center frequency of the wireless transmission is 5.3 GHz.

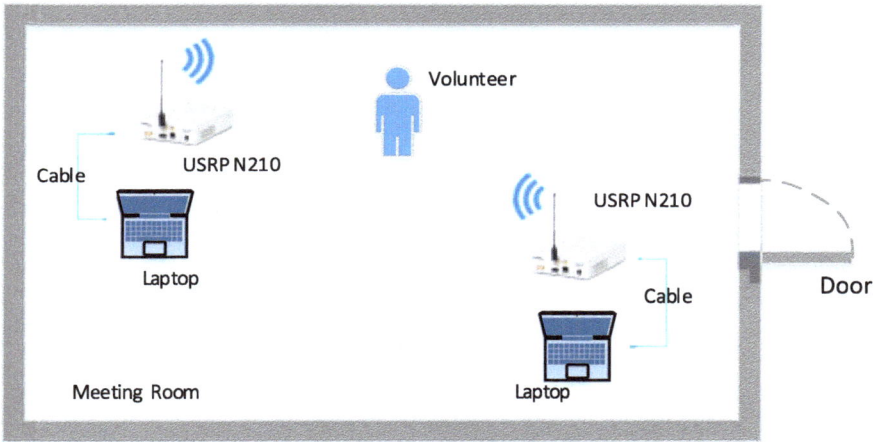

Fig. 2. Experiments environment and setting

Long Symbols Extraction The IEEE802.11a WLAN is a packet communication system [11]. Frame structure [6] is shown in Fig. 3, which consists of two parts, training sequence and OFDM Symbols; the training sequence contains designated short preamble and long preamble. Short preamble, a pattern that spans 16 samples and repeats ten times, which is used to detect the frame, and long preamble is used as frequency offset correction and symbol alignment, which is a pattern repeats 2.5 times, one of the purposes of the long preamble is symbol alignment, the intention of this process is detecting the start of signal. According to Fig. 3, when the start of signal is found, the long preamble can be extracted. The sample points of long preamble are related to the duration of long preamble and the sample rate:

$$N_{LP} = N_S \times T_S \tag{8}$$

where N_{LP} is the sample points of one long preamble and N_S is the sample rate; in this paper, we set N_S to 20 MHz, T_s is the length of long preamble patterns (two repeated sequence). Seen from Fig. 3, the length of long preamble is 8 μs, which contains a pattern that repeats 2.5 times. It is easy to get T_s, that equals to 6.4 μs. Each frame, we sample 128 points. Next, we talk about how to divide the continuous data into correct cycle while ignoring the minute error caused by participant in periodic action. In the process of data acquisition, we keep the normal communication,and transmit text. Meanwhile, people do periodic activities shown in Fig. 4. Obviously, periodic gestures cause cyclical effects on signals. Ideally, we can divide the serial data in fixed period. While, as we known, it is not able to avoid that some random difference or minute errors could be superposed in data. So proper method of period division produces vital influence on recognition performance.

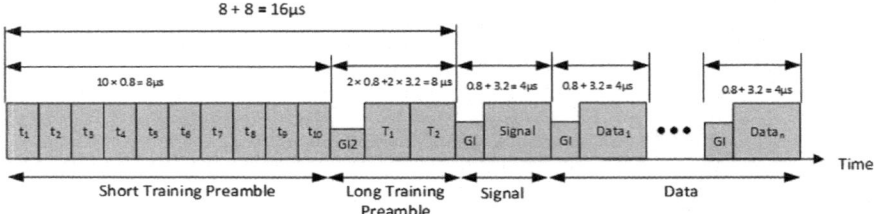

Fig. 3. OFDM training structure

Data Preprocessing Consider the long preamble $X_1, X_2, X_3, \ldots, X_i, \ldots,$ X_{n-2}, X_{n-1}, X_N, where X_i is a column vector of $1 \times N_{LP}$ that samples from long preamble of data frames, and N is the number of frame of one group body gestures (different gestures may have different N), we calculate the mean of X_i. Say x_i denotes the mean power of the X_i, then, let $\bar{X}_i$ ($i = 1, 2, 3, \ldots, N-2, N-1, N$) be the new vector formed by x_i:

$$\bar{X}_i = [x_1, x_2, x_3, \ldots, x_i, x_{N-1}, x_N] \tag{9}$$

Figure 5 shows the amplitude spectrum of nine body gestures. Inside, the horizontal axis is the frame index, and the vertical axis is the mean power of X_i (i.e. $\bar{X}_i$). The plots show the nine kinds of periodic body gestures defined in Fig. 4. It is shown that different gestures have completely different effects on the preamble periodically. The cycle of difference depends on the duration of body gesture, and the amplitude and the shape of wave are determined by the location and the range of the body.

Endpoint Detection A new method to detect the endpoint is proposed in this paper. Obviously, body gestures have some effects on preamble, so detecting the start and end point of gesture is vital. The steps of detection are in follows:

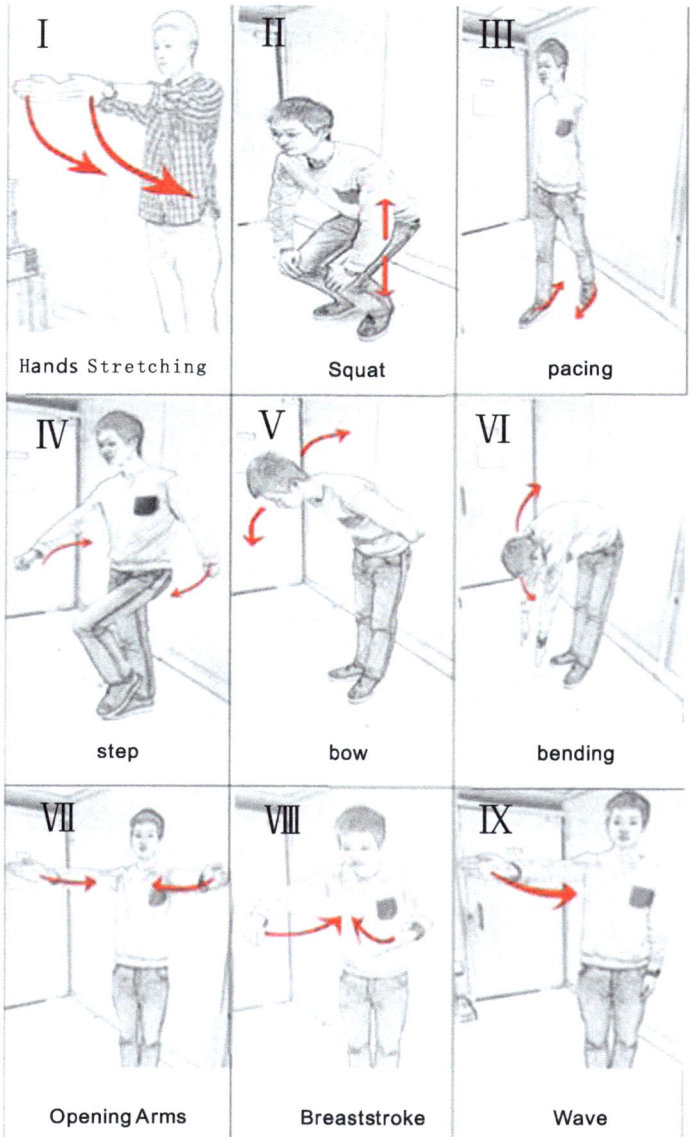

Fig. 4. Predefined dynamic body gestures

Step 1. Zero-mean need to be carried out on $\bar{X}_i$.

$$P_i = \bar{X}_i - \bar{X}_i^{mean} \tag{10}$$

where $P_i (i = 1, 2, 3, \ldots, N-2, N-1, N)$ denotes the zero-mean amplitude of $\bar{X}_i$, X_i^{mean} is the mean of the $\bar{X}_i$.

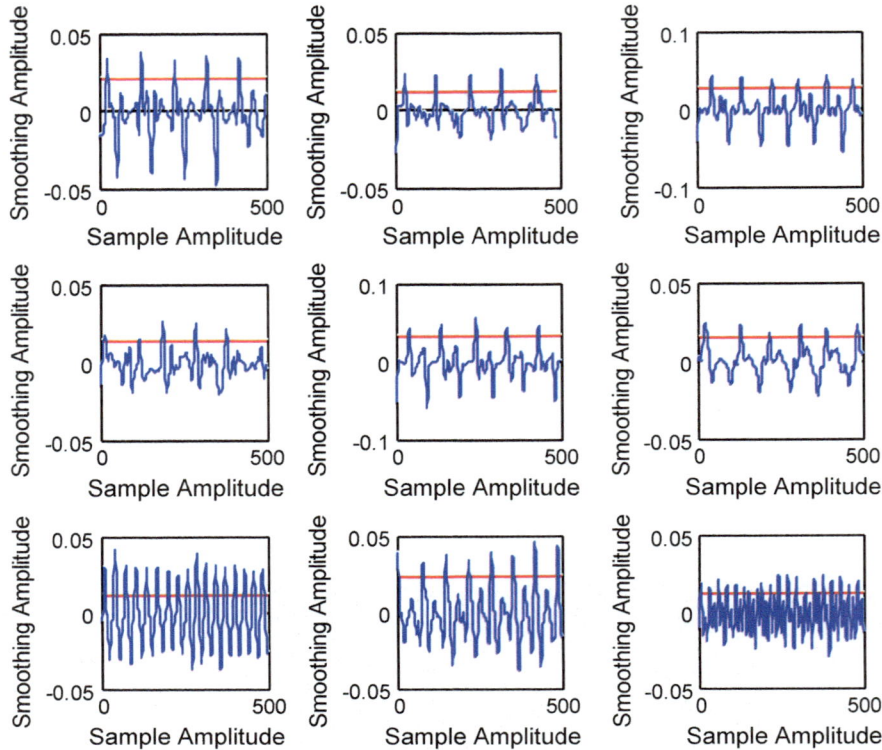

Fig. 5. The normalize amplitude spectrum of nine kinds of body gestures

The result is shown in Fig. 5.

Step 2. Make differential operation on P_i, and then, we let the differential operation result of the P_i is the Y_j

$$Y_i = P_i - P_{i-1} \tag{11}$$

Next, we calculate the $y_j{}^{threshhold}$:

$$y_j{}^{threshhold} = y_j{}^{mean} + \beta(y_j{}^{max} - y_j{}^{mean}) \tag{12}$$

where j $=1, 2, 3, \ldots ,7 ,8 ,9$, which denotes nine kinds of body gestures. $y_j{}^{mean}$ is the mean of Y_i, $y_j{}^{max}$ is the max of the Y_i. And β is the adjustable coefficient to make the cycle division reasonable.

Step 3. Calculate the mean of y_{i-1}, y_i and y_{i+1} then replace it with z_i, this operation needed to be done to eliminate some singular points (i.e.small peaks).

Step 4. Adjust β and find i that meets the following conditions:

$$\begin{cases} z_i - z_{i-1} > 0, z_{i-1} - z_{i-2} > 0, \\ z_i - z_{i+1} > 0, z_{i+1} - z_{i+2} > 0, \\ \qquad z_i > y_i{}^{threshhold} \end{cases} \tag{13}$$

where we need adjust β on the basis of the actual periods to find reasonable i. Later, we get a new vector V_j consists of the special i that meets above conditions (i.e., v_1 equals eligible i).
Say

$$V_j = [v_1, v_2, v_3, v_l, \ldots, v_{L-2}, v_{L-1}, v_L] \tag{14}$$

where L is the length of the V_j. The elements of this vector denotes the start and end point of one cycle. Using V_j, divide $\bar{X}_i$ to form a matrix M_j. The row of M_j denotes the sample numbers of one body gestures. Actually it is $L - 1$, the column of M_j is the period of one body gestures, we set the column is:

$$\max_{1 < l < L-1} \|v_{l+1} - v_l\| \tag{15}$$

Here, we get the whole samples of body gestures, which are shown in Fig. 6.

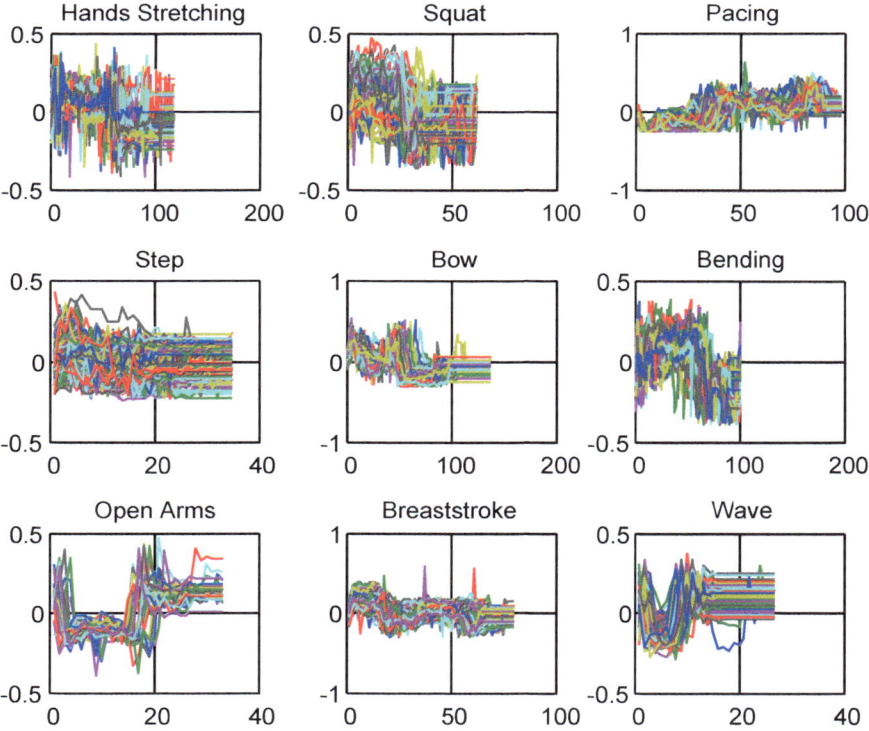

Fig. 6. Nine groups of samples, final samples of nine gestures

4   Experiment and Analysis

Experiment In Sect. 3, we get the human body gestures dataset, which consists of nine kinds of gestures shown in Fig. 4, every gesture contains 100 samples. Obviously, nine gestures have different cycle, considering facilitate subsequent process,we choose the median to make the dimension of all vectors become 100. That means the dataset is a matrix of 900×100, which denotes 900 samples. Besides, the label matrix is 900×9, is one-hot vector, which denotes the class of sample. The following processing, we use this dataset to train the CNN and ELM. In every process of training and testing, the train dataset is 540×100 and test dataset is 360×100.

In our experiment, the features produced by CNN in the last layers before the class assignment work are 48-dimensional, which are extracted from each 100-dimensional body gesture data. Considering the data size, our net selected one convolutional layer and one fully connected layer. The CNN was originally trained on nine kinds of body gestures with labels. Features are obtained by extracting activation values of the last hidden layer. The extracted features for each sample are then used as input to ELM for classification. The flow chart of the whole processing is shown in Fig. 7. In the process of training the CNNs, MSE loss function is used:

$$\min C(\widehat{Y}, Y) = \|\widehat{Y}^2 - Y^2\| = \sum_i (\widehat{y} - y)^2 \tag{16}$$

iteration result in train dataset is shown in Fig. 8.

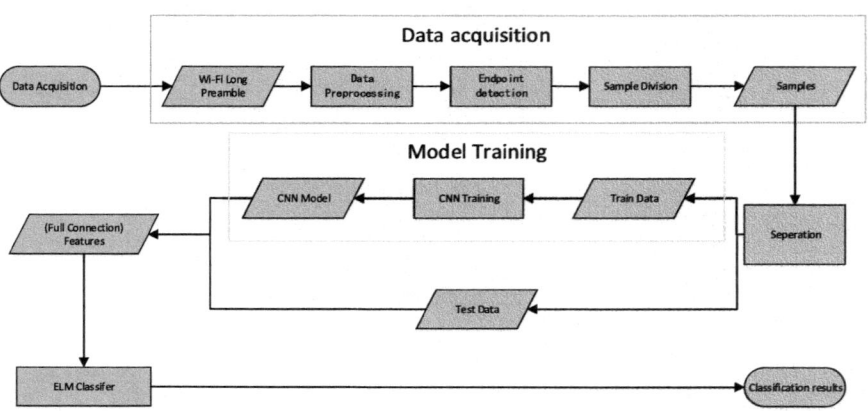

Fig. 7. Body gestures recognition framework

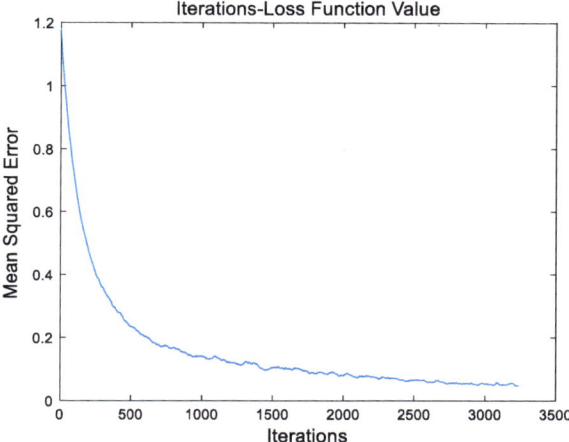

Fig. 8. World map

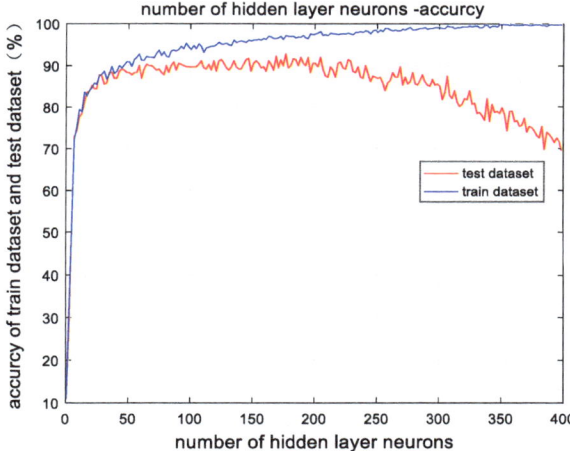

Fig. 9. Concrete and constructions

We choose randomly generated hidden nodes, sigmoid function is used as activation. The number of hidden neurons is determined through cross-validation. Figure 9 shows the process of it.

Compared with other classifiers and features extraction method, we also use them in our dataset. In the extraction features, we select ten kinds of statistics features manually [12]: power-delay-profile (PDP), mean excess delay (RMS), RMS delay spread, multipath components (MPCs) within 10 dB, number of MPCs that contain 60% of the received energy, max value of sample, mean of sample, standard deviation, kurtosis, and skewness. And support vector machine (SVM) is applied in our dataset, the result is shown in Table 1.

Table 1. Summary of classification accuracy (%) using different feature extraction methods and classifiers

	Hands stretching	Squat	Pacing	Step	Bow	Bending	Opening arms	Breast stroke	Waves	Mean accuracy
Manual&SVM	75.79	81.32	86.81	89.20	91.46	88.18	97.44	83.58	92.39	87.35
CNN&SVM	82.50	87.50	100.00	97.50	97.50	75.00	100.00	90.00	100.00	92.22
CNN&ELM	95.00	85.00	85.00	100.00	95.00	100.00	100.00	92.50	100.00	94.72

In Table 1, Manual&SVM means we selected features manually and SVM classifier is adopted. CNN&SVM stands for CNN is used to extract features and SVM classifier act as classifier. The last is what we proposed in this paper. The detailed recognition rates of the nine body gestures in Fig. 4 are shown in Table 2.

Table 2. Summary of classification accuracy (%) of different body gestures

	i	ii	iii	iv	v	vi	vii	viii	ix
i	95.00	0	0	0	0	2.50	0	2.50	0
ii	5.00	85.00	0	5.00	0	2.50	0	2.50	0
iii	2.50	0	85.00	0	0	0	10	2.50	0
iv	0	0	0	100.00	0	0	0	0	0
v	0	0	0	0	95.00	0	0	2.50	0
vi	0	0	0	0	0	100.00	0	0	0
vii	0	0	0	0	0	0	100.00	0	0
viii	0	0	0	0	0	0	0	92.50	0
ix	0	0	0	0	0	0	0	0	100.00

5   Conclusion

In this paper, we treated the problem of human body gestures recognition with Wi-Fi signals. Instead of using CSI to analyze the motion, we adopt the long preamble of Wi-Fi signal through the analysis of the communication process. Compared with CSI, long preamble is more primitive, and more features of motion could be collected. Aimed at long preamble, in the data processing stage, we proposed new motion capture and endpoint detection algorithm. And inspired by speech processing and image processing, instead of extracting features manually, CNN is used in this task. In contrast, we select ten statistic features of samples, and use the same classifier SVM, the result shows that the accuracy rate of CNN is higher 4.87%. Besides, in the small dataset, ELM has milder optimization constraints and less computational complexity, we choose ELM as the classifier in the end, we can find from Table 1, when the same feature vector input ELM and SVM, ELM has the better accuracy. The mean accuracy of SVM is 92.22%, while ELM is 94.75%. Meanwhile for SVM, the lowest recognition is 75%, while ELM is 85%. This means ELM is better than SVM in the body gestures recognition. Moreover, different gestures may have totally distinct effect on signals, which reflects on recognition rate. The nine body gestures we select consist of many parts of daily motions, this could promote the home care on the condition of respect of privacy.

Acknowledgments. We thank the anonymous for their thoughtful and constructive remarks that help me improve the quality of this paper.

References

1. Qifan P, Gupta S, Gollakota S, Patel S. Whole-home gesture recognition using wireless signals. Comput Commun Rev. 2013;43(4):485–6.
2. Adib F, Kabelac Z, Katabi D, Miller RC. 3D tracking via body radio reflections; 2013. p. 317–29.
3. Wang G, Zou Y, Zhou Z, Wu K, Ni LM. We can hear you with wi-fi!. In: International conference on mobile computing and Networking; 2014. p. 593–604.
4. Zhou G, Jiang T, Liu Y, Liu W. Dynamic gesture recognition with wi-fi based on signal processing and machine learning. In: IEEE global conference on signal and information processing; 2015. p. 717–21.
5. Liu CH. On the design of OFDM signal detection algorithms for hardware implementation. In: Global telecommunications conference, 2003. GLOBECOM '03. IEEE; 2003, vol. 2. p. 596–9.
6. IEEE. Part 11: wireless LAN medium access control (mac) and physical layer (phy) specifications. In: the 5-GHz band, IEEE; 2007. p. C1–1184.
7. Bloessl B, Segata M, Sommer C, Dressler F. An IEEE 802.11 a/g/p OFDM receiver for GNU radio; 2013.
8. Sainath TN, Kingsbury B, Saon G, Soltau H, Mohamed AR, Dahl G, Ramabhadran B. Deep convolutional neural networks for large-scale speech tasks. Neural Netw Official J Int Neural Netw Soc. 2015;64:39–48.
9. Ciresan DC, Meier U, Masci J, Gambardella LM, Schmidhuber J. Flexible, high performance convolutional neural networks for image classification. In: IJCAI 2011, Proceedings of the international joint conference on artificial intelligence, Barcelona, Catalonia, Spain, July; 2011. p. 1237–42.
10. Huang GB, Zhu QY, Siew CK. Extreme learning machine: theory and applications. Neurocomputing. 2006;70(1–3):489–501.
11. Sourour E, El-Ghoroury H, Mcneill D. Frequency offset estimation and correction in the IEEE 802.11a wlan. In: Vehicular technology conference, 2004. Vtc2004-Fall. 2004 IEEE; 2005, vol. 7. p. 4923–7.
12. Zhai S, Jiang T. Target detection and classification by measuring and processing bistatic UWB radar signal. Measurement. 2014;47(1):547–57.

Evaluation of Local Features Using Convolutional Neural Networks for Person Re-Identification

Shuang Liu[1,2(✉)], Xiaolong Hao[1,2], Zhong Zhang[1,2], and Mingzhu Shi[1,2]

[1] Tianjin Key Laboratory of Wireless Mobile Communications and Power Transmission, Tianjin Normal University, Tianjin, China
[2] College of Electronic and Communication Engineering, Tianjin Normal University, Tianjin, China
{shuangliu.tjnu,haoxiaolong17,zhong.zhang8848}@gmail.com
shimingzhu1@163.com

Abstract. In this paper, we mainly evaluate the influence of local features extracted by convolutional neural networks for person re-identification. Considering the variant body parts with different structural information, we divide the holistic person images into several parts and extract their features. Two kinds of aggregation methods are used to aggregate local features. Experiments on the challenging person re-identification database, Market-1501 database, show that the max aggregation is more effective for extracting the discriminative local features than the sum aggregation.

Keywords: Local features · Convolutional neural networks · Person re-identification

1 Introduction

Given a specified person image captured from one camera, person re-identification aims to spot the same person captured from other cameras [1]. Person re-identification plays a significant role in practical applications such as video surveillance, image retrieval, and so on [2,3]. Hence, person re-identification becomes gradually popular in the research field. Nevertheless, the performance of person re-identification is affected by many factors, such as different pose, various illumination, and so on [4]. In order to solve these problems, a large number of researchers have devoted themselves to the study of person re-identification.

Recently, the convolutional neural networks display potential for extracting the discriminative features for person re-identification [5,6] and gradually draw more attention in this domain. Up to now, the available methods mainly concentrate on extracting a discriminative features representation [7–9] and learning

Q. Liang et al. (eds.), *Communications, Signal Processing, and Systems*, Lecture Notes in Electrical Engineering 516,
https://doi.org/10.1007/978-981-13-6504-1_107

an effective measure method [10–12]. As for the former, global and local features are the mainstream algorithms. Some approaches utilize the whole person image to train the network model in order to obtain global representations. For instance, the method named the domain guided dropout algorithm was proposed by Xiao et al. [13], which utilizes convolutional neural networks to obtain a robust global features from multiple domains. To acquire the high-level discriminative features, Wu et al. [7] used deep convolutional networks and extremely small filters to extract features. Combining the strengths of the identification model and verification model, Zheng et al. [14] proposed a siamese network model to learn discriminative global features using three loss functions. Although the global features extracted the convolutional neural networks are effective for person re-identification, it ignores the local structure information. Therefore, under the guidance of the deep learning methods, some researchers have focused on mining local structure information and have achieved promising results. In order to extract robust local features, Yi et al. [15] split the holistic person images into three partial areas and trained three convolutional neural networks, respectively. Considering the various structure information in different positions, the Spindle Net captured micro-body information and learnt the local features completely. Sun et al. [16] presented a network called part-based convolutional baseline (PCB), which utilizes a simple uniform partition strategy to obtain local features consisting of several part-level features. As for the measure method, some methods, such as the Euclidean distance [11,17], the Mahalanobis distance [18], the cosine distance [12], and so on, are exploited to compute the similarity between person images.

The above-mentioned local features can achieve good results for person re-identification [15,16,19]. In this paper, we mainly focus on evaluating the impact of local features with different aggregation methods for person re-identification. In order to learn local features using the convolutional neural networks, we first split the holistic person image into several parts and allocate a person re-identification label for each part. Then, the output of fully connected layer is regarded as the features vector of each part. After extracting the features for each part, we utilize two kinds of methods to aggregate these local features for each person image. Finally, we evaluate the performance with different aggregation methods on the Market-1501 database [20].

The remain of this article is organized as follows. The detailed implementation is given in Sect. 2. In Sect. 3, we show the experimental results on the Market-1501 database and we make a conclusion in Sect. 4.

2 Approach

In this section, we first introduce the framework of convolutional neural networks and then present implementation details. Finally, two kinds of aggregation methods are introduced.

2.1   The Framework of Convolutional Neural Networks

There are many convolutional neural networks models, such as CaffeNet [21], VGG16 [22], ResNet-50 [23], and so on. We select ResNet-50 as the convolutional neural networks to fine-tune on our database as shown in Fig. 1. In order to apply the ResNet-50 network on the Market-1501 database, the number of neurons in the original fully connected layer is changed to 751, and the features f extracted from the fully connected layer are a 2048 dimensional vector.

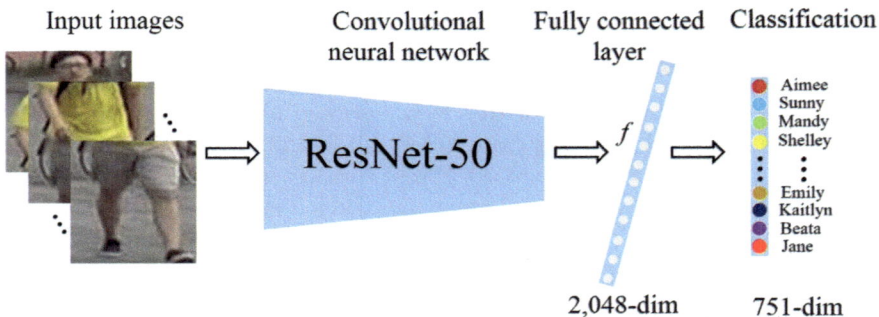

Fig. 1. The framework of convolutional neural networks.

2.2   Implementation Details

In order to extract the local features, we first divide each person image into three overlapped parts. Each part is then allocated a person re-identification label, which is the same as the original person re-identification label as shown in Fig. 2. Next, we employ these parts with the person re-identification labels to train the convolutional neural networks. After extracting the feature of each part from the fully connected layer, two aggregation approaches, i.e., sum aggregation and max aggregation, are used to aggregate the features of three parts, and the aggregated features are taken as the final features for each person image.

2.3   Two Kinds of Aggregation Methods

As shown in Fig. 3, since each person image is divided into three overlapped parts, we can obtain three partial features for each person image, i.e., f_1, f_2, and f_3. Afterward, two alternative methods are employed to aggregate the three partial features.

As for the sum aggregation, it aims to smooth the difference among all the parts. Figure 4 illustrates the fusion process. Using the sum aggregation, the local features can be aggregated into one global features. The aggregated features f_{sum} can be formulated as:

$$f_{sum} = f_1 + f_2 + f_3 \qquad (1)$$

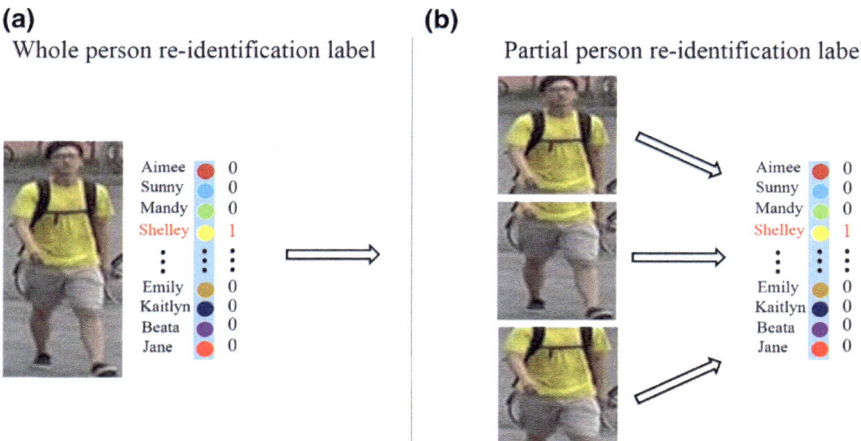

Fig. 2. a Is the whole person re-identification label, and **b** is the partial person re-identification label.

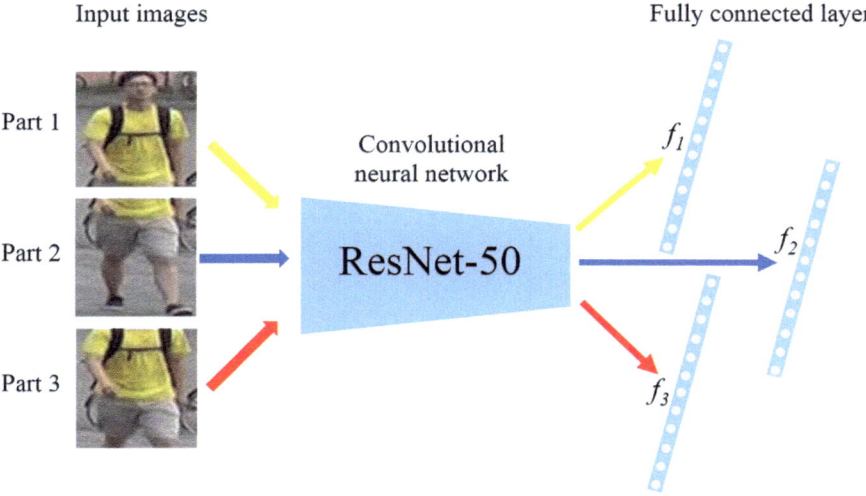

Fig. 3. Extracting three partial features for each person image.

where f_1, f_2, and f_3 are the partial features of a person image, respectively.

As shown in Fig. 5, the max aggregation method aims at maximizing the difference of the parts. In other words, the max aggregation method can maximize the detailed structure information for the person image. It can be formulated as:

$$f_{max} = \max(f_1, f_2, f_3) \tag{2}$$

where $\max(\cdot, \cdot, \cdot)$ indicates the maximum of three vectors in each dimension.

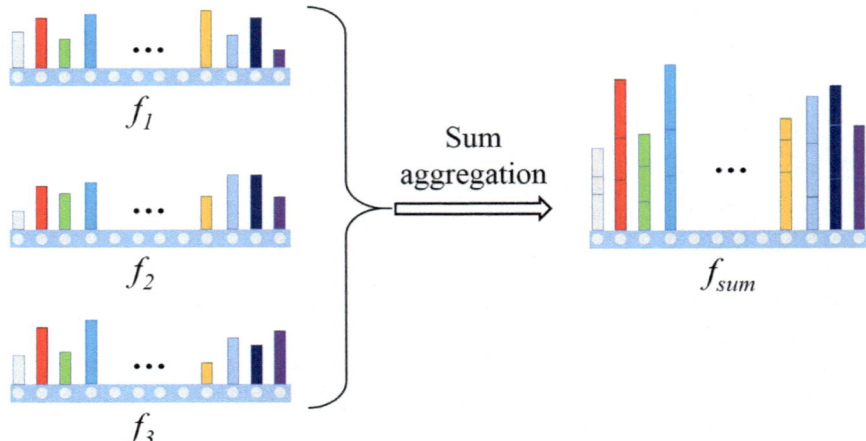

Fig. 4. Illustration of the sum aggregation.

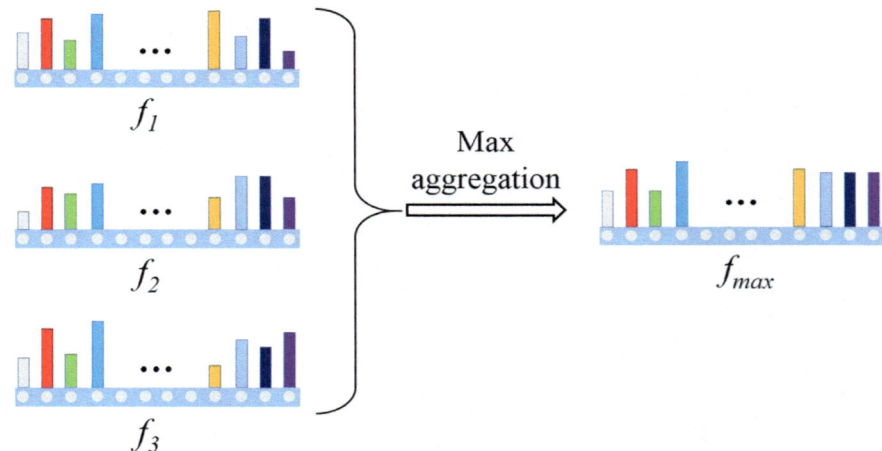

Fig. 5. Illustration of the max aggregation.

3   Experiments

In this section, we first briefly introduce the Market-1501 database [20] for person re-identification, and then evaluate the above-mentioned aggregation approaches on this database.

3.1   Market-1501 Database

Market-1501 database [20] is popular database for person re-identification. It utilizes six cameras to collect the person images in a campus and all the person images are divided into three parts, i.e., training part, test part, and disturbance part. Training part includes 12,936 images of 751 persons and each person has an average of 17.2 images. Test part contains 19,732 images of 750 persons. This database utilizes the deformable part model (DPM) [24] to automatically detect the pedestrians, which is very close to the realistic setting. During the evaluation process, we utilize the rank-i ($i = 1$, 5 and 10) accuracy and mAP algorithm to evaluate the experiment results.

3.2   Evaluation on Two Aggregation Methods

We evaluate two different aggregation approaches in this subsection. Table 1 lists the experimental results on the Market-1501 database. The max aggregation achieves the best results compared with the other method. Specifically, the max aggregation achieves 86.43% rank-1 accuracy and 67.78% mAP accuracy. Compared with the ResNet baseline, the max aggregation yields 3.74% rank-1 and 3.35% mAP improvements, respectively. The max aggregation achieves the best results due to two main reasons. On the one hand, the max aggregation is beneficial to excavation of the comprehensive structure information. On the other hand, utilizing the max aggregation, we can obtain the unique features of different persons. However, the sum aggregation is inferior to the ResNet baseline slightly. The reason is that the sum aggregation ignores some discriminative features, which decreases the performance for person re-identification.

Table 1. Comparison with two aggregation approaches on the Market-1501 database.

Methods	Rank-1	Rank-5	Rank-10	mAP
ResNet baseline	82.69	92.90	95.63	64.43
Sum aggregation	80.73	91.35	94.15	58.89
Max aggregation	**86.43**	**94.03**	**95.90**	**67.78**

4   Conclusion

In this paper, we have evaluated the influence of local features extracted by convolutional neural networks on the Market-1501 database. Specifically, we first divide the person image into three parts, and then allocate a person re-identification label for each part, which is the same as the original person label. After extracting the local features, we utilize two kinds of approaches to aggregate the partial features, respectively. Experimental results show that the max aggregation can yield better performance than the sum aggregation.

Acknowledgments. This work was supported by National Natural Science Foundation of China under Grant No. 61501327, No. 61711530240 and No. 61501328, Natural Science Foundation of Tianjin under Grant No. 17JCZDJC30600 and No. 15JCQNJC01700, the Fund of Tianjin Normal University under Grant No.135202RC1703, the Open Projects Program of National Laboratory of Pattern Recognition under Grant No. 201700001 and No. 201800002, the China Scholarship Council No. 201708120039 and No. 201708120040, and the Tianjin Higher Education Creative Team Funds Program.

References

1. Liao S, Hu Y, Zhu X, Li SZ. Person re-identification by local maximal occurrence representation and metric learning. In: IEEE conference on computer vision and pattern recognition; 2015. p. 2197–206.
2. Sathish PK, Balaji S. Person re-identification in surveillance videos using multi-part color descriptor. Int J Comput Appl. 2015;121(16):15–7.
3. Zhang R, Liang L, Zhang R, Wang M, Zhang L. Bit-scalable deep hashing with regularized similarity learning for image retrieval and person re-identification. IEEE Trans Image Process. 2015;24(12):4766–79.
4. Zhang Z, Wang C, Xiao B, Zhou W, Liu S. Action recognition using context-constrained linear coding. IEEE Sig Process Lett. 2012;19(7):439–42.
5. Zheng L, Zhang H, Sun S, Chandraker M, Yang Y, Tian Q. Person re-identification in the wild. In: IEEE conference on computer vision and pattern recognition; 2017. p. 1367–76.
6. Zheng L, Yang Y, Hauptmann AG. Person re-identification: past, present and future. arXiv preprint arXiv:1610.02984; 2016.
7. Wu L, Shen C, Hengel AVD. Personnet: person re-identification with deep convolutional neural networks. arXiv preprint arXiv:1601.07255; 2016.
8. Zhao H, Tian M, Sun S, Shao J, Yan J, Yi S, Wang X, Tang X. Spindle net: person re-identification with human body region guided feature decomposition and fusion. In: IEEE conference on computer vision and pattern recognition; 2017. p. 1077–85.
9. Zhang Z, Wang C, Xiao B, Zhou W, Liu S. Attribute regularization based human action recognition. IEEE Trans Inf Forensics Secur. 2013;8(10):1600–9.
10. Zhang Z, Wang C, Xiao B, Zhou W, Liu S, Shi C. Cross-view action recognition via a continuous virtual path. In: IEEE conference on computer vision and pattern recognition; 2013. p. 2690–7.
11. Ma B, Su Y, Jurie F. Local descriptors encoded by fisher vectors for person re-identification. In: European conference on computer vision; 2012. p. 413–22.
12. Zheng Z, Zheng L, Yang Y. Unlabeled samples generated by gan improve the person re-identification baseline in vitro. In: International conference on computer vision; 2017. p. 3774–82.
13. Xiao T, Li H, Ouyang W, Wang X. Learning deep feature representations with domain guided dropout for person re-identification. In: IEEE conference on computer vision and pattern recognition; 2016. p. 1249–58.
14. Zheng Z, Zheng L, Yang Y. A discriminatively learned cnn embedding for person reidentification. ACM Trans Multimedia Comput Commun Appl. 2017;14(1):1–20.
15. Yi D, Lei Z, Liao S, Li SZ. Deep metric learning for practical person re-identification. In: International conference on pattern recognition; 2014. p. 34–9.

16. Sun Y, Zheng L, Yang Y, Tian Q, Wang S. Beyond part models: person retrieval with refined part pooling (and a strong convolutional baseline). arXiv preprint arXiv:1711.09349; 2018.
17. Dikmen M, Akbas E, Huang TS, Ahuja N. Pedestrian recognition with a learned metric. In: Asian conference on computer vision; 2010. p. 501–12.
18. Xiang S, Nie F, Zhang C. Learning a mahalanobis distance metric for data clustering and classification. Pattern Recognit. 2008;41(12):3600–12.
19. Cheng D, Gong Y, Zhou S, Wang J, Zheng N. Person re-identification by multi-channel parts-based cnn with improved triplet loss function. In: IEEE conference on computer vision and pattern recognition; 2016. p. 1335–44.
20. Zheng L, Shen L, Tian L, Wang S, Wang J, Tian Q. Scalable person re-identification: a benchmark. In: IEEE international conference on computer vision; 2015. p. 1116–24.
21. Krizhevsky A, Sutskever I, Hinton GE. Imagenet classification with deep convolutional neural networks. In: Neural information processing systems; 2012. p. 1097–105.
22. Simonyan K, Zisserman A. Very deep convolutional networks for large-scale image recognition. In: International conference on learning representations; 2015.
23. He K, Zhang X, Ren S, Sun J. Deep residual learning for image recognition. In: IEEE conference on computer vision and pattern recognition; 2016. p. 770–8
24. Felzenszwalb PF, Girshick RB, Mcallester DA, Ramanan D. Object detection with discriminatively trained part-based models. IEEE Trans Pattern Anal Mach Intell. 2010;32(9):1627–45.

A Modulation Recognition Method Based on Bispectrum and DNN

Jiang Yu, Zunwen He[(⊠)], and Yan Zhang

School of Information and Electronics, Beijing Institute of Technology,
Beijing 100081, People's Republic of China
hezunwen@bit.edu.cn

Abstract. In this paper, we propose a new method for modulation recognition of received digital signals using bispectrum and AlexNet. The bispectrum analysis is used to generate the feature images, AlexNet, as a widely used deep neural network (DNN), is used as the classifier. It is able to classify six common digital communication signals, including 2ASK, 4ASK, 2FSK, 4FSK, 2PSK and 4PSK. Compared to the traditional decision-theoretic methods, the proposed method needs no prior information for the received signals. The numerical results indicate that this method is more robust and effective than the classical decision theory and its improved algorithm, particularly when the signal-to-noise ratio (SNR) is low. It is shown that the success rate of 90% can be achieved when the SNR is greater than or equal to 3 dB.

Keywords: AlexNet · Bispectrum · CNN · DNN · Modulation recognition

1 Introduction

With the rapid development of wireless communication technology in modern society, the communication environment has become more and more complicated. Under this complex environment, in order to ensure that the two parties can transmit information accurately and efficiently, the communication signal adopts different modulation formats. For the received signal, signal demodulation and other communication tasks can be completed only on the premise of correct recognition of signal modulation mode and parameters. Therefore, the recognition of modulated signals is very important. Modulation recognition refers to determining the modulation type of the received communication signal with unknown modulation information content [1].

The modulation recognition problem has been considered in many articles that are summarized in [2]. All modulation recognition algorithms considered, there are divided into two groups: one group deals with likelihood functions and the other is based on features extracted from received signals. The decision-theoretic method that deals with likelihood function uses the properties of the signal, and the mathematical formula is obtained after deriving the principle formula and distinguishing it from the appropriate threshold. Under the criterion of Bayesian minimum cost of misjudgment, this method ensures that the recognition result is optimal, but in practical applications, the

© Springer Nature Singapore Pte Ltd. 2020
Q. Liang et al. (eds.), *Communications, Signal Processing, and Systems*, Lecture Notes in Electrical Engineering 516,
https://doi.org/10.1007/978-981-13-6504-1_108

decision-theoretic method has problems such as too many parameters required in the recognition process and the computational expressions of the likelihood ratio function are complicated, and the calculation amount is quite large [2].

Feature-based (FB) methods generally consist of feature extraction subsystem and the function of the pattern recognition subsystem. The function of the feature extraction subsystem is to extract features from the received data, and the features are predefined so that the dimensions represented by the patterns are reduced. The role of the second subsystem is to distinguish the type of modulation. As the feature extraction and classifier play a fatal role in the FB methods, much work has done to get a better result. Different artificial neural network (ANN) architectures are used for classification [3, 4]. As for the feature extraction, there are usually used spectral feature set (e.g., Instantaneous amplitude, frequency and phase) [3], statistical feature set (higher-order statistics) [5–7] and wavelet transform [8]. The advantage of FB methods is that the basic algorithm is simple, the selected parameters are sensitive, and it is very applicable in the field of complex signal recognition. The disadvantage is that noise has a great influence on its performance [2].

Recently, deep learning (DL) has become the core technology of artificial intelligence (AI). DL shines in domains such as image, speech and natural language processing, because it is hard to characterize real-world language or images with rigid mathematical models. By using a cascade of nonlinear processing unit layers with high dimensional input data, DL is good at feature extracting and transforming. Furthermore, unsupervised learning of the features and representations of the data is fundamental to DL [9]. As DNN will extract the features and use them by itself, it will reduce the demand of prior knowledge of the received signal.

In this paper, we propose a new method, which use the bispectrum analysis to get the bispectrum features that generate feature images, and use the convolutional neural network (CNN) as the classifier. By using this method, we can identify six modulation types such as 2FSK, 4FSK, 2PSK, 4PSK, 2ASK and 4ASK. Our experimental results show that this method performs well in recognition rate.

The remainder of the paper is organized as follows: Sect. 2 describes the system model, the theory of bispectrum analysis and the CNN classifier. Simulation result of the six modulation types is presented in Sect. 3. Finally, conclusions are made in Sect. 4.

2   Proposed Method Based on Bispectrum and DNN

2.1   System Model

The system model is shown in Fig. 1 as follows.

As shown in Fig. 1, our system model is consist of three parts. The digital receiver receives the signal through the antenna, and then down-converts the IQ two-way signal, and segments the complex signal according to the sampling rate. We perform nonparametric direct bispectrum estimation on each signal to get bispectrum features to generate 3D images, and then select appropriate view of angle to obtain 2D images. In order to improve recognition accuracy, discarding axis metrics and number symbols for

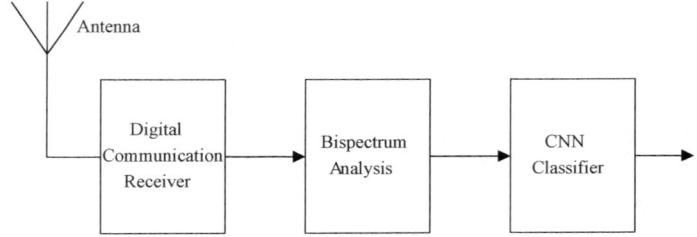

Fig. 1. System model

the generated images and selecting the white background as well. Set the image to 256 × 256 pixels, and use the folders as the image classification label, then put them into the CNN classifier to identify. The details are described as follows.

2.2 Signal Model

The initial phase of the carrier is assumed to zero, and the digital modulation signal with noise can be expressed as:

$$\widetilde{S}(t) = \overline{S}(t)\exp(j\omega_c t) + n(t) \tag{1}$$

where $\overline{S}(t)$ is the baseband signal, ω_c is the carrier frequency and $n(t)$ is the complex white Gaussian noise.

After down conversion, the ASK signal can be expressed as:

$$\overline{S}_{ASK}(t) = \sum_{i=1}^{N} \sqrt{S_i} p(t - iT_b) + n(t) \tag{2}$$

where T_b is the symbol cycle, $p(t)$ is the transmitted symbol waveform.

The FSK signal can be expressed as:

$$\overline{S}_{FSK}(t) = \sqrt{S} \sum_{i=1}^{N} e^{j(\omega_i t + \theta_i)} p(t - iT_b) + n(t) \tag{3}$$

where $\omega_i \in \{\omega_1, \omega_2, \omega_3, \ldots, \omega_M\}$, S is the signal power.

The PSK signal can be expressed as:

$$\overline{S}_{PSK}(t) = \sqrt{S} \sum_{i=1}^{N} e^{j\Phi_i} p(t - iT_b) + n(t) \tag{4}$$

where $\Phi_i \in \left\{\frac{2\pi}{M}(m-1), \ m = 1, 2, \ldots, M\right\}$, S is the signal power.

2.3   Bispectrum Analysis

Given higher-order cumulant $c_{kx}(\tau_1, \tau_2, \ldots, \tau_{k-1})$ of random sequence satisfy the next equation:

$$\sum_{\tau_1=-\infty}^{+\infty} \cdots \sum_{\tau_{k-1}=-\infty}^{+\infty} |c_{kx}(\tau_1, \tau_2, \ldots, \tau_{k-1})| < +\infty \tag{5}$$

then a k-order spectrum is the dimensional DFT of $k-1$ order cumulant, i.e.:

$$S_{kx}(\omega_1, \omega_2, \ldots, \omega_{k-1}) = \sum_{\tau_1=-\infty}^{+\infty} \cdots \sum_{\tau_{k-1}=-\infty}^{+\infty} |c_{kx}(\tau_1, \tau_2, \ldots, \tau_{k-1})| e^{-j\sum_{i=1}^{k-1} \omega_i \tau_i} \tag{6}$$

where $|\omega_i| \leq \pi, i = 1, 2, \ldots, k-1$, $|\omega_1 + \omega_2 + \cdots + \omega_{k-1}| \leq \pi$.

Bispectrum is the third-order spectrum, defined as:

$$B_x(\omega_1, \omega_2) = \sum_{\tau_1=-\infty}^{+\infty} \sum_{\tau_2=-\infty}^{+\infty} c_{3x}(\tau_1, \tau_2) e^{-j(\omega_1\tau_1 + \omega_2\tau_2)} \tag{7}$$

where $|\omega_i| \leq \pi$, $i = 1, 2$, $|\omega_1 + \omega_2| \leq \pi$.

$$c_{3x}(\tau_1, \tau_2) = E\{x^*(n)x(n + \tau_1)x(n + \tau_2)\} \tag{8}$$

where $*$ stands for the complex conjugate and $E\{\cdot\}$ denotes statistical expectation.

In general, two of the most popular conventional approaches to estimate the higher-order spectra are the indirect and direct methods, which may be seen as direct approximations of the definition of higher-order spectra [10], and in this paper we mainly adopt the latter bispectrum features of the 2ASK, 4ASK, 2FSK, 4FSK, 2PSK and 4PSK are illustrated in Fig. 2.

After that we have to select appropriate view of angle to obtain 2D images from 3D images as shown in Fig. 3. Due to the bispectrum symmetry, we set an observation angle of every $15°$ from $(-90°, 0°)$ to $(90°, 0°)$ with step of 2 dB from 0 to 8 dB according to the experimental setting in Sect. 3, generating 500 groups of samples per observation angle, of which 300 groups are used as training sets and 200 are used as validation sets, and use the same classifier as in Sect. 3. Experimental results show that the recognition rate of $(\pm 90°, 0°)$ and $(\pm 45°, 0°)$ is better than the other angles, and the recognition success rate between these four angles is almost the same, but $(-45°, 0°)$ requires a lower SNR when the recognition rate reaches 100%, so we use $(-45°, 0°)$ as the best viewing angle when generating 2D images from 3D images, and that may change when more signals are added for recognition. In order to improve recognition accuracy, discarding axis metrics and number symbols for the generated images and selecting the white background as well.

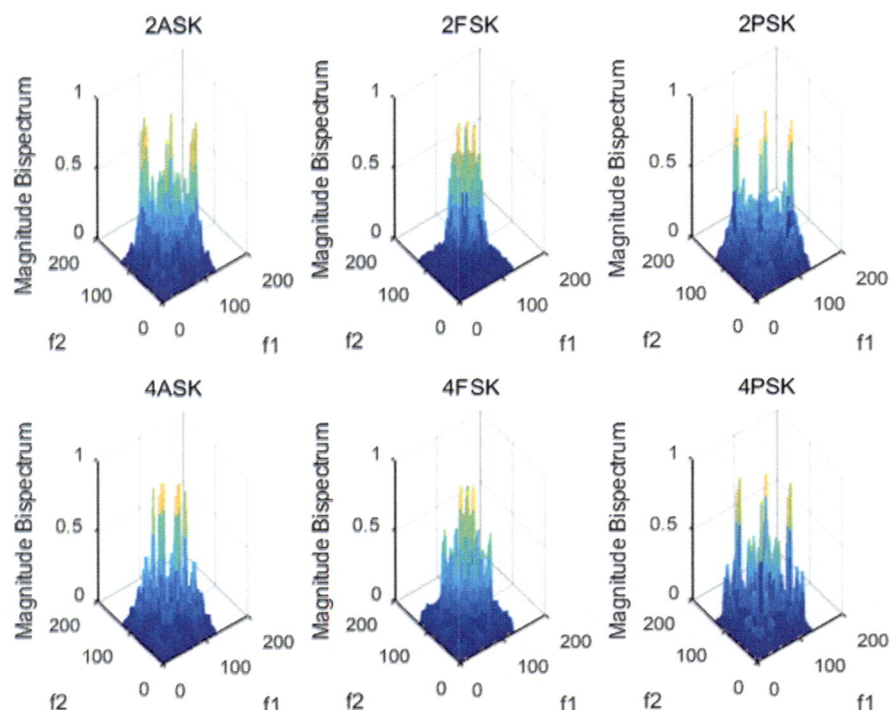

Fig. 2. Bispectrum features of the six modulated signals with SNR 5 dB

2.4   Deep Neural Network

Deep neural network (DNN) can express complex nonlinear relationships. There are some basic network architecture of DNN, such as fully connected feedforward neural network, convolutional neural network (CNN), recurrent neural network (RNN) and so on.

CNN is a famous model in architecture and it is developed from a fully connected feedforward network to prevent the rapid growth in the parameters when applied in image recognition. Generally, the basic architecture of CNN is to add pooling and convolutional layers before feeding into a fully connected network [11]. By taking most of the calculation, convolutional layer is the core part of CNN with the property of sparse connectivity and weight sharing. Convolutional layers pass the result to the next layer after applying a convolution operation to the input, and this operation involves the convolutional core. If the input image is fairly large, we would like to reduce the number of parameters with the pooling layer. The neurons in the pooling layer are grouped to compute for the mean value (average pooling) or maximum value (max pooling) to reduce the parameters before using the fully connected network. The fully connected layer is almost the same as in traditional neural networks that neurons have full connections to all activations in the previous layer.

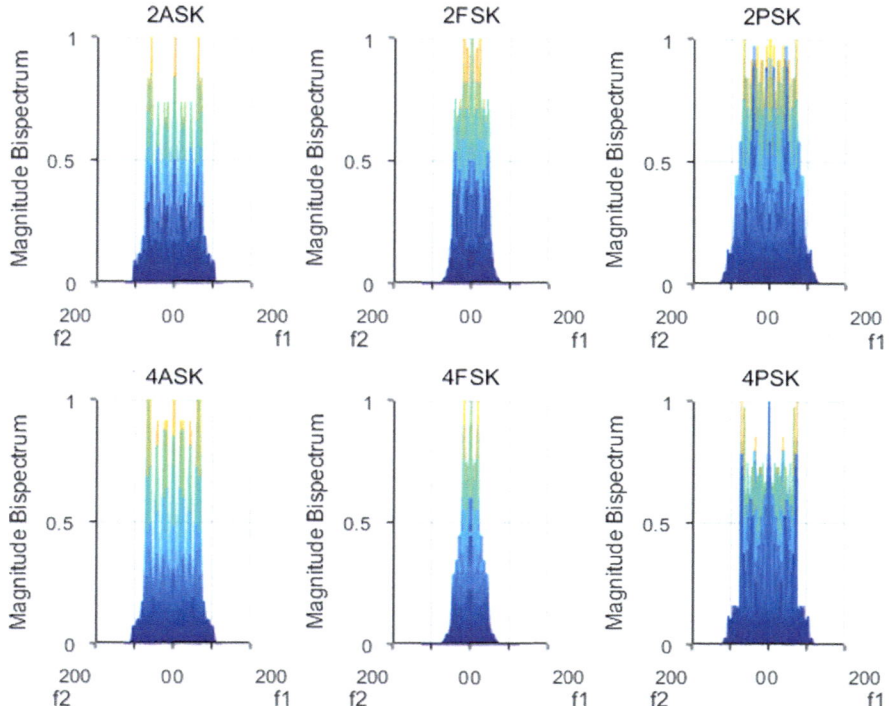

Fig. 3. Selected 2D images from 3D images as the input of the CNN classifier

AlexNet is consist of five convolutional layers and three fully connected layers with a 1000 ways softmax layer, and it proposes two primary methods that we can tackle with overfitting. One is "Dropout", the other is "Data Augmentation" [12]. To solve the vanishing gradient problem, AlexNet introduces a new activation function named rectified linear units (ReLU) to replace the classic sigmoid function.

As AlexNet performs well in image recognition, we choose it as the classifier, and the solver options of the AlexNet are set default while the base learning rate is 0.0001.

3 Performance Evaluation

3.1 Experimental Parameter Setup

In Sect. 2, we described how to generate classification images from received signals. In Sect. 3, we used simulation experiments to verify the validity of the proposed method. We use MATLAB 2016b communication signal toolbox to generate six kinds of simulation signals that are consistent with the parameters of [13], i.e., 2ASK, 4ASK, 2FSK, 4FSK, 2PSK and 4PSK.

The specific parameters of the simulation experiment were set as follows: the symbol rate was 500 bps, the carrier frequency was 2000 Hz, the sampling frequency was 12,000 Hz, the signal frequency offset was equal to the symbol rate, the carrier amplitude was 1, the number of symbols was 500, the SNR range from 0 to 20 dB, the noise is Gaussian white noise. For each signal, 500 signal samples were collected every 1 dB in a SNR of 0 to 20 dB, of which 300 were used for training set and 200 were used for testing set.

3.2 Comparison and Analysis

The comparison of the proposed method named bis and [13] are shown in Table 1 and Fig. 4.

Table 1. Average recognition rate of this method and that of [13]

SNR/dB	2ASK/%		4ASK/%		2FSK/%		4FSK/%		2PSK/%		4PSK/%		AVE/%	
	bis	[13]	bis	[13]	bis	[13]	bis	[13]	bis	[13]	bis	[13]	bis	[13]
0	81.60	21.50	80.90	0.80	90.30	0.00	90.00	0.00	82.30	0.00	80.30	0.00	84.23	3.72
2	89.60	96.15	85.30	97.80	91.30	14.40	91.30	14.20	87.30	0.00	84.60	0.00	88.23	37.10
4	94.30	100.0	94.60	100.0	97.30	34.10	98.30	71.70	93.60	50.00	92.60	99.10	95.12	75.80
6	98.90	100.0	97.30	100.0	100.0	90.90	100.0	93.30	98.60	99.30	97.60	100.0	98.73	97.25
8	100.0	100.0	100.0	100.0	100.0	100.0	100.0	100.0	100.0	100.0	100.0	100.0	100.0	100.0

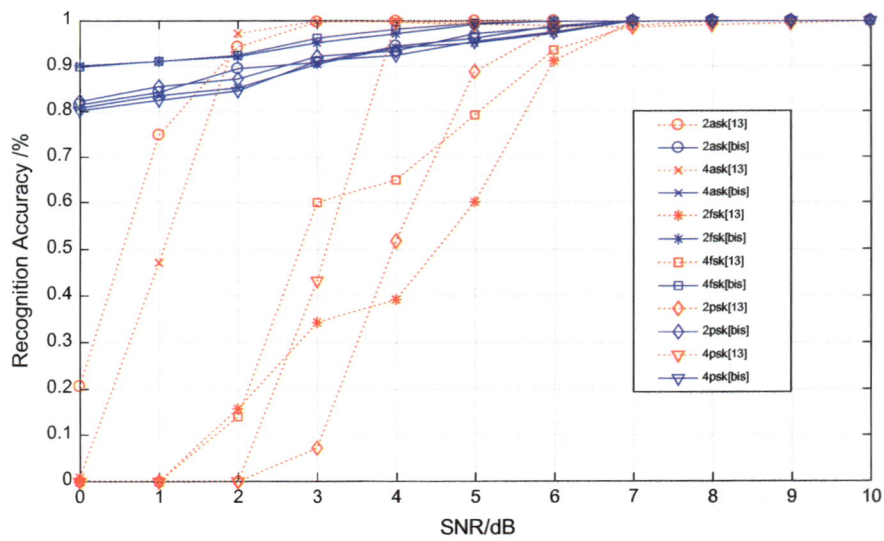

Fig. 4. Comparison of the recognition rate all six signals of the proposed method with [13]

Table 1 shows the comparison of the average recognition rate of the proposed method named bis and that of [13]. It can be seen from the table that the average recognition rate of this method is higher than that of [13] in 0–6 dB, which is more obvious at low SNR.

In Fig. 4, the blue solid line is the proposed method named bis and the red dotted line is the method of [13]. From Fig. 4, we can see that the proposed method generally performs better than that of [13] in distinguishing the 2PSK, 2FSK and 4FSK signals, especially in the low SNR. Furthermore, we can see that the proposed method can achieve a recognition success rate of approximately 100% when the SNR is greater than or equal to 7 dB, and success rate of 90% can be achieved when the SNR is greater than or equal to 3 dB, and success rate of 80% can be achieved when the SNR is no less than 0 dB. It performs much better than the algorithm in [4], which reaches 90% recognition rate when the SNR is 10 dB. Meanwhile, it has a better recognition effect compared with [13] when the SNR is low.

4 Conclusions

In this paper, we propose a new modulation recognition method based on bispectrum and DNN. The bispectrum analysis is used to generate the feature images, AlexNet, as a widely used DNN, is used as the classifier. The problem of image angle selection is analyzed and some useful conclusions are obtained. The numerical results show that compared with the classical decision theory and its improved algorithm, the proposed method performs better at low SNR, and success rate of 90% can be achieved when the SNR is greater than or equal to 3 dB, and success rate of 100% can be achieved when the SNR is no less than 7 dB.

Acknowledgments. This work was supported in part by National Nature Science Foundation of China under Grants No. 61201192 and the National High Technology Research and Development Program of China under Grants No. 2015AA01A708.

References

1. Azzouz EE, Nandi AK. Algorithms for automatic modulation recognition of communication signals. IEEE Trans Commun. 1998;46(4):431–6.
2. Dobre OA, Abdi A, Bar-Ness Y, et al. A survey of automatic modulation classification techniques: classical approaches and new trends. IET Comm. 2007;1(2):137–56.
3. Wong MLD, Nandi AK. Automatic digital modulation recognition using artificial neural network and genetic algorithm. Sig Process. 2004;84(2):351–65.
4. Park CS, Kim DY. Modulation classification of analog and digital signals using neural network and support vector machine. In: 4th international symposium on neural networks, ICNN2007, Nanjing, Jun 2007.
5. Lopatka J, Pedzisz M. Automatic modulation classification using statistical moments and a fuzzy classifier. In: International conference on signal processing, vol. 3, 61(24); 2000. p. 1500–6.
6. Ebrahimzadeh A, Ardeshir GR. A new signal type classifier for fading environments. J Comp Info Technol. 2007;15(3):257–66.
7. Wong MDL, Nandi AK. Efficacies of selected blind modulation type detection methods for adaptive OFDM systems. In: Proceedings of the international conference on signal processing and communication systems, ISCPSC2007; Dec 2007.

8. Liu J, Luo Q. A novel modulation classification algorithm based on daubechies5 wavelet and fractional fourier transform in cognitive radio. In: IEEE ICCT; 2012. p. 115–20.
9. Li D, Yu D. Deep learning: methods and applications. Found Trends Sig Process. 2014;7(3–4):197–387.
10. Nikias CL, Mendel JM. Signal processing with higher-order spectra. IEEE Sig Process Mag. 1993;10(3):10–37.
11. Lin Y, Tu Y, Dou Z et al. The application of deep learning in communication signal modulation recognition. In: 2017 IEEE/CIC international conference on communications in China (ICCC), Qingdao; Oct 2017.
12. Krizhevsky A, Sutskever I, Hinton G. ImageNet classification with deep convolutional neural networks. In: International conference on neural information processing systems, p. 1097–105 (2012).
13. Zhang DM, Wang X. Improved method of digital signals modulation identification based on decision theory. J Comput Appl. 2009;29(12):3227–30.

Image-to-Image Local Feature Translation Using Double Adversarial Networks Based on CycleGAN

Chen Wu$^{(\boxtimes)}$, Lei Li, Zhenzhen Yang, Peihong Yan, and Jiali Jiao

National Engineering Research Center of Communication
and Network Technology School of Science, Nanjing University
of Posts and Telecommunications, Nanjing 210003, China
wc915495936@live.com

Abstract. Image-to-image translation is a hot field in the machine learning with the emergency of the generative adversarial networks. Most of the latest models easily lead to changes in the overall image and overfitting when they are used to local feature translation. To address these limitations, this article adds a suppressor and proposes a double adversarial CycleGAN. The suppressor is added to suppress the change of images, and the suppressor and generator form a new adversarial relationship. We hope it will achieve Nash equilibrium that is the change of image focus on the local feature. Finally, a contrast experiment was conducted. In the case of image local feature transfer, the change of image is focused on the local features and the overfitting phenomenon can be well resolved.

Keywords: Local feature translation · Generative adversarial networks · Double adversarial CycleGAN

1 Introduction

The image-to-image translation [1] was defined as the task of translating one possible representation of a scene into another. The ideal of image-to-image translation starts with the Image Analogies [2]. Lots of researches in computer version, image processing, computational photography, and graphics have been done to learn the mapping between images. The image-to-image translation contains lots of applications such as style transfer, object transfiguration, season transfer, and photograph enhancement. In this paper, we focus on the local feature transfer. Recently, as the emergence of the generative adversarial networks [3], a lot of methods based on GANs have been proposed to deal with the image-to-image translation. Some methods [4,5] were proposed to overcome the

© Springer Nature Singapore Pte Ltd. 2020
Q. Liang et al. (eds.), *Communications, Signal Processing,
and Systems*, Lecture Notes in Electrical Engineering 516,
https://doi.org/10.1007/978-981-13-6504-1_109

difficulty that it is hard to get a large quantity of paired images. The dual learning approaches [6–10] have been further exploited to the mapping between source image domain and target image domain. Based on these theories, the researchers proposed some image-to-image translation models such as pix2pix [11], Cycle-GAN [10], StarGAN [12], and GeneGAN [13]. Compared to other generative model, GAN can produce clearer images and the state-of-the-art models could learn the mapping of two domains through the unpaired data. However, when these models are used to image-to-image local feature translation by unaligned data, it will easily lead to overall image changes and overfitting. The aligned data means the local feature that we want to transfer is in the same position of all pictures. We try to use the CelebFaces Attributes Dataset for CycleGAN to train the *glass* ↔ *noglass* translation model (local feature) and fined that it changes the tint of face slightly (see Fig. 1). The CelebFaces Attributes Dataset that we used is the aligned dataset. However, when we try to use the unaligned *horse* ↔ *zebra* data for CycleGAN, the tint of the generated images is changed obviously (see Fig. 4). The CycleGAN does a good job in image-to-image style translation. However, when we use it in image-to-image local feature translation, it will lead to global changes and overfitting of the images. The other latest models rely on data alignment and they did not try to solve the potential problems in image-to-image local feature translation. It is a big step from paired data to unpaired data to reduce the difficulty of data preprocessing. The StarGAN, GeneGAN, and ELEGAN use the aligned CelebFaces Attributes Dataset and the

glass → *noglass*   *noglass* → *glass*

Fig. 1. Comparison chart: The first column (1): original people wearing glasses. The second column (2): *glass* → *noglass* using CycleGAN. The third column (3): original people who do not wear glasses. The fourth column (4): *noglass* → *glass* using CycleGAN

Radboud Faces Database to train the modules. Learning the mapping between domains by aligned data is easier than using the unaligned data. When we use the CycleGAN to learn the mapping *glass* ↔ *noglass* by the aligned CelebFaces Attributes Dataset, the model works well and changes the tint of face a little bit. However, when we use the CycleGAN to learn the mapping *horse* ↔ *zebra* by the unaligned data, the model changes the tint of the images obviously and it is easy to overfit. If we need to train new attributes, it will spend us lots of time aligning dataset. The purpose of this paper is to find a way to make the image-to-image translation focus on the local feature by the unaligned data. Our goal is to learn the mappings G, F exactly between the domain X and domain $Y(G : X \rightarrow Y, F : Y \rightarrow X)$. The CycleGAN introduces an additional loss to inhibit the change of images. It was the technique of Taigman [14]

$$L_{identity}(G, F) = \mathbb{E}_{y \sim p_{data(y)}}[\|G(y) - y\|_1] + \mathbb{E}_{x \sim p_{data(x)}}[\|F(x) - x\|_1] \quad (1)$$

We try this method in local feature image-to-image translation, but this method is not good enough. The additional loss regularizes the generator directly; it means we hope the generator can change the images domain and inhibit the change of images at the same time. It is difficult for the generator to achieve the balance. In this paper, we propose a double adversarial network based on Cycle-GAN to improve the local feature image-to-image translation by the unaligned data. We let the $L_{identity}(G, F)$ as a suppressor instead of a regular. Therefore, the generator and the suppressor can train the G, F alternately. We find that the confrontational relationship between generator and suppressor can accelerate convergence.

2   Related Work

2.1   Generative Adversarial Networks

The GAN is a competition between two networks. The generator network receives a source of noise and exports the corresponding fake data samples. The discriminator network receives the fake data samples and the real data samples, and then it distinguishes the fake data samples and the real data samples. The aim of the generator is to fool the discriminator, and the aim of the discriminator is to distinguish samples better. The minimax objective function of GAN:

$$\min_G \max_D \mathop{E}_{x \sim p_r}[\log(D(x))] + \mathop{E}_{x \sim p_g}[\log(1 - D(x))] \quad (2)$$

where the p_r is the true data distribution and the p_g is the output distribution of generator. Firstly, the discriminator is trained to optimality. Secondly, when minimizing the objective function to minimize the Jensen–Shannon divergence between p_r and p_g, there are some problems with original GAN such as collapse problem, non-convergence problem, and vanishing gradient problem. Some methods were proposed to solve these problems such as DCGAN [15], LS-GAN [16], WGAN-gp [17]. CycleGAN uses the LS-GAN instead of the original GAN.

2.2   Cycle-Consistent Generative Adversarial Network

CycleGAN is a model algorithm based on the use of two generators and two discriminators to transform the distribution of two types of images based on generative adversarial network. The purpose of CycleGAN is to learn the mapping relationship between two data distributions X and Y. Assuming only the generator and discriminator, the generator does not learn the mapping relationship between the two data distributions very well. It is possible that all the images of one data distribution are mapped to one picture in another data distribution. Therefore, the cycle-consistent loss is proposed. The forward cycle consistency is $x \to G(x) \to F(G(x)) \approx x$, and the backward cycle consistency is $y \to F(y) \to G(F(y)) \approx y$ (see Fig. 2). The full loss function of CycleGAN is:

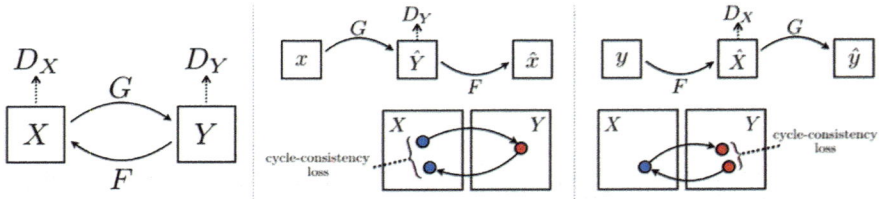

Fig. 2. CycleGAN tries to learn the mapping functions $G : X \to Y$ and $F : Y \to X$ by the loss function $L(G, F, D_{(X)}, D_{(Y)})$ combined by $L_{GAN}(G, D_Y, X, Y)$, $L_{GAN}(F, D_X, Y, X)$, and cycle-consistent loss.

$$L(G, F, D_X, D_Y) = L_{GAN}(G, D_Y, X, Y) + L_{GAN}(F, D_X, Y, X) + \lambda L_{cyc}(G, F) \tag{3}$$

$$L_{GAN}(G, D_Y, X, Y) = \mathbb{E}_{y \sim p_{data(y)}}[\log(D_Y(y))] + \mathbb{E}_{x \sim p_{data(x)}}[\log((1 - D_Y(G(x))))] \tag{4}$$

$$L_{GAN}(F, D_X, Y, X) = \mathbb{E}_{x \sim p_{data(x)}}[\log(D_X(x))] + \mathbb{E}_{y \sim p_{data(y)}}[\log((1 - D_X(G(y))))] \tag{5}$$

$$L_{cyc}(F, D_X, Y, X) = \mathbb{E}_{x \sim p_{data(x)}}[\|F(G(x)) - x\|_1] + \mathbb{E}_{y \sim p_{data(y)}}[\|F(G(y)) - y\|_1] \tag{6}$$

CycleGAN does a good job in unpaired image-to-image style translation such as season transfer and painting $\leftrightarrow$ photographs. However, the performance of CycleGAN in image-to-image local feature translation is not satisfied. It is easy to change the tint of the pictures, and it easily leads to the overfitting.

In the paper of CycleGAN, it adopts the technique of Taigman by a additional loss $L_{identity}(G, F) = \mathbb{E}_{y \sim p_{data(y)}}[\|G(y) - y\|_1] + \mathbb{E}_{x \sim p_{data(x)}}[\|F(x) - x\|_1]$. The purpose of the $L_{identity}(G, F)$ is to decrease the change between the image-to-image translation $G(x), F(y)$. It has a good performance in image-to-image style translation. However, when I try to use the new loss function in image-to-image local feature translation such as $horse \leftrightarrow horse$ or $glass \leftrightarrow noglass$, it does not work well.

3   Double Adversarial CycleGAN

When we try to transfer the $horses \rightarrow zebras$ by CycleGAN, the tint of the pictures is changed at the same time. The phenomenon of overfitting is occurred. Without the $L_{identity}(G, F)$, the CycleGAN is easy to lead to the overall change in the images. However, when we add the $L_{identity}(G, F)$ as a regular to the loss, we find that it is not easy for the generator to achieve the balance. It can even lead to the phenomenon that the local feature translation that should be changed is suppressed and the background area of the picture is changed. We can see that the translation of $horse \rightarrow zebra$ is more focused on local feature by the double adversarial CycleGAN (see Fig. 4). The experiment shows that the generated horse still has some zebra stripes (see Fig. 5). The generator is difficult to learn the mapping of $zebra \rightarrow horse$. When we add the regular term or the suppressor to the CycleGAN, it will lead to more zebra stripes. We have not yet solved this problem.

According to these problems, we take some methods to solve them. The previous method is to add the $L_{identity}(G, F)$ to the generator's loss function during the training process. It means that the $L_{identity}(G, F)$ loss is a regular term. When the generator tries to do image-to-image translation, it also limits the change of image-to-image translation at the same time. It is hard for the generator to learn the balance in one step. Inspired by the two-person zero-sum game, we let the $L_{identity}(G, F)$ as a suppressor and the generator can achieve the balance stably step by step (see Fig. 3). Therefore, the generator–discriminator and generator–suppressor constitute a double adversarial cycle-consistent loss. The objective function of suppressor:

$$min \quad S = \mathbb{E}_{y \sim p_{data(y)}}[\|G(y) - y\|_1] + \mathbb{E}_{x \sim p_{data(x)}}[\|F(x) - x\|_1]$$

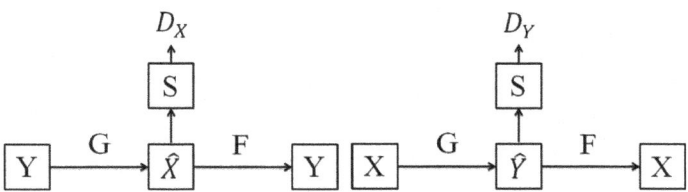

Fig. 3. The suppressor(S) is used to suppress the difference between the generated images and the original images.

4   Experiment

In this section, we compare these three methods: CycleGAN, CycleGAN + $L_{identity}(G, F)$, and double adversarial CycleGAN. As shown in Figs. 4 and 5, the CycleGAN model changes the tint of images obviously. The overall color of the image has become deeper, and generated ($horse \rightarrow zebra$) stripes are not clear. The CycleGAN + $L_{identity}(G, F)$ model improves the phenomenon of the tint change in image-to-image translation. However, the generated zebra ($horse \rightarrow zebra$) stripes are also not clear. From the fourth column of Fig. 4, we can see the generated zebra ($horse \rightarrow zebra$) stripes are clear and the tint of image is stable.

(1)   (2)   (3)   (4)

Fig. 4. Comparison chart ($horse \rightarrow zebra$): The first column (1): original images. The second column (2): $horse \rightarrow zebra$ using CycleGAN. The third column (3): $horse \rightarrow zebra$ using CycleGAN+$L_{identity}(G, F)$. The fourth column (4): $horse \rightarrow zebra$ using double adversarial CycleGAN

The Inception Score (IS) and Fréchet Inception Distance (FID) [18] are two ways to test the quality of generated pictures. We use the FID score to measure the distance between the real data and the generated data.

$$d^2((m, C), (m_\omega), C_\omega) = ||m - m_\omega||_2^2 + Tr(C + C_\omega - 2(CC_\omega)^{1/2}) \qquad (7)$$

(1)   (2)   (3)   (4)

Fig. 5. Comparison chart (*zebra → horse*): The first column (1): original images. The second column (2): *zebra → horse* using CycleGAN. The third column (3): *zebra → horse* using CycleGAN+$L_{identity}(G, F)$. The fourth column (4): *zebra → horse* using double adversarial CycleGAN

(m, C) and (m_ω, C_ω) are the means and covariance matrices of the real data distribution and fake data distribution. This table shows the FID of generated images. We can see the result of CycleGAN+$L_{identity}(G, F)$ is worse than Cycle-GAN. Double adversarial CycleGAN has a good performance in *horse → zebra*, and it is not suitable for the *zebra → horse*. However, both IS and FID are not suitable for our work. We need to measure the ratio of global and local changes in the image-to-image translation and the quality of the local features from generated pictures. We have not done this yet (Table 1).

Table 1. FID score

Name	CycleGAN	CycleGAN +$L_{identity}$ (G, F)	Double adversarial CycleGAN
FID(*zebra → horse*)	68.4	86.5	75.4
FID(*horse → zebra*)	69.0	65.6	51.9

This table shows the FID of generated images.

5 Conclusion

In this paper, we propose a double adversarial CycleGAN. Under the observation that only local part of the image should be modified in this task, we let the $L_{identity}(G, F)$ as a suppressor instead of regular term and it can accelerate convergence and achieve the balance more quickly. The confrontation between generator and suppressor can make the training more stable.

Acknowledgments. This work was supported by National Natural Science Foundation of China (Grant No. 61501251, 61373137, 61071167) and the Science Foundation of Nanjing University of Posts and Telecommunications Grant (NY214191).

References

1. Liu M-Y, Breuel T, Kautz J. Unsupervised image-to-image translation networks. CoRR, abs/1703.00848, 2017.
2. Hertzmann A, Jacobs CE, Oliver N, Curless B, Salesin DH. Image analogies. In: Proceedings of the 28th annual conference on computer graphics and interactive techniques, SIGGRAPH '01. ACM: New York, NY, USA, 2001. p. 327–40.
3. Goodfellow I, Pouget-Abadie J, Mirza M, Xu B, Warde-Farley D, Ozair S, Courville A, Bengio Y. Generative adversarial nets. In: Ghahramani Z, Welling M, Cortes C, Lawrence ND, Weinberger KQ, editors. Advances in neural information processing systems 27. New York: Curran Associates, Inc.; 2014. p. 2672–80.
4. Perarnau G, van de Weijer J, Raducanu B, Álvarez JM. Invertible conditional gans for image editing. CoRR, abs/1611.06355, 2016.
5. Zhu J-Y, Krähenbühl P, Shechtman E, Efros AA. Generative visual manipulation on the natural image manifold. In: Proceedings of European conference on computer vision (ECCV), 2016.
6. Xia Y, He D, Qin T, Wang L, Yu N, Liu T-Y, Ma W-Y. Dual learning for machine translation. CoRR, abs/1611.00179, 2016.
7. Shen W, Liu R. Learning residual images for face attribute manipulation. CoRR, abs/1612.05363, 2016.
8. Kim T, Cha M, Kim H, Lee JK, Kim J. Learning to discover cross-domain relations with generative adversarial networks. CoRR, abs/1703.05192, 2017.
9. Yi Z, Zhang H, Tan P, Gong M. Dualgan: Unsupervised dual learning for image-to-image translation. CoRR, abs/1704.02510, 2017.
10. Zhu J-Y, Park T, Isola P, Efros AA. Unpaired image-to-image translation using cycle-consistent adversarial networks. CoRR, abs/1703.10593, 2017.
11. Isola P, Zhu J-Y, Zhou T, Efros AA. Image-to-image translation with conditional adversarial networks. CoRR, abs/1611.07004, 2016.
12. Choi Y, Choi M-J, Kim M, Ha J-W, Kim S, Choo J. Stargan: Unified generative adversarial networks for multi-domain image-to-image translation. CoRR, abs/1711.09020, 2017.
13. Zhou S, Xiao T, Yang Y, Feng D, He Q, He W. Genegan: Learning object transfiguration and attribute subspace from unpaired data. CoRR, abs/1705.04932, 2017.
14. Taigman Y, Polyak A, Wolf L. Unsupervised cross-domain image generation. CoRR, abs/1611.02200, 2016.
15. Radford A, Metz L, Chintala S. Unsupervised representation learning with deep convolutional generative adversarial networks. CoRR, abs/1511.06434, 2015.

16. Qi G-J. Loss-sensitive generative adversarial networks on lipschitz densities. CoRR, abs/1701.06264, 2017.
17. Gulrajani I, Ahmed F, Arjovsky M, Dumoulin V, Courville AC. Improved training of wasserstein gans. CoRR, abs/1704.00028, 2017.
18. Heusel M, Ramsauer H, Unterthiner T, Nessler B, Klambauer G, Hochreiter S. Gans trained by a two time-scale update rule converge to a nash equilibrium. CoRR, abs/1706.08500, 2017.

Evaluation Embedding Features for Ground-Based Cloud Classification

Zhong Zhang[1,2](✉), Donghong Li[1,2], and Shuang Liu[1,2]

[1] Tianjin Key Laboratory of Wireless Mobile Communications and Power Transmission, Tianjin Normal University, Tianjin, China
{zhong.zhang8848,donghongli1139,shuangliu.tjnu}@gmail.com
[2] College of Electronic and Communication Engineering, Tianjin Normal University, Tianjin, China

Abstract. Ground-based cloud classification plays a vital important role in meteorological research. However, the existing methods perform well confined to one weather station. In this paper, we present a detailed introduction of two representative embedding features for ground-based cloud classification in various weather stations. The features are learned from the metric learning and the convolutional neural network (CNN), respectively. The two kinds of features are evaluated on two weather stations.

Keywords: Ground-based cloud classification · Embedding features · Metric learning · Convolutional neural networks

1 Introduction

Ground-based cloud classification has an important impact on meteorological studies [1]. However, the professional human observers are the main conductors to implement this task, resulting in mass consumption of human efforts and material resources. In order to address this issue, many researchers have focused on developing automatic techniques for ground-based cloud classification in recent years. Isosalo et al. [2] classified ground-based cloud images by utilizing local binary patterns (LBPs) and local edge patterns (LEPs), and LBPs had better performances than LEPs. Calbó and Sabburg [3] represented cloud images using pattern features based on the Fourier transformation and statistical features. Liu et al. [4] extracted both texture and structure features from the color cloud images. Liu et al. [5] proposed illumination-invariant completed local ternary patterns (ICLTP), which can effectively handle the illumination variations. They soon proposed the salient LBP (SLBP) [6] to capture descriptive cloud information. The desirable property of SLBP is the robustness to noise.

© Springer Nature Singapore Pte Ltd. 2020
Q. Liang et al. (eds.), *Communications, Signal Processing, and Systems*, Lecture Notes in Electrical Engineering 516,
https://doi.org/10.1007/978-981-13-6504-1_110

Recently, Ye et al. [7] first applied the convolutional neural network (CNN) to ground-based cloud classification, and they employed the Fisher Vector to encode the deep convolutional features. Shi et al. [8] concluded that the deep convolutional activation-based features outperformed traditional hand-craft features considerably for ground-based cloud classification. However, these methods are only suitable for one certain weather station. That means they can achieve desired classification accuracies on the certain dataset captured by the weather station, but obtain poor performances on other weather stations. In other words, these methods are not generalized to other weather stations, because the cloud images captured by them generally possess variations in illuminations, images resolutions, occlusions, camera settings, etc., as shown in Fig. 1. Therefore, we argue that training a classifier on one weather station, while test on another weather station.

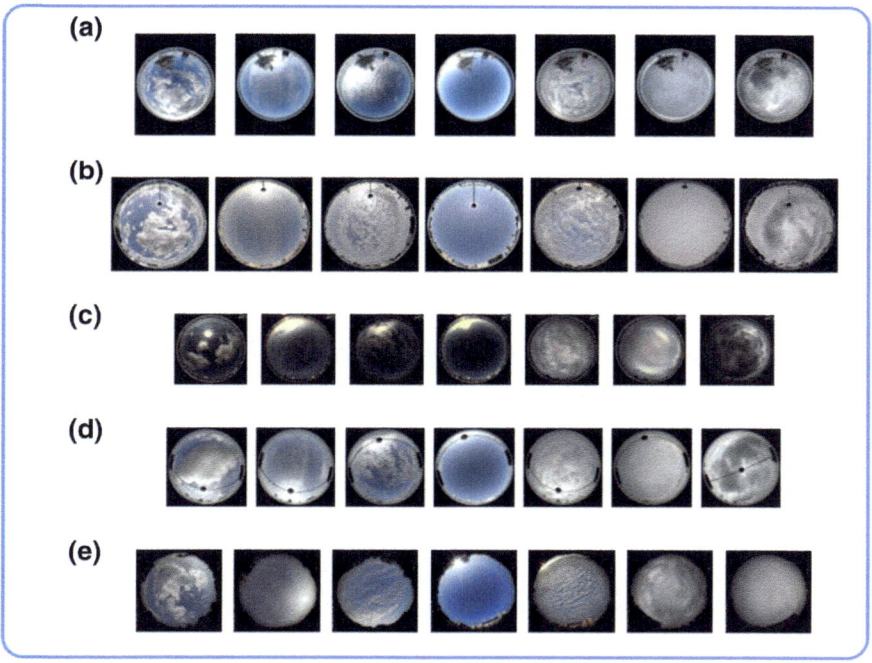

Fig. 1. Cloud samples are provided by various weather stations (**a**)–(**e**).

In this paper, we evaluate different cloud images provided by various weather stations based on two kinds of embedding features. The first one is to learn a subspace using metric learning methods. Metric learning has been successfully applied to the domain adaptation [9,10]. The metric learning is suitable for the situation that training on one weather station, while test on other weather station. The second one is the output of the fully connected (FC) layer of the CNN.

The CNN has achieved excellent performances in image classification [11–13], and it is a desirable choice to apply the CNN to ground-based cloud classification.

The remaining of the paper is organized as follows. Section 2 introduces the metric learning and CNN. Section 3 shows the experimental results of various weather stations based on the two embedding features. We conclude the paper in Sect. 4 the paper.

2   Approach

2.1   Metric Learning

First, the cloud images are divided into two sets of pairs where a pair consists of two cloud images. The first set S includes similar pairs where the two cloud images of each similar pair belong to the same cloud category. The second set D consists of dissimilar pairs where the two cloud images of each dissimilar pair belong to different cloud categories. The metric learning considers the relationships of pairs to learn a transformation matrix, which is used to project the cloud images into a subspace. Suppose that there is a pair (i, j), where $i \in \mathbb{R}^{d \times 1}$ and $j \in \mathbb{R}^{d \times 1}$ are from different weather stations. We learn a transformation matrix $H \in \mathbb{R}^{d \times m}$ which parameterizes the (squared) Mahalanobis distance:

$$d_M(i, j) = (i - j)^T M(i - j),\tag{1}$$

where $M = HH^T (M \in \mathbb{R}^{d \times d})$. If i and j belong to a similar pair, there is:

$$(i, j) \in S,\tag{2}$$

in the same way, if i and j belong to a dissimilar pair, there is:

$$(i, j) \in D.\tag{3}$$

Based on the Mahalanobis distance, in order to learn H, Xing et al. [14] proposed a metric that minimizes the distance between all similar pairs:

$$\min_{M} \sum_{(i,j) \in S} (i - j)^T M(i - j)$$

$$s.t. \sum_{(i,j) \in D} \sqrt{(i - j)^T M(i - j)} \geq 1,\tag{4}$$

$$M \geq 0.$$

The first constraint ensures that M does not collapse the dataset into a single point, and the second one ensures a valid metric.

Ghodsi et al. [15] proposed to minimize the squared distance between similar pairs, while maximize the squared distance between dissimilar pairs. Similarly, Globerson and Roweis [16] learned a metric by keeping images from the same category near each other and simultaneously far from the images from other categories. Weinberger et al. [17] utilize semidefinite programming to learn a metric for k-nearest neighbor (kNN) classification.

2.2   Convolutional Neural Network

The CNN model can learn the high nonlinear transformation for the input data. Thus, the features extracted from FC layer can be viewed as embedding features. The VGG-11 [13] model is representative, and therefore, we evaluate the VGG-11 model on the task of ground-based cloud classification. There are eight convolutional layers and three fully connected (FC) layers composing the VGG-11 model. The cloud images as the inputs of the VGG-11 model should be resized to 224 × 224 pixels. The filters of each convolutional layer are with a receptive field of 3 × 3 pixels, and the convolution stride is set to 1 pixel. Five max-pooling layers follow some of convolutional layers and are implemented by a sliding window of 2 × 2 pixels with stride 2. The first two FC layers both consist of 4096 neurons, and the last one includes 1000 neurons where each neuron denotes a category of the ImageNet dataset. However, when the VGG-11 model is fine-tuned on the ground-based cloud datasets, we change 1000 into the number of cloud categories, as illustrated in Fig. 2. We take the output of the second FC layer as features, which contain the overall spatial layout information. That means a cloud image is represented as 4096-dimensional vector.

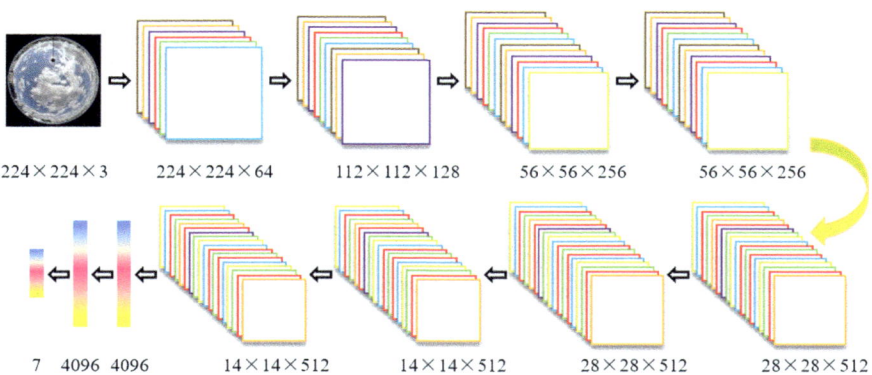

Fig. 2. Cloud image with the size of 224 × 224 × 3 is fed into the VGG-11 model, which results in a series of feature maps with different sizes. The final output is 7 neurons corresponding to the number of cloud categories.

3   Experiments

3.1   Datasets and Experimental Setup

The are two cloud datasets serving for the following evaluations. The Institute of Atmospheric Physics, Chinese Academy of Sciences, provides the first cloud dataset, named IAP_e. There are seven categories and 3533 cloud images in this dataset. The cloud images from the IAP_e dataset are 2272 × 1704 pixels. The second cloud dataset is named as CAMS_e with the number of 2491 cloud images, provided by Chinese Academy of Meteorological Sciences. Each cloud image has the resolution of 1392 × 1040. Since the two cloud datasets are captured from different weather stations, the IAP_e is quitely different from the CAMS_e in the aspects of camera settings, resolutions, occlusions, and illuminations, as shown in Fig. 3.

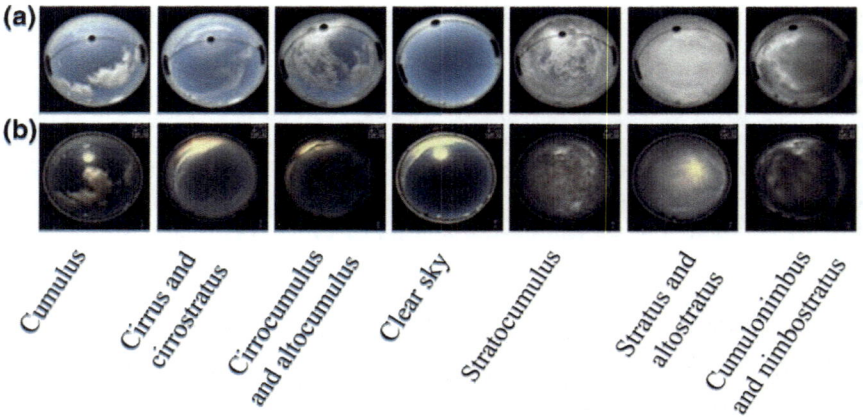

Fig. 3. **a** The IAP_e cloud dataset. **b** The CAMS_e cloud dataset.

All images from the two datasets are resized to 224 × 224 pixels. We extract LBP features with (P, R) equal to $(8, 1)$, $(16, 2)$, and $(24, 3)$, and then concatenate histograms of the three scales to form a feature vector with the size of $10 + 18 + 26 = 54$. The M is finally multiplied by the resulting feature vector to form a transformed feature vector for each cloud image. We select $k = 150$. When we fine-tune the VGG-11 [13] model on our cloud datasets, we set the batch size and the number of training epochs to 32 and 60, respectively. The learning rate is initialized as 0.0001 and then set to 0.00006 for the final 20 epochs. A half of cloud images is randomly selected from each category and consists of the training set, and the remaining as the test set. The final results represent the recognition accuracy averaged over ten randomly selected training/test sets. The nearest neighborhood classifier is used to classify the cloud images.

Table 1 lists the classification results of the two embedding features on different weather stations. It is obvious that the high classification results are achieved when training and testing on the same dataset. However, we obtain poor classification performances when training and testing on different datasets. Moreover, the CNN performs much better than the metric learning. To conclude, it is essential to make a further study for the ground-based cloud classification in various weather stations.

Table 1. Classification results (%) using different embedding features.

Datasets		Methods	
		Metric learning	CNN
IAP_e	IAP_e	86.93	92.74
CAMS_e	CAMS_e	89.42	93.81
IAP_e	CAMS_e	38.65	53.84
CAMS_e	IAP_e	36.81	52.07

4 Conclusion

In this paper, we have introduced two kinds of representative embedding features which are learned from the metric learning and the CNN, respectively. Then, we evaluated the two features on two weather stations. The experimental results show that the metric learning and the CNN perform poorly when training on weather station, while test on another weather station. In other words, they are unsatisfied with the ground-based cloud classification in different weather stations, and it is still eager to develop effective methods on this issue.

Acknowledgments. This work was supported by National Natural Science Foundation of China under Grant No. 61501327 and No. 61711530240, Natural Science Foundation of Tianjin under Grant No. 17JCZDJC30600 and No. 15JCQNJC01700, the Fund of Tianjin Normal University under Grant No.135202RC1703, the Open Projects Program of National Laboratory of Pattern Recognition under Grant No. 201700001 and No. 201800002, the China Scholarship Council No. 201708120039 and No. 201708120040, and the Tianjin Higher Education Creative Team Funds Program.

References

1. Chen Z, Zen D, Zhang Q. Sky model study using fuzzy mathematics. J Illum Eng Soc. 1994;23:52–8.
2. Isosalo A, Turtine M, Pietikäinen M. Cloud characterization using local texture information. In: Finnish signal processing symposium, Kuopio, Finland; 2007. p. 1–6.
3. Calbó J, Sabburg J. Feature extraction from whole-sky ground-based images for cloud-type recognition. J Atmos Ocean Tech. 2008;25:3–14.
4. Liu L, Sun X, Chen F, Zhao S, Gao T. Cloud classification based on structure features of infrared images. J Atmos Ocean Tech. 2011;28:410–7.
5. Liu S, Wang C, Xiao B, Zhang Z, Shao Y. Illumination-invariant completed LTP descriptor for cloud classification. In: International congress on image and signal processing, Chongqing, China; 2012. p. 449–53.
6. Liu S, Wang C, Xiao B, Zhang Z, Shao Y. Salient local binary pattern for ground-based cloud classification. Acta Meteorol Sin. 2013;27:211–20.
7. Ye L, Cao Z, Xiao Y. DeepCloud: ground-based cloud image categorization using deep convolutional features. IEEE Trans Geosci Remote. 2017;55:5729–40.

8. Shi C, Wang C, Wang Y, Xiao B. Deep convolutional activations-based features for ground-based cloud classification. IEEE Geosci Remote Sens Lett. 2017;14:816–20.
9. Ding Z, Fu Y. Robust transfer metric learning for image classification. IEEE Trans Image Process. 2017;26:660–70.
10. Saenko K, Kulis B, Fritz M, Darrell T. Adapting visual category models to new domains. In: European conference on computer vision, Heraklion, Crete, Greece; 2010. p. 213–26.
11. Krizhevsky A, Sutskever I, Hinton GE. Imagenet classification with deep convolutional neural networks. In: Advances in neural information processing systems, Lake Tahoe, Nevada, USA; 2012. p. 1097–105.
12. He K, Zhang X, Ren S, Sun J. Deep residual learning for image recognition. In: IEEE conference on computer vision and pattern recognition, Las Vegas, Nevada; 2016. p. 770–8.
13. Simonyan K, Zisserman A. Very deep convolutional networks for large-scale image recognition. In: International conference on learning representations, San Diego, California, USA; 2015. p. 1–14.
14. Xing EP, Jordan MI, Russell SJ, Ng AY. Distance metric learning with application to clustering with side-information. In: Advances in neural information processing systems, Vancouver and Whistler, British Columbia, Canada; 2003. p. 521–8.
15. Ghodsi A, Wilkinson DF, Southey F. Improving embeddings by flexible exploitation of side information. In: International joint conferences on artificial intelligence, Hyderabad, India; 2007. p. 810–6.
16. Globerson A, Roweis ST. Metric learning by collapsing classes. In: Advances in neural information processing systems, Vancouver and Whistler, British Columbia, Canada; 2006. p. 451–8.
17. Weinberger KQ, Blitzer J, Saul LK. Distance metric learning for large margin nearest neighbor classification. In: Advances in neural information processing systems, Vancouver and Whistler, British Columbia, Canada; 2006. p. 1473–80.

A Gradient Invariant DCT-Based Image Watermarking Scheme for Object Detection

Xiaocheng Hu$^{(\boxtimes)}$, Bo Zhang, Huibo Li, Jing Guo, Yunxiang Yang, Yinan Jiang, and Ke Guo

China Academy of Electronic Information Technology, Beijing 100041, China
675342900@qq.com

Abstract. In this paper, we proposed a novel DCT-based watermarking scheme for grayscale images, which utilizes the connection between discrete cosine transform (DCT) and the Histogram of Oriented Gradient (HOG) feature extraction operation. We embed messages into the low-frequency band, and correspondingly, an effective coefficients pair selection scheme is constructed. The proposed scheme not only maintains the superiority of compression robustness but also keeps good visual quality for the watermarked image. Moreover, the proposed method is insensitive to HOG feature extraction, which makes the watermarked image more suitable for further objection detection and recognition scenarios.

Keywords: Watermarking · Robustness · HOG feature extraction

1 Introduction

With the rapid development of mobile devices and social networks, digital images are widely captured and shared throughout the Internet, which draws several urgent issues relating to copyright protection and authentication during transmission. Digital image watermarking is a kind of data hiding method that has been extensively studied in recent decades.

Most of the image watermarking schemes satisfy one or more of the five requirements: imperceptibility, non-detectability, security, robustness, and capacity. Blind image watermarking methods generally belong to the spatial or transform domain. Many recent advances related are largely inspired by the manipulation of the domain transform of multimedia objects. Watermarks embedded in the transform domain are usually more robust and less perceptible, while their computational requirements tend to be higher than those spatial domain ones. The major domains used for image watermarking include discrete cosine transform (DCT) [1–3], discrete Fourier transform (DFT) [4–6], discrete wavelet transform (DWT) [7–9], and singular value decomposition (SVD) [10, 11].

© Springer Nature Singapore Pte Ltd. 2020
Q. Liang et al. (eds.), *Communications, Signal Processing, and Systems*, Lecture Notes in Electrical Engineering 516,
https://doi.org/10.1007/978-981-13-6504-1_111

DCT has proven particularly effective with regard to energy compaction and the ability to incorporate characteristics of human visual system (HVS) [12,13].

Recently in [14], Soumitra proposed a compromised scheme using the middle-frequency band DCT coefficients pairs, which are less vulnerable to modification and also the embedding capacity is guaranteed. Existing watermarking schemes mainly focus on robustness against lossy compression or geometric transformation attacks for message recovery, while before the message extraction procedure, kinds of image processing operations can be applied to the watermarked image. For example, the well-known Histogram of Oriented Gradients (HOG) feature descriptor is very popular for object detection and recognition in computer vision field. A modified watermarked image will affect the feature extraction and thus result in the decrease in detection and recognition performance. For some important applications such as pedestrian retrieval in a police or military system, slight accuracy decrease may involve fateful consequences. Therefore, watermarking scheme for images not interfering with further image processing is especially desired. In [15], Hou et al. proposed a recursive reversible data hiding method for color images, which keeps the gray version of the cover image unchanged, and thus, the feature extraction is not affected.

In this paper, we proposed a novel DCT-based watermarking scheme for grayscale images, which utilizes the connection between DCT transform and HOG feature extraction operation. We embed messages into the low-frequency band, and further, a novel coefficients pair selection scheme is constructed. The proposed scheme not only maintains the superiority of robustness but also keeps good visual quality for the watermarked image.

2   Previous Arts

In [14], Soumitra et al. proposed a block-based watermarking scheme using DCT coefficients pairs. First, the input $M \times N$ image I is divided into non-overlapping blocks of size $k \times k$. Then, each block I_b is transformed into the DCT domain through the following equation:

$$F_b(u,v) = \alpha(u)\alpha(v) \sum_{x=0}^{k-1}\sum_{y=0}^{k-1} I_b(x,y) \times cos\left[\frac{(2x+1)u\pi}{2k}\right]cos\left[\frac{(2y+1)v\pi}{2k}\right] \quad (1)$$

where

$$\alpha(u) = \begin{cases} \sqrt{1/k}, & u=0 \\ \sqrt{2/k} \ otherwise \end{cases} \quad and \quad \alpha(v) = \begin{cases} \sqrt{1/k}, & v=0 \\ \sqrt{2/k} \ otherwise \end{cases} \quad (2)$$

When computed, the DCT coefficients are collected according to the zigzag order. The method in [14] utilizes the size relation between the two coefficients in each pair. To be specific, while embedding message bit "1," if the first coefficient is not greater than the second, then swap their values, otherwise keep them unchanged; While for message bit "0," if the first coefficient is greater than the second coefficient, then swap there values, otherwise do nothing. After message embedding, the image block pixel values are reconstructed from the changed DCT coefficients through inverse DCT transform.

3 Proposed Scheme

3.1 Frequency Band Selection

Observing from (1), the computed DCT coefficients flow into three bands, namely the low-, middle-, and high-frequency bands. Generally, the energy of the image block mostly lies in the low-frequency band, and the middle- and high-frequency bands reflect the texture and edge details of the image block. DCT-based watermarking schemes usually utilize middle-band frequency for message embedding as it is less perceptible on modification, and the high-frequency band is rarely used because it is fragile to compression attacks.

The Histogram of Oriented Gradients (HOG) feature computes the horizontal and vertical gradients by simple high-pass filters and then, the direction of the gradient is resolved for each pixel, and later, the histogram of the gradient directions is counted for each block. HOG feature actually plays a high-pass filter role to the image block, such that modification on the high- or middle-frequency band will influence its computation. Seeing this connection, we propose to embed messages into the low-frequency band to minimize affection to the HOG features extraction. Moreover, we design a message size adaptive coefficients selection embedding scheme to decrease the perceptual distortion involved. The selected low-band DCT coefficients are depicted in Fig. 1, ranging from the second coefficient to the fifteenth coefficient according to the zigzag order.

3.2 Coefficient Pairs Selection

In [14], Soumitra et al. utilize the size relation between the two coefficients in each pair to embed message. For those pairs whose two DCT coefficients violate

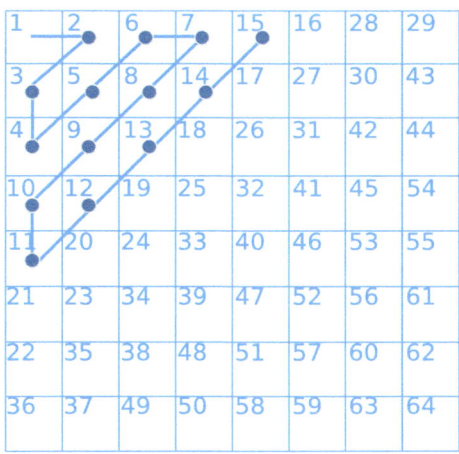

Fig. 1. Selected DCT coefficients of 8 × 8 image block following the zigzag order.

the order during embedding will swap their two coefficients, no matter how much they differ from each other.

As a matter of fact, whether the two coefficients in each pair swap their values or not, their absolute difference value stays invariant after message embedding. Utilizing this criterion, we can select coefficients pairs with small absolute difference to be embedded first to decrease the swapping distortion. Figure 2 shows the absolute difference value distribution between the two adjacent low-frequency band DCT coefficients in all the 8 × 8 image blocks for the Lena cover image. From Fig. 2, we can see that small absolute difference values between the two coefficients occurs more often than large absolute difference values, we can use a message length adaptive threshold to pick out close coefficient pairs for embedding. As a result, the coefficient pairs with large absolute difference values can be skipped, and the embedding distortion is better guaranteed.

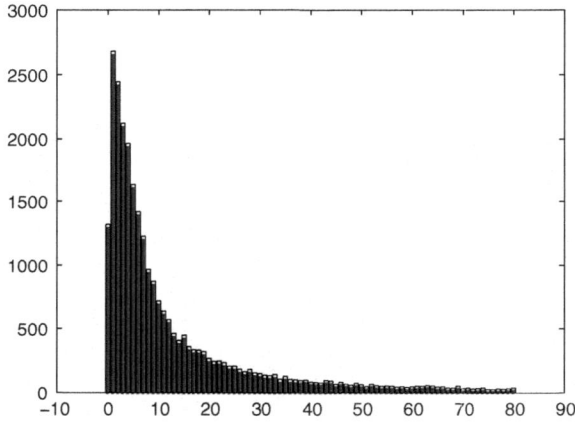

Fig. 2. Histogram of absolute difference value between the two adjacent DCT low-frequency band coefficients in all 8 × 8 image blocks for the Lena cover image.

3.3 Embedding and Extracting

To extract the message, we need to transfer the coefficient pairs filtering threshold to the receiver side. The overall message embedding and extraction procedure with adaptive coefficient pairs selection are described as follows.

Embedding

1. Divide the cover image into 8×8 blocks and apply discrete cosine transform(DCT) to each image block.
2. Define an initial threshold δ to filter all the 8×8 image blocks in the cover image, count all the adjacent DCT coefficient pairs whose absolute difference value is less equal than δ. If the total count is less than the message payload length, increase the δ threshold by 1 and repeat the filtering procedure until enough DCT coefficient pairs are selected to embed the message.
3. For each DCT coefficients pair in each 8×8 image block, check where the absolute difference value between them is less equal than the threshold δ, then a message bit can be embedded into this pair by:
   - While embedding message bit is "1," if the first coefficient is less than the second coefficient, then swap their values, else keep them unchanged.
   - While embedding message bit is "0," if the first coefficient is greater than the second coefficient, then swap their values, else keep them unchanged.
4. After embedding the message, apply the inverse discrete cosine transform to each image block to get the modified image block pixels, as a result, the whole watermarked image is obtained.
5. Actually, the threshold δ must be transmitted to the receiver side first, we simplify this through least significant bits (LSBs) replacement using the first several blocks in the cover image.

Extracting

1. Firstly, recover the threshold δ value from the LSBs in the first several blocks in the cover image.
2. Divide the cover image into 8×8 blocks and apply discrete cosine transform to each image block.
3. For each DCT coefficients pair in each 8×8 image block, check where the absolute difference value between them is less equal than the threshold δ, then a message bit can be extracted from this pair by:
   - If the first coefficient is greater than the second coefficient, then extract message bit "1."
   - If the first coefficient is less than the second coefficient, then extract message bit "0."
4. After extracting each message bit, the whole message payload is recovered.

4   Experimental Results

To evaluate the embedding performance of the proposed DCT coefficients pair selection scheme, we conduct several kinds of simulations on different types of cover images. The cover image of size 512×512 is shown in Fig. 3, which varies from smooth ones to those full of textures and edge components. In our proposed scheme, low-frequency DCT coefficients from the second position toward the fifteenth position following the zigzag scanning order for each 8×8 block are selected to embed messages. To demonstrate the effectiveness of the proposed scheme, comparisons between the proposed scheme and Soumitra et al.'s method in [14] are illustrated, in terms of visual imperceptibility, influence degree on HOG features, and also the compression robustness.

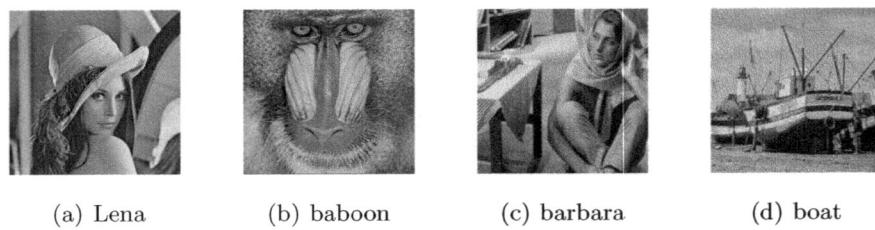

(a) Lena   (b) baboon   (c) barbara   (d) boat

Fig. 3. Test cover images of size 512×512.

4.1   Imperceptibility Comparison

Peak signal-to-noise ratio (PSNR) is largely used to evaluate the similarity between the cover image and the watermarked image, which is defined as:

$$PSNR = 10 \times \log_{10} \frac{255 \times 255}{MSE} \tag{3}$$

where the mean square error(MSE) is given as $MSE = \frac{1}{M \times N} \sum_{i=1}^{M} \sum_{j=1}^{N} (x_{ij} - x'_{ij})^2$. Here, the M and N represent the width and height of the image, and x_{ij} and x'_{ij} mean the pixel values of the cover image and the watermarked image.

We compare our proposed DCT coefficients pair selection scheme against Soumitra et al.'s method in [14] with respect to PSNR criterion. Comparison results under different embedding rates are shown in Fig. 4, from which we can see that even though the proposed scheme gains distinctly better PSNR performance in comparison against method in [14].

4.2   HOG Feature Influence

The HOG feature is widely used for object detection, especially for pedestrian detection. We embed message into the cover images using both our proposed scheme and Soumitra et al.'s method in [14] and extract their HOG features. Later, the two modified HOG features are compared to the original HOG feature extracted from the cover image before embedding. The HOG feature similarity measurement(FSM) is computed by:

$$FSM = \sqrt{\sum_{i=1}^{L} (f_i - f'_i)^2} \tag{4}$$

where f_i and f_i' are the feature elements of the cover image and the watermarked image. Note, here we stitch all the block HOG features together to form a feature vector $f_i, i = 1, \ldots, L$, and L is the overall vector length.

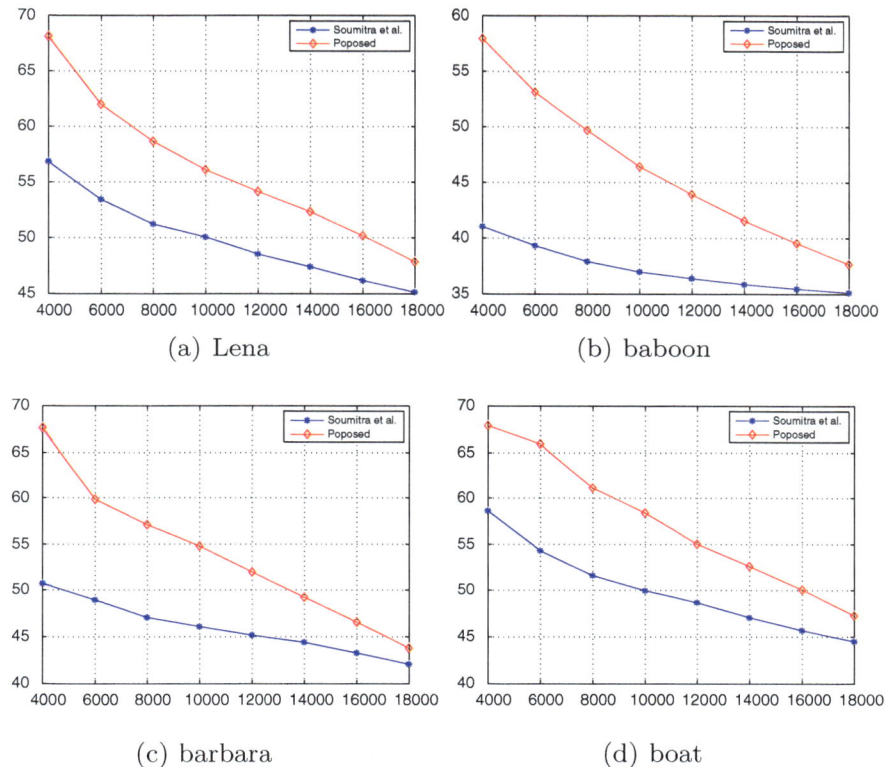

(a) Lena   (b) baboon

(c) barbara   (d) boat

Fig. 4. Imperceptibility performance comparisons with Soumitra et al.'s method [14].

The HOG feature similarity measurement (FSM) comparison results are depicted in Table 1. The results exhibit our proposed method's superiority for HOG-like features' invariability after message embedding and thus make it more suitable for object detection and recognition scenarios.

4.3   Compression Robustness

JPEG compression is one of the most popular compression attacks during watermarked image transmission. As the low-frequency band coefficients are more robust to compression operations than those middle- and high-frequency band coefficients, our proposed embedding scheme will expect better robustness to JPEG compression attack theoretically, in comparison with a method in [14]. For the sake of simplicity, the experimental comparisons are omitted.

Table 1. HOG feature similarity measurement (FSM) comparisons.

Image	Lena	Baboon	Barbara	Boat	Peppers	Airplane	Average
Soumitra et al.	9.945	11.267	8.001	10.184	8.575	8.275	9.375
Proposed	9.231	9.446	7.350	8.907	8.417	8.079	8.572

5 Conclusion

In this paper, we proposed a novel DCT-based watermarking scheme for grayscale images, which utilizes the connection between discrete cosine transform (DCT) and the histogram of Oriented Gradient (HOG) feature extraction operation. We embed messages into the low-frequency band, and correspondingly, a novel coefficients pair selection scheme is constructed. The proposed scheme not only maintains the superiority of compression robustness but also keeps good visual quality for the watermarked image. Moreover, the proposed method is insensitive to HOG feature extraction, which makes the watermarked image more suitable for further objection detection and recognition scenarios.

Acknowledgments. This work is supported by Beijing NOVA Program (Z181100006218041) and National key R&D program of China (2017YFC0820106).

References

1. Patra JC, Phua JE, Bornand C. A novel DCT domain CRT-based watermarking scheme for image authentication surviving JPEG compression. Digital Signal Process. 2010;20(6):1597–611.
2. Lin SD, Shie SC, Guo JY. Improving the robustness of DCT-based image watermarking against JPEG compression. Comput Stand Interfaces. 2010;32(1):54–60.
3. Hsu LY, Hu HT. Blind image watermarking via exploitation of inter-block prediction and visibility threshold in dct domain. J Vis Commun Image Represent. 2015;32(C):130–43.
4. Tao P, Eskicioglu AM. An adaptive method for image recovery in the DFT domain. J Multimedia. 2006;1(6):36–45.
5. Tsui TK, Zhang XP, Androutsos D. Color image watermarking using multidimensional fourier transforms. IEEE Trans Inf Forensics Secur. 2008;3(1):16–28.
6. Lang J, Zhang ZG. Blind digital watermarking method in the fractional fourier transform domain. Opt Lasers Eng. 2014;53(2):112–21.
7. Wang Y, Doherty JF, Dyck REV. A wavelet-based watermarking algorithm for ownership verification of digital images. IEEE Trans Image Process Publ IEEE Signal Process Soc. 2002;11(2):77–88.
8. Zhang G, Wang S, Wen Q. An adaptive block-based blind watermarking algorithm. In: International conference on signal processing, 2004. Proceedings. ICSP, Vol 3. New York: IEEE; 2004. p. 2294–7.
9. Liu N, Li H, Dai H, Guo D, Chen D. Robust blind image watermarking based on chaotic mixtures. Nonlinear Dyn. 2015;80(3):1329–55.

10. Chung K, Yang W, Huang Y, Wu S, Hsu Y. On SVD-based watermarking algorithm. Appl Math Comput. 2007;188(1):54–7.
11. Guo J, Zheng P, Huang J. Secure watermarking scheme against watermark attacks in the encrypted domain. J Vis Commun Image Represent. 2015;30:125–35.
12. Hernandez JR, Amado M, Perez-Gonzalez F. DCT-domain watermarking techniques for still images: Detector performance analysis and a new structure. IEEE Trans Image Process Publ IEEE Signal Process Soc. 2000;9(1):55–68.
13. Agarwal C, Mishra A, Sharma A. Gray-scale image watermarking using GA-BPN hybrid network. J Vis Commun Image Represent. 2013;24(7):1135–46.
14. Roy S, Pal AK. A blind DCT based color watermarking algorithm for embedding multiple watermarks. AEU - Int J Electron Commun. 2017;72:149–61.
15. Hou D, Zhang W, Chen K, Lin SJ, Yu N. Reversible data hiding in color image with grayscale invariance. IEEE Trans Circuits Syst Video Technol. 2018;99:1.

A Method for Under-Sampling Modulation Pattern Recognition in Satellite Communication

Tao Wen[✉] and Qi Chen

Battle Laboratory, Naval Command College, Nanjing 210016, China
1280691681@qq.com

Abstract. To solve the problem of reconnaissance and processing of broadband satellite communication signals, a kind of satellite communication signals BPSK/QPSK modulation pattern recognition method was put forward in this paper. This method deals with the satellite descending signal with BPSK/QPSK modulation in the under-sampling condition. Because the corrected spectrum of BPSK signal contains obvious crest, while QPSK signal does not contain this feature. The difference of the waveform characteristics is used to complete modulation pattern recognition. The simulation results show that this method can identify BPSK/QPSK modulation signals when SNR is greater than 1 dB. When the sampling points are reduced, the satellite communication signal under-sampling modulation pattern recognition method can still maintain good recognition performance.

Keywords: Satellite communication · Under-sampling · Modulation pattern recognition · Sparse reconstruction

1 Introduction

Satellite communication as a kind of high-frequency, large-broadband instant means of communication has the characteristics of wide coverage and signal transmission stability. It is an important means of battlefield communications at present. And noncooperative reception and processing of satellite communication signal are of great significance to grasp the trend of the space electromagnetic situation. The development of satellite communication anti-jamming technology, such as spectrum expansion, star processing, and limiting technology, presents a challenge to the noncooperative reception of satellite communication signals. However, the modulation pattern recognition process of satellite communication signals is not exactly the same as that of traditional communication signal modulation pattern recognition. First of all, the satellite communication signal has the characteristics of high frequency, large bandwidth, in noncooperative satellite communication signal processing, as a result of the signal sampling rate increase, the data quantity is big, have high requirements for ADC and data transmission. There are differences between satellite communication signal and traditional communication signal in transmission path. The transmission path of satellite communication has bigger longer noise influence , but also needs to take into

© Springer Nature Singapore Pte Ltd. 2020
Q. Liang et al. (eds.), *Communications, Signal Processing, and Systems*, Lecture Notes in Electrical Engineering 516,
https://doi.org/10.1007/978-981-13-6504-1_112

account the signal transmission in the atmosphere with the influence of the multipath effect, therefore, it is difficult to extract the characteristic parameters of the signal. In addition, it is an important characteristic parameter extraction method in the modulation pattern recognition of the satellite communication signal. The modulation types and parameters of satellite downlink communication signals can be obtained through modulation pattern recognition, and then useful information in satellite communication signals can be further obtained through demodulator.

Compressed sensing is an under-sampling acquisition and processing theory of signal, at the same time, the signal can be compressed by the acquisition system based on the compressed sensing theory framework [1]. Compressed sensing can retain the information in the original signal, the sampling rate is also far lower than that of the Nyquist sampling rate. Satellite communication signal has the characteristics of a large amount of data and difficult signal acquisition, and the problems need to be solved in under-sampling condition based on compressed sensing theory [2]. Therefore, this paper considers using under-sampling technology to get a small amount of sampling point satellite communication signal processing.

At present, there are two main types of signal modulated pattern recognition algorithms: likelihood ratio test and feature extraction algorithm. Literature 3 used in mixed likelihood ratio test (Hybrid likelihood ratio test , HLRT) method to complete BPSK/QPSK signal modulation recognition [3], the phase difference, and sequence of the signal is in the process of unknown parameters. This paper adopts maximum likelihood estimation, linear least squares estimation, and torque estimation method to estimate the unknown parameters. Furthermore, the whole process of HLRT recognition was completed. But the experimental environment was set in the additive white Gaussian noise (AWGN), which did not take into account the effect of the multipath effect, and the calculation was large. Feature extraction algorithms mainly include time–frequency method, wavelet transform method, and statistical extraction method [4]. Less prior information of the signal and low computational complexity are required in the feature extraction method, but it needs to solve the reliability problem of modulation style recognition performance. Literature 7 discussed several modulation pattern recognition algorithms based on feature extraction [5]. This method can not meet the requirements of satellite communication signal modulation recognition in sampling rate and signal-to-noise ratio.

BPSK/QPSK modulation is widely used in satellite communication due to its advantages of high-frequency utilization, strong anti-interference, and simple circuit implementation. A method for under-sampling modulation pattern recognition in satellite communication is proposed in this paper for the recognition of BPSK/QPSK modulation pattern recognition in satellite communication [6, 7]. This algorithm uses undersampling technology to reduce the sampling rate, and introduces the sparsity adaptive subspace tracking algorithm (SASP) in literature 9. As the reconstruction algorithm in the under-sampling technology, it improves the reconstruction accuracy and reduces the computational complexity at the same time [8]. Then, a method proposed in this paper for under-sampling modulation pattern recognition in satellite communication, using BPSK/QPSK signal differences of fixed frequency spectrum waveform characteristic parameters, completed the recognition of BPSK and QPSK modulation signal.

2   Satellite Communication Signal Under-sampling Modulation Recognition Based on Modified Spectral Feature Extraction

The overall processing flow of BPSK/QPSK signal modulated pattern recognition technology based on under-sampling technology is shown in Fig. 1.

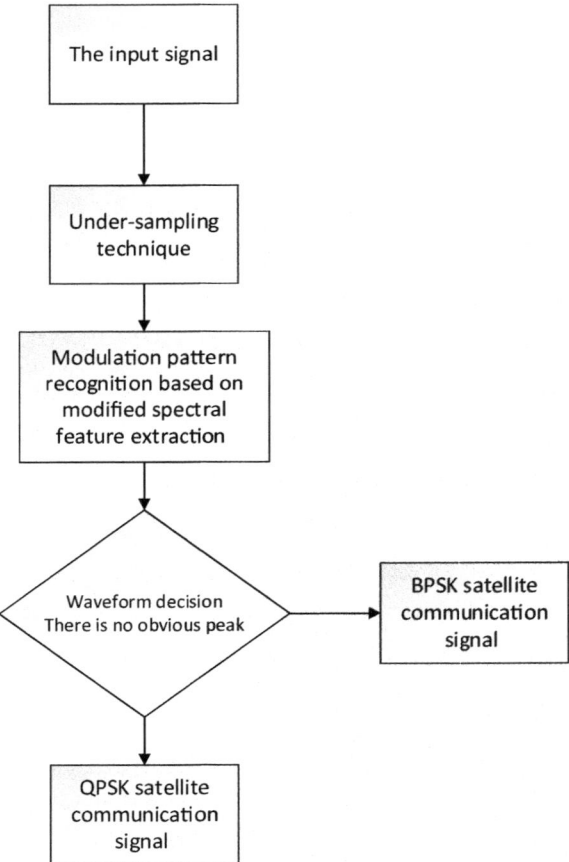

Fig. 1. BPSK/QPSK signal modulated pattern recognition technology based on under-sampling technology

As shown in the flow chart, this method is divided into three steps: first step, under-sampling technology processing, and the BPSK/QPSK satellite communication signal with less sampling points. In the second step, the characteristic parameter extraction part, the reconstruction signal extracted from the first step is extracted and corrected. The third step is to identify the modulation pattern by comparing the modified spectral signature waveform.

2.1 Under-sampling Technology

The theory of under-sampling technology is based on the theory of compressive sensing: The sparse signal is not sampled in the sparse signal with the sparse signal in different domains. That is, the signal is compressed at the same time of sampling, and the process of under-sampling is realized by removing most of the redundancy in the signal. The main steps of under-sampling technique are sparse representation, random measurement, and reconstruction algorithm. Signals in nature can be represented by specific sparse $\{\Psi\}$ basis basically.

The received BPSK/QPSK signal s can be regarded as a random process, which can be expressed as a linear combination of an orthonormal basis; signal s can be expressed as:

$$s = \Psi x \tag{1}$$

where x is the column vector of $N \times 1$, which is the projection coefficient $x = \langle s, \Psi_i \rangle$. Therefore, it can be seen that the projection coefficient x and the signal s are, respectively, the representation of the receiving signal in the Ψ domain and the time domain, and the two are equivalent expressions.

In the process of compression sampling, N visa number is expressed as M visa, which naturally reduces the amount of data sampled. Construct the matrix $M \times N$ of Φ, and the compression sampling value y can be expressed as:

$$y = \Phi s \tag{2}$$

where the row vector of Φ is composed of $\{\Phi_i\}_{j=1}^{M} (M < N)$; $y_j = \langle s, \Phi_i \rangle$, $j = 1, \ldots, M$.

Figure 2 shows the schematic diagram of the basic principle of compression sampling.

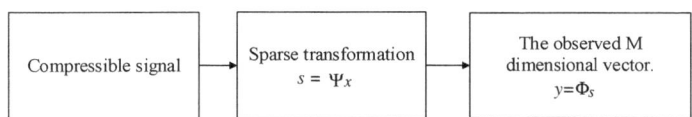

Fig. 2. Basic principle block diagram of compressed sampling

In addition, in order to ensure that the information in the original signal is not lost, the CS measurement matrix should satisfy conditions of the restricted isometry property (RIP). In general, random observation matrices composed of random elements can meet the RIP conditions with a relatively large probability, for example, Gaussian distribution and Bernoulli distribution.

2.1.1 Sparse Representation

The observation matrix adopted in this paper can be equivalent to matrix form:

$$\Phi = HP \tag{3}$$

where P is the orthogonal sparse matrix and H is the random measurement matrix. In this paper, the discrete Fourier transform (DFT) is used as the orthogonal sparse matrix, and the positive transformation formula is as follows:

$$X(k) = \sum_{n=0}^{N=1} x(n)e^{-j\frac{2\pi}{N}kn} = \sum_{n=0}^{N=1} x(n)W_N^{kn} \tag{4}$$

where $n,\ k = 0, 1, \ldots, N-1$, $W_N = e^{-j2\pi/N}$.

This process can also be expressed through matrix operations:

$$
\begin{bmatrix} X(0) \\ X(1) \\ X(2) \\ \vdots \\ X(N-1) \end{bmatrix}
=
\begin{bmatrix}
1 & 1 & 1 & \cdots & 1 \\
1 & W_N^1 & W_N^2 & \cdots & W_N^{N-1} \\
1 & W_N^2 & W_N^4 & \cdots & W_N^{2(N-1)} \\
\vdots & \vdots & \vdots & \ddots & \vdots \\
1 & W_N^{N-1} & W_N^{2(N-1)} & \cdots & W_N^{(N-1)(N-1)}
\end{bmatrix}
\begin{bmatrix} x(0) \\ x(1) \\ x(2) \\ \vdots \\ x(N-1) \end{bmatrix}
\tag{5}
$$

After orthogonalization, the matrix form of discrete Fourier transform is obtained:

$$
P = \frac{1}{\sqrt{N}}
\begin{bmatrix}
1 & 1 & 1 & \cdots & 1 \\
1 & W_N^1 & W_N^2 & \cdots & W_N^{N-1} \\
1 & W_N^2 & W_N^4 & \cdots & W_N^{2(N-1)} \\
\vdots & \vdots & \vdots & \ddots & \vdots \\
1 & W_N^{N-1} & W_N^{2(N-1)} & \cdots & W_N^{(N-1)(N-1)}
\end{bmatrix}
\tag{6}
$$

2.1.2 Random Measurement

The structure of the measurement matrix in the compressive sensing theory determines whether the information in the signal can be effectively collected. Normally, the measurement matrix should satisfy RIP measurement matrix properties to ensure that the sampling signal energy is unchanged before and after. At the same time, it need to be considered that the matching problem between the measurement matrix and measurement matrix reconstruction algorithm to ensure that the reconstruction effect optimization.

H is the Gaussian random measurement matrix adopted in this paper, and an $M \times N$-size matrix H is constructed, so that each element in H is independent of the mean value of 0, and the variance is the Gaussian distribution, namely

$$H \sim N(1, \frac{1}{M}) \tag{7}$$

Gaussian measurement matrix is most widely used in the sparse reconstruction technology of measurement matrix. When meet the conditions of the $M \geq cK \log(N/K)$, the matrix which has strong randomness will meet great probability of RIP conditions [9].

2.1.3 Reconstruction Algorithm

Sparsity adaptive subspace tracking (SASP) uses a new sparse estimation method to obtain the initial estimate of sparsity and then updates the estimation by iteration. The weak matching principle is used in each iteration to select new atomic, again through the subspace tracking to improve result and reconstruction signals. The method will be sampling the sparse degree of adaptive subspace tracking to accomplish the signal reconstruction.

The sparse degree of adaptive subspace tracking (sparsity adaptive subspace pursuit, SASP) algorithm steps:

Input: M dimension observation vector y, dimension measurement matrix
Output: reconstructed signal $\hat{x}$, $\hat{x}_{\Gamma^n} = \arg \min \|y - \Phi \Gamma^n \hat{x}_{\Gamma^n}\|_2^2$. The other elements in $\hat{x}$ are zero
(1) $g^0 = \Phi * y$, $K_0 = 1$
(2) $\Gamma^0 = \{|g_i^0|$ The maximum index value before K_0th$\}$
(3) If $\|\Phi_{\Gamma^0} * y\|_2 < \frac{1-\delta_K}{\sqrt{1+\delta_K}} \|y\|_2$, then $K_0 = K_0 + 1$; repeat (2)
(4) $r^0 = \min \|y - F_{G^0} x_{G^0}\|_2^2$
(5) $n = 1$
(6) $g^n = \Phi * r^{n-1}$
(7) $\hat{\Gamma} = \Gamma^{n-1} \cup \{|g_i^n|$ The maximum index value before $|\Gamma^{n-1}|$th$\}$
(8) $\hat{x}_{\hat{\Gamma}} = \arg \min \|y - \Phi_{\hat{\Gamma}} x_{\hat{\Gamma}}\|_2^2$
(9) $\tilde{\Gamma} = \{|x_{\hat{\Gamma}}|$ The maximum index value before $|\Gamma^{n-1}|$th$\}$
(10) $\tilde{r} = \min \|y - \Phi_{\hat{\Gamma}} x_{\hat{\Gamma}}\|_2^2$
(11) If $\|\tilde{r}\|_2^2 < \|r^{n-1}\|_2^2$, $\Gamma^n = \tilde{\Gamma}$, $r^n = \tilde{r}$, $n = n + 1$, repeat (6)
(12) When the correlation between atoms and residuals is less than a fixed value, otherwise $\hat{\Gamma} = \Gamma^{n-1} \cup \left\{ i : |g_i^n| \geq \alpha \max_{2 \leq j \leq N} |g_j^n| \right\}$, repeat (6)

In the algorithm, the first four steps are estimated for the sparse degree, where Γ^0 is the initial estimation set and r^0 is the residual. As the main part of the iterative algorithm, steps 5–12, where n is the number of iterations, the stopping condition of the algorithm iteration is that the correlation between the atoms and residuals in step 12 is less than some fixed value.

At this point, it is possible to reconstruct signal $\tilde{x}(n)$, $n = 1, 2$, in less than the conditions of Nyquist sampling rate. Figures 3 and 4 are the simulation waveforms of sparse reconstruction process of BPSK/QPSK modulated signals when sampling points are 1000 and the sampling rate is 1/4. The results show that BPSK/QPSK signals can be sparse represented by discrete Fourier transform, and the original signal can be recovered by the sparse adaptive subspace tracking (SASP) reconstruction algorithm, and the reconstruction error is less than 10%.

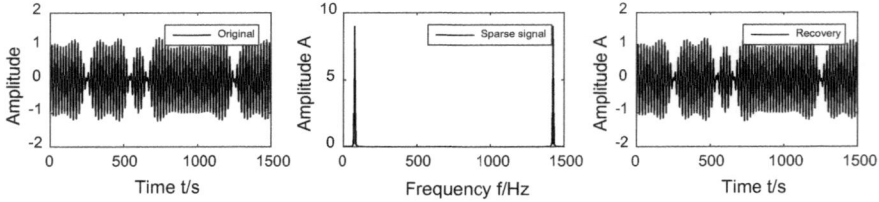

Fig. 3. Sparse reconstruction simulation waveform of BPSK signal

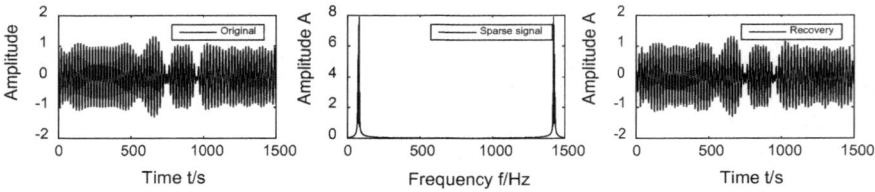

Fig. 4. Sparse reconstruction simulation waveform of QPSK signal

2.2 BPSK/QPSK Satellite Communication Signal Under-sampling Modulation Pattern Recognition

2.2.1 BPSK/QPSK Signal Model

The mathematical model of BPSK/QPSK modulation signal can be expressed as:

$$s(n) = A\exp(j[\phi(n)]),\ 0 \le n \le N - 1 \tag{8}$$

where $s(n)$ represents BPSK/QPSK modulation signal, A is the modulation signal amplitude, $\phi(n)$ is the phase function, and N is the number of sampling points. For phase modulation, the phase function $\phi(n)$ reflects the different modulation styles.

$$\phi(n) = 2\pi f_0 n\Delta t + \pi d_i(n) + \theta \tag{9}$$

The multipath effect of signal in transmission process is an important problem in satellite communication and reconnaissance. The multipath effect is mainly caused by multipath reflection in communication signal transmission path. The influence of synchronous timing error and incomplete matched filtering of receiver can also be equivalent to multipath model [10]. The discrete equivalent model of multipath fading channel can be expressed as:

$$h(n) = h_0 e^{j\theta_0} \delta(n) + \sum_{k=1}^{L} h_k e^{j\theta_k} \delta(n-k) \tag{10}$$

In the above equation, the number of channels in the channel is $L + 1$, the amplitude gain of h_0 main path, and the phase shift factor of θ_0, while h_k and θ_k are, respectively, the amplitude gain and phase shift factor of the KTH multipath.

This paper considers the simulation experiment scenario for mobile satellite communication in a wide area. The direct signal is basically unblocked, the direct wave envelope is a fixed value. The signals containing the direct waves and the diffuse components subject to Rayleigh distribution are subject to the rice distribution. At this point, the satellite channel is a linear time-varying channel, and the response varies with time. The effect of this interference on useful signals is multiplicative interference. Suppose the effective signal is $s(n)$, $w(n)$ as multiplicative interference, $e(n)$ as the additive white Gaussian noise, received signal $x(n)$ can be expressed as:

$$x(n) = s(n) \cdot w(n) + e(n) \tag{11}$$

2.2.2 Correction of Spectral Feature Extraction

Through the under-sampling technique, the reconstructed signal $\tilde{x}(n)$ of input signal $x(n)$ is obtained. The next step is to extract the characteristic parameters of the reconstructed signal.

First, the signal is squared, namely

$$y(n) = \tilde{x}^2(n) \tag{12}$$

If the observed signal is BPSK signal, then $y(n)$ is converted into $y(n)$ sine wave with phase 2θ and carrier $2f_0$. If it is QPSK signal, it degenerates into BPSK signal [11, 12]. At this time, the modulation pattern recognition of BPSK/QPSK signal was translated into the recognition of the sinusoidal affected BPSK signal.

The frequency form of sine wave signal is concentrated in a certain frequency. BPSK signal spectral waveform is subject to sinc function distribution, bandwidth is affected by the symbol rate. Therefore, the former on the spectrum distribution is single frequency on a separate line, while the latter should be on the spectrum distribution peak shape distribution; therefore, this difference can be used for BPSK/QPSK signal modulation style recognition.

For the reconstructed signal $\tilde{x}(n)$, discrete Fourier transform (DFT) is taken and the model is taken, and $\widehat{X}(n)$ is defined as the modified spectrum:

$$\widehat{X}(n) = |\text{DFT}[y(n)]| = \left| \sum_{n=0}^{N-1} W_N^{-nk} \right|, \ 0 \le k \le N-1 \qquad (13)$$

where $W_N = e^{j2\pi/N}$.

Figure 5 shows the modified spectra of BPSK/QPSK two signal squares, respectively, and the simulation conditions are: Gaussian noise intensity is 3 dB, the carrier frequency is 500 Hz, the symbol rate is 100 bps, element number is 100 bit, the modulation modes are BPSK and QPSK. Figure 5 shows that the BPSK signal is reduced to sine wave by square transformation. The QPSK signal is degraded to BPSK signal by square transformation. When the BPSK signal and QPSK signal are represented in the modified spectrum, it can be seen that the BPSK signal has a single spectral line component, and QPSK signal is without a single component of the spectrum. After removing the single line component, BPSK signal peak is not obvious, and QPSK signal has obvious peak value.

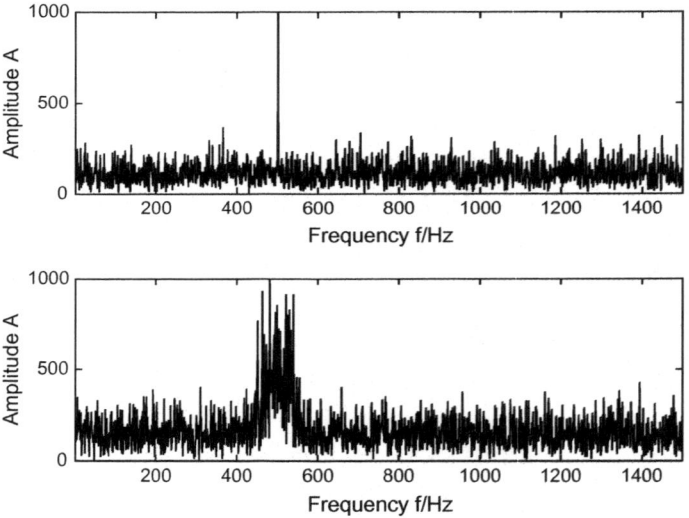

Fig. 5. Comparison of the corrected spectrums after the square of BPSK/QPSK

2.3   Modulation Pattern Recognition

BPSK signal transforms back a sine signal through the square transformation, QPSK signal through the square transforms back into Sinc function. Because of the effect of channel noise and multipath fading effect, after receiving the under-sampling of signal $x(n)$, the modified spectrum characteristic waveform $R(n)$ of the extraction $x(n)$ can be expressed as:

$$R(n) = \Omega(n) + N(n) \tag{14}$$

where $\Omega(n)$ is the effective characteristic waveform, and $N(n)$ is the noise waveform. When the input signal is BPSK signal, $R(n)$ contains a single spectral line, and the waveform is shown as a single spectral line distributed in the noise waveform. When the input is QPSK signal, $\Omega(n)$ still contains the single spectral line, but also contains the sinc function waveform. At this time, the modified spectrum characteristic waveform $R(n)$ shows a single spectral line and peak amplitude at the same time distributed in the noise waveform.

Therefore, in order to use the characteristics of the BPSK/QPSK signal waveform to realize the modulation pattern recognition, first remove the single line, the two kinds of characteristic waveforms due to the characteristics of the QPSK signal waveform contains obvious peak value, then sentence by setting the threshold for BPSK/QPSK signal modulation style recognition.

3 Simulation Analysis

Let's say the input signal $s(n)$ style for BPSK and QPSK modulation. After the transmission of the mobile satellite channel, the output signal $x(n)$ obeys the rice distribution and contains additive white Gaussian noise through under-sampling based on compression perception processing, and reconstruction signal $\tilde{x}(n)$ is obtained by less sampling point. The simulation experiment will analyze the performance of the satellite communication signal under-sampling modulation recognition method proposed in this paper under different conditions. The recognition rate of BPSK signal is $P(I_0)$, the recognition accuracy of QPSK signal for $P_t = (P(I_1) + P(I_2))/2$, has the performances evaluation for simulation of BPSK using 13 Barker code, QPSK with 14 Taylor code, number of simulation experiments/group is set to 100 times.

3.1 Under-sampling Modulation Pattern Recognition Simulation Experiment Under Different SNR Environment

In the process of satellite communication signal processing, the influence of damping and noise in the channel is a must to consider. Therefore, in order to test the performance of the method in satellite communication channel, a simulation experiment was carried out in conditions of different SNRs. The simulation condition is: carrier frequency is 75 MHz, element width is 800 ns, sampling rate $W = 1/2$, simulate the multipath effect, les channel attenuation factor intensity to 2 dB, gaussian noise intensity change interval [−8 dB, 8 dB], the step length is 1 dB.

The simulation results are shown in Fig. 6; under the influence of the multipath effect, under-sampling satellite communication signal modulation pattern recognition method in the signal-to-noise ratio of 3 dB complex still has 87% of the recognition accuracy in noisy environment; BPSK/QPSK modulation signal can be stabilized when SNR is greater than −1 dB.

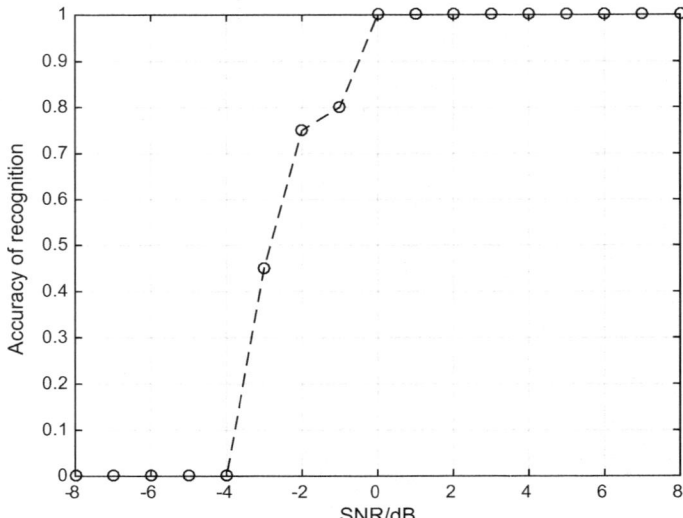

Fig. 6. Recognition rate of satellite communication signal under-sampling modulation pattern recognition method in different SNR environments.

3.2   Analysis of Under-sampling Modulation Pattern Recognition Performance

In order to verify the effect of the method in this paper under the condition of under-sampling. The performance analysis of under-sampling modulation pattern recognition was carried out. The simulation conditions is: the carrier frequency is 75 MHz; element width is 800 ns; les channel factor intensity is 2 dB; gaussian noise intensity is 2 dB.

Figure 7 shows the simulation results. The traditional sampling frequency conforms to the modulation pattern recognition method of Nyquist sampling theorem, and the sampling points are 1000, 2000, 3000, and 4000, respectively. Simulation results are shown in Fig. 7b, the modulation pattern recognition algorithm based on the technology of under-sampling recognition performance under different conditions of sample points, sampling points respectively in 1000, 2000, 3000, 2000, and other conditions of the same recognition performance experiment. The simulation results show that when the sampling points are reduced, the recognition accuracy of the unsampled modulation pattern recognition method can maintain the recognition performance better when the SNR changes.

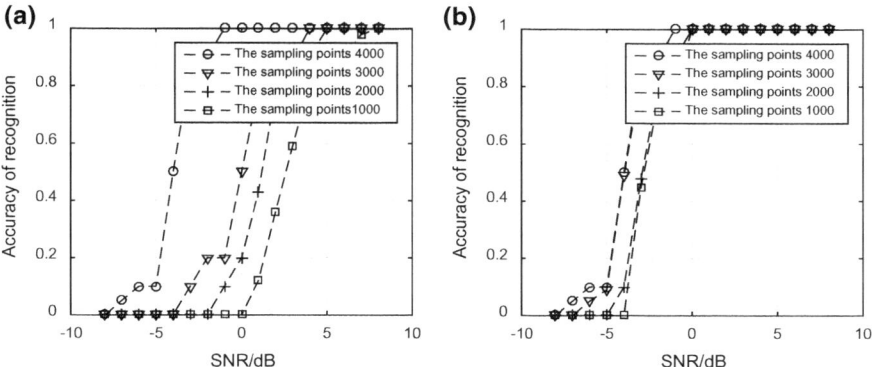

Fig. 7. Performance comparison of recognition rate under different sampling points

4 Conclusion

In this paper, a method is proposed to identify the under-sampling modulation based on feature parameter extraction. This method mainly aims at BPSK/QPSK modulation pattern recognition of communication signals. The methods in this paper considers the influence of the satellite communication channel multipath effect joins the additive gaussian noise at the same time to join the multiplicative noise, simulates the les channel attenuation phenomenon, in the process of signal source generated. In the signal sampling phase, sparse reconstruction of under-sampling technique is used to sample and reconstruct the signal with a lower sampling rate to solve the problem of a large sampling difficulty of satellite communication data. Finally, the spectrum characteristic waveform of the reconstructed signal was extracted, according to the characteristic difference of corrected spectral waveforms, and the BPSK/QPSK signal modulation pattern recognition was realized. The simulation results show that this method can be applied to the modulation pattern recognition in satellite communication channel environment and has a strong robustness.

Fund Project. National Natural Science Foundation of China (youth project): No. 61501484; Research and Development Fund of Naval Engineering University (Science and Technology [2016] No. 66), accounting subject: 425517K170; 425517K167.

References

1. Xiang Q, Cheng B, Yu J. Hierarchical porous CdS nanosheet-assembled flowers with enhanced visible-light photocatalytic H_2-production performance. Appl Catal B Environ. 2013;138–139:299–303.
2. Huan X, Safta C, Sargsyan K, et al. Compressive sensing with cross-validation and stop-sampling for sparse polynomial chaos expansions. arXiv preprint arXiv:1707.09334. 2017.
3. Sejdić E, Orović I, Stanković S. Compressive sensing meets time–frequency: an overview of recent advances in time–frequency processing of sparse signals. Digital Sig Process. 2017.

4. Sur P, Chen Y, Candès EJ. The likelihood ratio test in high-dimensional logistic regression is asymptotically a rescaled chi-square. arXiv preprint arXiv:1706.01191. 2017; Xu JL, Su W, Zhou M. Likelihood-ratio approaches to automatic modulation classification. IEEE Trans Syst Man Cybern, Part C (Appl Rev). 2011;41(4):455–69.

5. Gong W, Huang K. An image-feature based method for feature extraction of intra-pulse modulated signals. Electron Opt Control. 2008;4:016.

6. Guo-Bing H, Yu L. Signal intrapulse modulation recognition algorithm based on sine wave extraction. Comput Eng. 2010;36(13):21–3.

7. Ming J, Guobing H. Intrapulse modulation recognition of radar signals based on statistical tests of the time-frequency curve. In: 2011 international conference on electronics and optoelectronics (ICEOE). IEEE; 2011, 1, p. V1-300–V1-304.

8. Yang L, Guobing H. An improved BPSK/QPSK signal modulation pattern recognition algorithm. Telecommun Eng, 2017;57(8).

9. Zefang XU, Shunlan LIU. Adaptive regularized subspace pursuit algorithm. Comput Eng Appl. 2015;51(3):208–11.

10. Mishali M, Eldar YC. From theory to practice: sub-Nyquist sampling of sparse wideband analog signals. IEEE J Sel Top Sig Process. 2010;4(2):375–91.

11. Simon MK, Alouini MS. Digital communication over fading channels. Hoboken: Wiley; 2005.

12. Guobing H, Xu J, Li Y, et al. Modulation radar signal analysis and processing technology of. Boston: Addison Wesley Publishing; 2014.

Sequential Modeling for Polyps Identification from the Vocal Data

Fangqi Zhu[1], Qilian Liang[1], and Zhen Zhong[2(✉)]

[1] Department of Electrical Engineering, University of Texas at Arlington, 416 Yates
St., Arlington, TX 76010, USA
fangqi.zhu@mavs.uta.edu, liang@uta.edu
[2] Department of Otolaryngology Head and Neck Surgery, Peking University First
Hospital, Beijing 100034, China
{Zhong_zhen,Zhen_zhong}@sina.com

Abstract. Given the revival of neural networks and its recent impact
in other disciplines and record-breaking performances in a variety of
applications, in this paper, we employed a deep sequential model for
polyps detection from the vocal data. Previous research of acoustic signal
recognition (ASR) has focused on hand-crafted machine learning fashion,
such as Mel-frequency cepstral coefficients with hidden Markov model
and Gaussian mixture model. The deep model demonstrates its flexibility
and potential to outperform the traditional methods, and we expand its
scope on medical symptom identification. The mapping between the raw
vocal signal and the symptom recognition is established, and we show
that we can achieve a good recognition accuracy, which may appear to
clinical diagnosis in the near future.

Keywords: Vocal features · Polyps · Sequential model · LSTM

1 Introduction

Classical model for time-series modeling is mainly based on statistical analysis, state-space description, and dynamical causality. From the statistical point
of view, a Cox proportional hazard regression model for predicting stroke and
heart disease was proposed [1]. The milestone work of Rabiner summaries the
development of hidden Markov Model, which embedded the observation with
hidden layer and used the transformation in the hidden states for inference [2].
A review of Wiener–Akaike–Granger–Schweder (WAGS) influence, state-space
method, and dynamic casuality method for fMRI data is discussed [3]. Although
all these three kinds of methods make a huge progress on some specific tasks,
they rely on lots of hand-crafted details and assumptions, which may not always
be satisfied in the real applications.

The emergence of deep learning has a great impact on the medical and biological fields. One of the classical but still challenging research branches is how to use

© Springer Nature Singapore Pte Ltd. 2020
Q. Liang et al. (eds.), *Communications, Signal Processing,
and Systems*, Lecture Notes in Electrical Engineering 516,
https://doi.org/10.1007/978-981-13-6504-1_113

the neural network to achieve sequential modeling and inference. For example, a fuzzy logic-induced feed-forward neural network for sequential classification is proposed to handle the inherent uncertainty [4]. The DeepBind framework, which contains convolution and pooling layer for its neural network, achieves good performance on the DREAM5 evaluation for DNA and RNA sequence specification [5].

A more suitable network structure for time-series modeling is the recurrent neural network, especially the long short-term memory (LSTM) network [6–8]. It exhibits impressive performance in numerous sequence-based tasks such as speech recognition, acoustic modeling of speech, knowledge graph, and machine translation. LSTM is also a flexible and extensible structure, and different variations of the LSTM have been proposed in terms of different aspects such as saving the computation and memory cost, increasing the adaption for longer sequences, and tailored to fit more tasks [9]. The Granger causality analysis with $l1$-norm constraints on the weights of neighboring input periods leads to a lasso regularization on the LSTM network [10], which optimizes the structure of the LSTM to save computation and memory resources.

Some previous interdisciplinary work between RNN and pathophysiologic analysis has validated RNN's ubiquitous property. For instance, the gated recurrent unit network, one of the simplified versions of LSTM, has been employed for early detection of the heart failure onset [11]. In this paper, the sequential modeling for polyps identification from the vocal data is investigated. Twelve respondents, four of which have throat polyps and the rest are healthy (no throat polyps), are recorded for producing two vowels with high sampling rate. We use time-frequency analysis on these data and they are trained with sequential deep learning model. The main contributions of this paper are as follows:

- A concrete real vocal dataset for polyps identification is provided, and an easy-to-employ temporal frequency transform is utilized.
- Deep sequential learning model with dropout regularization is employed on the two specific temporal frequency features.
- Outstanding performance compared to previous work [12], which provides promising approach for clinical application.

The rest of the paper is arranged as follows: In Sect. 2, the vocal experiment and dataset description are demonstrated for formulating the problem. In Sect. 3, the deep sequential model is employed to learn the coefficients of time frequency analysis and achieve the polyps identification. In Sect. 4, the performance of the training will be demonstrated, and finally, we conclude in Sect. 5.

2   Vocal Experiment and Dataset Description

One of the typical symptoms among adults is the vocal cord polyps, which is the result of an acute injury (such as yielding and shouting really loud) and or several other causes (such as cigarette smoke). The original diagnosis of this problem is from the pathological viewpoint, and it is highly depended on the expert knowledge, which is not suited to the urgent requirement of self-diagnosis currently. In principle, sounds generated by human beings are filtered by the structure of the vocal tract which comprises of parts like tongue, teeth, etc., which is nonlinear processing comprising of both physical and aerodynamic effects. This kind of structure leads to what kind of sound will come out.

The recording of the voice for automatic diagnosis of the problem is achieved based on the speech analysis. We collect the data that are about the discrete vocal samples of recording twelve respondents, in which four of them have throat polyps and the rest are healthy (no throat polyps). In the study, the vowels /a:/ and /i:/ are collected and each of them lasts for approximately 10 seconds. The samples of vowels within the same scenarios are snipped off a longer vowel sample to choose the stable part of the whole sample for processing. The sampling rate of the vocal data is 192,000, and bits per sample is 32. The details of the vocal dataset are summarized in Table 1. The time and frequency of vocal data for four cases (/a:/ polyps, /a:/ non-polyps, /i:/ polyps and /i:/ non-polyps) are shown in Fig. 1.

Table 1. Vocal dataset summary

Parameters	Values
No. of patients	12 (4 with polyps, 8 without polyps)
Age range	21–50
Sample rate	192,000
Samples range	1,157,200–1,344,000
Bits per samples	32
Channels	1
Normalized data	FALSE

As shown in Fig. 2, it is difficult to find features or correlation directly in the time domain, but we are able to distinguish the difference in the frequency domain. Therefore, we resort to display the latent features in the temporal frequency domain. The main components in the frequency domain lie within 0–2000 Hz, and we employ a rectangle window to filter out the rest of the part and adopt the short-time Fourier transform (STFT) to form a spectrogram. Such sampled window clipped STFT defined over the region $m \in [0, R-1]$ is given by [13]:

$$X_{STFT}[k, lL] = X_{STFT}\left(e^{j2\pi k/N}, lL\right) = \sum_{m=0}^{R-1} x[lL - m]w[m]e^{-j2\pi km/N}, \quad (1)$$

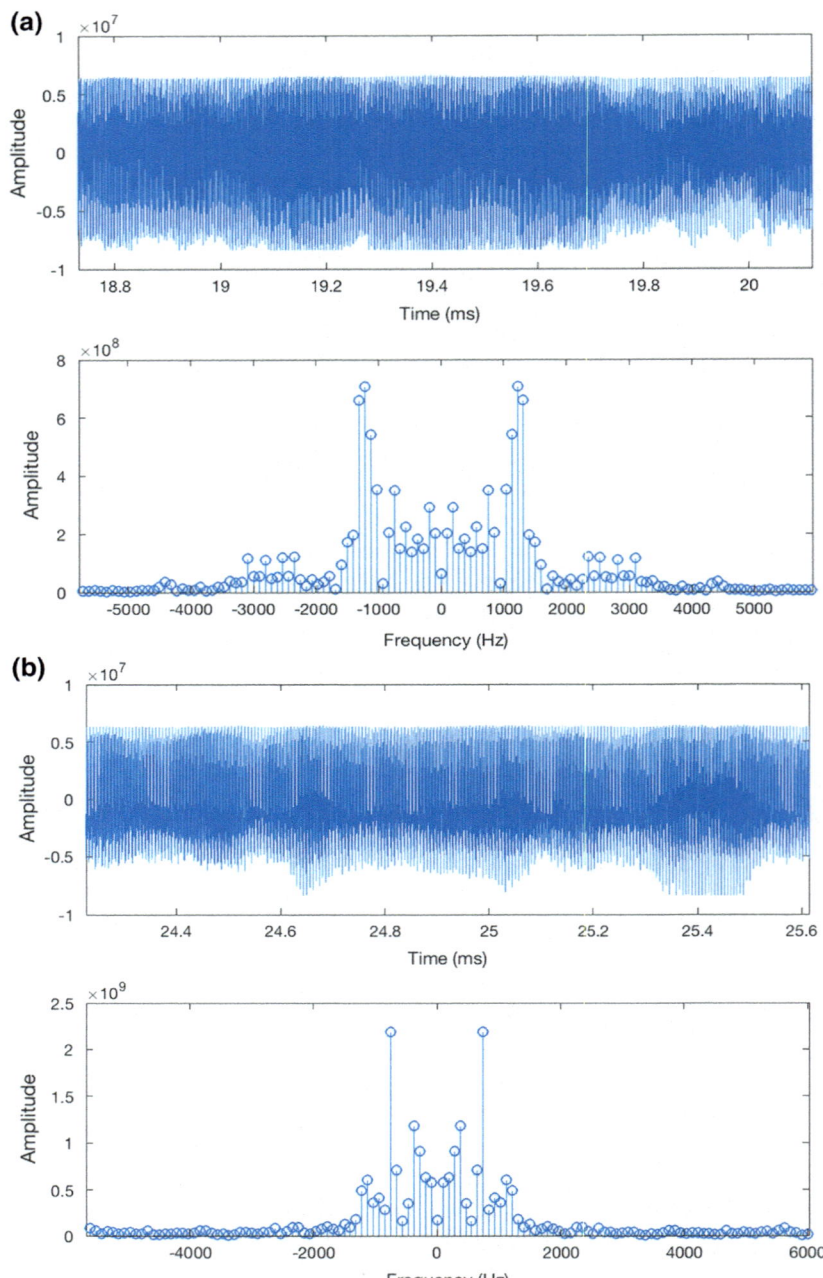

Fig. 1. Time and frequency of the vocal signal for vowels, **a** /a:/ normal; **b** /a:/ abnormal(polyps); **c** /i:/ normal; **d** /i:/ abnormal(polyps)

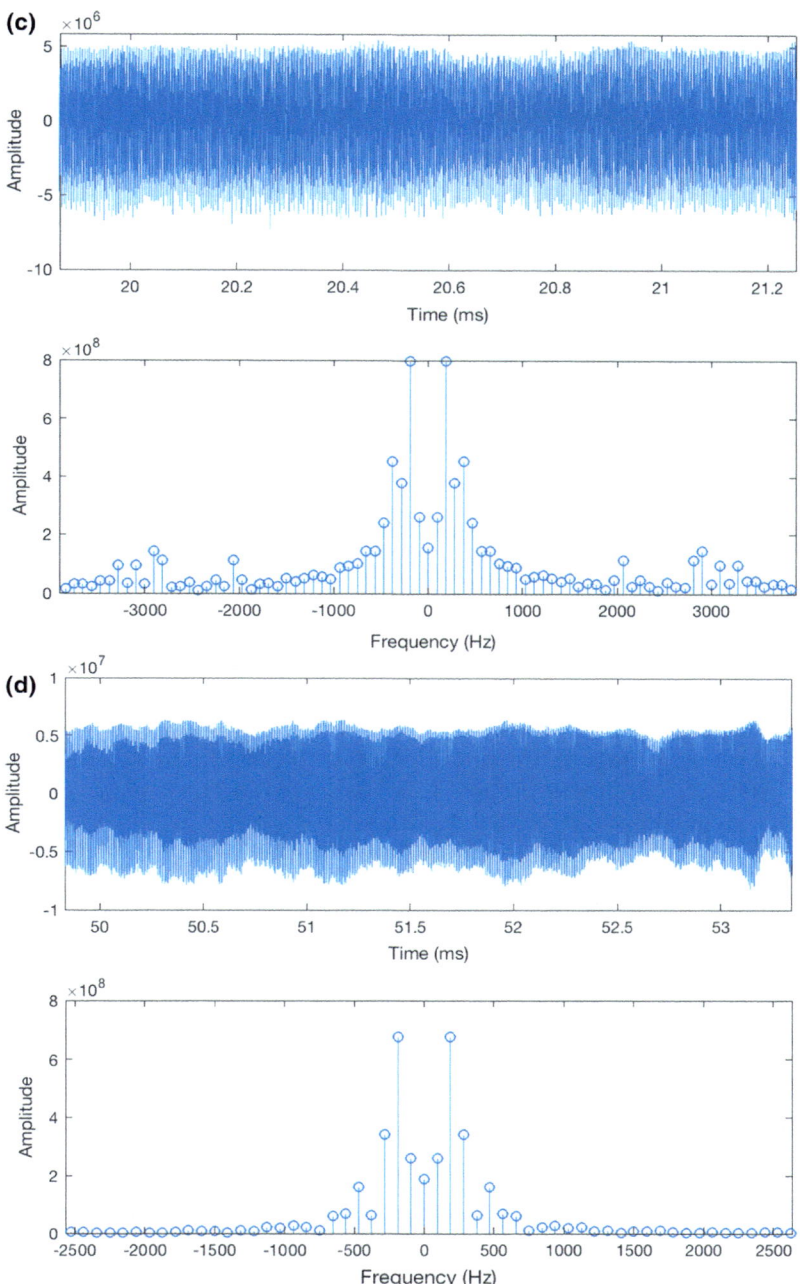

Fig. 1. (*continued*)

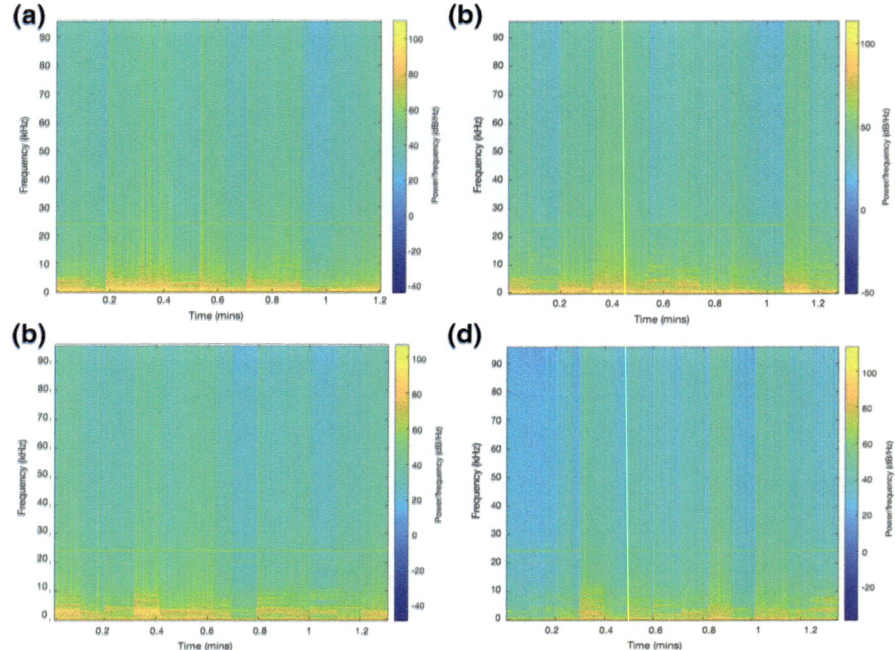

Fig. 2. Spectrogram (STFT) of the vocal signal for vowels, **a** /a:/ normal; **b** /a:/ abnormal(polyps); **c** /i:/ normal; **d** /i:/ abnormal(polyps)

Another widely adopted feature of ASR is the Mel-frequency cepstral coefficient (MFCC), which was proposed by Davis and Mermelstein in the 1980s.[1] The main steps are as follows:

- Apply the Mel filterbank to the power spectra, and sum the energy in each filter. $Y(i) = \sum_{k=0}^{N/2} \log | s(n) | H_i \left(k\frac{2\pi}{N} \right)$. ($S(n)$ is the DFT of the input signal, H_i is the band filter, N is the frame length, and N' is the number of points in short-term DFT with zero padding)
- For each frame, calculate the periodogram estimate of the power spectrum.
- Frame the signal into short frames and take the Fourier transform
- Keep DCT coefficients 2–13, and discard the rest (can be adjusted accordingly).
- Take the logarithm of all filterbank energies, and take the DCT of the log filterbank energies.

3   Deep Sequential Modeling for Classification

LSTM is an efficient tool to selectively read, write, and forget the sequential input by leveraging the gated control mechanism. The LSTM framework that is

---

[1] We follow the implementation from https://github.com/jameslyons/ python_speech_features.

adopted in our model contains the five modules. The input module is inspired by the requirement to choose what kind of information that we want to update, which can be expressed as follows:

Since for an RNN, it is difficult to own a very long memory and also it has constraints due to the problem of vanishing and exploding of the gradients, it is necessary to introduce a mechanism to selectivity lose some information which has less contribution to the learning task.

$$i_t = \sigma(W_i s_{t-1} + U_i x_t + b_i) \tag{2}$$

$$f_t = \sigma(W_f s_{t-1} + U_f x_t + b_f) \tag{3}$$

$$o_t = \sigma(W_o s_{t-1} + U_o x_t + b_o) \tag{4}$$

$$\tilde{s}_t = \phi(W(o_t \odot s_{t-1}) + U x_t + b) \tag{5}$$

$$s_t = f_t \odot s_{t-1} + i_t \odot \tilde{s}_t \tag{6}$$

where the σ is the sigmoid function $\sigma(x) = \frac{1}{1+e^{-x}}$ and ϕ is the hyperbolic tangent function ("tanh") $\phi(x) = \frac{e^x - e^{-x}}{e^x + e^{-x}}$. W_* is the weight matrix from the input to the hidden layer, U_* is the recurrent state transition matrix, b_* is the bias vector, and $\odot$ stands for the element-wise product.

For the cost function, the well-known cross-entropy function is a proper choice for this target detection and data retrieval task, since we can model both the problems as classification-like problems. Given M input–target pairs $\mathbf{x}^{(i)}, y^{(i)}, i = 1, \ldots, M$ and the estimation of the distribution for the ith sample $\hat{y}^{(i)}$, the discrete binary cross-entropy can be expressed as follows:

$$L(\mathbf{x}, y) = -\sum_i^M \left(y^{(i)} \log \hat{y}^{(i)} + (1 - y^{(i)}) \log(1 - \hat{y}^{(i)}) \right) \tag{7}$$

4   Performance Analysis

The datasets contain STFT coefficients and MFCC coefficients for both vowel /a:/ and vowel /i:/, for normal and abnormal situations. For both of the two data, they are separately imported into the LSTM networking and will be split as 60% : 20% : 20% for training, validation, and testing. The amount of data for STFT are larger than that of MFCC, and the performance of the two features is compared. The implementation of the whole program is accomplished by leveraging the Keras 2.0 with TensorFlow backend. The program is running on the Imac i7 4 GHz with 16 GB RAM.

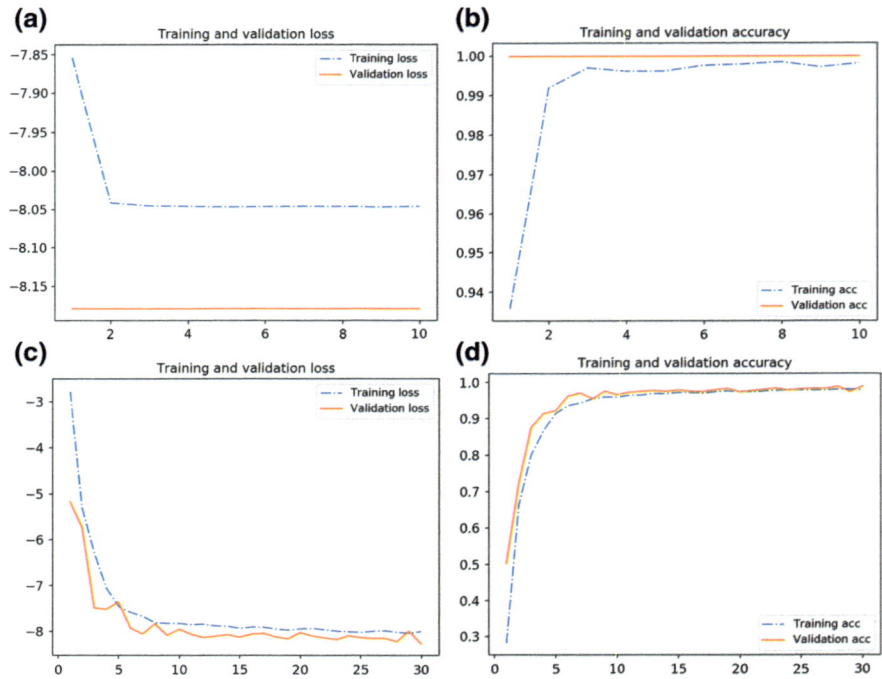

Fig. 3. Training and testing performance of the deep sequential model of **a** loss of STFT features; **b** accuracy of STFT features; **c** loss of MFCC features; **d** accuracy of MFCC features

A 32-unit LSTM is employed with dropout ratio of 0.4. A dense layer with hyperbolic tangent activation function is built on top of it. The Adam optimization method is adopted for iterations. The training process contains 10 epochs with minibatch size of 32. The random seed is set so that we do not need to run multiple times and average (it is still able to run average as well). The training performance and validation performance for both STFT and MFCC are shown in Fig. 3. The training accuracy approximates stable after 6 epochs and the loss approximates the lower bound after 2 epochs for STFT, while for MFCC, since the datasets are much smaller than STFT, it shows oscillation but eventually achieves a good performance, which shows the efficiency of the LSTM network.[2] The testing loss and accuracy are shown in Tables 2 and 3 for both the STFT features and MFCC features.

---

[2] The corresponding code will be uploaded to Github for open source purpose.

Table 2. Training, validation, and testing performance of the deep sequential model of STFT features

Performance index	vowel /a:/	vowel /i:/
STFT training accuracy	0.9983	0.9998
STFT training loss	−8.0454	−8.0449
STFT validation accuracy	1.0000	1.0000
STFT validation loss	−8.1784	−8.1784
STFT test accuracy	1.0000	1.0000
STFT test loss	−7.9814	−7.9784

Table 3. Training, validation, and testing performance of the deep sequential model of MFCC features

Performance index	vowel /a:/	vowel /i:/
MFCC test accuracy	0.9896	−0.9778
MFCC test loss	−8.1018	−7.8884
MFCC training accuracy	0.9794	0.9774
MFCC training loss	−8.0009	−7.8938
MFCC validation accuracy	0.9886	0.9796
MFCC validation loss	−8.2706	−7.9664

5   Conclusion

In this paper, the polyps detection based on the vocal samples is solved by leveraging the deep sequential model. Two types of preprocessing methods, i.e., the STFT and MFCC, are adopted to project the raw data on new domain to demonstrate the inherent information. The problem is formulated as the binary classification task, and the LSTM network with Adam optimization is able to achieve good performance on the binary classification task, with high accuracy and fast convergence.

Acknowledgments. This work was supported in part by NSFC under Grant 61771342, 61731006, 61711530132, and Tianjin Higher Education Creative Team Funds Program.

References

1. Wang TJ, Massaro JM, Levy D, et al. A risk score for predicting stroke or death in individuals with new-onset atrial fibrillation in the community: the framingham heart study. J Am Med Assoc. 2003;290(8):1049–56.
2. Rabiner LR. A tutorial on hidden Markov models and selected applications in speech recognition. Proc IEEE. 1989;77(2):257–86.
3. Roebroeck A, Seth AK, Valdes-Sosa P. Causal time series analysis of functional magnetic resonance imaging data. In: NIPS mini-symposium on causality in time series, 2011. p. 65–94.

4. Zhu F, Liang J. Soil moisture retrieval from UWB sensor data by leveraging fuzzy logic. IEEE Access, 2018. https://doi.org/10.1109/ACCESS.2018.2840159.
5. Alipanahi B, Delong A, Weirauch MT, et al. Predicting the sequence specificities of DNA-and RNA-binding proteins by deep learning. Nat Biotechnol. 2015;33(8):831–9.
6. Hochreiter S, Schmidhuber J. Long short-term memory. Neural comput. 1997;9(8):1735–80.
7. Hinton G, Deng L, Yu D, et al. Deep neural networks for acoustic modeling in speech recognition: the shared views of four research groups. IEEE Signal Process Mag. 2012;29(6):82–97.
8. Sutskever I, Vinyals O, Le QV. Sequence to sequence learning with neural networks. Adv Neural Inf Process Syst, 2014, p. 3104–12.
9. van der Westhuizen J, Lasenby J. The unreasonable effectiveness of the forget gate. arXiv preprint arXiv:1804.04849, 2018.
10. Tank A, Cover I, Foti NJ, et al. An interpretable and sparse neural network model for nonlinear granger causality discovery. In: Accepted by NIPs time series workshop, 2017.
11. Choi E, Schuetz A, Stewart WF, et al. Using recurrent neural network models for early detection of heart failure onset. J Am Med Inf Assoc. 2016;24(2):361–70.
12. Zhong Z, Jiang T, Zhang W, et al. Analyzing speech of patients with vocal polyps based on channel parameters and fuzzy logic systems. Comput Math Appl. 2011;62(7):2834–42.
13. Mitra SK, Kuo Y. Digital signal processing: a computer-based approach. New York: McGraw-Hill Higher Education; 2006.

Audio Tagging With Connectionist Temporal Classification Model Using Sequentially Labelled Data

Yuanbo Hou[1(✉)], Qiuqiang Kong[2], and Shengchen Li[1]

[1] Beijing University of Posts and Telecommunications, Beijing, China
hyb@bupt.edu.cn
[2] Centre for Vision, Speech and Signal Processing, University of Surrey, Guildford, UK

Abstract. Audio tagging aims to predict one or several labels in an audio clip. Many previous works use weakly labelled data (WLD) for audio tagging, where only presence or absence of sound events is known, but the order of sound events is unknown. To use the order information of sound events, we propose sequentially labelled data (SLD), where both the presence or absence and the order information of sound events are known. To utilize SLD in audio tagging, we propose a convolutional recurrent neural network followed by a connectionist temporal classification (CRNN-CTC) objective function to map from an audio clip spectrogram to SLD. Experiments show that CRNN-CTC obtains an area under curve (AUC) score of 0.986 in audio tagging, outperforming the baseline CRNN of 0.908 and 0.815 with max pooling and average pooling, respectively. In addition, we show CRNN-CTC has the ability to predict the order of sound events in an audio clip.

Keywords: Audio tagging · Sequentially labelled data (SLD) · Convolutional recurrent neural network (CRNN) · Connectionist temporal classification (CTC)

1 Introduction

Audio tagging aims to predict an audio clip with one or several tags. Audio clips are typically short segments such as 10 s of a long recording. Audio tagging has many applications in information retrieval [1], audio classification [2], acoustic scene recognition and industry sound recognition [3].

Many previous works of audio tagging rely on strongly or weakly labelled data. In strongly labelled data [4], each audio clip is labelled with both tags and onset and offset times of sound events. Labelling the strongly labelled data is time-consuming and labour expensive, so the size of strongly labelled dataset is often limited to minutes or a few hours [4]. Additionally, the onset and offset time of some sound events are ambiguous due to the fade-in and fade-out effect [5]. In fact, many audio datasets contain only tags, without the onset and offset times of sound events. This is referred to as weakly labelled data (WLD) [5]. In WLD, only the presence or absence of sound events is known, the occurrence sequence of sound events is not known. These weaknesses limit the use of strongly labelled data and WLD.

© Springer Nature Singapore Pte Ltd. 2020
Q. Liang et al. (eds.), *Communications, Signal Processing, and Systems*, Lecture Notes in Electrical Engineering 516,
https://doi.org/10.1007/978-981-13-6504-1_114

To avoid the weakness of strongly labelled data and WLD and use order information of sound events, we propose sequentially labelled data (SLD). This idea is
inspired by the label sequences in speech recognition [6]. In SLD, the tags and order of
tags are known, without knowing occurrence time of tags. SLD reduces the workload
of data annotation and indicates the order of tags in WLD. The order information of
events will benefit tasks like acoustic scene analysis and environment recognition.
Figure 1 shows an audio clip and its strong, sequential and weak tags.

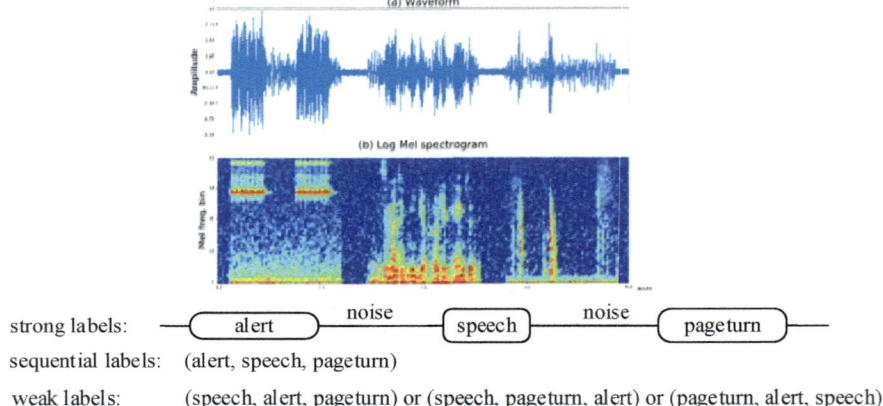

Fig. 1 From top to bottom: **a** waveform of an audio clip containing three sound events: "alert",
"speech" and "pageturn"; **b** log Mel spectrogram of (**a**); strong labels, sequential labels and weak
labels of the audio clip

To utilize the SLD in audio tagging, we propose to use CTC technique to train a
CRNN (CRNN-CTC). CTC is a learning technique for sequence labelling with RNN
[7], which has achieved great success in speech recognition [6]. In fact, CTC is an
objective function that allows RNN to be trained for sequence-to-sequence tasks,
without requiring any prior alignment between the input and target sequences. In
training, CTC computes the total probability of input sequences and sums over all
possible alignments. CTC allows train an RNN without any prior alignment (i.e. the
starting or ending times of each sound event), hence, even without strong labels, it is
sufficient to do audio tagging with SLD based on CTC model.

There are two contributions in this paper. First, in audio tagging, we propose SLD,
which reduces the workload and difficulties of data annotation in strong labels, and
indicates the order of tags in weak labels. Second, to utilize SLD in audio tagging, we
propose CRNN-CTC compare its performance with other common CRNN models in
previous works. This paper is organized as follows, Sect. 2 introduces related works.
Section 3 describes CRNN baseline. Section 4 describes CRNN-CTC with SLD.
Section 5 describes dataset, experimental set-up and results. Section 6 gives
conclusions.

2   Related Work

Audio classification and detection have obtained increasing attention in recent years. There are many challenges for audio detection and tagging such as DCASE 2013 [3], DCASE 2016 and DCASE 2017 [4].

In previous works in audio classification and tagging, Mel-frequency cepstrum coefficient (MFCC) and Guassian mixture model (GMM) is widely used in baseline system [3]. Recent methods include deep neural networks (DNNs), convolution neural networks (CNNs) and RNN [2], with inputs varying from Mel energy, spectrogram, MFCC to constant Q transform (CQT).

Many methods described above rely on the bag of frames (BOF) model [5]. BOF is based on an assumption that tags occur in all frames, which is however not the case in practice. Some audio events like "gunshot" only happen a short time in audio clip. State-of-the-art audio tagging methods [8] transform waveform to the time-frequency (T-F) representation. T-F representation is treated as an image which is fed into CNNs. However, unlike image where the objects usually occupy a dominant part of the image, in an audio clip audio events only occur a short time. To solve this problem, some attention models [8] for audio tagging and classification are applied to attend to the audio events and ignore the background sounds.

3   CRNN Baseline in Audio Tagging

CRNN has been successfully used in audio tagging [8]. First, the waveforms of the audio are transformed to time-frequency (T-F) representation such as Mel spectrogram. Next convolutional layers are applied on the T-F representation to extract high-level features. Then, bidirectional gated recurrent units (BGRU) are adopted to capture the temporal context information. Finally, the output layer is a dense layer with the sigmoid activation function since it is a multi-class classification problem [5], the sigmoid activation function to predict probability of each sound event in the audio clip. Inspired by the good performance of CRNN in audio tagging [8], we use CRNN as our baseline system in this paper.

An audio clip from real life may contain more than one sound event, as environmental sound is often a mixture audio that comes from multiple sound sources simultaneously. Thus, the audio tagging task is a multi-label classification problem, and a binary decision is made for each class. In training phase, the binary cross-entropy loss is applied between the predicted probability of each tag and the ground-truth tag in an audio clip. The loss can be defined as:

$$E = -\sum_{n=1}^{N} (P_n \log Q_n + (1 - P_n) \log(1 - Q_n)) \tag{1}$$

where E is the binary cross-entropy, Q_n and P_n denote the predicted and reference tags sequence of the n-th audio clip, respectively. N is the batch size.

In CRNN baseline, clip level probability of tags can be obtained from the last layer. However, there is no frame level information of each event in it. To obtain the probability of each event at each frame, a dense layer with the number of event classes, following the BGRU layer, as shown in Fig. 2. These frame level predictions can be used for sound event detection. To map the frame level tags to clip level tags, pooling layer was used. In training, the clip level predictions are compared against the weak labels of the audio clip to compute the loss function of model.

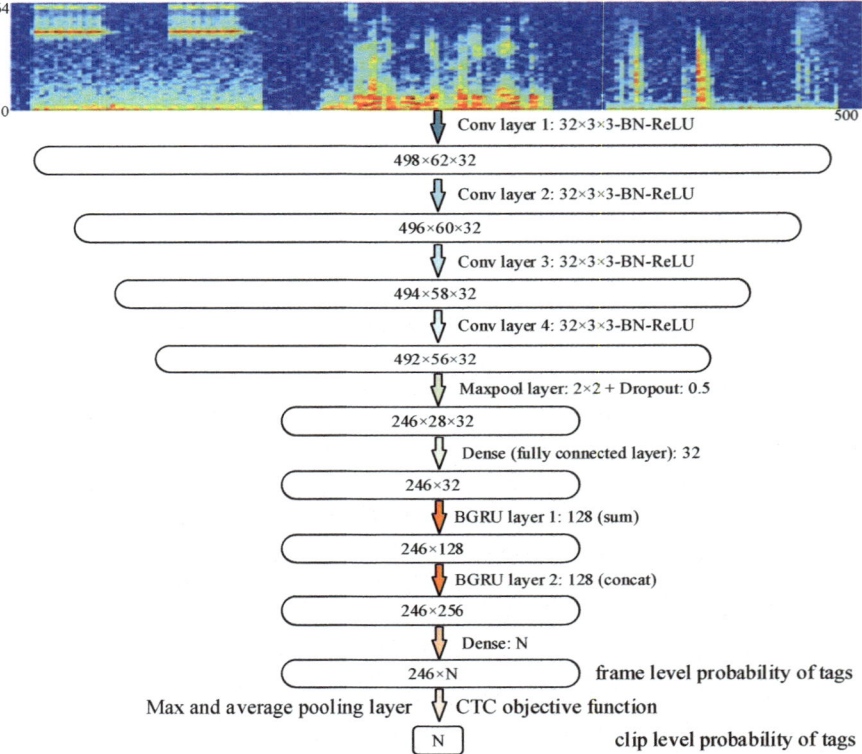

Fig. 2 Model structure. BN: Batch normalization. ReLU: Rectified linear unit. For baseline, CRMP and CRAP, $N = 16$. For CRNN-CCT, $N = 17$ ($16 + 1$), the extra "1" indicates the blank label

There are two pooling operations in Fig. 2, max pooling and average pooling. For CRNN with max pooling (CRMP) and CRNN with average pooling (CRAP), pooling performs down-sampling along time axis and transforms the frame level probability of tags to clip level tags, respectively. Max pooling and average pooling as way of aggregation have been successfully used [8].

4   CRNN-CTC in Audio Tagging

As discussed before, strongly and weakly labelled data have their own drawbacks in audio tagging, so we propose sequentially labelled data (SLD) and use CRNN-CTC to detect presence or absence of several sound events in SLD.

4.1   Sequentially Labelled Data

Let $\mathcal{D}$ be a set of training examples drawn from audio dataset. Input space $\mathcal{X} = (\mathbb{R}^n)$ is the set of all sequences of n dimensional vectors. Target space $\mathcal{Z} = L$ is the set of all sequences of labels over audio events. In general, we refer to elements of L as label sequences or labelling [7]. Each example in $\mathcal{D}$ consists of a pair of sequences $(\mathbf{x}, \mathbf{z})$. The target sequence $\mathbf{z} = (z_1, z_2, \ldots, z_Q)$ is at most as long as input sequence $\mathbf{x} = (x_1, x_2, \ldots, x_T)$, i.e. $Q \leq T$. Since, the input and target sequences are not generally the same length, there is no a priori way of aligning them [7]. In the label sequence $\mathbf{z}$, the tags of the audio clip and sequence of tags are known, without knowing their occurrence time, that is, there are no starting/ending times of sound events. We refer to audio data labelled by label sequence as sequentially labelled data (SLD).

In essence, SLD is a weakly labelled data with events sequence information. In audio tagging using SLD, we can use the model like CRNN described in Sect. 3. However, there is no order information of sound events in predictions of baseline, CRMP and CRAP. So we propose CRNN-CTC in audio tagging using SLD.

4.2   CRNN-CTC in Audio Tagging Using SLD

CTC has achieved great success in speech recognition [6, 7]. In this section, we will show how to use CTC technique to train a CRNN in audio tagging using SLD.

CTC is a learning technique for sequence labelling, it shows a way for training RNN with label unsegment sequences. CTC redefines the loss function of RNN [7] and allows RNN to be trained for sequence-to-sequence tasks without requiring any prior alignment (i.e. starting or ending time of sound events) between the input and target sequences. Thus, it is sufficient to train a CRNN using SLD with CTC. Given $y_t(k)$ is probability of observing label k at time t output by the last recurrent layer in CRNN, and z_t is the ground-truth label, conventional loss function of RNN for a sequence X of length T is $L = -\sum_{t=1}^{T} \log y_t(z_t)$, which is the negative logarithm of the joint probability of desired label sequence and its alignment. In audio tagging, we are only interested in label sequence, not the ground-truth alignment.

CTC gives a solution to how to marginalize out the alignment. First, CTC adds an extra "blank" label (denoted by "–") to original label set L [7]. Then, it defines a many-to-one mapping β that transforms the alignment (i.e. the sequence of output labels at each time step, also called a path [7]) to label sequence. The mapping β removes repeated labels from the path to a single one and then removes the "blank" labels. For example, $\beta(C - AT-) = \beta(-CC - -ATT) = CAT$, that is, path "$C - AT-$" and "$-CC - -ATT$" both map to the label sequence "CAT".

The CTC objective function is defined as the negative logarithm of total probability of all paths that map to the ground-truth label sequence. The total probability can be found using dynamic programming algorithm [7] on the trellis shown in Fig. 3. On the x-axis is time steps, on the y-axis is "modified label sequence", target label sequence with blank labels added to the beginning and end and inserted between every pair of labels. Given the length of modified label sequence is L and l_i denotes i-th label. An effective path may start at either l_1 or l_2 and end at l_{L-1} or l_L. At each time step, the path may (i) stay at the same label; (ii) move to the next label; (iii) move to the label after the next if it is not a blank label different from the current label. Let $\alpha_t(s)$ be the total probability of $l_{1:s}$ at time t. Assuming conditional independence between $y_t(k)$ (i.e. probability of observing label k at time t) across time steps, the $\alpha_t(s)$ can be calculated as follows:

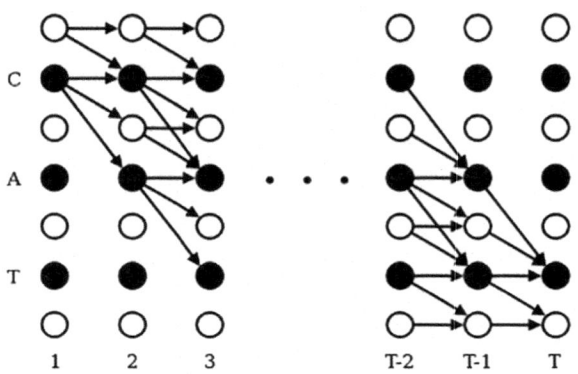

Fig. 3 Trellis for computing CTC objective function [7] applied to the example labelling "CAT". Black circles represent labels, white circles represent blanks. Arrows signify allowed transitions

$$\alpha_1(s) = \begin{cases} y_1(l_s) & s \leq 2 \\ 0 & s > 2 \end{cases} \tag{2}$$

$$\alpha_t(s) = [\alpha_{t-1}(s) + \alpha_{t-1}(s-1) + \delta_s \alpha_{t-1}(s-2)]y_t(l_s), t > 1 \tag{3}$$

where $\delta_s = 1$ if $l_s \neq l_{s-2}$, and terms that go past the start of the modified label sequence are zero. The sum of total probability of paths that map to original label sequence is $\alpha_T(L-1) + \alpha_T(L)$, and its negative logarithm is CTC loss function.

To decode the CTC output, we use the simple best path decoding [7] in this paper. This method is to select the label with the maximum probability at each frame, reduce adjacent repeating labels to a single one and remove the blank labels.

The output of CTC model is directly a label sequence corresponding the audio clip. The detailed structure of CRNN-CTC was shown before in Fig. 2.

5 Experiments and Results

We use the audio events in DCASE 2013 [3] to make SLD and evaluate the proposed method. There are 16 kinds of sound events in DCASE 2013 includes: *alert, clearthroat, cough, doorslam, drawer, keyboard, keys, knock, laughter, mouse, pageturn, pendrop, phone, printer, speech* and *switch*. We remixed these sound events to 10 s audio clips totalling 7.1 h, where each audio clip contains no overlapped three or several sound events mixed with noise background.

For experimental set-up, fourfold cross-validation was used for model selection and parameter tuning. Dropout, batch normalization and early stopping criteria are used in training phase to prevent over-fitting. The model is trained for maximum 1000 epochs with Adam optimizer with learning rate of 0.001.

The results are evaluated by precision, recall and *F*-score [9]. To calculate these metrics, we need to count the number of: true positive (TP), false negative (FN) and false positive (FP). Precision (*P*), recall (*R*) and *F*-score are defined as:

$$P = \frac{\text{TP}}{\text{TP} + \text{FP}}, \quad R = \frac{\text{TP}}{\text{TP} + \text{FN}}, \quad F = \frac{2\text{P} \cdot \text{R}}{\text{P} + \text{R}}. \tag{4}$$

To evaluate the true positive rate (TPR) versus false positive rate (FPR), the receiver operating characteristic (ROC) curve was used [9]. AUC score is the area under the ROC curve which summarizes the ROC curve to a single number. Larger *P*, *R*, *F*-score and AUC indicates better performance. The AUC score of audio tagging is shown in Table 1; CRAP, CRMP and CRNN-CTC outperform baseline system. CRNN-CTC achieves an averaged AUC of 0.986.

Table 2 shows the averaged statistic including precision, recall, *F*-score and AUC over 16 kinds of sound events, and CRNN-CTC performs better than other models. Figure 4 shows the frame level predictions of models on example audio clip. In Fig. 4, CRNN-CTC predicts the tag sequence of audio clip, typically as a series of spikes [7]. Although the spikes align well with the actual position of sound events in audio clip, there is no time span information about these events.

In Fig. 4, CRMP produces wide peaks, indicating the onset/offset times of each event. That shows max pooling has ability to locate audio events, while average pooling seems to fail. The reason may be max pooling encourages the response for a single location to be high, for similar audio events which can obtain similar features. While average pooling in CRAP encourages all response to be high, the difference features of each event make it difficult to locate audio events.

Table 1 AUC of audio tagging

	alert	clearthroat	cough	doorslam	drawer	keyboard	keys	knock	laughter	mouse	pageturn	pendrop	phone	printer	speech	switch	Average
Baseline	0.609	0.627	0.674	0.691	0.690	0.569	0.702	0.816	0.617	0.668	0.693	0.662	0.654	0.862	0.550	0.625	0.669
CRAP	0.737	0.948	0.792	0.804	0.895	0.811	0.864	0.971	0.783	0.587	0.759	0.809	0.715	0.910	0.800	0.850	0.815
CRMP	0.959	0.970	0.915	0.875	0.953	0.735	0.918	0.973	0.883	0.835	0.892	0.936	0.892	0.985	0.887	0.922	0.908
CRNN-CTC	0.968	1.0	1.0	0.977	1.0	0.959	0.972	1.0	1.0	0.995	0.990	0.972	1.0	0.995	0.990	0.965	0.986

Table 2 Averaged stats of audio tagging

	Precision	Recall	*F*-score	AUC
Baseline	0.687	0.371	0.482	0.669
CRAP	0.847	0.647	0.733	0.815
CRMP	0.933	0.827	0.877	0.908
CRNN-CTC	0.983	0.975	0.98	0.986

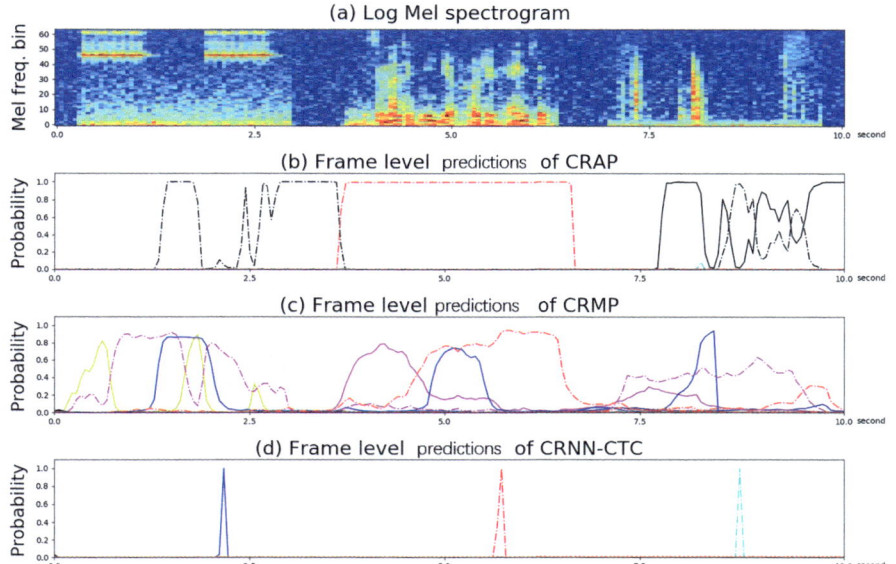

Fig. 4. A frame level predictions of CRAP (**b**), CRMP (**c**) and CRNN-CTC (**d**). The ground-truth tag is "alert, speech, pageturn". Peaks are annotated with corresponding tag

6 Conclusion

In this paper, we analyse the weakness of strongly and weakly labelled data, then propose SLD. To utilize SLD in audio tagging, we propose CRNN-CTC. In CRNN-CTC, CTC layer maps frame level tags to clip level tags, similar to the pooling layer. So we compare them. Experiments show CRNN-CTC outperforms CRAP, CRMP and baseline. The frame level predictions of models in Fig. 4 show CRNN-CTC predicts the presence/absence and tag sequence of events in the audio clip well.

References

1. Guo, G, and S. Z. Li. Content-based audio classification and retrieval by support vector machines. IEEE Press, 2003.
2. Xu Y, Kong Q, Wang W and Plumbley MD. "Largescale weakly supervised audio classification using gated convolutional neural network," arXiv preprint arXiv:1710.00343, 2017.
3. Stowell D, Giannoulis D, Benetos E, Lagrange M, and Plumbley MD. "Detection and classification of acoustic scenes and events," IEEE Transactions on Multimedia. 17(10): 1733–1746, 2015.
4. Mesaros A, Heittola T, et al. Dcase 2017 challenge setup: Tasks, datasets and baseline system, in Workshop on DCASE 2017, Munich, Germany, 2017.
5. Kong Q, Xu Y, Wang W and Plumbley MD. A joint separation-classification model for sound event detection of weakly labelled data, in IEEE Int. Conf. on Acoustics, Speech and Signal Processing (ICASSP), Apr. 2018.
6. Graves A and Jaitly N. Towards end-to-end speech recognition with recurrent neural networks, in Proc. of ICML, 2014.
7. Graves A and Gomez F, Connectionist temporal classification:labelling unsegmented sequence data with recurrent neural networks, in ICML, 2006, pp. 369–376.
8. Xu Y, Kong Q, Huang Q, Wang W and Plumbley MD. Attention and localization based on a deep convolutional recurrent model for weakly supervised audio tagging, in INTERSPEECH, 207, pp. 3083–3087.
9. Bhavna K, Jain K and Sharma SK. Estimation of Area under Receiver Operating Characteristic Curve for Bi-Pareto and Bi-Two Parameter Exponential Models. Open Journal of Statistics 4.1(2014):1–10.

Implementation of AdaBoost Face Detection Using Vivado HLS

Sanshuai Liu, Kejun Tan$^{(\boxtimes)}$, and Bo Yang

Information Science and Technology College,
Dalian Maritime University, Dalian, Liaoning Province, China
`tankejun@dlmu.edu.cn`

Abstract. For the problem that Adaptive Boosting (AdaBoost) face detection algorithm is slowly implemented on the embedded platform by software, this paper adopts the method of the full hardware acceleration. The intellectual property (IP) core of AdaBoost algorithm is designed by Vivado high-level synthesis (HLS), which may reduce the development difficulty and shorten the development cycle. The design adopts the serial–parallel structure to accelerate face detection and uses several methods of optimizing hardware resource. The face detection algorithm is implemented on the Zedboard platform and achieves the purpose of real-time detection.

Keywords: AdaBoost · Face detection · HLS · Real-time detection

1 Introduction

Face detection refers to the process of determining whether there are faces in the image, and their positions, and sizes. The face detection problem is originally proposed as the positioning link of the automatic face recognition system. Due to its application value in the areas of security access control and visual inspection, it has become an independent topic that has received widespread attention [1].

As face detection becoming more and more important in many fields, researchers have tried various methods to achieve face detection. Software implementation implemented on the embedded platform is slow [2]. Therefore, some scholars have studied the hardware acceleration of AdaBoost algorithm; for example, Shi Yuehua used Verilog HDL to implement the face detection algorithm [3]. In this paper, AdaBoost algorithm intellectual property (IP) core is designed by HLS and implemented on the Zedboard platform. HLS could convert C, C++, and other codes into hardware description language [4], and the generated IP cores can be directly connected with register-transfer-level (RTL) design in Vivado. Compared to software implementations, this method has a faster detection speed. Compared to the hardware description language implementations, the algorithm is easier to implement and the development cycle is shorter.

© Springer Nature Singapore Pte Ltd. 2020
Q. Liang et al. (eds.), *Communications, Signal Processing, and Systems*, Lecture Notes in Electrical Engineering 516,
https://doi.org/10.1007/978-981-13-6504-1_115

2   Face Detection Algorithm

2.1   Haar Features and Integral Image

The Haar features reflect the gray pixel subtraction module in an image and are mainly used to describe human face feature, and the Haar features used in this paper are shown in Fig. 1a. For facilitating the calculation of Haar feature, Viola and Jones introduced the concept of integral image [5]. The definition of the integral image is:

$$SAT(x,y) = \sum_{x'<x, y'<y} I(x',y') \tag{1}$$

The value of $I(x',y')$ represents the pixel value of the point (x',y'), and $SAT(x,y)$ is the pixel sum of all pixels in the upper-left corner of the point (x,y), as shown in Fig. 1b. Figure 1c shows the incremental calculation algorithm which is used to quickly calculate the Haar features. The sum of pixels in the rectangular shaded region could be obtained by adding and subtracting the values of the integral image of the four vertex positions. The calculation formula is as follow:

$$Sum(D) = SAT(x,y) + SAT(x+w, y+h) - SAT(x+w, h) - SAT(x, y+h) \tag{2}$$

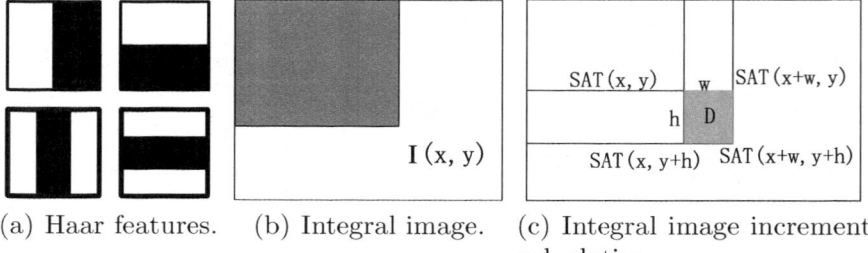

(a) Haar features.   (b) Integral image.   (c) Integral image increment calculation.

Fig. 1. Haar features and integral image.

2.2   Face Detection Process

Scaling the image and scaling the classifiers are two methods of face detection [6]. The method of scaling the classifiers only needs to calculate the integral image once, and it needs to cache the integral image. It takes less time, but it consumes a large amount of memory resources. The detection process is shown in Fig. 2a. The method of scaling the image slides a fixed size detection window in the integral image, while the size of detection window in the method of scaling the classifiers is changed. The memory resources consumption of scaling the image is less, because it requires to cache the raw image, but each time the image is scaled the integral image needs to calculate again. Therefore, scaling the image method is used in this paper. The specific detection process is shown in Fig. 2b.

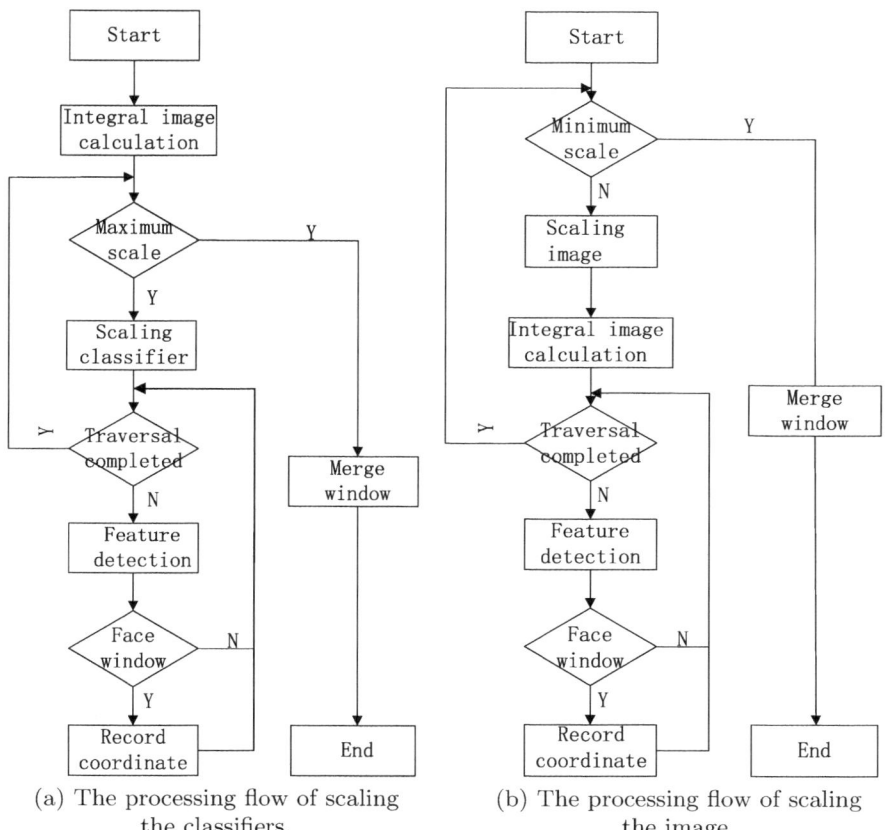

(a) The processing flow of scaling
the classifiers.

(b) The processing flow of scaling
the image.

Fig. 2. Processing flow of AdaBoost.

The process of scaling the image face detection firstly judges whether the image is scaled to the minimum scale that is the size of training sample and, if not, continues to scale the image. The square integral image and the integral image are used to obtain the image variance which could correct the illumination of the image. The detection window traverses in the integral image to find face window. If a face window is detected, the coordinate and the scaling ratio of face window are recorded. When the integral image traversal is over, it is judged whether the image is scaled to the smallest scale that means the end of the detection. The face windows are merged to obtain the correct coordinate position after the detection is completed.

3   Hardware Design and Optimization

3.1   Face Detection Module

Feature detection module is the core module in the face detection. The classifier used in the module is the front face detection classifier provided by the Open Source Computer Vision (OpenCV) library. The Haar features in the classifier we used have only two rectangular features and three rectangular features, and we encoded them. It can be determined what kind of feature is used by identifying the encoding, which could save memory resources compared to all default to three rectangular features (the third rectangular data in two rectangular features is filled with zeros). Firstly, the module of feature detection reads the stored rectangular data of the Haar features of the classifier, including the rectangular starting coordinates, widths, lengths, and weights. The sum of pixels in the rectangular region can be calculated according to formula (2). The sum of pixels in each rectangular area is multiplied by itself related weight and then added. The sum is compared with the product of the corresponding weak classifier threshold and the variance within the window. If it is greater than the product, it takes the right value, and if it is less than the product, it takes the left value. The results of the weak classifiers accumulated compared with the threshold of the strong classifier; if it is greater than the threshold, it means that the strong classifier is passed and the next strong classifier will be detected; otherwise, it is not a face window, and the next candidate window will be detected.

3.2   Serial and Parallel Structure Design

The classifier has 22 strong classifiers, each of which could filter out a certain amount of candidate windows. The filtered results of top 5 strong classifiers are tested in 50 different face pictures as shown in Fig. 3. It could be seen that the filtered increase of the top 3 strong classifier is larger, and the filtered increase in the fourth stage and the fifth stage gradually decreases. Since the top 3 strong

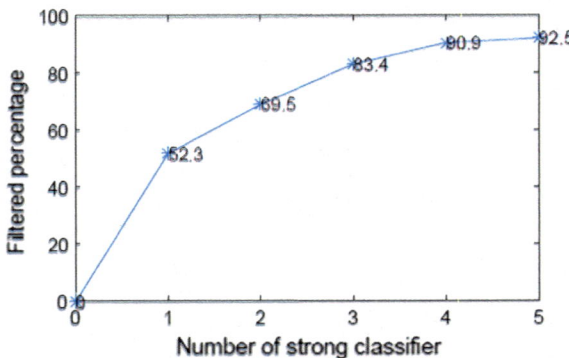

Fig. 3. Filtered rate of strong classifier.

classifiers could filter out 83.4% of non-face windows, taking into account the relationship between resources and speed, the top 3 strong classifiers use a parallel structure to design, and the latter 19 strong classifiers use a serial structure to design, as shown in Fig. 4. The weak classifiers of the top 3 strong classifiers are all expanded, and a parallel structure could be obtained. If any one of top 3 strong classifiers detects failed, the next window is directly detected. If the detection of the parallel module is successful, the detection of the serial module will be continued. In the serial module, only the strong classifier at current could continue to detect the next strong classifier. Otherwise, it will detect the next window. If the candidate window passes all strong classifiers, the window is a face window, the coordinate of which will be recorded, and the next window continues to be detected.

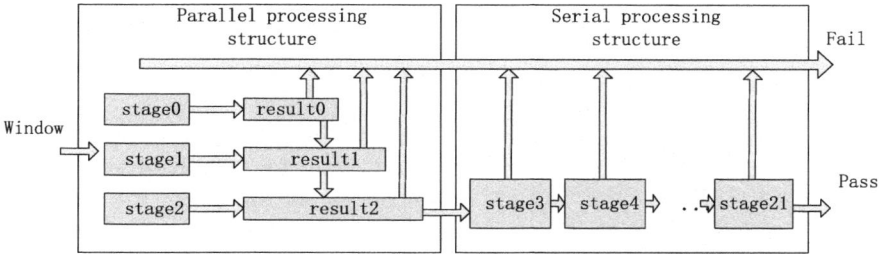

Fig. 4. Serial and parallel structure.

3.3 Data Bit Qualification and Type Conversion

The calculation of the integral image could be easily obtained according to the incremental calculation method of the integral image, but the integral image needs to cache when the window traverses. Limiting the bit width of data can save block random access memory (BRAM) resource consumption. For example, the maximum of the 320×240 image's integral image is $320 \times 240 \times 255 = 19,584,000$, so at least 25 bits are needed to store the data of the integral image. The maximum of the squared integral image is the maximum of the integral image moved left 8 bits, so storage requires at least 33 bits.

Floating-point calculation is used in the bilinear interpolation scaling image and the scaling calculation. Fixed-point data is defined in Vivado HLS, which could set the length of integer part and decimal part according to the required accuracy. As shown in Table 1, solution 1 is the resource consumption of the fixed-point data, the bit width of the integer part is 10, the bit width of the decimal part is 16, and solution 2 is the resource consumption report of the floating-point-type data. Solution 1 is compared with solution 2, the look-up table (LUT) resource is reduced about 15%, and other hardware resources are also reduced. Since the floating-point-type data consumes lots of resources, the thresholds, the left-value parameters, the right-value parameters of the weak classifiers, and the thresholds of the strong classifiers are expanded 4096 times to become integer data.

Table 1. Comparison of resource consumption

	Solution 1	Solution 2
BRAM_18K	197	197
DSP48E	54	62
FF	44424	48715
LUT	30978	39069

3.4 Multiplication Optimization

Multiplication calculation takes much time and occupancies digital signal processing (DSP) resources. Multiplication calculation can be split into shift operations which take less time, so the multiplication calculation in the code could be replaced with shift operations. It not only saves time, but also reduces the use of DSP resources.

4 Experiment Result

The experiment is implemented on the Zedboard platform, which includes a dual-core ARM Cortex A9 and more than 5 million programmable logic elements. In the experiment, the image size is 320×240, and the scaling is 1.2. The method of scaling the image is bilinear interpolation, and the system operating frequency is 100 MHz.

The experiment selected 100 images which contain face and no face in different backgrounds. The rate of false detection is connected to the background, we may reduce the false detection rate by adjusting the algorithm of window selection, and the detection rate will decrease. Comprehensively, adjusting the window screening algorithm that when two or more scales are continuously passed detection, the candidate face windows are determined to be face windows, and the accuracy rate of the experimental detection is 95%. Software implementation of the face detection algorithm is on the Zedboard platform processing system (PS), and it takes about 3377 ms in the case of having face in the image and 3265 ms in the case of having no face. The hardware implementation is 93 and 82 ms, respectively, and the hardware acceleration effect is more than 30 times, achieving the purpose of real-time detection. In some specific applications, the distance between the person and the sensor is often relatively fixed, such as driver fatigue detection. We may fix scale in several scaling sizes, which could greatly reduce the traversal window, so the speed can reach about 10 ms.

References

1. Liang L, Ai H, Zhang B. A survey of human face detection. Chin J Comput. 2002;25(5):449–58.
2. Zhu M, Lu X, Lu H. Transplantation and optimization of AdaBoost face detection algorithm on DSP. Comput Eng Appl. 2014;50(20):197–201.
3. Zhang Z, Feng Z, Wei S. Implementation of face detection system based on AdaBoost algorithm on embedded system. Inf Technol. 2008;07:167–70.
4. Xilinx Inc. High-level-synthesis v17.1 [EB/OL]. URL https://www.xilinx. com/support/documentation/sw_manuals/xilinx2018_1/ug902-vivado-high-level-synthesis.pdf.
5. Viola P, Jones M. Rapid object detection using a boosted cascade of simple features. In: Computer society conference on computer vision and pattern recognitionl, vol. 1; 2001. p. I-511–8.
6. Acasandrei L, Barriga A. Accelerating Viola-Jones face detection for embedded and SoC environments. In: ACM/IEEE international conference on distributed smart cameras, Ghent; 2011. p. 1–6.

Research on Rolling Bearing On-Line Fault Diagnosis Based on Multi-dimensional Feature Extraction

Tianwen Zhang[⊠]

College of Information and Communication Engineering,
Harbin Engineering University, Harbin, China
{287133770,1325069385}@qq.com

Abstract. In the paper, a novel rolling bearing fault diagnostic method was proposed to fulfill the requirements for effective assessment of different fault types and severities with real-time computational performance. Firstly, multi-dimensional feature extraction is discussed. And secondly, a gray relation algorithm was used to acquire basic belief assignments. Finally, the basic belief assignments were fused through Yager algorithm. The related experimental study has illustrated the proposed method can effectively and efficiently recognize various fault types and severities.

Keywords: Rolling element bearing · Pattern recognition · Gray relation algorithm · Yager algorithm

1 Introduction

Rolling bearing as an important part is widely used in almost all types of rotating machinery, such as gas turbine, steam turbine and diesel engine. Rolling bearing failure is one of the main causes of failure and damage of rotating machinery and leads to huge economic losses [1–3]. Among the many fault diagnosis approaches for bearings, vibration-based diagnostic methods have received much attention in the past few decades [4, 5]. Bearing vibration signal contains a wealth of information on mechanical health status. This also makes it possible to extract the dominant features that characterize the mechanical health state from vibration signals through signal processing techniques [6]. Currently, many signal processing techniques have been applied to bearing off-line fault diagnosis. However, due to many nonlinear factors (e.g., stiffness, friction, clearance, etc.), bearing vibration signals (especially in a faulted condition) will exhibit nonlinear and unsteady character [7]. In addition, the measured vibration signal contains not only information about the operating conditions associated with the bearing itself, but also information about a large number of other rotating components and structures in the plant equipment [8]. Due to the usually large background noise, slight bearing fault information is easily submerged in background noise and difficult to extract. Therefore, conventional time-domain and frequency-domain methods do not

© Springer Nature Singapore Pte Ltd. 2020
Q. Liang et al. (eds.), *Communications, Signal Processing,
and Systems*, Lecture Notes in Electrical Engineering 516,
https://doi.org/10.1007/978-981-13-6504-1_116

allow for an accurate assessment of the bearing health status [9]. In this paper, a novel rolling bearing fault diagnostic method was proposed to fulfill the requirements for effective assessment of different fault types and severities with real-time computational performance.

2 Methodology

2.1 Multi-dimensional Feature Extraction

Firstly, multi-dimensional feature extraction on the basis of Entropy characteristics, Holder coefficient characteristics were proposed for extracting health status feature vectors from the bearing vibration signals, respectively.

2.1.1 Entropy Characteristics

Entropy is a crucial concept in information theory and is a measure of information uncertainty of the signal distribution and the complexity of the signal [10].

Suppose the bearing vibration signal is f. The signal is first sampled and discretized into a discrete signal sequence $f(i), i = 1, 2, 3, \ldots, n$, where n is the total number of discrete signals. Perform FFT transform as follows:

$$F(k) = \sum_{i=0}^{n-1} f(i) \exp\left(-j\frac{2\pi}{n}ik\right), \quad k = 0, 1, \ldots, n-1 \tag{1}$$

After obtaining the signal spectrum, calculate the energy of each point:

$$E_k = |F(k)|^2 \tag{2}$$

Calculate the total energy value of each point:

$$E = \sum_{k=0}^{n-1} E_k \tag{3}$$

Shannon entropy E_1 and exponential entropy E_2 can be defined as follows:

$$E_1 = -\sum_{i=1}^{n} P_i \log_e P_i \tag{4}$$

$$E_2 = \sum_{i=1}^{n} P_i e^{1-P_i} \tag{5}$$

The entropy characteristics $[E_1, E_2]$ are taken as dominant feature vectors for rolling element bearing fault pattern recognition.

2.1.2 Holder Coefficient Characteristics

The Holder coefficient algorithm evolves from the Holder inequality [11, 12]. Holder coefficient can be used to measure the similar degree of two sequences.

Holder coefficient of these two discrete signals is obtained as follows:

$$H_c = \frac{\sum_{i=1}^{n} f_1(i) f_2(i)}{\left(\sum_{i=1}^{n} f_1^p(i)\right)^{1/p} \cdot \left(\sum_{i=1}^{n} f_2^q(i)\right)^{1/q}} \tag{6}$$

where $0 \leq H_c \leq 1$.

Calculate the Holder coefficient value H_1 between the bearing vibration signal sequence $f(i)$ and the rectangular signal sequence $s_1(i)$.

$$H_1 = \frac{\sum_{i=1}^{n} f(i) s_1(i)}{\left(\sum_{i=1}^{n} f^p(i)\right)^{1/p} \cdot \left(\sum_{i=1}^{n} s_1^q(i)\right)^{1/q}} \tag{7}$$

In the same way, we obtain the Holder coefficient value H_2 between the vibration signal sequence $f(i)$ and the triangular signal sequence $s_2(i)$.

$$H_2 = \frac{\sum_{i=1}^{n} f(i) s_2(i)}{\left(\sum_{i=1}^{n} f^p(i)\right)^{1/p} \cdot \left(\sum_{i=1}^{n} s_2^q(i)\right)^{1/q}} \tag{8}$$

The Holder coefficient characteristics $[H_1, H_2]$ are taken as dominant feature vectors for rolling element bearing fault pattern recognition.

2.2 Diagnostic Procedure

Totally, the process of the proposed method for rolling bearing on-line fault diagnosis is as follows.

Step 1: The vibration signals from the object bearing are sampled under different healthy status, including normal operating condition and conditions with different fault types and severities, to establish the knowledge base (i.e., the recognition template).

Step 2: The health status feature vectors are extracted from the sample knowledge base through the multi-dimensional feature extraction based on entropy characteristics $[E_1, E_2]$, Holder coefficient characteristics $[H_1, H_2]$, respectively.

Step 3: The sample knowledge base is established based on the fault symptom (i.e., the extracted fault feature vectors) and the fault pattern.

Step 4: The health status feature vectors extracted based on bearing vibration signals to be identified are input into the gray relation algorithm (GRA) to acquire basic belief assignments (i.e., BBA1, BBA2, BBA3) and fuse the basic belief assignments through Yager method and output the diagnostic results (i.e., fault types and severities).

3 Experimental Validation

In the paper, the rolling bearing vibration signals for testing are from Case Western Reserve University Bearing Data Center [13]. The fault types contain outer race fault, the inner race fault, and the ball fault, and the fault diameters, i.e., fault severities, contain 28 mils, 21 mils, 14 mils and 7 mils. An accelerometer is installed on the motor drive end housing with a bandwidth of up to 5000 Hz, and the vibration data for the test bearing under different fault patterns is collected by a recorder. The bearing vibration data used for analysis is obtained under the motor speed of 1797 r/min and load of 0 hp (Table 1).

Table 1 Description of experimental data set

Health status condition	Fault diameter (mils)	The number of base samples	The number of testing samples	Label of classification
Normal	0	10	40	1
Inner race fault	7	10	40	2
	14	10	40	3
	21	10	40	4
	28	10	40	5
Ball fault	7	10	40	6
	14	10	40	7
	28	10	40	8
Outer race fault	7	10	40	9
	14	10	40	10
	21	10	40	11

Where the abscissa axis E_1 represents the Shannon entropy, and ordinate axis E_2 represents the exponential entropy.

Where the abscissa axis H_1 represents the Holder coefficient with the rectangular sequence as the reference sequence, and ordinate axis H_2 represents the Holder coefficient with the triangular sequence as the reference sequence.

From Figs. 1 and 4, it is interesting to see that the dominant fault feature vectors extracted from the bearing vibration signals with different fault types and severities through the multi-dimensional feature extraction based on entropy characteristics, Holder coefficient characteristics show apparent differences, respectively. The knowledge base (i.e., the recognition template) for GRA is established based on the fault symptom (i.e., the extracted feature vectors) and the fault pattern (i.e., the known fault types and severities). The fault feature vectors extracted based on the testing rolling bearing vibration signals to be recognized input to GRA, and the diagnostic results (i.e., fault types and severities) are output after the fusion of BBAs, shown in Table 2 (Figs. 2 and 3).

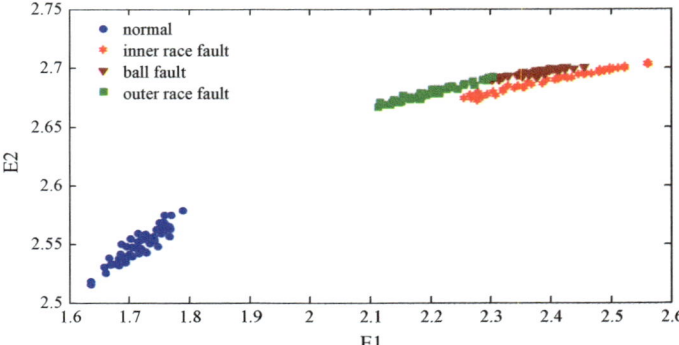

Fig. 1 Entropy characteristics of a random selected sample from normal operating condition and various fault conditions with fault diameter 7 mils

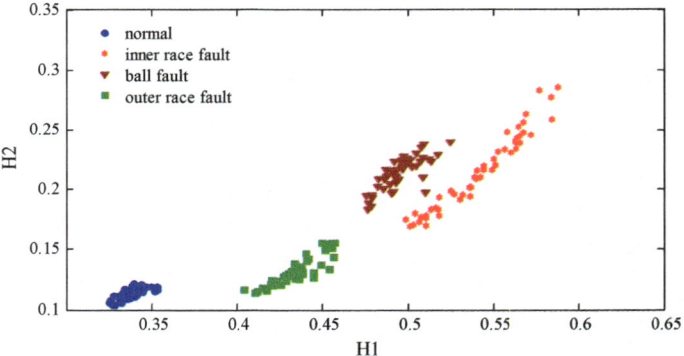

Fig. 2 Holder coefficient characteristics of a random selected sample from normal operating condition and various fault conditions with fault diameter 7 mils

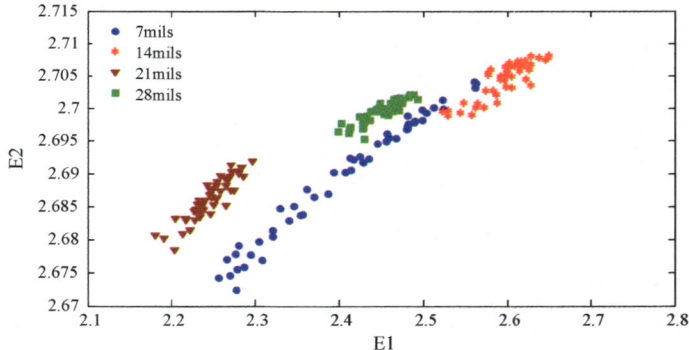

Fig. 3 Entropy characteristics of a random selected sample from inner race fault condition with various severities

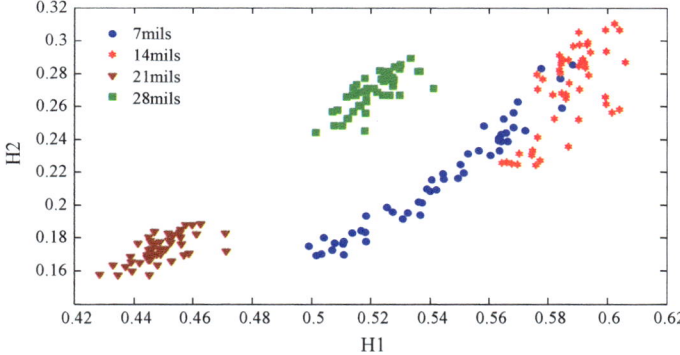

Fig. 4 Holder coefficient characteristics of a random selected sample from inner race fault condition with various severities

The diagnostic results from Table 2 show that the detecting success rate for bearing faulty conditions can reach 100%, with the total fault pattern recognition success rate almost 99.09%, which shows a certain improvement in diagnostic accuracy compared with the methods from references [14]. The time cost of the proposed method through a laptop computer with a 4.0 GHz dual processor for one test case is only 0.016 s.

Table 2 Diagnostic results by the proposed method compared with results from Ref. [14]

Label of classification	The number of testing samples	The number of misclassified samples		Testing accuracy (%)	
		[14]	Proposed	[14]	Proposed
1	40	0	0	100	100
2	40	0	0	100	100
3	40	0	2	100	95
4	40	3	0	92.5	100
5	40	0	0	100	100
6	40	2	0	95	100
7	40	3	2	92.5	95
8	40	3	0	92.5	100
9	40	0	0	100	100
10	40	0	0	100	100
11	40	4	0	90	100
In total	440	15	4	96.59	99.09

4   Conclusion

In the paper, a novel rolling bearing fault diagnostic method was proposed to fulfill the requirements for accurate assessment of different fault types and severities with real-time computational performance. The related experimental study has illustrated the following conclusions: The diagnostic results by the proposed approach show that the diagnostic success rate for bearing faulty conditions can reach 100%, with the total diagnostic success rate almost 99.09%; the proposed approach can improve the fault diagnostic accuracy compared with the existing expert systems, and it is very suitable for on-line bearing fault diagnosis.

Acknowledgments. This work is supported by the National Natural Science Foundation of China (61771154) and funding of State Key Laboratory of CEMEE (CEMEE2018K0104A).

Meantime, all the authors declare that there is no conflict of interests regarding the publication of this article.

We gratefully thank of very useful discussions of reviewers.

References

1. Hecke BV, Qu Y, He D. Bearing fault diagnosis based on a new acoustic emission sensor technique. J Risk Reliab. 2015;229(2):105–18.
2. Jiang L, Shi T, Xuan J. Fault diagnosis of rolling bearings based on marginal fisher analysis. J Vib Control. 2014;20(3):470–80.
3. Xu J, Tong S, Cong F, Zhang Y. The application of time-frequency reconstruction and correlation matching for rolling bearing fault diagnosis. ARCHIVE Proc Inst Mech Eng Part C J Mech Eng Sci 1989–1996, vols. 203–210. 2015;229(17):3291.
4. Lin Y, Wang C, Ma C, Dou Z, Ma X. A new combination method for multisensor conflict information. J Supercomputing. 2016;72(7):2874–90.
5. Zhang X, Hu N, Hu L, Chen L, Cheng Z. A bearing fault diagnosis method based on the low-dimensional compressed vibration signal. Adv Mech Eng. (2015);7(7).
6. Zhang DD. Bearing fault diagnosis based on the dimension–temporal information. ARCHIVE Proc Inst Mech Eng Part J. J Eng Tribol 1994–1996, vols. 208–210. 2011;225 (8):806–13.
7. Vakharia V, Gupta VK, Kankar PK. A multiscale permutation entropy based approach to select wavelet for fault diagnosis of ball bearings. J Vib Control. 2014;21(16):3123.
8. Tiwari R, Gupta VK, Kankar PK. Bearing fault diagnosis based on multi-scale permutation entropy and adaptive neuro fuzzy classifier. J Vib Control. 2015;21(3):461–7.
9. Sun W, Yang GA, Chen Q, Palazoglu A, Feng K. Fault diagnosis of rolling bearing based on wavelet transform and envelope spectrum correlation. J Vib Control. 2013; 19(6):924–41.
10. Li J, Guo J (2015) A new feature extraction algorithm based on entropy cloud characteristics of communication signals. In: Mathematical problems in engineering. p. 1–8.
11. Li J. A new robust signal recognition approach based on holder cloud features under varying SNR environment. KSII Trans Internet Inf Syst. 2015;9(12):4934–49.

12. Li J. A novel recognition algorithm based on holder coefficient theory and interval gray relation classifier. KSII Trans Internet Inf Syst. 2015;9(11):4573–84.

13. The Case Western Reserve University Bearing Data Center, http://csegroups.case.edu/bearingdatacenter/pages/download-data-file. Accessed 11 Oct 2015.

14. Li J, Cao Y, Ying Y, Li S. A rolling element bearing fault diagnosis approach based on multifractal theory and gray relation theory. PLoS ONE. 2016;11(12):1–16.

Multi-pose Face Recognition Based on Contour Symmetric Constraint-Generative Adversarial Network

Ning Ouyang[1,2], Liyuan Liu[2], and Leping Lin[1,2(✉)]

[1] Key Laboratory of Cognitive Radio and Information Processing,
Ministry of Education, Guilin University of Electronic Technology,
Guilin 541004, Guangxi, China
ouyangning@guet.edu.cn, lin_leping@163.com
[2] School of Information and Communication, Guilin University of Electronic
Technology, Guilin 541004, Guangxi, China
342520927@qq.com

Abstract. In order to address the impact of large-angle posture changes on face recognition performance, we propose a contour symmetric constraint-generative adversarial network (CSC-GAN) for the multi-pose face recognition. The method employs the convolutional network as the generator for face pose recovery, which introduces the global information of the constrained pose recovery of positive face contour histogram. Meanwhile, the original positive face is used as the discriminator, and the symmetric loss function is added to optimize the learning ability of the network. The positive face with gesture recovery is obtained by striking the balance between training of the generator and discriminator. Then we employed the nearest neighbor classifier to identify. The experimental results show that CSC-GAN obtained good posture reconstruction texture information on the multi-pose face reconstruction. Compared with the traditional deep learning method and 3D method, it also achieves higher recognition rate.

Keywords: Generative adversarial network · Pose recovery · Face contour · Symmetric loss

1 Introduction

In recent years, face recognition [1–4] technology is a hot topic in machine vision research. However, the performance of face recognition depends largely on the influence of various factors such as lighting, posture, and time span. Currently, the posture change is the most challenging face recognition problem. Asthana et al. [5] used the view-based active appearance model (VBAAM) to project images of non-positive faces into an aligned 3D face model for normalizing the pose. With the rapid development of deep learning in recent years, face posture reconstruction based on deep learning has also made great progress. Kan et al. [6] proposed stacked progressive auto-encoders (SPAEs) for face changes caused by different poses. It uses shallow stepping auto-encoding to gradually reconstruct a non-positive face image into a positive face image,

© Springer Nature Singapore Pte Ltd. 2020
Q. Liang et al. (eds.), *Communications, Signal Processing, and Systems*, Lecture Notes in Electrical Engineering 516,
https://doi.org/10.1007/978-981-13-6504-1_117

but the partial details of the positive face image recovered are not clear. Later, Zhu et al. [7] proposed a multi-view perceptron (MVP) deep network, which employed multi-layer perceptions to obtain better reconstruction images. In 2014, Goodfellow et al. [8] proposed a generative adversarial network (GAN). This model consists of a generator and a discriminator, where the generator generated new sample data, and the discriminator evaluated the authenticity of the generated data. Generative adversarial network is an unsupervised learning model, which effectively solves a series of data generation problems. Meanwhile, it has more self-learning ability in feature extraction, which makes the image output from the generator more realistic.

Various factors such as lighting, posture, and occlusion will have a certain impact on the recognition performance of realistic faces. For large-angle pose reconstruction issues, a method based on contour symmetric constraint-generative adversarial network (CSC-GAN) is proposed to reconstruct and recognize a multi-face poses. This method competes each other by generated network and adversarial network, optimizes, and trains iteratively until the output images are close to real people's faces. Experimental results show that the proposed method can not only eliminate the attitude error in the large-angle attitude reconstruction, but also enrich the detail of the reconstructed positive face. Compared with other face recognition methods, the recognition rate is also greatly improved.

2 Contour Symmetric Constraint-Generative Adversarial Network Model

The network model uses a convolution neural network as a generator in a generative adversarial network, the positive face contour histogram was introduced to constrain the restoration of the positive face to ensure the global contour quality of the reconstructed image, and the data output from this section serves as the generator part of the overall network. The adversarial network uses the real face data as a discriminator to distinguish the authenticity and filter the face data reconstructed in the resolution generator, in order to achieve a minimized synthesis error to introduce a symmetrical loss function. When the network is continuously trained against the generator and discriminator, the entire network is balanced when no real data or data can be generated; the output data is a real image that can be confused. Figure 1 shows the convolution generation adversarial network model. SF is the composite picture output by the generator, and GT is the true face data of the sample.

2.1 Generated Network

In the generated network, the image of any pose under normal lighting is processed through the convolution pool, and the output positive face image is used as the output of the generated network. It can be seen from Fig. 1 that the face picture input in the generation network is $w \times h$, and any gesture under normal lighting is x^0. The generated network consists of two 5×5 convolution layers and two 3×3 pooled layers. W_i^1 is the characteristic map produced by the first layer convolution, and W_j^2 is the characteristic map produced by the second layer convolution. V^1 and V^2 are the first

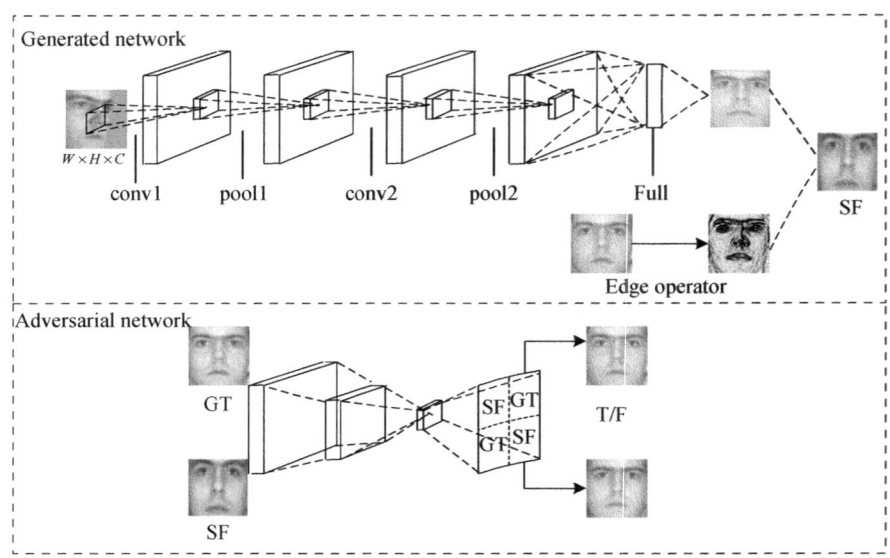

Fig. 1. Contour symmetric constraint-generative adversarial network

and second levels of matrix pooling, so the feature map formula of image x^0 through two-layer convolution network is as follows:

$$x_i^2 = \sum_{j=1} V^2 \sigma \left(W_j^2 V^1 \sigma \left(W_i^1 x^0 \right) \right) \tag{1}$$

where W is the feature map of the weight matrix map, σ indicates activation function; select RELU function here. Positive face image output normal lighting is reconstructed positive face y_1. The extraction of the feature map x_i^2 is performed through the W weight matrix for W^3 reconstruction as follows:

$$y_1 = \sigma \left(W^3 x_i^2 \right) \tag{2}$$

Generated network is based on convolution network-based reconstruction of the positive face, using gradient descent to continuously update the network parameters' back-propagation; Δ represent intermediate variables, and γ learning rate is 0.0001. The updated parameters W_{k+1}^i:

$$\Delta_{k+1} = 0.9\Delta_k - \gamma \frac{\partial E}{\partial W_k^i} \tag{3}$$

$$W_{k+1}^i = \Delta_{k+1} + W_k^i \tag{4}$$

In the process of face reconstruction with a large angle, the restored positive face often results in the discontinuity of the gray value of the edge due to the partial loss of the overall information feature, and the recognition effect is not good. The edge information histogram can reflect the local discontinuity of the image, along the edge of the pixel intensity to ensure the integrity of the overall contour, and better noise resistance can effectively highlight the edge direction information. In order to ensure that the faces reconstructed at large angles are more realistic and the features extracted are more abundant, after generating a positive face image recovered from the network, a positive face contour histogram is added to constrain the quality of the global features. The central position of the image is represented by any pixel (i, j) of the positive face image $f(x, y)$, and the gradient of the center pixel of the given window in the m and n directions is calculated. Obtain the edge points of the image coordinates $S_{m,n}$ through S:

$$S = \arctan\left(\frac{S_m}{S_n}\right)\sqrt{S_m^2 + S_n^2} \qquad (5)$$

Convolution of the image output from θ and convolution networks to obtain a reconstructed positive face image:

$$y_i = \sigma\left(W^3 S \otimes x_i^2\right) \qquad (6)$$

Finally, obtain the cost function of convolution network after positive face contour constraint, X^0 represents the input image set $\{x_i^0\}$, $\overline{Y} = \{\overline{y_i}\}$ is the original image, and $Y = \{y_i\}$ is a reconstructed image:

$$E(X^0; W) = \left\|Y_1 - \overline{Y}\right\|^2 \qquad (7)$$

2.2   Adversarial Network

The adversarial network is similar to a comparative analysis, and the authenticity of the reconstructed image I^P output from the network output is compared with that of the real data I^F until an image similar to the real data is identified. That is, the true and false samples are distinguished, and the two-dimensional matrix values that are opposed to the network output are converted into a probability value. The network is trained by using a separate alternate iterative update method. Remember that the data generated by the input against the network is $G_{\tau_G}(I^P)$ and the real data is I^P. First, the generated network G is fixed and trained adversarial network D to maximize the accuracy of the confrontation discrimination. Secondly, the fixed adversarial network D is trained to generated network G, so that the accuracy of confrontation discrimination is minimized. This results in the final destination function, which is 1 when the output value is true and 0 when the output value is false. Therefore, when training the network, the adversarial network D is updated twice and the parameters of the generated network G are updated once:

$$\min_{\tau_G} \max_{\tau_D} = E_F\left(X^0; W\right) \log D_{\tau_D}\left(I^F\right) + E_P\left(X^0; W\right) \log\left(1 - D_{\tau_D}\left(G_{\tau_G}\left(I^P\right)\right)\right) \qquad (8)$$

3   Minimize Loss Function

In the generation network, in order to ensure the naturalness of the generated face, a parameter g is proposed to update the loss function to minimize the synthesis error. Reconstruct the image as the dataset $I_n^P = Y\{y_1, y_2, y_3, \ldots\}$. The real data $I_n^F = \overline{Y}\{\overline{y_i}\}$ is the face positive face image. L_{syn} is the total loss function. The parameter $\hat{\tau}_G$ is updated by:

$$\hat{\tau}_G = \frac{1}{N} \arg\min \sum_{n=1}^{N} \{L_{\text{syn}}(G_{\tau_G}(I_n^P), I_n^F)\} \tag{9}$$

The total loss function L_{syn} that minimizes the synthesis error is composed of real data discriminating losses and generating symmetry losses of the reconstructed image. $I_{x,y}^P$ is the reconstructed sample pixel level, $I_{x,y}^{gt}$ is the real image pixel level, and L_r is the error loss of training real sample reconstruction:

$$L_r = \frac{1}{w \times h} \sum_{x=1}^{w} \sum_{y=1}^{h} \left| I_{x,y}^P - I_{x,y}^{gt} \right| \tag{10}$$

In the traditional generation adversarial network training learning, the problem of degeneration of generator samples may occur, the same sample points are always generated, and the sample face tends to shrink as the number of iterations increases. According to the characteristics of face symmetry, the width of all input images is covered by half, and the feature points are gradually described by reconstructing the absolute value of the subtracted sample image, which ensures the symmetry of the reconstruction of the visible part and the masked part. Especially for large-angle reconstructions, the asymmetry of the local feature reconstruction is very obvious. Introducing the symmetric loss L_{sl} can effectively constrain the symmetry of the positive face after reconstruction of different poses, and it also has a face detail that is reconstructed and better richness:

$$L_{\text{sl}} = \frac{1}{w/2 \times h} \sum_{x=1}^{w/2} \sum_{y=1}^{h} \left| I_{x,y}^P - I_{w-(x-1),y}^P \right| \tag{11}$$

L_{cee} is the cross-entropy loss function used to limit hidden activation functions. λ_1 and λ_2 are coefficients of balanced penalty terms, and take the empirical values $\lambda_1 = 0.001$, $\lambda_2 = 0.03$. The final loss function is a weighted loss function whose purpose is to achieve the minimum synthesis error:

$$L_{\text{syn}} = L_r + \lambda_1 L_{\text{sl}} + \lambda_2 L_{\text{cee}} \tag{12}$$

4 Experimental Results and Analysis

The paper verifies the effectiveness of the proposed algorithm on the Multi-PIE [9] face image library. The database contains 337 people with a total of 754,204 face images in different poses, lighting, and expressions; each person contains 15 postures, and 20 different lighting are included in the same posture. One of the subsets of the Multi-PIE database was selected, which included 11 poses in the range of $-75°$ to $+75°$, and $15°$ between poses. The selected face data is normal expressions under normal lighting. The image size of alignment clip is set to 64×64, the batch size is 10, and the learning rate is 10^{-4}.

Most of the previous multi-pose face–face reconstruction models aim to solve the problem of attitude reconstruction in the $±45°$ range. In recent years, it is generally believed that it is difficult for a model to recover postures larger than $±60°$. However, this paper provided a solution for recovering postures with large deflection angles. In the experiment, Local Gabor binary pattern histogram sequence (LGBP) [10], Stacked progressive auto-encoders (SPAE) [6], Convolutional restricted boltzmann machines (CRBM) [11], Fully automatic Ensemble Gabor Fisher Classifier (FA-EGFC) [12], Deep Convolutional Neural Networks (DCNN) [1] et al. Act recognition rate comparison experiment. From Fig. 2, it can be seen intuitively that the positive face image reconstructed by the CSC-GAN method can restore the texture information of the original image better. Lines 1, 3, and 5 are different gestures for different faces in $-75°$ to $+75°$. Lines 2, 4, and 6 are face–face images reconstructed by this method when corresponding to different angles. It can be seen from the above that the positive face of the face reconstructed by this method under the large angle of $+75°$ still maintains the symmetry and visual clarity of the original image.

Figure 3 shows the comparison between the method and GAN, DCNN, MVP, and other deep learning methods when the face pose angle is $+75°$. From the visual observation, it can be seen intuitively that when CSC-GAN reconstructs a large-angle gesture face, facial symmetry is better than other methods, making facial information more natural. In particular, the comparison with the common GAN network also verifies the importance of adding the symmetric loss function in this paper. In addition, facial features such as eyebrows, eyelids, and beards are more comprehensive than DCNN and MVP models, and face reconstruction is more abundant, ensuring more face features.

Table 1 compares the recognition rate of different face reconstruction methods on Multi-PIE face database. It can be seen that when the attitude angle is very small, such as $±15°$, the recognition rate of each method is not significantly different. When the angle reaches $±45°$, the method of this paper is slightly lower than that of DCNN. However, with the gradual increase of the angle, when the angle is greater than $±60°$ CSC-GAN compared with other methods, the recognition rate has a greater improvement.

| 75° | 60° | 45° | 30° | 15° | -15° | -30° | -45° | -60° | -75° |

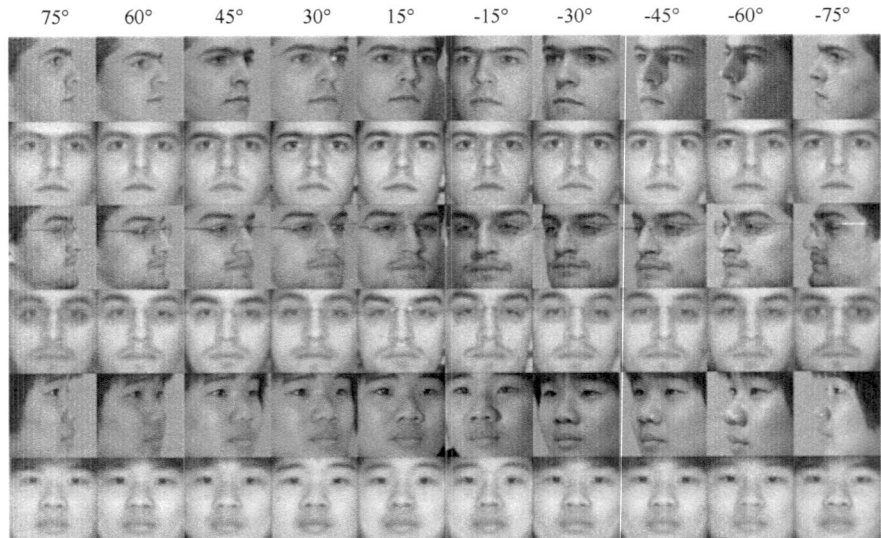

Fig. 2. Face reconstructed by CSC-GAN at various angles 75° to −75°

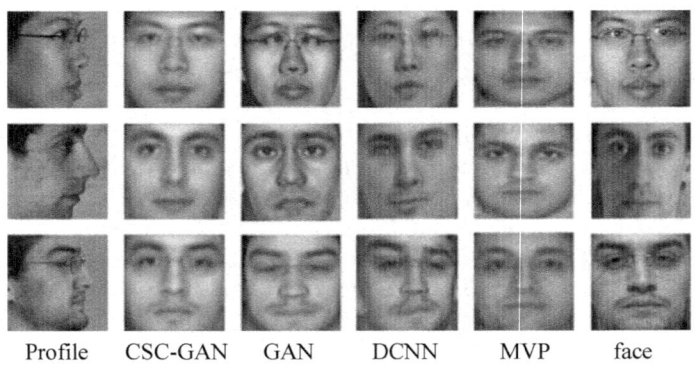

Profile   CSC-GAN   GAN   DCNN   MVP   face

Fig. 3. Comparison of CSC-GAN and other reconstruction models at 75°

Table 1. Comparison of recognition rates of different face reconstruction methods on Multi-PIE face database

Methods	−75°	−60°	−45°	−30°	−15°	15°	30°	45°	60°	75°
LGBP	10.2	20.2	37.7	62.5	77.0	83.0	59.2	36.1	22.6	11.3
VBAAM	30.2	52.3	74.1	91.0	95.1	95.7	89.5	74.8	54.0	32.4
CRBM	40.1	69.4	80.3	90.5	94.9	96.4	88.3	75.2	65.8	44.5
SPAE	60.4	72.2	84.9	92.6	96.3	95.7	94.3	84.4	77.6	68.2
FA-EGFC	70.5	80.6	87.4	95.0	99.3	99.0	92.9	85.2	81.8	70.8
MVP + LDA	72.4	80.2	89.6	96.5	100	99.0	94.5	90.8	79.8	70.2
DCNN + LDA	74.8	82.6	95.6	98.8	100	99.3	98.2	97.8	80.4	73.2
SC-GAN	**82.6**	**86.2**	95.2	**99.4**	100	**99.8**	**99.4**	97.4	**85.4**	**82.4**

5 Conclusions

We present a method based on contour symmetric constraint-generative adversarial network (CSC-GAN) to reconstruct and recognize a variety of face poses; this method uses a convolution network as a generator to introduce a positive face contour histogram to constrain the global quality and to confront the real data positive face pose. Symmetrical loss is added during the attitude reconstruction to achieve a better posture recovery effect. Experiments show that this method not only eliminates the effect of pose on face recognition performance, but also improves the quality of face reconstruction to a certain extent. It also maintains a good recognition rate after large-angle pose reconstruction and achieves experimental expectations. Future research work will take into account various factors such as lighting and expression to reconstruct the human face and integrate the influence in the actual environment to further improve the recognition rate.

Acknowledgments. This work is partially supported by the following foundations: the National Natural Science Foundation of China (61661017); the China Postdoctoral Science Fund Project (2016M602923XB); the Natural Science Foundation of Guangxi province (2017GXNSFBA198212, 2016GXNSFAA38014); the Key Laboratory Fund of Cognitive Radio and Information Processing (CRKL160104, CRKL150103, 2011KF11); Innovation Project of GUET Graduate Education (2016YJCXB02); the Scientific and Technological Innovation Ability and Condition Construction Plans of Guangxi (159802521); the Scientific and Technological Bureau of Guilin (20150103-6).

References

1. Zhu Z, Luo P, Wang X, et al. Deep learning identity-preserving face space. In: Proceedings of the IEEE international conference on computer vision. Darling Harbour, Sydney; 2013.
2. Taigman Y, Yang M, Ranzato MA, Wolf L. Closing the gap to human-level performance in face verification. In: CVPR, Colombia; 2014.
3. Sun Y, Wang X, Tang X. Deep learning face representation from predicting 10,000 classes. In: CVPR, Colombia; 2014.
4. Taigman Y, Yang M, Ranzato MA, Wolf L. Web-scale training for face identification. arXiv:1406.5266; 2014.
5. Asthana A, Marks TK, Jones MJ, Tieu KH, Rohith M. Fully automatic pose-invariant face recognition via 3D pose normalization. In: ICCV, Barcelona, Spain; 2011.
6. Kan M, Shan S, Chang H, et al. Stacked progressive auto-encoders (spae) for face recognition across poses. In: Proceedings of the IEEE conference on computer vision and pattern recognition; 2014. p. 1883–90.
7. Zhu Z, Luo P, Wang X, et al. Multi-view perceptron: a deep model for learning face identity and view representations. In: Advances in Neural Information Processing Systems; 2014. p. 217–25.
8. Goodfellow I, Pouget-Abadie J, Mirza M, Xu B, et al. Generative adversarial nets. In: NIPS, Montreal, Canada; 2014.
9. Gross R, Matthews I, Cohn J, et al. The CMU multi-pose, illumination, and expression (multi-PIE) face database. CMU Robotics Institute. TR-07-08, Tech. Rep; 2007.

10. Zhang W, Shan S, Gao W, Chen X, Zhang H. Local Gabor binary pattern histogram sequence (LGBPHS): a novel non-statistical model for face representation and recognition. In: ICCV, Beijing, China; 2005.
11. Huang GB, Lee H, Learned-Miller E. Learning hierarchical representations for face verification with convolutional deep belief networks. In: CVPR, Rhode Island, America; 2012.
12. Li S, Liu X, Chai X, Zhang H, Lao S, Shan S. Morphable displacement field based image matching for face recognition across pose. In: ECCV, Florence, Italy; 2012.

Flight Target Recognition via Neural Networks and Information Fusion

Yang Zhang[✉], Zhenzhen Duan, Jian Zhang, and Jing Liang

University of Electronic Science and Technology of China
Information and Communication Engineering, Chengdu, China
xixrui@yeah.net

Abstract. The purpose of this research is to increase the target recognition rate by means of neural networks and feature fusion. We analyze the performance of different recognition methods (Bayesian classifier, support vector machine (SVM), and neural networks) based on high-resolution range profile (HRRP). The result shows the superiority of neural networks to Bayesian classifier and SVM in classification. We apply multi-source feature fusion to target recognition based on neural networks. The results show that, in certain cases, the target recognition ratio using fusion feature is higher than that of HRRP only.

Keywords: Target recognition · Neural networks · Information fusion

1 Introduction

High-resolution range profile (HRRP) is the coherent summations of complex echoes from target scatters in each range cell [1]. It represents the projection of target scattering centers on radar line of sight. Since HRRP is easy to be obtained and contains much structure information of targets, it has attracted intensive attentions from the radar automatic target recognition (RATR) community [2,3].

After a long term of research, the US Army Missile Command thought that target recognition based on HRRPs is the most promising identification method in the existing technology [4]. This viewpoint made a rapid development of this method. Zyweck and Bogner [5] proposed a method to solve the problem of aspect sensitivity of HRRP. It uses multi-template matching to carry out target recognition with HRRPs of different aspect. Du et al. [6] proposed a method for calculating the Euclidean distance in higher-order spectra feature space. It avoids calculating the higher-order spectra directly. Botha proposed an identification method based on neural network in 1996 [7]. He thought that, compared with other recognition mehtods, neural networks can extract more unknown features from HRRP and provide a very high target correct recognition rate if the amount of training data is large enough. However, the neural networks he used are simple

© Springer Nature Singapore Pte Ltd. 2020
Q. Liang et al. (eds.), *Communications, Signal Processing,
and Systems*, Lecture Notes in Electrical Engineering 516,
https://doi.org/10.1007/978-981-13-6504-1_118

perceptrons. More complex and useful neural networks are not applied because of the lack of theory or hardware performance.

Now, with the development of hardware technology, deep neural networks have applied in various fields. In 2012, Geoffrey Hinton [8] won the competition of ImageNet image recognition by using deep neural network. He reduced the error rate of "top 5" to 15.315%. The same year, Graves A. proposed improved long short-term memory (LSTM) networks [9], which can effectively process the timing issues. For HRRP, deep neural networks can extract more features from it for identification and LSTM networks can process the dependence between sampling points. We believe the target recognition ratio based on HRRP can be improved via deep neural networks.

However, the information of target provided by HRRP is limited. Although the deep neural networks can extract more features, the correlation of these features is too high to improve the recognition rate further. We need to use more other features to do that. Nowadays, the reconnaissance system is mostly composed of multi-source sensor networks. The whole system can work effectively with the fused information obtained by different sensors (visible light sensors, infrared sensors, and radar sensors). It is believed that the target recognition ratio can be improved via the fusion feature of different sensors.

In this paper, we analyze the performance of different recognition methods based on HRRP. The result shows the superiority of neural networks to Bayesian classifier and SVM in classification. Meanwhile, we apply multi-source feature fusion to target recognition based on neural networks. We establish a multilayer perceptron to carry out target recognition with fusion feature obtained by color images, grayscale images, and HRRPs. The results show that, in sunny day, the target recognition ratio using fusion feature is higher than using HRRP only.

The structure of this paper is organized as follows. In Sect. 2, we propose the methods of getting fusion feature of images and HRRPs. Section 3 introduces the neural networks. In Sect. 4, we conduct the training of neural network and the simulation of target recognition. Finally, Sect. 5 summarizes our investigation.

2   Feature Fusion

First of all, we need to extract some certain features from the images and HRRP of target to form the fusion feature.

2.1   HRRP

HRRP cannot be used for target recognition or feature extraction directly. We must do some processing about it. The amplitude of HRRP obtained by radar sensor is defined as:

$$R = [r(1), \ r(2), \ldots, r(N)] \tag{1}$$

N is the number of the distance units.

The processing of HRRP is power transformation and normalization. Botha et al. [7] think these processing can improve the target recognition rate. Power transformation is defined as:

$$H(n) = r(n)^\alpha \quad 0 < \alpha < 1 \tag{2}$$

$H(n)$ is the amplitude of HRRP's sample after power operation.

The next step is normalization. After that, we can calculate the length and the change rate according to [10]. Then, we can obtain a feature vector:

$$F_{radar} = [L, \, t] \tag{3}$$

L is the length of target, and t is the change rate.

2.2 Image Feature Extraction

We can obtain color images and grayscale images of target from visible light sensor and infrared sensor, respectively. It is difficult to use the entire images for fusion. So, we need to extract some typical features from these images.

For color image, we extract its color moments [11] for target recognition. The mathematical definitions of three color moments are:

$$u_i = \frac{1}{N} \sum_{j=1}^{N} p_{i,j} \tag{4}$$

$$\sigma_i = \left(\frac{1}{N} \sum_{j=1}^{N} (p_{i,j} - u_i)^2 \right)^{\frac{1}{2}} \tag{5}$$

$$s_i = \left(\frac{1}{N} \sum_{j=1}^{N} (p_{i,j} - u_i)^3 \right)^{\frac{1}{3}} \tag{6}$$

$p_{i,j}$ represents the probability of the appearance of pixel with a gray level of j in color channel i. N represents the number of pixels in the image.

According to three color channels of RGB image, we can obtain a feature vector of color moments:

$$F_{color} = [u_R, \, \sigma_R, \, s_R, \, \mu_G, \, \sigma_G, \, s_G, \, \mu_B, \, \sigma_B, \, s_B] \tag{7}$$

For grayscale image, we can extract several features based on gray histogram as follows:

$$m = \sum_{i=0}^{L-1} z_i p(z_i) \tag{8}$$

$$\sigma = \sqrt{\sum_{i=0}^{L-1} (z_i - m)^2 p(z_i)} \tag{9}$$

$$e = -\sum_{i=0}^{L-1} p(z_i) \log_2 (p(z_i)) \qquad (10)$$

m, σ, and e represent the average brightness of texture, the average contrast of texture, and the entropy of gray histogram, respectively. L is the number of the gradation, z_i represents the ith gradation and $p(z_i)$ represents the probability of the appearance of pixel with a gradation of z_i.

Then, we can obtain another feature vector for fusion:

$$F_{gray} = [m,\ \sigma,\ e] \qquad (11)$$

Finally, we can get the fusion vector:

$$F_{fusion} = [F_{gray},\ F_{radar},\ F_{color}] \qquad (12)$$

3   Neural Network

After getting the features, we begin to introduce the neural network. Here, we only briefly introduce two types of neural networks used in the research.

3.1   Multilayer Perceptron

Multilayer perceptron is the most common neural network. It is composed of input layer, hidden layers, and output layer. The structure of multilayer perceptron is shown in Fig. 1.

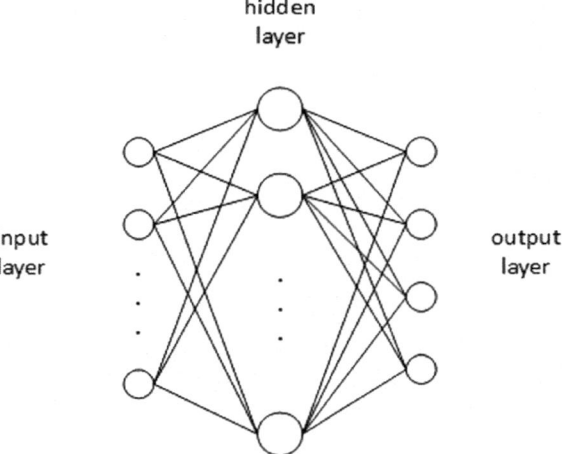

Fig. 1. The structure of multilayers perceptron

For multilayer perceptron, we make the entire HRRP or feature vectors as input data. Output is set as the types of targets. We can change the number of features to be extracted by changing the number of hidden layers or the nodes in each layer.

3.2 LSTM Network

LSTM network is a variant of recurrent neural network. It can be used for learning long-term dependence information [9]. Each LSTM unit is composed of cell, input gate, output gate, and forget gate. The structure is shown in Fig. 2.

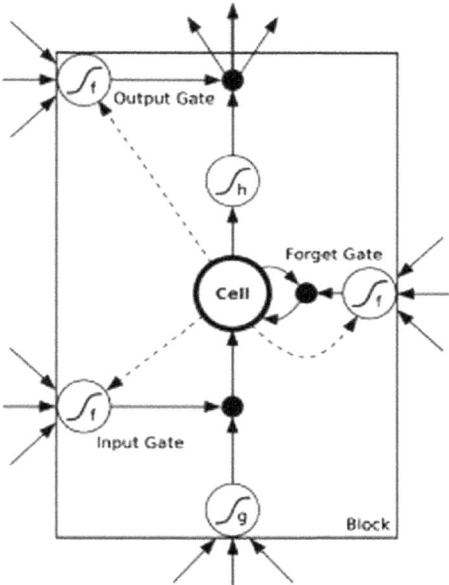

Fig. 2. The structure of LSTM networks

In LSTM unit, the first step is to compute what information should be thrown away from the cell state according to the output of last state and current input. The next step is to compute the new information which should be stored in the cell state. After that, updating the old cell state. Finally, computing the output based on current input, current cell state, and the output of last state.

HRRP is a variant of radar echo. And it can be considered as a time series vector. We believe it is reasonable to use LSTM networks to conduct recognition based on HRRP.

4 Simulation

The echo data in the simulation obtained by the radar simulation software BSS. In this research, F-15, Tu-16, and AH-64 are selected as targets to be recognized. We get 300 sets of HRRPs, color images, and grayscale images for each kind of aircraft. The attitude angle is transformed from 3 to 30° with the step size of 3°. The number of sampling points of each HRRP is 320.

After processing the data, we can acquire HRRP and F_{fusion} of targets. Then, the rest work is divided into two parts:

1. Performance analysis of different recognition methods based on HRRP (simple perceptron, multilayer perceptron, LSTM networks, Bayesian classifier, and SVM).
2. Performance analysis of recognition method based on fusion feature.

4.1 Performance Analysis of Different Recognition Method

This part is to analyze the performance of five different recognition methods. First, we build three different neural networks (LSTM networks, multilayer perceptron, and simple perceptron). The parameters of them are shown in Tables 1, 2, and 3.

Table 1. Parameters of LSTM networks

Time step	Input size	Hidden size	Hidden layer	Output size
16	20	64	1	3

Table 2. Parameters of multilayer perceptron

Input size	W_1	W_2	W_3	W_{out}
320	320×160	160×64	64×16	16×3

Table 3. Parameters of simple perceptron

Input size	W_1	W_{out}
320	320×160	160×3

In Tables 2 and 3, W_i is the shape of weight matrix of hidden layer i. And W_{out} is the shape of weight matrix of output layer.

HRRPs are used as training data of three neural networks. For LSTM networks, We need to reshape the HRRP to (16,20) to fit the input size and time step. The number of training times for all neural networks is 100,000, batch size is 20, the loss function is cross entropy and the optimizer is Adam optimizer.

Then, we add a Bayesian classifier [10] and a SVM to compare the performance of different recognition methods based on HRRP.

In testing, each time we change the SNR, we will conduct 600 tests and calculate the average recognition rate of three aircrafts. The curves of SNR and recognition rate are shown in Fig. 3.

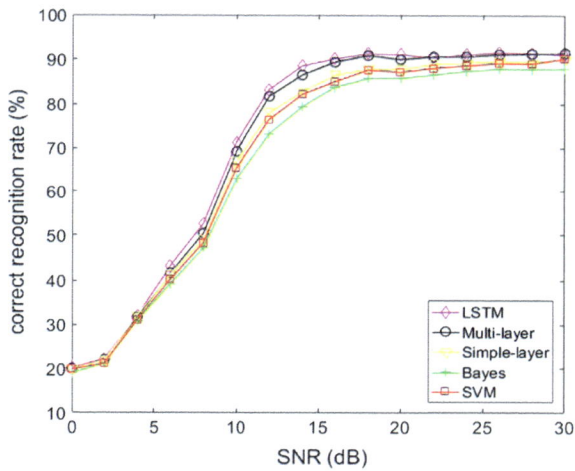

Fig. 3. The recognition rate of different methods based on HRRP

Figure 3 shows the performance of five different classifiers based on HRRP. It is obvious that neural networks are superior to other classifiers. Because they can extract more features from HRRP to conduct target recognition. From neural networks, simple perceptron has the worst performance because of the fewer nodes, fewer layers, and fewer features. So, we believe that the performance can be improved by adding the number of layers and nodes. We can see that the recognition rate based on LSTM networks is very close to that of multilayer perceptron. Although LSTM networks have advantages in timing processing, multilayer perceptron have more layers or nodes and can extract more features from the target. So, for target recognition rate, they almost have the same performance.

4.2   Performance Analysis of Recognition Method Based on Fusion

It can be learned from the above that neural networks can provide higher target correct recognition rate than that of SVM and Bayesian classifier. In this part, we establish another multilayer perceptron to conduct target recognition based on fusion vectors F_{fusion}. The parameters are shown in Table 4.

Table 4. Parameters of multilayer perceptron based on fusion feature

Input size	W_1	W_2	W_3	W_{out}
14	14×64	64×64	64×16	16×3

In testing, we simulate a rainy scene by adding Gaussian blur filter to the parts of images. And we make the rest of the images as test data on sunny day. Then, we can obtain two sets of F_{fusion} to conduct target recognition. Meanwhile, we add two neural networks based on HRRP (LSTM networks and multilayer perceptron) to analyze the performance. The testing results are shown in Figs. 4 and 5.

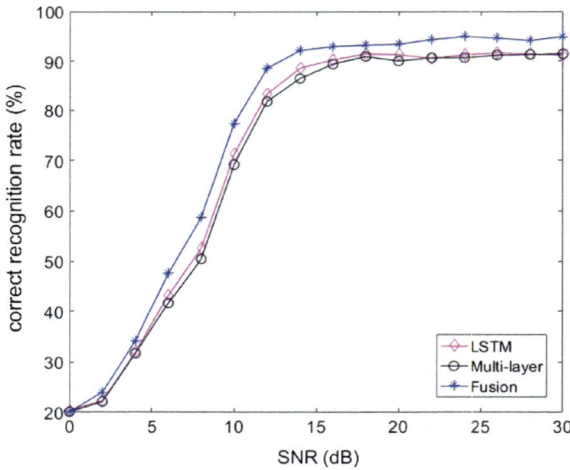

Fig. 4. The recognition rate of different methods in sunny day

From Fig. 4, we can see that the correct recognition rate based on fusion is better than that of HRRP only when the SNR changes from 8 to 13 dB. It is believed that, on sunny day, color images and grayscale image can provide important information of targets. Using fusion feature is conducive to the recognition.

Figure 5 shows the performance of method based on fusion is worse than the others on rain day. Because in this case, the resolution of image is too low to contribute to the recognition. The features extracted from images are useless and the effective features are only the length and the change rate extracted from

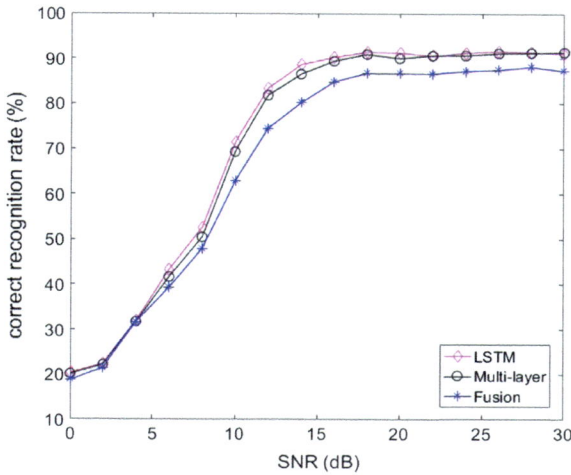

Fig. 5. The recognition rate of different methods in rainy day

HRRP. It is difficult to get a high recognition rate based on only two features. So, it is proved that the strategy of recognition should change with the environment.

5 Conclusion

In this paper, we analyzed the performance of different recognition methods based on HRRP. The result shows the superiority of neural networks to SVM and Bayesian classifier. The next, we established a multilayer perceptron to carry out target recognition with fusion feature obtained by color images, grayscale images, and HRRPs. The results show that, in sunny day, when the resolution of the images is high, the target recognition ratio using fusion feature features is higher than using HRRP only.

Acknowledgments. This work was supported by the National Natural Science Foundation of China (61671138, 61731006), and was partly supported by the 111 Project No. B17008.

References

1. Smith CR, Goggans PM. Radar target identification. IEEE Antennas Propag Mag. 1993;35(2):2738.
2. Zhang XD, Shi Y, Bao Z. A new feature vector using selected bispectra for signal classification with application in radar target recognition. IEEE Trans Sig Proces. 2001;49(9):1875–85.
3. Kim DH, Seo DK, Kim HT. Efficient radar target recognition using the music algorithm and invariant features. IEEE Trans Antennas Propag. 2002;50(3):325337.

4. Webb AR. Gamma mixture models for target recognition. Pattern Recogn. 2000;33(12):2045–54.

5. Zyweck A, Bogner RE. Radar target classification of commercial aircraft. IEEE Trans Aerosp Electron Syst. 1996;32(2):598–606.

6. Du L, Liu H, Bao Z, Xing M. Radar HRRP target recognition based on higher order spectra. IEEE Trans Sig Process. 2005;53(7):2359–68.

7. Botha EC, Barnard E, Barnard CJ. Feature-based classification of aerospace radar targets using neural networks. Amsterdam: Elsevier Science Ltd.; 1996.

8. Montavon G, et al. Explaining nonlinear classification decisions with Deep Taylor decomposition. Pattern Recogn. 2016;65:211–22.

9. Graves A. Long short-term memory. Supervised sequence labelling with recurrent neural networks. Berlin Heidelberg: Springer; 2012. p. 1735–80.

10. Yang Z, et al. Flight recognition via HRRP using fusion schemes; 2016.

11. Stricker MA, Orengo M. Similarity of color images. In: Storage and retrieval for image and video databases III; 1995. p. 381–92.

Specific Emitter Identification Based on Feature Selection

Yingsen Xu[1(✉)], Shilian Wang[1], and Luxi Lu[2]

[1] College of Electronic Science and Engineering,
National University of Defense Technology, Changsha, China
591297993@qq.com, wangsl@nudt.edu.cn
[2] National Key Laboratory of Science and Technology
on Blind Signal Processing, Chengdu, China

Abstract. For the high dimension of fingerprint feature set in the process of specific emitter identification (SEI), feature selection method is utilized to reduce the feature dimension and improve individual recognition rate. This paper adopted the filter feature selection in four ways: MIFS, mRMR, CMIM, and JMIM fingerprint feature set of high-dimensional feature selection and combined with PCA dimensionality reduction algorithm to minimize the feature dimension. The simulation results show that feature selection is feasible in individual recognition of the radiation source and can be effectively combined with dimension reduction algorithm.

Keywords: Specific emitter identification · Feature selection ·
Dimension reduction

1 Introduction

Due to design tolerance, the manufacture error, equipment aging and the influence of environmental factors, there must be some differences in hardware equipment between different radiation sources, and these differences are shown in the signal that there are subtle features that do not affect the detectability of information transmission. Through fingerprint identification, a certain amount of information can be obtained if the captured enemy signal cannot be decrypted. The technology of communication radiation source identification includes the following steps fine feature extraction, feature processing, and feature classification.

Xu [1] uses the square integral bispectra (SIB) method to extract the radiation source fingerprint characteristics. Literature [2] combined empirical mode decomposition (EMD) and Hilbert spectrum analysis to analyze signal characteristics at time-scale and frequency scale; Bertoncini et al. [3] established fingerprint characteristic quantity by using multi-scale wavelet transform and statistical feature quantity and

© Springer Nature Singapore Pte Ltd. 2020
Q. Liang et al. (eds.), *Communications, Signal Processing,
and Systems*, Lecture Notes in Electrical Engineering 516,
https://doi.org/10.1007/978-981-13-6504-1_119

conducted individual fingerprint identification experiments for different RF tags; some scholars used intrinsic time-scale decomposition (ITD) [4] to decompose signal extraction characteristics. The subtle feature extraction method mentioned above based on the signal itself, to achieve high resolution at the same time increased the feature dimension. The existence of feature redundancy may result in a large amount of classification calculation, and at the same time, there may be interference problems between various features. These questions, using the feature selection process, characteristics on the one hand, reduce the feature dimension, on the other hand, reduce the characteristics of redundancy, is helpful to improve the efficiency of emitter individual fingerprint identification.

Feature selection is a work of selecting the most effective features of a set of features to reduce the dimension of feature space [5]. Dash [6] describes the basic framework of feature selection in the literature, as shown in Fig. 1. Feature selection based on evaluation criteria can be divided into filter, wrapper, and embedded [7]. Filter method uses the evaluation criteria to ensure the relationship between the characteristics and classification and reduce the correlation between features and characteristics. Wrapper method evaluates each of the selected feature subsets, to choose the optimal subset. Filter method using the statistical characteristic of all the learning samples, fast at the same time, may be subjected to the imbalance of the small sample and the sample problem, the wrapper method based on the follow-up study to evaluate feature subset, classification effect is best, but is not suitable for large data sets. Because the feature dimension extracted by the feature extraction method mentioned above is sharp, the filter method is chosen.

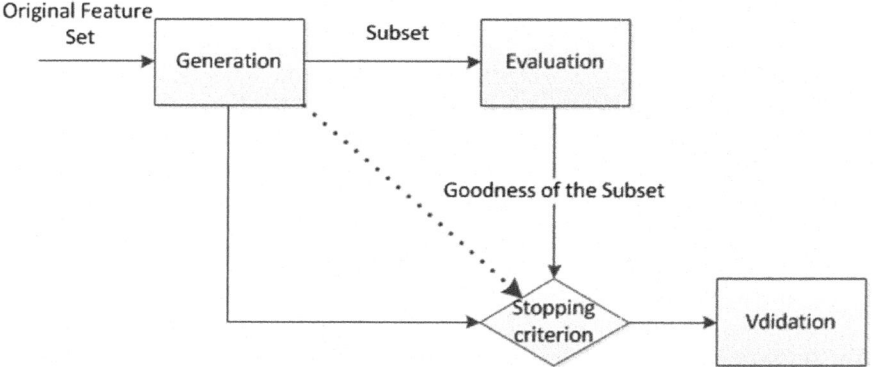

Fig. 1. Basic framework of feature selection

2   Feature Selection Methods Based on Filter

The main idea of the filter method is to evaluate each characteristic property, distribution of weight, and then choose the feature subset. The method needs to follow the principle to maximize the principle of classification effect and minimize redundancy. According to the evaluation content, it can be divided into four categories: distance measurement, information measurement, dependent measurement, and consistent measurement. In this paper, we choose evaluation feature based on information measurement which usually uses the information gain (IG) mutual information (MI). Information gain is defined as a priori uncertainty and expectations of the difference between a posteriori uncertainty, it can effectively select key characteristics, eliminate irrelevant features. Mutual information describes the interdependence between two random variables.

2.1   Related Theory

The method used in this paper is based on the measurement of mutual information. Suppose there are two random variables, one of which provides information on another random variable, that is, mutual information. This section mainly introduces some theoretical knowledge related to mutual information and subsequent methods.

The entropy of a discrete random variable $f = \{x_1, x_2, \ldots, x_N\}$ is denoted by $H(f)$ defined as:

$$H(f) = -\sum_{i=1}^{N} p(x_i) \log p(x_i) \tag{1}$$

where $p(x_i)$ is the probability mass function. For any two variables f_1 and f_2, the conditional entropy of the variable f_1 given f_2 is defined as:

$$H(f_1|f_2) = -\sum_{f_1}\sum_{f_2} p(f_1|f_2) \log p(f_1|f_2) \tag{2}$$

And the MI is the amount of uncertainty in f_1 due to the knowledge of f_2, it is defined as:

$$I(f_1;f_2) = \sum_{f_1}\sum_{f_2} p(f_1,f_2) \log \frac{p(f_1,f_2)}{p(f_1)p(f_2)} \tag{3}$$

MI represents the interdependence between the variables. If the mutual information is larger, it means that the correlation between two random variables is larger, and vice versa. We can also say:

$$I(f_1;f_2) = H(f_1) - H(f_1|f_2) = H(f_1) + H(f_2) - H(f_1f_2) \tag{4}$$

2.2 Process of Feature Selection

The feature selection method used in this paper adopts the heuristic search strategy. Specifically, the search in the state space evaluates the location of each search, gets the best location, and then searches from this location to the target. In this way, we can omit a large number of unnecessary search paths and improve the efficiency. The disadvantage is that the most complete subset cannot be obtained.

The filter feature selection method based on mutual information measurement focuses on selecting the appropriate information measurement function. There are many different forms of functions, but the purpose is the same, that is, the selected feature subset has the greatest correlation with the category, and the selected features have the least correlation with each other. The general form of measurement function is [8]:

$$J(f_i) = \alpha \cdot g(f_i, S, C) - \delta \tag{5}$$

where C is the category set, f_i is candidate feature, and S is a subset of selected features.

The basic process of the filter methods is shown in Table 1.

Table 1. Basic process of the filter methods

Initialize: Set $F = \{f_1, f_2, \ldots, f_D\}$, $S = \{\}$
Step 1: Compute $I(f_i; C) i = 1, 2, \ldots, D$
Step 2: Select $f_i = \arg \max(I(f_i; C))$, set $F \leftarrow F \backslash \{f_i\}$ and $S \leftarrow \{f_i\}$
Step 3: Select the next feature f_i according to the $J(f)$, set $F \leftarrow F \backslash \{f_i\}$ and $S \leftarrow S \cup \{f_i\}$
Step 4: Repeat the third step until $|S| = k$
Output: $S = \{f_1, f_2, \ldots, f_k\}$

2.3 MIFS and mRMR

MIFS (mutual information feature selection) and mRMR (minimal redundancy and maximal relevance) are selected as the selection methods of fingerprint features of radiation sources, mainly considering that their calculation is simple, rapid, and easy to realize [9]. The information measurement function of the MIFS algorithm considers the candidate feature's contribution to the classification and takes the MI between the candidate feature and the selected feature as the penalty factor. The MIFS evaluation function is:

$$J(f_i) = I(f_i : C) - \beta \sum_{f_k \in S} I(f_i; f_k) \tag{6}$$

where β is regulation constant in the interval [0.5, 1].

Compared with MIFS method, mRMR method is more cautious about redundant computation between the features. The idea is that when the quantity of selected features increases, the regulating coefficient in the evaluation function should be relatively reduced. The mRMR evaluation function is:

$$J(f_i) = \frac{I(f_i; C)}{\frac{1}{|S|^2} \sum_{f_k \in S} I(f_i; f_k)} \tag{7}$$

2.4   CMIM and JMIM

CMIM (conditional mutual information maximization) determines the importance of candidate features through the dependence degree of candidate features and categories under the condition that the selected feature set is known. The evaluation function of CMIM is:

$$J(f_i) = \arg\max_{f_i \in F} I(f_i; C|S) \tag{8}$$

JMIM (joint mutual information maximization) is similar to CMIM, but there are different ways to define the importance of candidate features. Specifically, there are different ways to select features that can represent the selected subset. The evaluation function of JMIM is:

$$J(f_i) = \arg\max_{f_i \in F} \left(\min_{f_k \in S}(I(f_i, f_k; C)) \right) \tag{9}$$

Venn diagram is used to illustrate the ways in which these two methods can be used to select features that represent selected subsets. $I(f_i, f_k; C)$ is the union of areas 1, 2, and 3; $I(f_i; C|f_k)$ is area 3 in Fig. 2. Because the calculation cost of MI is high and the number of features is large, a single selected feature which maximizes the $I(f_i; C|f_k)$ is selected to replace the entire selected subset in the CMIM method. Formula (8) can be written as:

$$J(f_i) = \arg\max_{f_i \in F} \left(\max_{f_k \in S}(I(f_i; C|S)) \right) \tag{10}$$

Unlike the CMIM method, the JMIM method first finds the f_k that minimizes $I(f_i, f_k; C)$ and then selects the f_i.

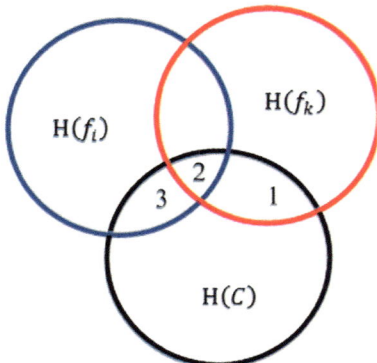

Fig. 2. Venn diagram illustrating the relation between feature and class

3 Experiments and Results

This article extracted different fingerprint based on the same set of simulation signals used the above four feature selection methods to reduct the dimension of characteristic fingerprint in order to validate the feasibility of the emitter signal features of fingerprint feature selection.

We consider the simulation based on the steady-state characteristics of the radiation source signal and divide the radiation source transmitter into several modules [10]; each module will affect the signal to generate different distortion. In this paper, the I/Q orthogonal modulator module and power amplifier module are considered.

We add the gain imbalance and orthogonal error of the orthogonal modulator into the simulation model and use Taylor series to model the power amplifier module. Set the Taylor series to the third order: $y(n) = \varepsilon_1 x(n) + \varepsilon_2 x(n)^2 + \varepsilon_3 x(n)^3$.

Five types of transmitters are constructed. Each type of transmitter generates 600 sample signals, including 300 training samples and 300 test samples. Each sample contains 200 randomly generated QPSK modulation symbols. The transmitter simulation coefficient is shown in Table 2.

Table 2. Parameter caused signal distortion

	T1	T2	T3	T4	T5
Gain imbalance	0.1	0.13	0.15	0.17	0.19
Orthogonal error (angle)	4.2	3.3	3.9	3.6	3
ε_1	1.521	1.181	0.820	1.000	0.987
ε_2	−0.104	−0.001	−0.001	0.109	0.014
ε_3	0.0021	0.0015	0.0011	−0.0019	0–0.002

In a noise environment where the signal-to-noise ratio is 10 dB, we took 2000 consecutive points from each sample to run the experiments. In the first experiment, we used EMD and ITD to decompose signals, respectively [11], and get five layers from each component signal, we extract each of the signal component samples of permutation entropy, approximate entropy, and fuzzy entropy as the fingerprint characteristics, forming a 40 story feature set. The feature set of 60 layers was extracted by rectangular double spectrometry in the second experiment. On the basis of experiment 1, PCA dimensionality reduction was added. All three experiments were identified by SVM classifier.

In the first experiment, the recognition probability of radiation source is 82% when feature selection is not carried out. Figure 3 shows the change of recognition probability along with the quantity change of selected feature. As can be seen from the figure, the curves of the four methods are basically similar and achieve the best recognition rate when the value of k is in the interval [20, 25]. Meanwhile, the curves of CMIM method and JMIM method are slightly better than those of mRMR method and MIFS method.

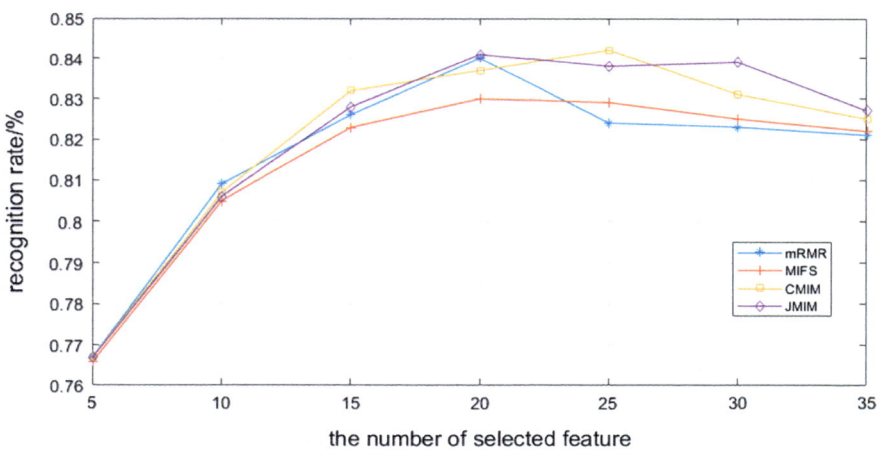

Fig. 3. Recognition rate in the first experiment

Figure 4 shows the recognition probability in experiment 2, along with the change of feature selection number. It can be seen that CMIM method and JMIM method have obvious advantages. From the analysis on evaluation function, mRMR method and MIFS method may ignore some features which have high values. Figure 5 shows the relationship between the f_i and f_k in some particular case. When the value of the penalty factor (area of $I(f_i; f_k)$) is close to the value of $I(f_i; C)$, those two feature selection methods cannot select the candidate feature.

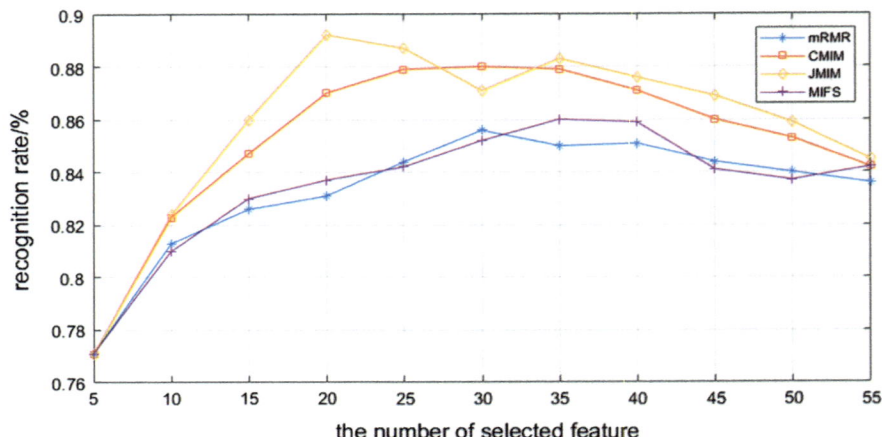

Fig. 4. Recognition rate in the second experiment

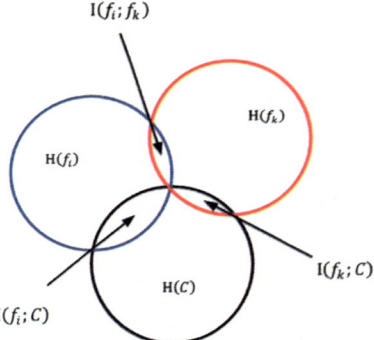

Fig. 5. Venn diagram illustrating the relation between the f_i and f_k in some particular case

The simulation results in the third experiment are shown in Fig. 6. The feature subset which includes 20 feature selected by JMIM method was sent into the PCA dimension reduction function. The experimental results show that the feature dimension reduction can be effectively combined with feature selection and bring the advantages that improve emitter individual identification probability and further reduce the subset dimension.

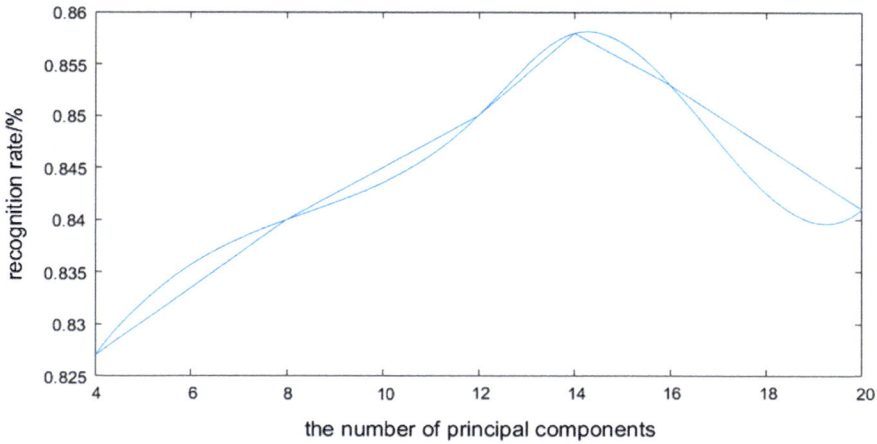

Fig. 6. Recognition rate in the third experiment

4   Conclusion

Considering the problem of large feature dimension would be met in the process of SEI, feature selection is helpful in reducing the feature dimension and improving individual recognition rate. Moreover, the feature selection can be effectively combined with feature dimension reduce function and bring a better performance.

References

1. Xu S, Huang B, Xu Z, et al. A new feature vector using local surrounding-line integral bispectra for identifying radio transmitters. In: International symposium on signal processing and its applications. IEEE; 2007. p. 1–4.
2. Yuan Y, Huang Z, Wu H, et al. Specific emitter identification based on Hilbert-Huang transform-based time-frequency-energy distribution features. IET Commun. 2014;8(13):2404–12.
3. Bertoncini C, Rudd K, Nousain B, et al. Wavelet fingerprinting of radio-frequency identification (RFID) tags. IEEE Trans Industr Electron. 2012;59(12):4843–50.
4. Frei MG, Osorio I. Intrinsic time-scale decomposition: time-frequency-energy analysis and real-time filtering of non-stationary signals. Proc Math Phys Eng Sci. 2007;463(2078):321–42.
5. 边肇祺, 张学工. 模式识别(第二版). 清华大学出版社; 2000.
6. Dash M, Liu H. Feature selection for classification. Intell Data Anal. 1997;1(1–4):131–56.
7. Blum AL, Langley P. Selection of relevant features and examples in machine learning. Artif Intell. 1997;97(1–2):245–71.
8. 刘华文. 基于信息熵的特征选择算法研究. 吉林大学; 2010.
9. Somol P, Haindl M, Pudil P. Conditional mutual information based feature selection for classification task. In: Congress on pattern recognition, Iberoamerican conference on

progress in pattern recognition, image analysis and applications. Berlin: Springer; 2007. p. 417–26.

10. 黄渊凌, 郑辉. 通信辐射源指纹产生机理及其仿真. 电信技术研究. 2012;1:1–12.

11. 谢阳, 王世练, 张尔扬,等. 基于差分近似熵和EMD的辐射源个体识别技术研究. 全国信号和智能信息处理与应用学术会议专刊; 2016.

Nonlinear Dynamical System Analysis for Continuous Gesture Recognition

Wenjun Hou and Guangyu Feng$^{(\boxtimes)}$

School of Digital Media and Design Arts, Beijing University of Posts
and Telecommunications, Beijing 100876, China
kinsney@bupt.edu.cn

Abstract. Extracting applicable features from continuous gesture is uneasy since it shows up as a nonlinear dynamic system with a spatial–temporal pattern. This paper introduces a continuous gesture recognition framework that analyzes, models, and classifies the nonlinear dynamics of gestures based on chaotic theory. In this system, the trajectories of finger joints are captured as the discrete observations of nonlinear dynamic system, which defines the feature matrix of gestures by reconstructing a phase space through employing a delay-embedding scheme, the properties of the reconstructed phase space are captured in terms of dynamic and metric invariants that include Lyapunov exponent, correlation integral, and fractal dimension. Finally, we extract a feature matrix for training several classifiers with relatively few samples and get best accuracy of around 96.6% to prove our assumption that the nonlinear dynamics of continuous gesture can be approximated by a particular type of dynamical system for classification.

Keywords: Continuous gesture recognition · Human computer interaction · Feature extraction · Chaotic theory

1 Introduction

Mid-air gesture interaction, as a common way of communication in the real world, has advantages of being potentially natural, efficient and multidimensional in the field of human–computer interaction. This interactive model contains not only the mouse-based panning operations of a desktop coordinate system, but also other unlimited operations with different postures. The requirement of natural interaction is met from the aspects of both multiple dimensions and intuition. Especially in the virtual reality (VR) environment, the mid-air gesture interaction is able to free users from the wearable devices, so that they can enjoy a more immersive experience. Recent years have witnessed the progress of the graphics

© Springer Nature Singapore Pte Ltd. 2020
Q. Liang et al. (eds.), *Communications, Signal Processing,
and Systems*, Lecture Notes in Electrical Engineering 516,
https://doi.org/10.1007/978-981-13-6504-1_120

processing algorithms, many hardwares have the ability to capture the trajectories of the skeletons with built-in units. Specifically, LeapMotion and Kinect can identify human's bare hands and obtain the position of finger joints in real time. There have been more and more attention to the mid-air gesture interaction; the priority lies in the semantic processing of a large amount of dynamic spatial–temporal information that is mostly metaphoric, ambiguous, and personalized. The performance of conventional continuous gestures recognition algorithms are mainly affected by such four factors:

1. A large number of degrees of freedom (DOF); The human hand is estimated to possess 27 DOF, with each pose and motion modeled by many more state variables.
2. Nonlinearity. The movement of gestures is inherently nonlinear, which is hard to extract accurate features for describing the motion in mathematical terms.
3. Noise. The appearance of dynamical gesture seems irregular due to different scales, shelter, orientations, luminosity, and other factors.
4. Different input dimensions. Gestures with same meaning are usually captured as time series of different length, while traditional machine learning methods require unified input dimensions.

In the early time, many researchers have borrowed models from speech recognition such as HMM [1], conditional random fields (CRFs) [2] and dynamic time scheduling (DTW) [3]. Among them, the HMM and CRF mainly employ a probability distribution model to establish a adaptive standard of target gestures and unintended gestures, but their temporal modeling capabilities are not good enough for dynamic spatial information. Although DTW methods have the advantage of low computational complexity, their sensitivity to the noise and outliers will reduce the robustness. Besides, descriptive variables such as tangential angular change [4] and coordinate of center point [5] are mostly used in spatial feature extraction, their abilities to characterize the movement are rather weak due to unknown governing equations of the system. There are also researchers focusing on gesture strokes and proposing a combination of new features such as left and right sector trajectory features, but they are too dependent on orientation, distance and other spatial features of strokes, and the differences between some similar gestures with confusing strokes such as five with S are hard to tell [6]. Recently, there have been growing interests in convolutional neural network (CNN) [7] and recurrent neural network (RNN) [8]. The hardware revolution has greatly increased the speed of computers and helped the original complex algorithms of deep learning to get a finer result than in the past. According to the results and analysis of the 2016 ChaLearn [9], the best recognition accuracy of isolated gesture recognition has improved from 56.90 to 67.71%, the Mean Jaccard Index (MJI) of continuous gesture recognition has improved from 0.2869 to 0.6103. However, their massive computing power does not pay off in the practical system, this reveals the fact that there is still a lack of effective modeling methods in feature engineering.

The aim of this paper is to provide a modeling method of continuous gesture recognition based on chaotic theory. The hypothesis is made that the irregularly sampled data of continuous gestures can be approximated with a particular type of dynamic system, and the characterization of this nonlinear dynamics will help with the establishment of feature vectors. The remainder of this paper is organized as follows: Sect. 2 presents the methodology of main proposals and its solutions. The proposed algorithm is validated experimentally on a set of mid-air alphabetic gestures in Sect. 3. Future work and conclusion are given in Sect. 4.

2 Chaotic Analysis of Continuous Gesture Recognition

When performing a particular alphabetical gesture, the trajectory we draw with bare hand usually contains not only the stroke itself, but also those unintended transition movement. When and how participants begin and end their gestures depend on their own habits, which make the pattern recognition more difficult. Besides, the underlying mechanism is hard to characterize in mathematical terms, descriptive statistics such as position of average coordinate points or orientation angle are not sufficient for characterizing since they make no physical sense without a detailed mathematical knowledge of the underlying dynamics.

The novelty of this paper stems from assuming a particular type of dynamical system to approximate the nonlinear dynamics of gesture and describe the features by discussing the structure and stability of this system. In other words, rather than letting data speak for itself about numbers of independent variables and unknown parameters, we try to make assumptions about it and fit the experimental data to the model by finding the parameters that best explains the patterns.

2.1 Strange Attractor

Assuming that there is a determinism present in the seemingly stochastic dynamics of gestures, which can be used to extract rich information for identification and classification. In a mathematical field of dynamical systems, the attractor is used to denote the set of numerical values toward which system tends to evolve, when the trajectory in an attractor with similar initial conditions tend to move apart with increasing time, there is a strange attractor characterizing this chaotic system with sensitive dependence on initial conditions. The next issue to be considered is the invariants of system's attractor, which also decides the properties of the constructed phase space for further classification. There are three major features in discussion—correlation integral, fractal dimension, and maximal Lyapunov exponents. The correlation integral is the mean probability that the states at two different times are close, while the fractal dimension measures the change in the density of phase space with respect to the neighborhood radius and characterizes the geometric structure of a strange attractor. The Lyapunov exponents describe how trajectories on the attractor move under the evolution of the system dynamics and the uncertainty about the future state. A chaotic process is generated by a nonlinear deterministic system with at least one positive Lyapunov exponent, which as a sufficient condition for stability also helps us filter the training data. As these exponents quantify the exponential rate at which nearby trajectories separate from each other while moving on the attractor, the maximal Lyapunov exponent is chosen as a dynamical invariant of the attractor to measure the exponential divergence of the nearby trajectories in the phase space.

2.2 Dynamic Reconstruction

Dynamic reconstruction describes a state of mapping from one-dimensional signal to an n-dimensional signal, which provides the dynamical modeling of a time series for an unknown n-dimensional system to capture the underlying dynamics. A fundamental method in dynamic reconstruction theory is a geometric theorem called the delay-embedding proposed by Takens [10], which reconstructs a phase space equivalent to the original phase space by delaying coordinate values. To form a matrix with each row representing a point in the reconstructed phase space (RPS), delayed copies of the original time series are used to slide a window of length m through the time series, and stack the m-dimensional vectors into the matrix with m noting embedding dimension and t noting delay parameter [11].

$$X = \begin{pmatrix} x_0 & x_t & x_{2t} & \cdots & x_{(m-1)t} \\ x_1 & x_{t+1} & x_{2t+1} & \cdots & x_{1+(m-1)t} \\ x_2 & x_{t+2} & x_{2t+2} & \cdots & x_{2+(m-1)t} \\ x_3 & x_{t+3} & x_{2t+3} & \cdots & x_{3+(m-1)t} \end{pmatrix} \tag{1}$$

There are two main methods to estimate a proper embedding delay: sequence correlation method and phase space geometry. The autocorrelation is a kind of sequence correlation method, which reduces the correlation between the reconstructed time series after selecting the delay time by the autocorrelation function to make the sequence dynamics less lost as much as possible. However, this method is essentially a linear concept, suitable for judging linear correlation rather than a nonlinear system. Based on this, Fraser and Swinney suggest using the first local minimum of the mutual information between the delayed and non-delayed time series, effectively identifying a value of delaying time for which they share the least information [12]. The false nearest neighbor method is used for finding the optimal embedding dimension m [13]. From the geometric point of view, the chaotic time series is the projection of the high-dimensional phase space on one-dimensional space with distorted trajectory, while the false neighbor points are those independent points which turn out be adjacent after projection in high-dimensional space and make chaotic time series irregular. With the increase of the embedded dimension m, the false neighbor points will be gradually removed and the trajectory of the chaotic motion will be restored to regular one. This reveals the fact that if points are sufficiently close in a reconstructed phase space, then they should remain close during a forward iteration, a phase space point that does not fulfill this criterion has a false neighbor. Figure 1 shows the three-dimensional projection of the reconstructed phase space for the chosen values of t and m.

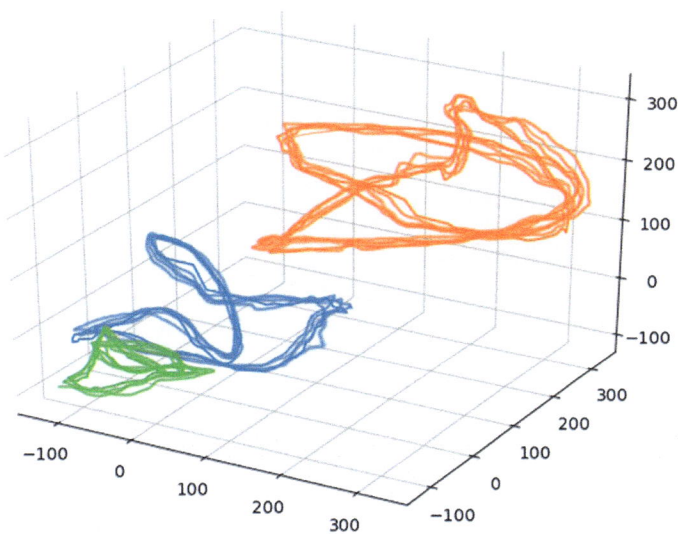

Fig. 1. Three-dimensional projection of reconstructed phase space

2.3   Correlation Factors

The correlation factors include correlation integral and fractal dimensions, they two characterize the metric structure of the attractor. Of them, the correlation integral measures the number of points within a neighborhood of radius, averaged over the entire attractor as

$$C(\varepsilon) = \frac{2}{N \times (N-1)} \sum_{i=1}^{N} \sum_{j=i+1}^{N} H(\varepsilon - \|\boldsymbol{x}_i - \boldsymbol{x}_j\|), \qquad (2)$$

where $\boldsymbol{x}_i$ are points on the attractor, $H(x)$ is the Heaviside function and ε is the value of radius. When set a specific value of ε to compute $C(\varepsilon)$, we can get the correlation integral as a feature vector. In our training, we go through several gesture samples to observe the stability of data with varying radius, and finally set 50 for a distinguishing distribution. The correlation function of the attractor, denoted as $C(q, r)$, describes the probability that any two points on the attractor are separated by a distance r for some integer q, indicated as

$$C(q, r) = r^{(q-1)D_q}. \qquad (3)$$

For $q = 2$ and taking the logarithm of both sides, the resulting dimension

$$D_q = lnC(\varepsilon)/ln\varepsilon \qquad (4)$$

is correlation dimension we need and bounds the degrees of freedom required to describe the system.

3   Experiment Result

3.1   Tracking Results of Gestures

In our experiment, we choose eleven continuous alphabetic gestures for classification viz 'A', 'B', 'E', 'F', 'H', 'I', 'K', 'P', 'R', 'X', and 'Z'. With the initial coordinate data captured from LeapMotion, we extract each nonlinear trajectory, respectively, in x, y, z dimensions from finger joints movement, and draw the approximate coordinate function after smoothing (shown in Fig. 2).

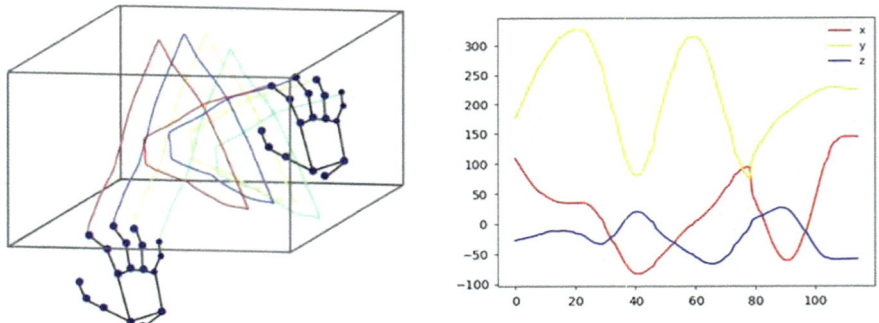

Fig. 2. The trajectory of mid-air gesture

3.2   Classification Results

The performances of the feature extraction are evaluated using three classifiers viz k-NN, SVM, and random Forest. In our experiment, fivefold cross-validation is conducted to analyze the performance of each classier with different parameters, the dataset is equally divided into five subsets, among them four subsets are used for training, the other is for testing. The framework we proposed put a lot of concentration on feature engineering, which improves the performance of the model and even achieves a accurate result from a simple dataset. Our dataset has 200 gestures of which 160 are used for training and remaining 40 are used for testing. Finally, the average accuracy of cross-validation is calculated and shown in Table 1. The three-dimensional projection of feature vector is shown in Fig. 3 with good clustering performance.

Table 1. The average recognition rate of each alphabetic gesture in three classifiers

Gesture	K-NN (%)	Random forest (%)	SVM (%)	N/G (%)
A	95	97.5	97.5	2
B	87.5	92.5	80	1
E	97.5	100	95	1.5
F	97.5	100	95	2.5
H	92.5	92.5	92.5	3
I	95	100	97.5	3.9
K	100	100	100	2.5
P	90	92.5	80	3
R	85	90	80	4.5
X	85	100	95	9.5
Z	92.5	100	92.5	7.8
Overall	91.8	96.6	91.6	3.7

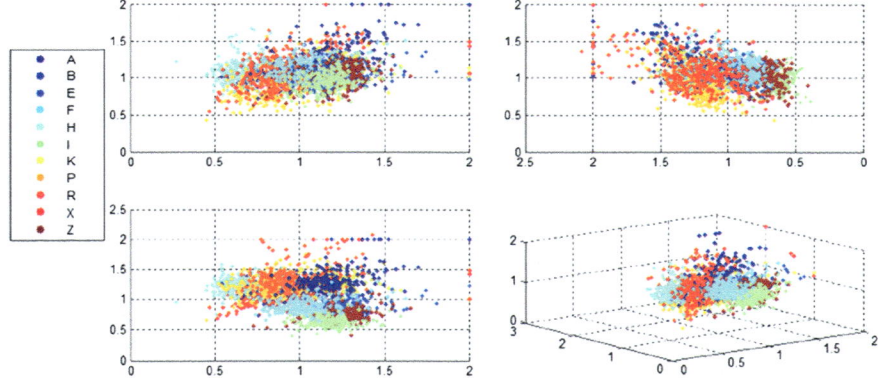

Fig. 3. Distribution of chaotic factors in feature matrix

From the average recognition classification accuracy of several alphabets' gesture, we can draw a conclusion that the algorithm we proposed for feature extraction applies to these three popular classification algorithm efficiently. Specifically in case of individual's gesture, the alphabets' gesture 'F', 'H', 'I', 'K', and 'Z' achieve the best performance of about 100%, they all share a common character of presenting as sequential movement, while for alphabets' gesture 'B', 'P', and 'R' with certain curves, it is uneasy for LeapMotion to capture the real trajectory without compensation, which put some constrains and effect on recognition, this will help with our further design of some interactive gestures. The experimental results suggest that our model provides pretty good recognition rate for most of the gestures; however, there is still reduction in overall accuracy due to misclassification of few similar gestures, such as D with P, E with F, and I with L, which is also in line with our expectations and indirectly illustrates the effective characterization of chaotic variables.

4   Conclusion

There are very few studies on the application of chaotic theory in dynamical gesture recognition. In this paper, we have attempted to provide some insights into feature extraction to recognize continuous alphabetical gestures, the recognition rate of around 93% in three popular classifiers have proven our assumption that these dynamical variables can characterize gesture motions in part and help us with the further study from a new perspective of chaotic theory. However, these three indicators we extract through phase space construction can only characterize the approximate chaotic movement tendency to some extent, which means our classifiers still cannot tell the difference between similar motions, there are still many valid feature variables to be explored. So our future work will include finding more indicators that characterize the gesture motion precisely to improve the recognition performance and design easy-to-identify interactive gesture considering its chaotic structure.

References

1. Yang Z, Narayanan SS. Modeling dynamics of expressive body gestures in dyadic interactions. IEEE Trans Affect Comput. 2017;8(3):369–81.
2. Yang HD, Sclaroff S, Lee SW. Sign language spotting with a threshold model based on conditional random fields. IEEE Trans Pattern Anal Mach Intell. 2009;31(7):1264–77.
3. Celebi S, Aydin AS, Temiz TT, Arici T. Gesture recognition using skeleton data with weighted dynamic time warping; 2013. p. 620–5.
4. Vo DH, Huynh, HH, Nguyen TN, Meunier J. Automatic hand gesture segmentation for recognition of Vietnamese sign language; 2016. p. 368–73.
5. Lu W, Tong Z, Chu J. Dynamic hand gesture recognition with leap motion controller. IEEE Sig Process Lett. 2016;23(9):1188–92.
6. Singha J, Misra S, Laskar RH. Effect of variation in gesticulation pattern in dynamic hand gesture recognition system. Neurocomputing. 2016;208:269–80.
7. Wang P, Li W, Liu S, Gao Z, Tang C, Ogunbona P. Large-scale isolated gesture recognition using convolutional neural networks. In: ArXiv e-prints; 2017.
8. Chai X, Liu Z, Yin F, Liu Z, Chen X. Two streams recurrent neural networks for large-scale continuous gesture recognition; 2016. p. 31–6.
9. Wan J, Escalera S, Escalante HJ, Baro X, Guyon I, Allik J, Lin C, et al. Results and analysis of ChaLearn LAP multi-modal isolated and continuous gesture recognition, and real versus fake expressed emotions challenges; 2017.
10. Takens F. Detecting strange attractors in turbulence. Berlin, Heidelberg: Springer; 1981. p. 366–81.
11. Oselio B, Hero A. Dynamic reconstruction of influence graphs with adaptive directed information; 2017.

12. Fraser AM, Swinney HL. Independent coordinates for strange attractors from mutual information. Phys Rev A. 1986;33(2):1134–40.
13. Brown L, Abarbanel J, Brown D, Kennel J, Kennel C. Determining embedding dimension for phase-space reconstruction using a geometrical construction. Phys Rev A. 1992;45(6):3402–4311.

Feature Wave Recognition-Based Signal Processing Method for Transit-Time Ultrasonic Flowmeter

Yanping Mei[1(✉)], Chunling Zhang[1], Mingjun Zhang[1], and Shen Wang[2]

[1] City Institute, Dalian University of Technology, Dalian, China
meiyanp@dlut.edu.cn
[2] Dalian Hui Ming Instrument Co., Ltd., Dalian, China

Abstract. In order to improve the measuring precision and stability of transit-time ultrasonic flowmeter as well as the locating accuracy of datum point for ultrasonic received signal, a feature wave recognition-based signal processing method is proposed in this study, which derives from analyzing the cause of errors in conventional threshold approach. By introducing a phase-shifted pulse into the ultrasonic excitation one, a feature wave with different period and phase is consequently produced in the ultrasonic received signal and recognized using a high-precision TDC chip according to the period of the received signal at first. Then the datum point of the received signal is accurately located with regard to the relationship between the position of feature wave and the initial position of the received signal so that the transit time of ultrasonic signal is finally measured. The following experiments focusing on a real-world problem demonstrate that the proposed method can effectively reduce the measurement errors caused by the amplitude change of the received signal. Such an approach is greatly beneficial for improving the precision of measurement along with the stability of the ultrasonic flowmeter.

Keywords: Ultrasonic flowmeter · Transit time · Signal processing · Feature wave recognition · Measuring precision

1 Introduction

Ultrasonic flowmeter (UF), with its non-contact characteristic and superiority on measuring accuracy, has become one of the most widely applied flow measurement instruments [1]. And the transit-time ultrasonic flowmeter (TTUF) is always the first alternative for solving practical problems regarding the fast response, easy installation, and low maintenance costs.

Considering the relationship between the transit-time difference caused by upstream and downstream and the velocity when ultrasonic pulse transmits through the flowing medium, the TTUF can calculate the flow rate, in which the transit time of ultrasonic signal plays a pivotal role for the flow rate measurement [2–4]. Such a parameter is mainly affected by the measurement precision of transit time and the detection accuracy of ultrasonic received signal. As for the former, it can be solved by utilizing time-to-

© Springer Nature Singapore Pte Ltd. 2020
Q. Liang et al. (eds.), *Communications, Signal Processing,*
and Systems, Lecture Notes in Electrical Engineering 516,
https://doi.org/10.1007/978-981-13-6504-1_121

digital converter (TDC), of which the measuring precision is typically up to picosecond level nowadays [5]. Bearing this in mind, the accurate detection of ultrasonic received signal becomes the most important factor affecting for the flow rate measurement. Conventional threshold method is commonly deployed for detecting the ultrasonic received signal, whereas measurement error is always produced by signal attenuation or noise disturbance. As such, various solutions are proposed to solve this problem. For instance, a digital signal processing algorithm using wavelet threshold for de-noising is proposed by Meng et al. [6]. A variable delay-time filter method is proposed by Zhao et al. [7]. A variable threshold-based zero-crossing detection signal processing method to determine a feature wave as well as the zero-crossing point and the transit time is presented by Wang et al. [8]. A dual-threshold method involving the determination on the abnormality of the signal is reported by Chen et al. [9].

In this paper, aiming at the measurement error and poor stability of the conventional threshold method caused by the noise in favor of practical application, considering the fact that the amplitude of the received signal could be changed easily while its period and phase are not changed significantly by the noises, a feature wave recognition-based signal processing method is proposed in this study. By means of introducing a phase-shifted pulse into the specific position of ultrasonic excitation one, a received signal with feature wave is consequently produced, of which the feature wave can be recognized by a high-precision TDC chip according to its period. Based on the position relationship with the feature wave, the datum point of the received signal is accurately located which leads to an accurate detection of ultrasonic received signal. Such an approach essentially reduces the measurement errors comparing with the one of threshold-based methods caused by the amplitude change of the received signal. The following experimental results show that the proposed method exhibits superior performance over the conventional methods, and the real-world application further demonstrates its practicability.

2 Analysis of Signal Processing

2.1 Error Analysis of Conventional Threshold Method

The principle of conventional threshold method is shown in Fig. 1. If the amplitude of the received signal reaches the preset threshold voltage, the detection circuit will trigger the zero-crossing detection and produce the stop pulses at the same time, then the ultrasonic transit time will be measured. However, when the amplitude of the received signal attenuates by disturbance, the stop pulse will have at least one periodic error because the datum point is changed. As such, the measurement error of transit time will be produced.

It can be depicted that the main problem for conventional threshold method is the detection error of the received signal. Therefore, it is of great significance to improve the accuracy on locating the datum point of ultrasonic received signal regardless of the varied amplitude caused by the disturbance.

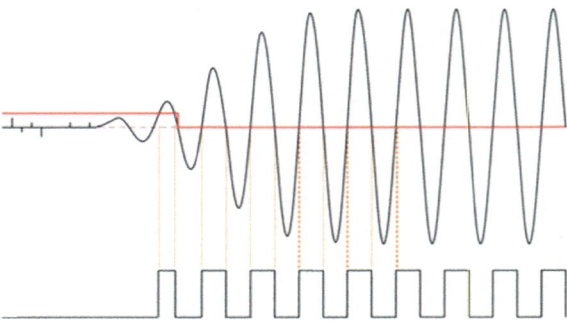

Fig. 1. Conventional threshold method

2.2   Feature Wave Recognition-Based Signal Processing Method

According to the basic principles of underwater acoustics [10], the vibration produced on its surface when the transducer receives the pulse signal can be described as

$$F = F_m \sin(\omega t + \varphi) \tag{1}$$

where F_m, ω, and φ denotes the amplitude, angular frequency, and phase angle of vibration. The received signal shares the same period and phase with the ones of (1), but its amplitude is related to the velocity of the vibration which can be calculated as

$$u_2 = \frac{2P_1 S}{Z_M} \tag{2}$$

where P_1 denotes the acoustic pressure of incident wave, S and Z_M refers to the section area and mechanical impedance of receiver transducer, respectively. When the acoustic waves are interrupted by disturbance during propagation, the S in (2) is correspondingly reduced which will cause the amplitude attenuation of the received signal, whereas the related period and phase will not be influenced. Taking such characteristic into account, the feature wave recognition-based signal processing method is proposed. The detailed elaboration as well as analysis is as follows.

A normal excitation pulse and its received signal are shown in Fig. 2. If a 180° phase-shifted pulse is introduced into the excitation one, its received signal will be changed as shown in Fig. 3. It can be depicted by comparing the two cases that the period and phase of the received signal in specific position are varied by a 180° phase-shifted excitation pulse, where a feature wave is introduced into the specific position of the received signal. Considering the favorable recognition of feature wave and the sensitivity and vibration speed of the utilized transducer with a resonant frequency of 1 MHz, an ideal received signal with higher initial amplitude and distinct feature wave can be obtained if the 180° phase-shifted pulse is inserted after the third excitation pulse as shown in Fig. 3.

Take a transducer with a resonant frequency of 1 MHz as an example, its received signal with feature wave is shown in Fig. 4. After the zero-crossing detection, the

(a) **(b)**

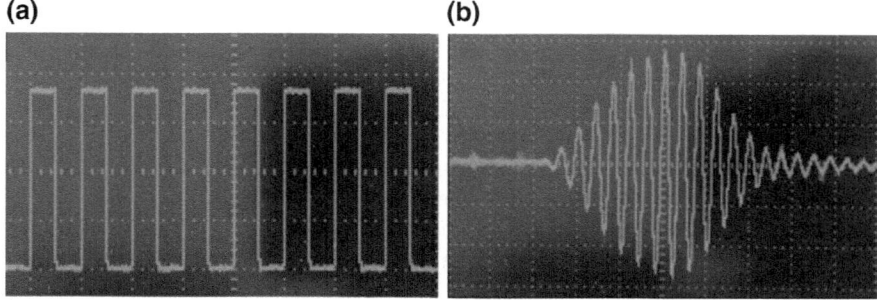

Fig. 2. a Normal excitation pulse and **b** received signal

(a) **(b)**

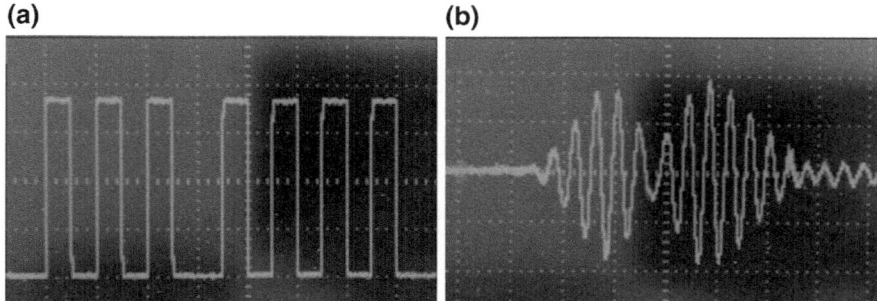

Fig. 3. a Excitation pulse with 180° phase-shifted and **b** received signal with feature wave

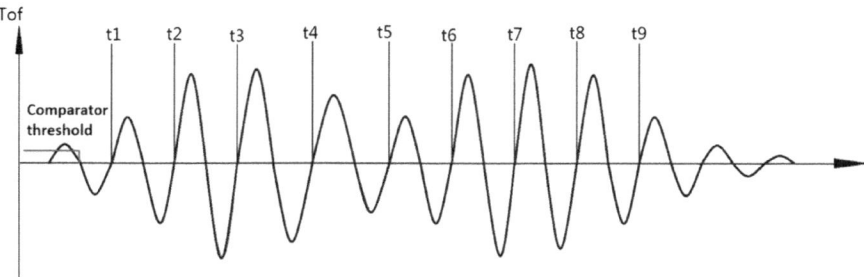

Fig. 4. Received signal with feature wave

periods of the received signal can be measured, i.e., $t2 - t1$, $t3 - t2$, $t4 - t3$, $t5 - t4$, $t6 - t5$,..., and the period change rate is defined as,

$$S_{\_p} = \frac{T_n}{T_s} \quad (n = 1, 2, 3, \ldots) \tag{3}$$

where T_n is the period of the nth received signal after zero-crossing detection; T_s is the standard resonant period of the transducer used.

A threshold value of the period change rate S_{std} is preset according to the characteristics of the transducer and the practical experience. If $S\_p < S\_{std}$, it can be depicted there are no changes in the periods of the received signal. Else if $S\_p \geq S\_{std}$, then the periods of the received signal are considered change, and the first change received signal wave is identified as the feature wave. The position relationship between the feature wave and the initial wave of the received signal can be determined by the position of phase-shifted pulse signal in the excitation pulse. Thus, the datum point of the received signal can be located accurately.

3   System Tests and Comparative Study

3.1   System Hardware Structure

Figure 5 gives the system hardware structure of TTUF method on the basis of the feature wave recognition. EFM32G880F128 is used to be the controller of the system. The high-precision TDC chip MAX35101 is used to transmit and receive the ultrasonic signal as well as measure the transit time of ultrasonic signal. The phase-shifted pulse drive circuit is responsible for generating an excitation pulse with fixed sequence. The dual-channel SPDT analog switch TS5A23159 is adopted to the transfer switch for the ultrasonic transmit circuits and the receive circuits.

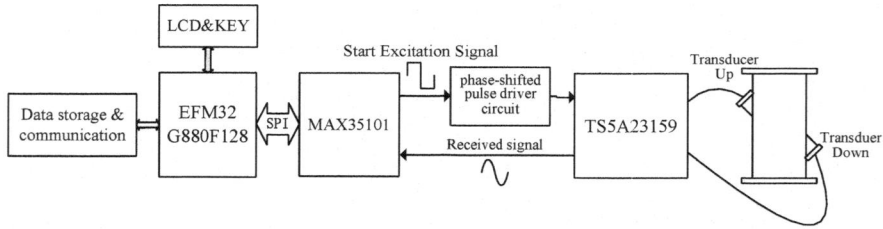

Fig. 5. System hardware structure diagram

3.2   Transit-Time Measurement

The accurate measurement of transit time requires the requirement on TDC chip is not only on the high-precision, but also having the capability of recording multiple times stop pulse for supporting the feature wave recognition method. The measurement precision of MAX35101 is up to 20 ps, and it can continuously record six stop pulses at one time, i.e., five periods of the received signal can be measured continuously, which the recognition of feature wave can be completed within one-time measurement scan.

When the measurement stars, MAX35101 emits the ultrasonic excitation pulses with no phase-shifted pulse in order to measure the T_s in (3) firstly, then it reemits one excitation pulse to the phase-shifted pulse drive circuit as well as generate a start pulse. The phase-shifted pulse drive circuit immediately generates a pulse sequence containing ten excitation pulses in which the first three pulses have 180° phase difference with the latter seven ones. When the signal containing feature wave is received, six stop pulse signals are produced by MAX35101 with which five consecutive periods of the received signal and the S_p are calculated. Comparing the S_p and the S_{std} presented, the corresponding feature wave is identified.

According to the position of the phase-shifted pulse introduced in the excitation pulse, it can be determined that the third wave before the feature wave is the initial received signal wave. Consequently, the datum point of the received signal is located and the transit time of ultrasonic is measured.

3.3 Comparative Study

In order to verify the effectiveness of the feature wave recognition method presented in this paper, two comparison experiments are conducted to the TTUF based on feature wave recognition method (TTUF-FWR) and the TTUF based on conventional threshold method (TTUF-CT): the transit-time measurement experiment on simulated condition and the flow rate measurement experiment on flow test platform.

3.3.1 Transit-Time Measurement Experiment

This experiment is conducted to test the influence for the precision and stability of transit-time measurement when the amplitude of the received signal changes. The transducers with the resonant frequency of 1 MHz are installed on the TTUF with DN50 in the 16 °C static clean water, and the voltage of excitation pulse is 3.0 V, the threshold value of comparator is set to 30 mV. The amplitude attenuation of the received signal is simulated by adjusting the resistance of resonant resistor which series with the transducers. Table 1 shows the results of measurement.

Table 1. Results of transit time measurement with different methods

Maximum amplitude of received signal (mV)	TTUF-FWR		TTUF-CT	
	Downstream transit time (µs)	Upstream transit time (µs)	Downstream transit time (µs)	Upstream transit time (µs)
400	48.275–48.276	48.275–48.276	48.628–48.629	48.628–48.629
300	48.281–48.282	48.281–48.282	49.633–49.634	49.633–49.634
200	48.286–48.287	48.286–48.287	50.631–50.633	50.631–50.633
<150	Low jitter, measurable		Large jitter, more than 2 µs	

The results show that the transit times measured by TTUF-FWR basically keep the same regardless the amplitude changes of the received signal, but the ones measured by TTUF-CT produce about 2 µs errors. When the amplitude of the received signal is lower than 150 mV, a data jitter occurs in the both flowmeters, which is smaller and measurable for the TTUF-FWR and greater than 2 µs for the TTUF-CT which cannot work normally. It can be seen that the TTUF-FWR has higher precision and stability of measurement than the TTUF-CT when the amplitude of the received signal changes.

3.3.2   Flow Rate Measurement Experiment on Flow Test Platform in Real Working Condition

In order to comparing the practical effect of flow rate measurement, this experiment is conducted on the flow test platform of Dalian Hui Ming Instrument Co., Ltd. Figure 6 shows the experimental layout. The excitation pulse voltage of the transducers with 1 MHz resonance frequency is 3.0 V, and the threshold value of comparator is set to 30 mV. The water in the pipe is added amount of sediment in order to reflect the practical situation that the bubbles and silt mixed in the water will cause the amplitude change of the received signal when the pump runs. The photograph of the experiment is shown in Fig. 7.

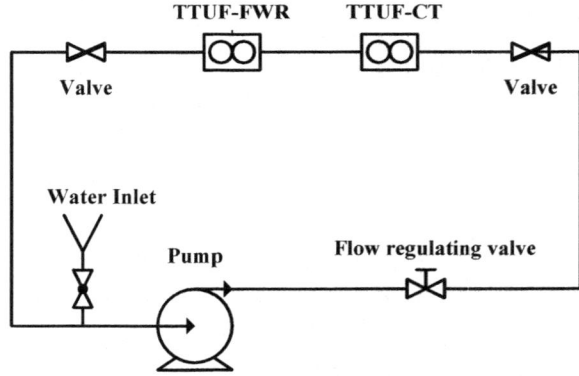

Fig. 6. Layout of experimental devices

The flow rate measurement experiments are conducted at three different flow rates of 1.5, 15, and 45 m³/h, and the measurements are done 10 times every 5 s at each flow rate. The experimental results including the repeatability calculated are shown in Table 2.

Table 2 shows that the measurement repeatability's of TTUF-FWR are all better than TTUF-CT at every flow rate, which indicates the more effectiveness of the feature wave recognition method. Because the accurate flow rate cannot be calibrated in this experiment, the minimum flow rate of 10 sets of measured data at 45 m³/h is, respectively, selected as the standard flow rates to calculate the measurement errors of the two flowmeters as shown in Table 3.

Fig. 7. Experimental photograph

Table 2. Measurement of flow rate

Group number	TTUF-FWR			TTUF-CT		
	Flow rate 1.5 m³/h	Flow rate 15 m³/h	Flow rate 45 m³/h	Flow rate 1.5 m³/h	Flow rate 15 m³/h	Flow rate 45 m³/h
1	1.523	15.619	45.561	1.492	15.961	44.784
2	1.562	15.895	46.024	1.593	16.334	45.436
3	1.422	15.724	46.352	1.651	15.805	45.825
4	1.475	15.636	46.684	1.454	15.529	46.391
5	1.395	15.525	46.236	1.328	15.035	46.632
6	1.613	15.149	45.853	1.466	15.297	45.457
7	1.453	15.266	45.352	1.354	15.536	45.203
8	1.537	15.749	45.213	1.473	15.154	44.565
9	1.569	15.817	45.942	1.629	14.897	44.239
10	1.637	15.725	46.204	1.626	14.662	44.354
Repeatability	0.081	0.238	0.461	0.115	0.550	0.825

According to Table 3, the maximum measurement error of TTUF-FWR is 3.253%, whereas the one of TTUF-CT is 5.409%. Although the measurement error mentioned above cannot accurately reflect the real measurement error, it proves that the feature wave recognition method has a better performance in accuracy.

The two comparing experiments show that the TTUF-FWR is superior to the TTUR-CT in the accuracy of measurement, the repeatability, the stability, and the measurement precision because of decreasing the measurement error of transit time caused by the amplitude changed of the received signal.

Table 3. Measurement error statistics

Group number	TTUF-FWR		TTUF-CT	
	Flow rate 45 m³/h	Measurement error (%)	Flow rate 45 m³/h	Measurement error (%)
1	45.561	0.770	44.784	1.232
2	46.024	1.794	45.436	2.706
3	46.352	2.519	45.825	3.585
4	46.684	3.253	46.391	4.864
5	46.236	2.263	46.632	5.409
6	45.853	1.416	45.457	2.753
7	45.352	0.307	45.203	2.179
8	45.213	0	44.565	0.737
9	45.942	1.612	44.239	0
10	46.204	2.192	44.354	0.260

4   Conclusions

Aiming at the measurement precision and stability of TTUF, a feature wave recognition-based signal processing method is proposed in this study. By measuring the period of feature wave which has a specific position in the received signal, the datum point of ultrasonic received signal is accurately located. The zero-crossing detection error of threshold comparator caused by the amplitude change of the received signal is effectively reduced so that the measurement accuracy of transit time is further improved. The following experimental study demonstrates that the proposed method performs better on both precision and stability comparing with the conventional approach, which has a high feasibility and practical value for real application.

References

1. Yu Y, Zong GH, Ding FL. Comparison of flow rate calculation method for ultrasonic flow measurement. J Beijing Univ Aeronaut Astronaut. 2013;39(1):37–41.
2. Svilainis L, Dumbrava V. The time-of-flight estimation accuracy versus digitization parameters. Ultrasound. 2008;63(1):12–7.
3. Wanderson ES, Edson DCB. Development and signal processing of ultrasonic flowmeters based on transit time. In: IEEE international conference on industry applications; 2016. p. 1–7.
4. Wang XF. Research on the key technologies of transit time ultrasonic gas flowmeters. Doctoral Dissertation. Dalian: Dalian University of Technology; 2011.
5. Rajita G, Nirupama M. Review on transit time ultrasonic flowmeter. In: 2016 2nd international conference on control, instrumentation, energy & communication; 2016. p. 88–92.
6. Meng H, Wang H, Li MW. High-precision flow measurement for an ultrasonic transit time flowmeter. In: International conference on intelligent system design and engineering application; 2010. P. 823–6.

7. Zhao WG, Jiang YF, Huang CC. A new ultrasonic flowmeter with low power consumption for small pipeline applications. In: IEEE international instrumentation and measurement technology conference; 2016. p. 1–6.
8. Wang W, Xu KJ, Fang M, Zhu WJ, Shen ZW, Wang G, Wang B. Study of a signal processing method for gas ultrasonic flowmeter. J Electron Measur Instrum. 2015;29 (9):1365–73.
9. Chen J, Yu SS, Li B, Fan CY. Signal processing based on dual-threshold of ultrasonic flow meter. J Electron Measur Instrum. 2013;27(11):1024–33.
10. Gu JH, Ye XQ. Foundation of underwater acoustics. Beijing: Defence Industry Press; 1981.

Realization of Unmanned Cruise Boat for Water Quality

Zhongxing Huo$^{(\boxtimes)}$, Yongjie Yang, and Yuelan Ji

School of Electronics and Information, Nantong University,
Nantong 226019, China
1254588282@qq.com, yang.yj@ntu.edu.cn

Abstract. In order to solve the problems of difficult wiring, poor flexibility, and high cost in aquaculture water quality monitoring, an unmanned water quality monitoring cruise ship with water quality monitoring device was built. The ship navigated automatically on the surface of the water according to the set course and collected water quality data during the voyage, which saved a lot of resources in this innovative way. There are two modes of operation for cruise ships: manual mode and autopilot mode. In the manual mode, the user can realize the manual operation of the ship through the remote controller and can set the autopilot path of the cruise ship. In the autopilot mode, the cruise ship moves automatically according to the preset path.

Keywords: Cruise boat · Water quality detection · Data acquisition · Automatic navigation

1 Introduction

Aquaculture water is the environment that fishes depend on for survival, so the good and bad culture environment is directly related to the growth and development of cultured fish, which determines the yield and quality of aquatic products [1–3]. The normal online water quality monitoring system, which usually replaces manual water quality monitoring system, adopts the scheme of arranging a number of nodes on the surface of the water. The system has the advantages of good real-time performance and simple operation, but the price of each set of sensors is very expensive, which is beyond the economic affordability of some farmers.

This paper designed an intelligent cruise boat for water quality monitoring, which only needs to carry a set of water quality detection sensors, saving a lot of equipment costs. At present, most of the researches on the water quality monitoring boat are manually controlled by the shore personnel, so it is impossible for the boat to run automatically for a long time.

© Springer Nature Singapore Pte Ltd. 2020
Q. Liang et al. (eds.), *Communications, Signal Processing, and Systems*, Lecture Notes in Electrical Engineering 516,
https://doi.org/10.1007/978-981-13-6504-1_122

2   Overall Design of System

The system block diagram of the scheme is shown in Fig. 1. The cruise boat is connected to the satellite positioning system, an electronic compass, the remote controller, and the PC. The satellite positioning system is used to obtain the longitude and latitude of the position information of cruise boat, and the electronic compass is used to get boat's course.

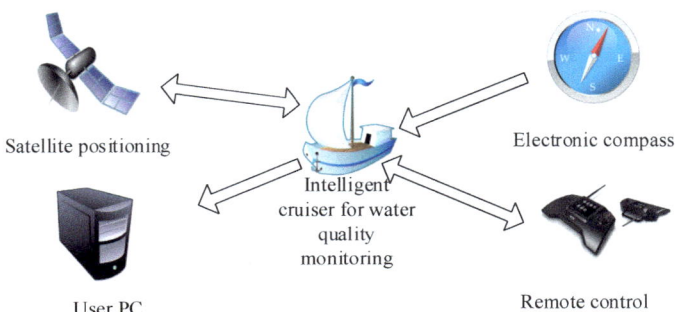

Fig. 1. Intelligent cruise boat system for water quality monitoring

The remote controller is used to communicate with the cruise boat. Through it, users can manually control the cruiser to move. They can also set up automatic navigation routes for cruisers, which need to avoid obstacles on the surface of the water. After completion, the remote control is required to set up the cruise boat into autopilot mode, and then the cruise boat continuously collects data on water quality parameters during the voyage. During autopilot, the user triggers a forward, left, right, or backward operation, the cruise boat exits autopilot mode.

3   Hardware Design of Cruise Boat for Water Quality Monitoring

In this paper, the circuit is divided into two parts: boat cruise and remote control.

3.1   Hardware Design of Cruise Boat

In order to realize intelligent cruising, the cruise boat carries the Beidou module, the electronic compass module, the steering gear, the motor drive circuit, the 433 MHz communication module, and the display module. The circuit diagram is shown in Fig. 2 [4]. The cruiser adopts a 32-bit STM32F407 single-chip microcomputer with high performance and high-performance price ratio. The frequency of operation is as high as 72 MHz, and the code execution efficiency is high.

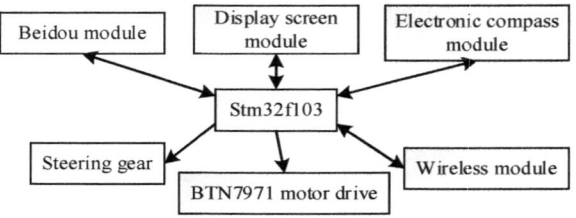

Fig. 2. Hardware block diagram of the front-end device

The geographic position of the cruise boat is indicated by latitude and longitude, which is obtained by the Beidou module UM220-III [5]. The positioning accuracy of this module is about 2.5 m CEP. The module uses serial port to send data to STM32.

The electronic compass adopts the GY-273 module, the measuring range is ±1.3–8 Gauss, and the MCU can communicate with the module through the IIC protocol to obtain the angle between the prow of cruising boat and the geomagnetic North Pole. Combined with the output information of the above two modules, the motion state of the cruise boat can be determined.

The wireless module uses the SX1278 module under the 433 MHz frequency to communicate with the wireless remote control provided by the cruise boat [6].

The OLED carried on the cruise boat is used to display the boat's information, mainly including the current longitude, latitude, heading angle and so on [7].

3.2   The Hardware Design of Remote Controller

The remote controller is used to interact with the user and the interactive interface is displayed on the display screen. Finally, the wireless module under the frequency of 433 MHz transmits the data to the cruise boat. The hardware block diagram is shown in Fig. 3.

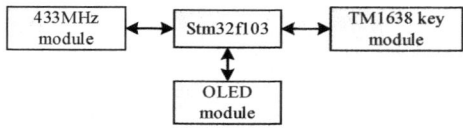

Fig. 3. Hardware block diagram of remote controller

The key module uses TM1638 chip to reduce the input/output occupation of microprocessor and simplify the complexity of program. In the design of the remote controller, the keyboard is used to receive user information and LED displays the state of the button pressed by the user. The digital communication of the module is designed by serial port, which only occupies three input/output ports of microprocessor, and the circuit connection is simple.

4 Software Design of Intelligent Cruise Boat for Water Quality Monitoring

4.1 Programming of Remote Control

The function realization of cruise boat needs to cooperate with the remote control, shown in Fig. 4, which needs to issue manual command, set path command, and autopilot command to cruise boat. The program first initializes the required hardware by using the function Board_Init(), including the 433 MHz wireless module, the key module and the display module. The Input_deal() function then enters to detect the input state of the user, assembles different data packets according to the different keys pressed, sends them out through the Send_packet() function, and displays the operation carried out by the current user on the display screen.

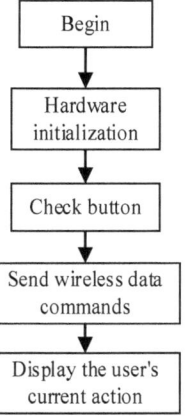

Fig. 4. Flowchart of remote control software

The remote control command packet format includes the following: "forward" button corresponding to "forward" is the cruise boat's advance command; "left turn" button corresponding to "left" is the cruise boat assembly command; "right turn" button corresponding to "right" for cruiser right-turn command; the "back" button corresponds to "backward" for the cruise boat's back command. These are manual driving orders. The "setting place" button corresponds to "Set1" as the first point command to set the path, and the "clear all" button corresponds to "clear" to clear the set location command, and the "autopilot" button corresponds to the "auto" as the autopilot command. When the user sets the path, the current set point mark is displayed back on the display screen, and the settings are automatically added 1. The schematic diagram of the remote control keys is shown in Fig. 5.

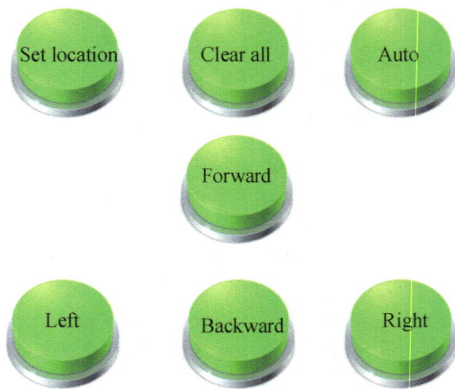

Fig. 5. Schematic diagram of remote control keys

4.2   Design of Autopilot Software for Cruise Boat

The embedded program is written by KEIL software and C language, and the flow of autopilot software is shown in Fig. 6 [8–10]. After the software is started, the wireless module, the Beidou module, the electronic compass module, the display screen module, the steering gear, the motor drive circuit and the stm32 internal peripherals are initialized by Board_Init() function.

The wireless module receives command data from the cruise boat's remote controller, which is divided into three types: the "manual pilot" command, the "set the path" command, and the "autopilot" command. The Get_lc(&hello) function performs different programs according to different commands, and if it is a "manual pilot" command, the cruiser can move back and forth from side to side.

If it is the "set path" command, the program first reads the coordinate data of the cruiser's location, then the stm32flash_write() function records it to the inner flash of the STM32 and displays the coordinate point label on the display screen.

If the "autopilot" command, then immediately enter autopilot mode, the program first read the electronic compass and Beidou satellite data, and then determine whether the cruiser's current coordinates are near the coordinates of the destination. According to the precision of the Beidou module, if the distance is less than 2.5 m, it means to reach the destination. If the cruise boat reaches its destination, the next stored coordinate is used as the destination. Then, according to the above information, the angle of the cruise boat need to be adjusted is calculated. Finally, the calculation results are input into the PID algorithm, and the output of the PID algorithm is used to control the cruiser movement [11–13]. Through the continuous cycle of the above process, the trajectory of the cruise boat is continuously adjusted, and then the cruise boat intelligent cruise is realized.

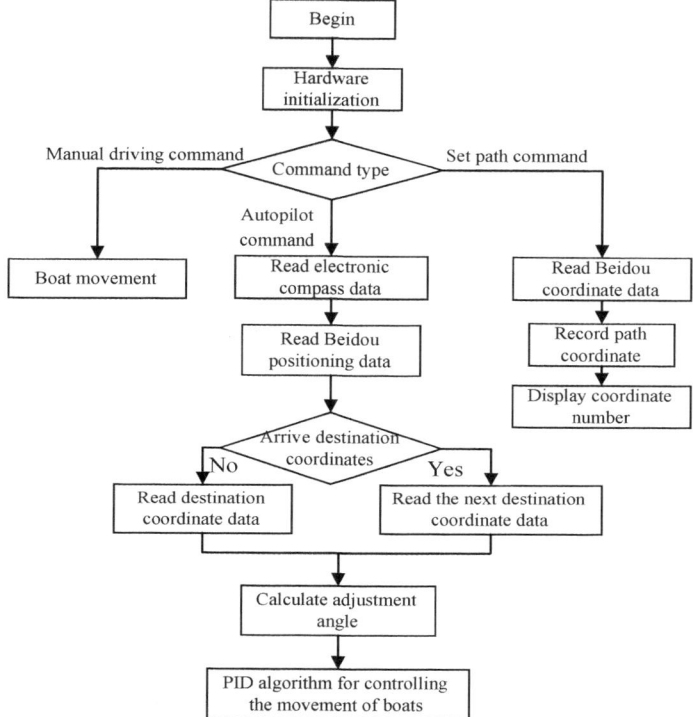

Fig. 6. Flowchart of cruise boat autopilot software

4.3   Software Design for Environmental Information Collection

When a cruise boat controlled by autopilot software is navigating on the surface, the data of water quality parameters collected by the cruise boat will be continuously uploaded, and the software flowchart is shown in Fig. 7. The program first initializes the 485 interface and the GPRS module by using Board_Init(), in which the 485 interface uses the baud rate of the 9600 bps, and each device has a unique address for each device on the 485 bus, according to which each sensor can be distinguished [14]. And then the program reads the data from each sensor. After initialization, when the cruiser is sailing on the water, the program sends the collected data to the user's PC through Gprs_send() function.

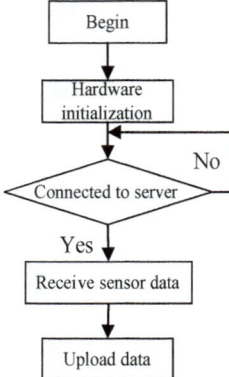

Fig. 7. Flowchart of environment information acquisition software

5 System Test

The test uses three target points to form a triangle, and each point is 10 m apart. After the target point is set, the cruiser is controlled into autopilot mode by the remote controller. The actual path is shown in Fig. 8.

Fig. 8. Schematic diagram of cruise boat operation path

In the course of the experiment, the distance of each cruiser from the target point was recorded four times, and the data was sent to the user PC by using the GPRS of the data acquisition part. The result is shown in Table 1.

The cruiser can reach every target point correctly, and the distance error is less than 2.5 m. This precision can meet the requirements on the surface of the water, can be used, and has a good application prospect.

Table 1. Cruise boat navigation record (unit: m)

Serial number	Distance from point 1	Distance from point 2	Distance from point 3
1	1.0	1.6	1.3
2	1.2	2.1	1.3
3	1.9	0.8	1.1
4	1.4	0.8	1.9

6 Conclusions

In this paper, a kind of intelligent cruise boat with water quality parameters is realized, and the water quality parameters information of the whole water area is collected in real time by a novel "multi-point one boat" method, which saves a lot of material resources and can meet the needs of aquaculture for water quality measurement.

Acknowledgments. This work was supported by Postgraduate Research and Practice Innovation Program of Jiangsu Province (KYCX17-1921) and First phase project of Jiangsu University Brand Specialty Construction Project (PPZY2015B135). In addition, it was completed under the support of Nantong University-Nantong Intelligent Information Technology Joint Research Center Open Topic (KFKT2017B05).

References

1. Shah MR, Lutzu GA, Alam A, et al. Microalgae in aquafeeds for a sustainable aquaculture industry. J Appl Phycol. 2017;1:1–17.
2. Mo WY, Man YB, Wong MH. Use of food waste, fish waste and food processing waste for China's aquaculture industry: needs and challenge. Sci Total Environ. 2017;635:613–4.
3. Tai H, Ding Q, Li D, et al. Design of an intelligent PH sensor for aquaculture industry. IFIP Adv Inf Commun Technol. 2017;347:642–9.
4. Yang B. Design and implementation of intelligent home wireless gateway based on STM32. In: International conference on information science and control engineering. IEEE computer society; 2017. p. 258–60.
5. Goncharova I, Lindenmeier S. Compact satellite antenna module for GPS, Galileo, GLONASS, BeiDou and SDARS in automotive application. IET Microwaves Antennas Propag. 2018;12(4):445–51.
6. Zhou L, Sun S, Zhang Y, et al. Long-distance running test system based on 433 MHz wireless module. In: IEEE international conference on communication technology. IEEE; 2016. p. 339–43.
7. Kurban M, Gündüz B. Physical and optical properties of DCJTB dye for OLED display applications: experimental and theoretical investigation. J Mol Struct. 2017;1137:403–11.
8. Mustakerov I, Borissova D. A framework for development of e-learning system for computer programming: application in the C programming language. J e-Learn Knowl Soc. 2017;13(2):89–101.
9. Kim JH, Whang IH. Augmented three-loop autopilot structure based on mixed-sensitivity H_∞ optimization. J Guidance Control Dyn. 2017;4:1–6.

10. Graham DM. An AutoPilot platform for high-resolution light-sheet microscopy. Lab Anim. 2017;46(2):25.
11. Kong H, Fang Y. Neural network PID algorithm for a class of discrete-time nonlinear systems. Int J Online Eng. 2018;14(2):103.
12. Dideriksen JL, Feeney DF, Almuklass AM, et al. Control of force during rapid visuomotor force-matching tasks can be described by discrete time PID control algorithms. Exp Brain Res. 2017;235(2):1–13.
13. Pradhan PC, Sahu RK, Panda S. Firefly algorithm optimized fuzzy PID controller for AGC of multi-area multi-source power systems with UPFC and SMES. Eng Sci Technol Int J. 2016;19(1):338–54.
14. Xie Y, Yu M, Fu J, et al. A hazmat transportation monitoring system based on global positioning system/beidou navigation satellite system and RS485 bus. In: International congress on image and signal processing, biomedical engineering and informatics. IEEE; 2017. p. 1059–63.

Improved K-Means Clustering for Target Activity Regular Pattern Extraction with Big Data Mining

Guo Yan[1(✉)], Lu Yaobin[1], Ning Lijiang[1], and Wang Jing[2]

[1] Nanjing Institute of Electronic Technology, Guorui Road 8#, Nanjing, China
9821078@qq.com
[2] Troop, PLA, Beijing 66132, China

Abstract. The traditional target activity regular pattern extraction methods replay previous target tracks, activities of the specified target are manually analyzed by checking all the tracks on map. This paper adopts big data mining technology to solve the problem of automatically extracting target classic tracks and converts the original pure manual map analysis into system automatic track extraction. This method greatly reduces the operation intervention of classic track extraction, which can reduce the 3–4 manual days to 3–4 h.

Keywords: Big data mining · K-means clustering · Target activity regular pattern

1 Introduction

The traditional method for target activity regular pattern extraction analyzes the activity of a specified target in a manual manner by replaying previous trajectory data. It is necessary to manually analysis the target trajectory according to time, region, country, mission to form a classic trajectory, and then sum up the target activity pattern. Therefore, there is a lack of effective methods for automatic extraction of target activity patterns.

Big data mining technology [1] uses non-traditional machine learning tools to process massive amounts of structured and unstructured data, to extract data relationships that are unknown, such as group data analysis (cluster analysis), unusual data monitoring (abnormal monitoring), and relationship mining. Using big data mining technology, intelligent processing similar to human intelligence can be performed on the previous target trajectory data to extract target activity patterns.

Target activity regular pattern extraction with big data mining technology, whose input time, region, and other conditions can be specified by requirements, utilize the K-means clustering previous big target trajectory mining algorithm to extract classic tracks, utilize geographic grid to merge the classic tracks into classic track patterns. This technology solves the problem of automatically extracting interested target from big amount of track data and converts the classic manually selection method to automatic extraction with little human participation. This change for track regular pattern extraction is fundamental, which greatly reduces the workload of operator.

© Springer Nature Singapore Pte Ltd. 2020
Q. Liang et al. (eds.), *Communications, Signal Processing, and Systems*, Lecture Notes in Electrical Engineering 516,
https://doi.org/10.1007/978-981-13-6504-1_123

2 Extraction of Target Activity Patterns Based on Big Data Mining

2.1 Target Classic Track Data Mining Method

This method utilizes big data mining technology to analyze the intrinsic correlation among the large amount of target track data. According to the machine number, time, area information, machine learning algorithms, such as data cleaning, data clustering are used to excavate track pattern, then the classic tracks are shown to combat personnel for combat target rapid analysis. The flowchart for target classic track data mining method is shown in Fig. 1, including data cleaning, track extraction, track clustering, track merging, classic track generation, and classic track display. The core of this flowchart is to track clustering.

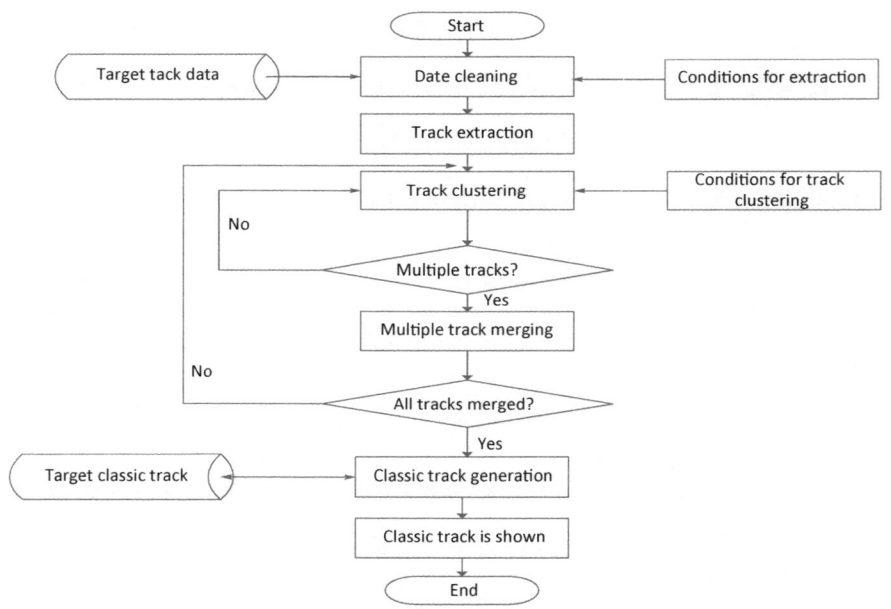

Fig. 1. Target activity pattern extraction with big data mining

(1) Data cleaning

The data for data mining is complex track data, which cover planes, ships, missiles, and satellites. The data sources include ground-based air defense radars, airborne radars, shipborne radars, and space-based radars, the sensor types include passive radar and optical equipment. Due to the different data sources, detection mechanisms, transmission mechanisms, the consistency and validity of the data should be considered. Data cleaning is necessary for data anomalies eliminating, normalization processing, and data formatting.

The track data is stored in the big data platform with near-line relational database and off-line distributed file system, where the track data within three months is stored in database, and the track data more than three months is stored with specified format in file system. During data cleaning, the data range is determined according to the conditions of track extraction, then the related track data is read into memory of big data platform, and the track data is cleaned.

(2) Track extraction

Using human–computer interaction window, the track extraction conditions, such as type of plane, start and stop time of specified target, area range (up-left and down-right corner of rectangular area, custom polygon area), country information, are properly set.

When the track is extracted, the big data platform acquires data components on demand. According to the different track extraction conditions set by warfighters, the related track data is parallel and quickly extracted into big data platform, and sorted by types, start and stop time, area range, and country.

(3) Track clustering

Track clustering is the core of the target activity pattern with big data mining. It mainly refines the analysis and processing of the large amount of unordered track data and classifies the different tracks with different shapes.

In this process, Euclidean position distance constraint [2] is improved as multiple track feature distance constraint for K-means clustering [3]. Euclidean position distance constraint performs the nearest neighbor calculation based on Euclidean distance of track point position. The computational complexity is low, and it is suitable for large-scale track clustering. Multiple track feature distance constraint performs the track feature Hausdorff distance [4], which includes track point position, velocity, attributes, course, and so on. The multidimensional features are used, and the clustering accuracy is better than the classic Euclidean position constraint.

The steps of K-means clustering algorithm is as follow:

Input: track data $C_1, C_2, \ldots, C_n$, clustering initial value k	
Output: classic track clustering results $x_1, x_2, \ldots, x_k$	
Algorithm steps	Step 1: Initialize k classic track seeds for track clustering. Extract n track data $C_1, C_2, \ldots, C_n$ from big data memory database, select random k tracks as the initial clustering seeds of the classic track clustering
	Step 2: Distribute track point data samples to the nearest center vector and construct disjoint clusters. For the remaining objects, according to their similarity with these cluster centers, they are assigned to their most similar clusters: $\sum_{i=1}^{n} \min_{j \in \{1,2,\ldots,k\}} (x_i - p_i)^2$
	Step 3: Use the sample mean in each cluster as the new cluster center (average of all objects in the cluster): $x = \frac{1}{|C_i|} \sum_{x \in C_i} x$
	Step 4: Repeat steps 2 and 3 until the cluster center no longer changes. Repeating this process until the standard measure function E to start to

(*continued*)

(continued)

Input: track data $C_1, C_2, \ldots, C_n$, clustering initial value k
Output: classic track clustering results $x_1, x_2, \ldots, x_k$

---

converge, and then get k clusters, making the clusters themselves as compact as possible, and separating the clusters as much as possible. The standard deviation function is used here as the standard measure function ad follow:

$E = \sum_{i=1}^{k} \sum_{x \in C_i} |x - x_i|^2$

The above K-means track clustering method based on Euclidean distance can achieve rapid clustering processing with massive target track data, but the accuracy of clustering is limited. In order to improve the clustering accuracy, K-means clustering algorithm based on multidimensional track features can be considered in this process. The algorithm steps are the same as the above method, but the distance calculation is calculated using Hausdorff as follows:

$$\delta_H(TR_A, TR_B) = \max_{P_A \in TR_A} \left\{ \min_{P_A \in TR_B} \{dist(P_A, P_B)\} \right\} \tag{1}$$

In the above equation, $dist(P_A, P_B)$ is the Euclidean distance considering the position between the two points, $\delta_H(TR_A, TR_B)$ is the maximum distance between TR_A and TR_B. Where from the middle point to the middle point and represents the degree of similarity with. Where TR_i is multidimensional track point sequence as follows:

$$TR_i = \{P_{i1}, P_{i2}, \ldots, P_{ij}, \ldots, P_{im}\} \tag{2}$$

P_{ij} is multidimensional track point vector as follow:

$$P_{ij} = [\text{label} \quad \text{longitude} \quad \text{latitude} \quad \text{altitude} \quad \text{attribute} \quad \text{velocity} \quad \ldots \quad \text{course}] \tag{3}$$

Which contains the target number, longitude, latitude, altitude, attributes, speed, velocity, etc. of the track point P_{ij}.

(4) Multiple tracks merging

The results of track clustering are merged according to the type of target. For aircraft and ship targets, tracks are sorted by model, number, time, area, and country.

(5) Classic track generation

For the collection of target tracks obtained by the multiple tracks merging, the classical track is extracted using the geographic grid technology. For the target tracks belonging to the same category, the track points are divided by a suitable grid width, and multiple tracks falling within a grid are replaced by one classic track point. The connectivity of the grid connects the classic track points to form a classic track.

(6) The classic track displaying

According to the model, time, area, country, etc. that the warfighter pay attention to, the classic tracks are displayed on the map.

2.2 General Machine Learning Algorithm Model

Target activity regular pattern extraction needs big data platform to implement data cleaning, clustering analysis. Therefore, a general machine learning hardware and software platform is necessary.

This paper develops a universal machine learning management tool to realize multisource data training and real-time data prediction. The tool encapsulates seven types of algorithms such as classification, regression, integration, and clustering, among which anomaly detection and clustering are applied to target activity regular pattern extraction. The universal machine learning algorithm model library is shown in Table 1.

Table 1. Universal machine learning algorithm model library

Algorithm classification	Algorithm name	Algorithm applicability
Anomaly detection	Support vector machine	Detecting unusual or abnormal data
Clustering	Enhanced K-means, orthogonal partition clustering	Finding natural grouping
Classification	Logistic regression, Bayes classification, support vector machines, decision trees	Predicting specific results
Regression	Ridge regression, lasso regression, elastic network	Prediction for continuous numerical results
Integration	Binary decision tree, gradient promotion	Random forest data
Feature extraction	Non-negative matrix factorization	Generate new attributes as a linear combination of existing attributes

3 Verification Results

3.1 Lightweight Big Data Platform

To build a lightweight platform for the target activity regular pattern extraction, a big data hardware platform with blade server I9000 + disk array, core software components such as Hive, Spark, Yarn, and Zookeeper is constructed. The hardware and software configurations are shown in Table 2.

Table 2. Big data platform hardware and software environment

Server	There are 4 servers, 3 for constructing a fleet, and 1 for building a stand-alone test environment. Blade server I9000 + disk array
Operation system	Linux
Distributed storage system	HDFS
Distributed computing framework	Spark computing framework
Distributed resource management system	YARN
Distributed collaborative component	Zookeeper
Distributed message queue	Kafka
Data migration tools	Sqoop, kettle
Data processing algorithm library	Mahout, Mlib
Cluster monitoring deployment tool	CM

3.2　Target Classic Track Clustering Mining Results

In order to verify the validity of the target classic track extraction, the track data of an airport in 2013 was used, and the data of one day is selected for cluster mining. Using the above process, multiple tracks of similar shapes are grouped into clusters from a large number of unrelated tracks, and multiple groups of tracks of different shapes are divided into clusters. Figure 2a shows the five track clusters of interest, and Fig. 2b, c shows the two of the five tracks.

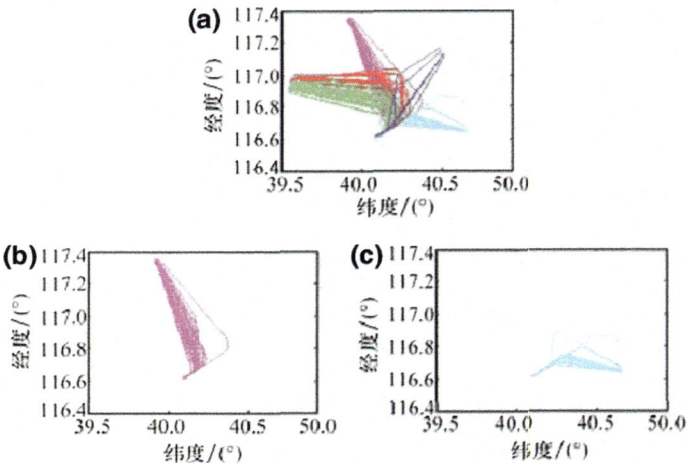

Fig. 2. Target classic track clustering results

4 Conclusion

Target activity regular pattern extraction technology, according to the time, area, and country constraint, utilizes the improved track-feature-based K-means clustering method for big data mining in large amount of previous tracks. This technology greatly reduces the operation intervention of classic track extraction, which can reduce the 3–4 manual days to 3–4 h. This technology has a wide range of application, such as radar netting system and edge sea defense system.

References

1. Liang JY, Qian YH, Li DY, et al. Theory and method of granular computing for big data mining. China Sci Inf Sci. 2015;45(11):1355–69.
2. Xu T, Li YX, Lv ZP. Track clustering based on distance from the track point metod. Syst Eng Electron Technol. 2015;37(9):2198–204.
3. Wang ZF, Pan Q, Lang L, et al. Dynamic track clustering algorithm based on subtractive clustering. Syst Simul J. 2009;21(16):5240–3.
4. Chen H, Zhang BY, Chen Y. Research on multi-hypothesis tracking with adaptive depth adaptation. Syst Eng Electron Technol. 2016;38(9):2000–7.

PSO-RBF Small Target Detection in Sea Clutter Background

ZhuXi Li, ZhenDong Yin$^{(\boxtimes)}$, and Jia Shi

School of Electronics and Information Engineering, Harbing Institute of
Technology, Harbin 150006, China
51280684@qq.com, yinzhendong@hit.edu.cn,
13182169819@163.com

Abstract. Target detection in the background of sea clutter is an important part
of sea surface radar signal processing. The traditional detection of weak targets
in sea clutter is based on the statistical characteristics of sea clutter, which does
not reflect the intrinsic dynamics of sea clutter. Therefore, the detection results
are not ideal. Based on the chaotic characteristics of sea clutter, this dissertation
reconstructs the space structure of the sea clutter and proposes an improved
particle swarm optimization (PSO) algorithm based on adaptive time-varying
weights and local search operators. This method was applied to the optimization
learning of the parameters of the radial basis function (RBF) neural network
kernel function. The method was validated by using McIX University in Canada
to measure the sea clutter data with the target in the Dartmouth area using IPIX
radar. The results showed that the PSO-RBF algorithm in the background of
chaotic sea clutter has good predictability. Compared with the general radial
basis neural network, the improved algorithm not only has fast convergence
speed but also has high error accuracy.

Keywords: Particle swarm optimization (PSO) · Radial basis function (RBF) ·
Target detection · Neural network · Sea clutter

1 Introduction

In recent years, the small target detection technology in the sea has attracted more and
more attention. Sea clutter is an echo signal that radar shines on the sea surface and is
very vulnerable to external environmental factors such as waves, wind speed, and wind
power. Since sea clutter can seriously affect the detection of small targets on the sea
(including low-altitude flying planes, small warships, floating ice on the sea, navigation
buoys, etc.), the study of sea clutter characteristics is of great significance.

The research of sea clutter based on chaos theory can reflect the inherent nonlinear
dynamics of sea clutter [2]. The introduction of nonlinear theory and neural network
technology has improved the detection performance of small sea targets. Many scholars
have done a lot of related works. Optimization of RBFNN center points, variances, and
number of hidden layer nodes by genetic algorithm [3]. An algorithm based on dual-
constraint least square support vector machine is introduced to detect the weak signal in
the background of chaotic sea clutter [4]. Based on the study of normalized RBF neural

Q. Liang et al. (eds.), *Communications, Signal Processing,
and Systems*, Lecture Notes in Electrical Engineering 516,
https://doi.org/10.1007/978-981-13-6504-1_124

network and least squares support vector machine (LSSVM) nonlinear prediction of sea clutter, a prediction method of LSSVM-coupled image lattice is proposed [5].

In this dissertation, the sea clutter is reconstructed by phase space, and an improved PSO algorithm based on adaptive time-varying weights and local search operators is proposed. This algorithm is applied to the optimal learning of kernel function parameters in RBF neural network. Algorithm flowchart is shown in Fig. 1. The simulation results show that the PSO-RBF algorithm has good predictability in the chaotic sea clutter background. Compared with the general radial basis neural network, the improved algorithm not only has fast convergence speed but also has high error precision.

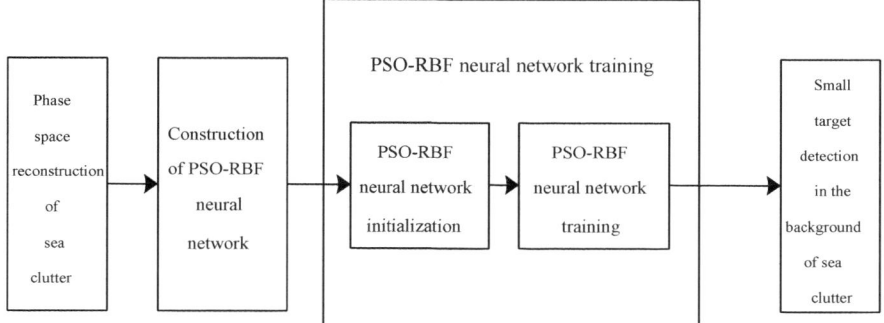

Fig. 1. PSO-RBF target detection flowchart

2   Phase-Space Reconstruction Model

Phase space is a multidimensional space that can reflect the inherent dynamic structural characteristics of the system. For a nonlinear system, the internal structure is usually more complex, and the phase space has a higher dimension. Therefore, it is difficult to determine the dimension. In practical applications, for a given one-dimensional time series, it is usually extended to a three-dimensional or even higher-dimensional space in order to extract the information contained in the time series. This process is called phase-space reconstruction. Its essence is to reconstruct a one-dimensional time series into an m-dimensional phase-space vector:

$$y(t + m\tau) = F(y(t), y(t + \tau), \ldots, y(t + (m - 1)\tau)) \tag{1}$$

where m is the embedding dimension, τ is the delay time, and F is the nonlinear function. Phase-space reconstruction's main steps are as follows:

Step 1: Calculate the embedding dimension m. The GP algorithm is a simple and relatively mature method for calculating the embedding dimension.

Step 2: Calculate the delay time τ. The autocorrelation function method is a commonly used calculation method that is relatively simple and mature.

3   PSO-RBF Neural Network Target Detection Model

3.1   The Basic Principle of Particle Swarm Optimization

Particle swarm optimization is a simulation of bird flocking behavior. The global search is achieved through the collective cooperation and competition between the particles. All particles have a fitness determined by the function being optimized, and a speed determines the direction and distance of their movement. During the iteration of the PSO algorithm, the particle updates its speed and position by tracking two extreme values (individual extremum p_{best} and global extremum g_{best}). When these two optimal values are found, the particles update their speed and position according to the following formula:

$$v_{id}^{k+1} = v_{id}^k + c_1\text{rand}()\left(p_i^k - x_{id}^k\right) + c_2\text{rand}()\left(p_{gd}^k - x_{id}^k\right) \tag{2}$$

$$x_{id}^{k+1} = x_{id}^k + v_{id}^{k+1} \tag{3}$$

where c_1 and c_2 are learning factors; v_{id} is the speed of the particle, $v_{id} \in [-v_{\max}, v_{\max}]$, $v_{\max}$ is a constant, set by the user to limit the speed of the particle; rand() is a random number between [0, 1].

The improved PSO algorithm proposed in this paper: In iterative formula (2) adding self-knowing time-varying weights w_i, and in iterative formula (3) adding α. Its mathematical description is as follows:

$$\begin{aligned} v_{id}^{k+1} &= v_{id}^k + c_1\text{rand}()\left(p_i^k - x_{id}^k\right) + c_2\text{rand}()\left(p_{gd}^k - x_{id}^k\right) \\ &= \left\{w_{\text{end}} + (w_{\text{start}} - w_{\text{end}})\exp\left[-15\left(\frac{k}{k_{\max}}\right)^3\right]\right\}v_{id}^k \\ &\quad + c_1\text{rand}()\left(p_i^k - x_{id}^k\right) + c_2\text{rand}()\left(p_{gd}^k - x_{id}^k\right) \end{aligned} \tag{4}$$

$$x_{id}^{k+1} = x_{id}^k + v_{id}^{k+1}\alpha = x_{id}^k + v_{id}^{k+1}\text{rand}()\theta \tag{5}$$

where w_i is in the range of $(w_{\min}, w_{\max})$; $k_{\max}$ is the maximum number of iterations set by the algorithm; rand() is a random number between [0, 1]; $\theta = [\text{rand}() + 0.5]$ (where [] is a rounding operation).

3.2   RBF Neural Network

RBF neural network is a kind of feed-forward back-propagation network constructed based on function approximation theory. It has strong ability to classify and approximate arbitrary nonlinear continuous functions. The structure of RBF neural network is shown in Fig. 2.

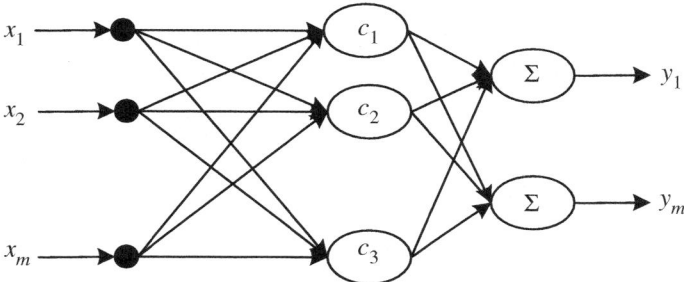

Fig. 2. RBF neural network structure

3.3 PSO-RBF Algorithm Flowchart

PSO-RBF algorithm flowchart is shown in Fig. 3.

4 Simulation

4.1 Simulated Sea Clutter Prediction

Assuming that in the case of a turbid background signal, the grazing echoes for the presence of sea clutter and noise are described by the following mathematical model:

$$\begin{cases} H_0 : x(n) = c(n), & 1 \leq n \leq N_T \\ H_1 : x(n) = s(n) + c(n), & 1 \leq n \leq N_T \end{cases} \tag{6}$$

where $x(n)$ is the observation time series, $s(n)$ is the target signal, $c(n)$ is the chaotic background signal, N_T is the total number of points of the observation sequence signal, H_0 represents the signal that does not contain the target, and H_1 represents the signal that has the target.

This paper uses Lorentz turbidity time series to simulate sea clutter. System parameters $\sigma = 16, b = 4, r = 45.92$. The initial value $x = -1, y = 0, z = 1$. And using the 4th-order Runge–Kutta method with an integration step of 0.01 to solve the Lorenz equation.

Set a single periodic target signal $s(n)$ as:

$$s(n) = 0.01 \sin(2\tau f), \quad f = 0.3 \tag{7}$$

Figure 4 shows the spectrum of a single periodic signal. SNR $= -75.98$ dB. Phase-space reconstruction of the mixed time series, using PSO-RBF neural network to predict the mixed time series with rectangular target signal. Forecast results are as shown in Fig. 5, and forecast error is as shown in Fig. 6.

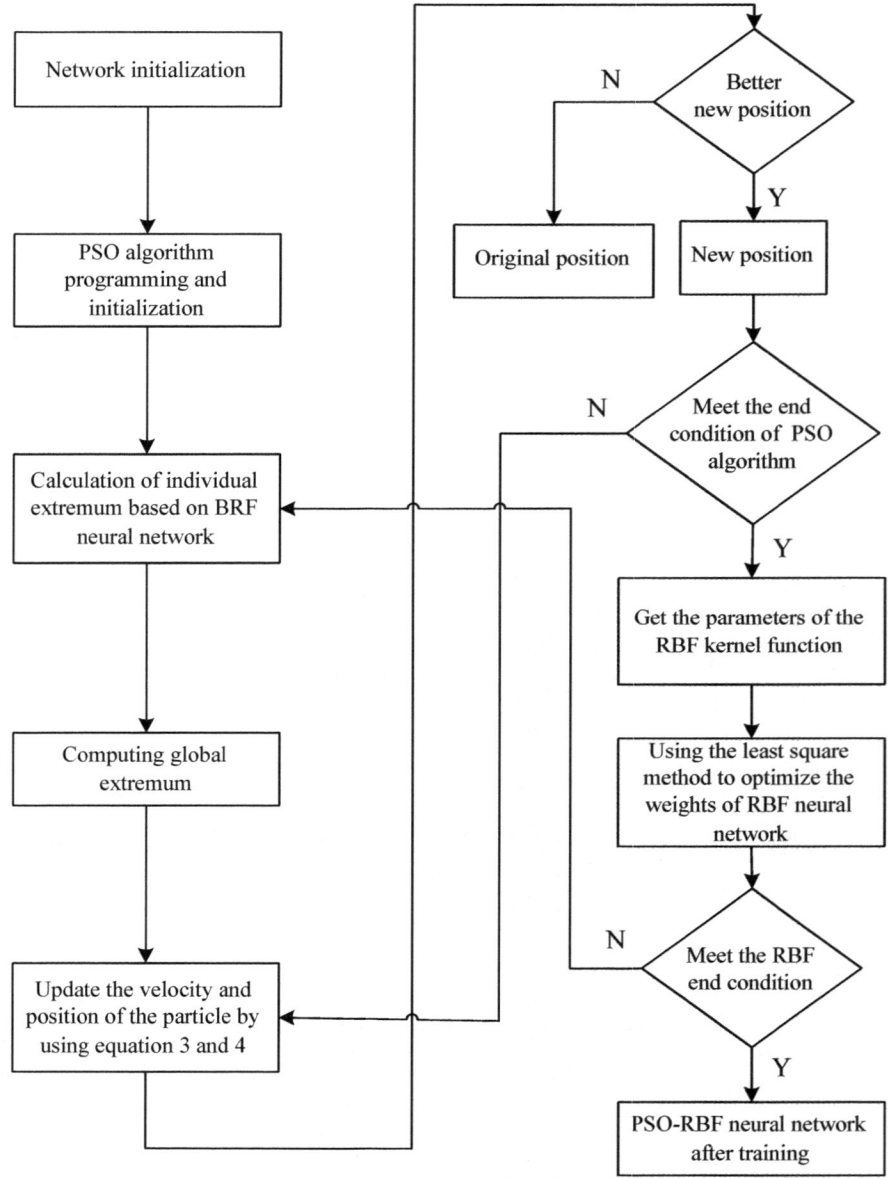

Fig. 3. PSO-RBF train algorithm flowchart

From Fig. 6, it can be seen that the frequency spectrum peaks at $f = 0.3$, thus enabling the detection of periodic signals.

From Table 1, it can be seen that the improved PSO-RBF can achieve faster convergence and better accuracy.

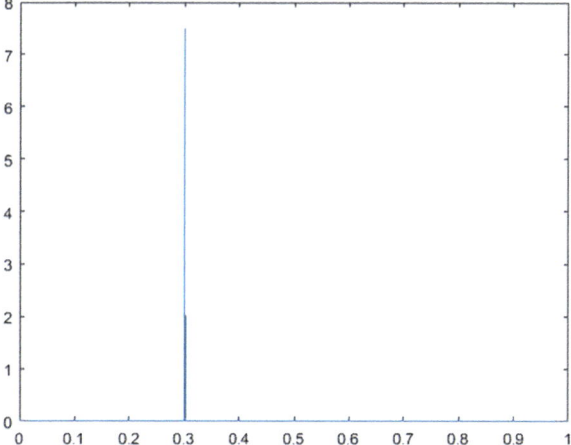

Fig. 4. Spectrum of a single periodic signal

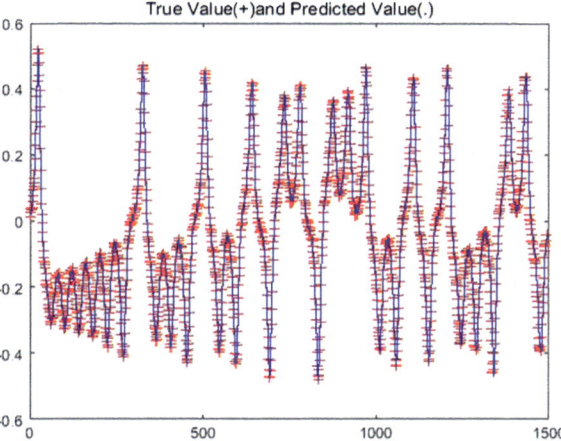

Fig. 5. Prediction value and real value of chaotic time series with a single periodic signal

4.2   IPIX Radar-Measured Sea Clutter Prediction

Select the #18 sea clutter data of the IPIX radar data set and select the sampling points of the 10,000 distance sea clutter data as the test sample. Using the RBF neural network trained to achieve prediction accuracy, the results are shown in Fig. 7.

It can be seen from Fig. 7 that there is a target at sampling point 1800-2200 and sampling point 8700-8800, which is consistent with the actual results (Table 2).

The experimental simulation shows that for the measured sea clutter data without weak targets, it can use RBF neural network and PSO-RBF neural network to perform single-step prediction because of its chaotic characteristics. For the measured sea clutter

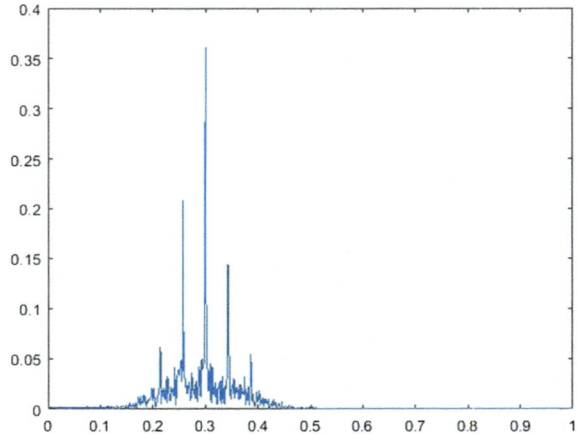

Fig. 6. Spectrum of prediction error of a single periodic signal

Table 1. Performance comparison of three algorithms

Algorithm name	RBF	Classic PSO-RBF	Improved PSO-RBF
Train time (s)	883.6	678.2	258.5
RMSE	0.032	0.183	0.000323

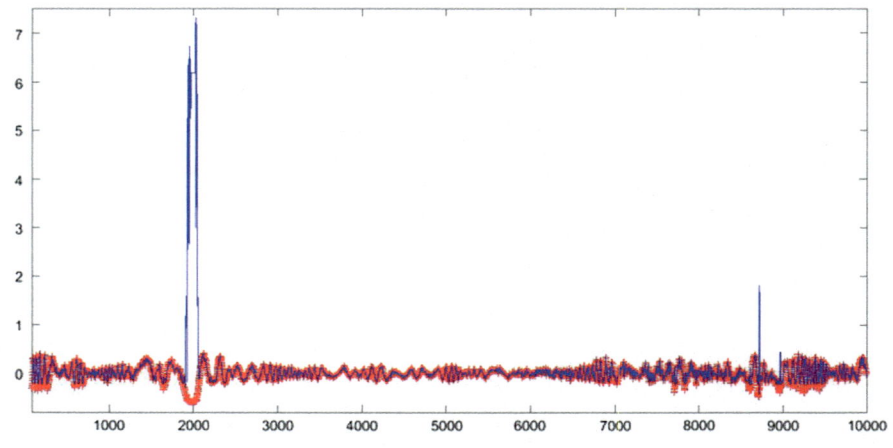

Fig. 7. IPIX radar-measured true value and predicted value sea clutter data

Table 2. Mean square error comparison of three algorithms

Algorithm name	RBF	Classic PSO-RBF	Improved PSO-RBF
RMSE	0.047	0.0143	0.00234

data with weak targets, the presence of the weak target destroys its inherent chaotic characteristics, which causes the prediction error to be larger than that of the pure sea clutter data. Therefore, we can effectively detect the weak target of the sea by comparing the turbidity characteristics of sea clutter by comparing the size of the prediction error. According to the comparison of prediction errors, it can be concluded that the improved PSO-RBF neural network detection is better.

5   Conclusion

In this paper, based on the analysis of particle swarm optimization, an improved particle swarm optimization algorithm is proposed for the PSO algorithm to fall into local minimum defects. The improved algorithm is applied to the hidden layer node parameters of the RBF neural network, namely the data center and the width. In the selection through simulation and IPIX radar data tests, it is verified that the improved PSO-RBF neural network has a better effect on weak target detection under sea clutter background.

References

1. Roy LP, Kumar RVR. Accurate K-distributed clutter model for scanning radar application. IET Radar Sonar Navigation. 2010;4(2):158–67.
2. Hennsssey G, Leung H, Yip PC. Sea-clutter modeling using a radar-basis-function neural network. IEEE J Oceanic Eng. 2001;28(3):358–72.
3. Leung H, Dubash N, Xie N. Detection of small objects in clutter using a GA-RBF neural network. IEEE Trans Aerosp Electron Syst. 2002;38(1):98–118.
4. Hongyan JT. Weak signal detection based on dual constraint least squares support vector machine in Chaotic Sea clutter background. Acta Physica Sinica. 2010.
5. Si W, Tong N, Wang Q. Small target detection algorithm research in sea clutter background. Sig Process. 2014.
6. Wang F. Sea clutter fractal characteristic analysis, modeling and small target detection. Doctoral Dissertation. Harbin Engineering University; 2009.
7. Si W, Tong N, Wang Q. Small target detection algorithm in the sea clutter background. Sea clutter processing and target detection based on nonlinear analysis. Doctoral Dissertation. Dalian Maritime University; 2008.
8. Li D. Characteristic detection method for small and medium targets in sea clutter. Master Thesis. Xidian University; 2016.
9. Xing H, Xu Y. Neural network based on the detection of weak signals in chaotic background; 2007.

An Identity Identification Method
for Multi-biometrics Fusion

Yingli Wang, Yan Liu, Hongbin Ma$^{(\boxtimes)}$, Xin Luo, and Danyang Qin

Electronic Engineering College, Heilongjiang University,
XueFu Road. 74, Harbin 150080, China
`mahongbin@hlju.edu.cn`

Abstract. In order to improve the reliability and security of biometric-based identity authentication system and reduce the risk of unauthorized access caused by forgery feature attacks, this paper proposes a method for identifying the identity of visitors. The method is based on D-S evidence theory. The palm print and palm vein are used as authentication features. Firstly, the same collection device is used to collect palm print and palm vein images under different wavelengths of light source and extract the HOG features of the image; then, use the one-vs-one multi-classification method of SVM to classify different individuals, and finally, using the D-S fusion strategy at the decision-making level to improve the security and accuracy of the identity authentication system. Through many experiments, the recognition rate of decision-making layer fusion is above 98%, which confirms the effectiveness of the proposed method.

Keywords: Multiple biometrics · D-S fusion · Identification

1 Introduction

In the existing data protection system, identity authentication is the first line of defense. At present, the mainstream identity authentication method is biometric authentication technology, including fingerprint, palm print, iris, vein, and so on. In recent years, bionic technology has developed rapidly. The emergence of items such as fingerprint caps makes identity identification schemes based on a single biometric no longer secure. However, it is difficult for criminals to obtain raw data of multiple biometrics at the same time, which provides new ideas for improving the reliability and security of existing biometric-based identification systems. In the existing feature fusion scheme, fusion is mainly performed at the following four levels: original data layer fusion, feature layer fusion, matching layer fusion, and decision layer fusion. It is very important to choose reasonable biometric features and fusion strategies. In a complex interactive environment, it is worthwhile to improve the ability of biometric-based identity authentication

© Springer Nature Singapore Pte Ltd. 2020
Q. Liang et al. (eds.), *Communications, Signal Processing,
and Systems*, Lecture Notes in Electrical Engineering 516,
https://doi.org/10.1007/978-981-13-6504-1_125

systems to defend against counterfeit feature attacks while ensuring recognition rates. Combined the above issue, this paper proposes a method for identifying the identity of visitors. The method is based on D-S evidence theory, and the palm print and palm vein are used as authentication features. Firstly, the same collection device is used to collect palm print and palm vein images under different wavelengths of light source and extract the HOG features of the image; then, use the one-vs-one multi-classification method of SVM to classify different individuals, and finally, using the D-S fusion strategy at the decision-making level to improve the security and accuracy of the identity authentication system.

2   Related Knowledge

Multi-feature fusion is the application of information fusion technology in the field of pattern recognition. The fusion of the decision-making layer belongs to the top-level fusion method. According to the result of single feature recognition, the fusion decision is made according to a certain distribution principle, and finally, the identity authentication will be output [1].

In the process of multi-feature fusion, there are some uncertain factors. The D-S evidence theory has good fuzzy inference performance for the uncertainty evidence. Xu and Yu [2] pointed out that the D-S method provides strong theoretical basis for the expression and synthesis of uncertain information. Chang et al. [3] used the D-S evidence theory to fuse the features of the image such as color and texture. Experiments have shown that it has a good recognition effect. In the field of identity authentication, researches of using D-S evidence theory for multi-feature fusion are rare.

In order to ensure the identification efficiency of the identity authentication system in the actual complex environment, this paper applies the D-S evidence theory fusion algorithm to the decision-making layer. Firstly, the HOG algorithm be used to extract the eigenvalues of a single biological image on the training data set. Then, the recognition subsystem be used to train. Next, the parameter correction set be used to test the recognition accuracy rate of the feature, which will be considered as the BPA of D-S evidence theory, and the different BPA is used for decision-level fusion recognition. Finally, the recognition result and rate of the system after feature fusion will be obtained.

The decision output $f(x)$ of the standard SVM is a hard output, which means that only x is or does not belong to a certain class, and does not provide the probability of belonging to this class. The Dempster synthesis rule first needs to determine the basic probability setting of each piece of evidence. The recognition accuracy rate of each biometric in this paper is a piece of evidences BPA. Therefore, the SVM posterior probability proposed by Platt will be used as the BPA for each evidence.

After learning the training data set of a single feature, in order to ensure the objectivity of the BPA assignment, a parameter-corrected data set will be used to obtain more reliable position parameters, scale parameters, and recognition accuracy α of the feature-corresponding classifier. The C_i refers to the user's category number, and i is the number of categories. The definition of the palm print feature recognition result is the first proof body, the palm vein feature recognition result is the second proof body, and the BPA function corresponding to the jth proof body can be described in the following form:

$$m_j(C_1, C_2, \ldots, C_i, \ldots, C_N) = \begin{pmatrix} p_{11}a_1, p_{12}a_1, \ldots, p_{1N}a_1 \\ \cdots \\ p_{j1}a_j, p_{j2}a_j, \ldots, p_{jN}a_j \end{pmatrix}, j = \{1, 2, \ldots\} \quad (1)$$

The degree of uncertainty of the jth evidence body is described in the following form:

$$m_j(\Theta) = 1 - \alpha_j \quad (2)$$

According to the Dempster synthesis rules and calculating the belief function and the plausibility function of the two evidence bodies, the reliability of the evidence to identify all the propositions of the framework and the uncertainty of the evidence are obtained.

3 Experiments and Analysis

3.1 The Source of Image Database

In order to verify the effectiveness of the proposed method, it is necessary to do the simulation experiment. This paper uses the database of palm print and palm vein images provided by the Hong Kong Polytechnic University as experimental data.

In this paper, 40 people were randomly selected from the above image library as the internal staff of a small company. The palm print image and the palm vein image were combined one by one. One person has eight palm print images and eight palm vein images. Four palms were randomly selected from one feature. The four of palm print images and the four of palm vein images are used as training data set to train the model proposed in this paper. The next two of them are selected as parameter correction data set to correct the probability of the single feature SVM classifier output and obtain the classification accuracy rate. The last two of them are used as test data set to obtain the accuracy of the fusion method proposed in this paper. The evaluation index of the proposed method is correct recognition rate (CRR).

3.2 Results and Analysis

In order to verify the recognition rate of individual features, experiments were performed on test data sets used to test palm print and palm vein. The construction of BPA function of palm print and palm vein evidence body depends on the classification accuracy rate. According to the experiment, the recognition rate of palm print is 85%, and the palm vein recognition rate is 90%. They are

Table 1. Comparison of the D-S fusion performance

The name of experiment	The numbers of test images	The right numbers of the results	CRR in (%)	The max time of one sample in(s)
Palm print	80	68	85	1.091947
Palm vein	80	72	90	1.533456
D-S fusion	80	79	98.75	5.365985

brought into the D-S fusion rule. The method proposed in this paper was tested using the test data set. Table 1 shows the recognition results of the D-S fusion method based on the palm print and palm vein proposed in this paper.

By analyzing the above experiment, we can draw the following conclusions:

(1) The traditional HOG algorithm can extract palm print features with a recognition rate of 85%, but it can be easily forged. Therefore, using only the palm print feature does not guarantee that the system is safe.
(2) Since we are not sure whether it is a fake palm print, we can combine the palm print and vein features to improve the safety of the system. The experimental results show that the proposed D-S fusion strategy has good performance for joint authentication of two features. A higher recognition rate demonstrates the effectiveness of the method.

4   Conclusion

In this paper, the D-S evidence theory is applied to the authentication method based on palm print and palm vein. The experimental results show that the method is simple and effective. It can not only use the authentication based on the palm vein to resist the forgery feature attack, but also realize the high recognition rate authentication system and provide an effective scheme with certain engineering significance. However, there are some deficiencies in this paper. For example, the number of test samples is not enough, and the performance of the method under the large-scale sample set is not verified. The recognition performance of palm vein needs to be further improved. The next step is to extract the texture features of the palm vein using other feature extraction methods (such as LBP) to improve the recognition rate of the palm vein.

References

1. Zhang L, Liang T. Adaptive multi-modality biometric fusion based on classification distance scores. J Comput Res Develop. 2018;1:151–62.
2. Xu C, Yu W. A review of theory and application of dempster-shafer's evidence reasoning method. Pattern Recogn Artif Intell. 1999;12(4):424–30.
3. Chang C, Xiaoyang Y, Guang Y. The research of image retrieval based on multi feature ds evidence theory fusion. Int J Sig Process Image Process Pattern Recogn. 2016;9(1):51–62.

Iced Line Position Detection Based on Least Squares

Yanwei Wang[1(✉)] and Jiaqi Zhen[2]

[1] College of Mechanical Engineering, Harbin Institute of Petroleum,
Harbin 150080, China
xianxinyue@163.com
[2] College of Electronic Engineering, Heilongjiang University,
Harbin 150080, China

Abstract. Ice and snow have a detrimental effect on the transmission and distribution lines. In the calculation of the ice thickness of the ice-covered transmission lines by the image method, the center position of the target transmission lines is crucial for the detection result. The detection of ice-covered transmission lines is affected by external factors, resulting in images that are mostly inclined ice-covered images. Therefore, it is necessary first to straighten the wire and then by fitting the image rotation technique.

Keywords: Ice-covering transmission lines · Least squares · Position detection

1 Introduction

In the ice detection system, the thickness of the ice coating is required to be detected due to different environmental conditions [1, 2]. In calculating the ice thickness of the target transmission line, we use the pixel difference in the vertical direction of the edge of the transmission line instead of the thickness of the ice coating. However, due to problems such as the angle of shooting, the power lines obtained after dividing the image are often inclined. So before calculating the thickness of the ice coating, we need to straighten the power line by image rotation technology [3, 4]. First, measure the tilt angle of the transmission line. Here, use the straight line fitting technique to obtain the tilt angle of the wire ice coating image, and then use the image rotation technique to straighten the inclined wire.

This work was supported by the National Natural Science Foundation of China under Grant No. 61501176, Heilongjiang Province Natural Science Foundation (F2018025), University Nursing Program for Young Scholars with Creative Talents in Heilongjiang Province (UNPYSCT-2016017), The postdoctoral scientific research developmental fund of Heilongjiang Province in 2017 (LBH-Q17149).

Q. Liang et al. (eds.), *Communications, Signal Processing, and Systems*, Lecture Notes in Electrical Engineering 516,
https://doi.org/10.1007/978-981-13-6504-1_126

2 Fitting Method of Ice Coating Line

If the number of sample points is a set of data, through a set of experimental data, according to the principle of the smallest sum of squared residuals, look for an analytic function, which is the least squares method, for a given set of data $(x_i y_i)(i = 0, 1, \ldots, m)$. It is required to find a function $y = s * (x)$ in function class $\Phi = \{\Phi_0, \Phi_1, \ldots, \Phi_n\}$, so that the sum of squared errors

$$\|\delta\|^2 = \sum_{i=0}^{m} \delta_i^2 = \sum_{i=0}^{m} [s * (x_i) - y_i]^2 \tag{2.1}$$

tends to be the smallest, ie:

$$\|\delta\|^2 = \min_{s(x) \in \Phi_i} \sum_{i=0}^{m} [s * (x_i) - y_i]^2 \tag{2.2}$$

polynomial in (2.1):

$$s(x) = a_0 * \Phi_0(x) + a_1 * \Phi_1(x) + \cdots + a_n * \Phi_n(x) \, (n < m) \tag{2.3}$$

is the least squares method of curve fitting.

The coefficients $a_1, a_2, \ldots, a_n$ in the Eq. (2.2) can be obtained by solving the problem of the minimum value of (2.1). The final result $s(x)$ found is generally a high degree polynomial.

3 Detection of Ice-Covered Wires by Least Squares Straight Line Fitting Method

Combine the approximation parameter method to give an angle arbitrarily (according to our experimental experience, recommended as 70) give a maximum error according to the accuracy requirement, and then take the first two points according to the measured experimental data. Use the least squares method (for the first case, it is actually two points to determine a straight line) to fit its model (straight line), calculate the coordinates of its starting point and end point, and the connection between the next point and the starting point. The angle α of the fitted line: Assume that the starting point of the fitted line is $S(X_S, Y_S)$, the ending point is $E(X_E, Y_E)$, and the next point is $N(X_N, Y_N)$, then Lack, the distance between points S_E is

$$|S_E| = \sqrt{(Y_S - Y_E)^2 + (X_S - X_E)^2} \tag{2.4}$$

the distance between points S and N is

$$|S_N| = \sqrt{(Y_S - Y_N)^2 + (X_S - X_N)^2} \tag{2.5}$$

the distance between points E and N is

$$|E_N| = \sqrt{(Y_E - Y_N)^2 + (X_E - X_N)^2} \tag{2.6}$$

the angle obtained by the cosine theorem is

$$\alpha = \cos^{-1} \frac{|S_E|^2 + |S_N - E_N|^2}{2|S_E S_N|} \tag{6.7}$$

It is judged whether the angle α is smaller than a given angle. If α is less than a given value, then this point N is added to the previous point column to redo the least squares line fit until α is greater than the given value, and the resulting model is the model of the first straight line. Repeat the above-mentioned fitting and judgment with the last point on the straight line and the next point until all the points are fitted and obtain the models of all the segmented lines. Then use these models to calculate the intersection of adjacent lines. Get their intersections, which are the end points (start or end) of the fitted straight line segment. Once the model has been obtained, the error analysis of the model can be performed, that is, whether the obtained model satisfies the error requirement (Figs. 1, 2, and 3; Table 1).

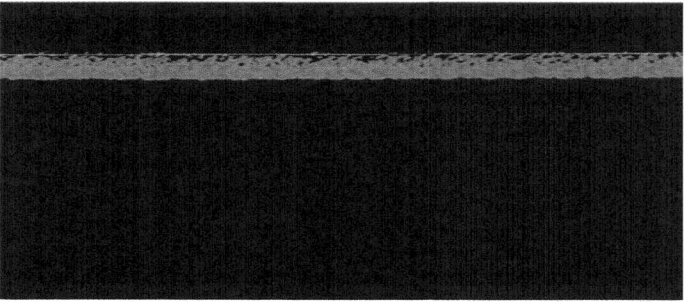

Fig. 1 Linear fitting of the first conductor

The experimental results show that the classification accuracy of color-based image feature extraction is between 80 and 90, and even 95% accuracy can be achieved. The more the number of samples, the higher the accuracy.

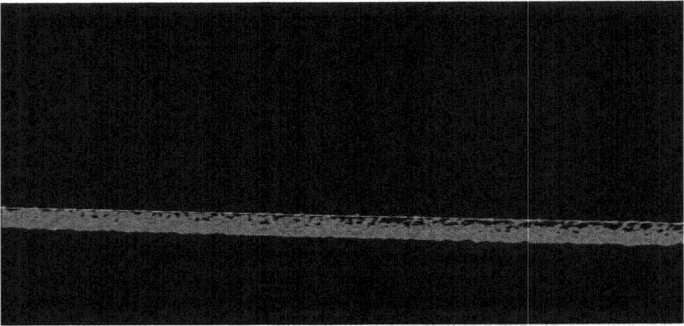

Fig. 2 Linear fitting of the second conductor

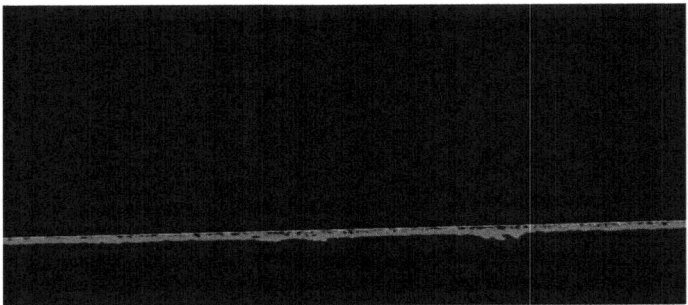

Fig. 3 Linear fitting of the third conductor

Table 1 Analysis

Wire	Slope
Ice coated wire 1	−0.5512
Ice coated wire 2	4.8472
Ice coated wire 3	−3.9271

4   Conclusions

This paper proposes a method for realizing the position detection of ice-covered wires by means of least squares fitting. The method has the advantages of simple and rapid adjustment of adjusting the position of the ice-covered wire, measuring the inclination angle of the wire by the image method, and adjusting the inclination angle. The test result shows that the method can effectively collect the vertical direction of the wire edge in the ice-covered image. The pixel difference provides a reliable basis for calculating the thickness of the ice.

References

1. Yano K. Studies of Icing and Ice-snow Accretion on Abies Mariesii. Bull Yamagata Univ Nat Sci. 1986;11:227–47.
2. Dai L, Huang H, Chen Z. Ternary sequential analytic optimization algorithm for SVM classifier design. Asian J Inf Technol. 2005;3:2–8.
3. Liao FM. Analysis of image processing and recognition technology of transmission line wire icing. South Agric Mach. 2017;48:94–5.
4. Gao J, Shen L, Xu YS, et al. Study on the detection of thickness and shortage diameter of ice coating based on sequence alignment method. Yunnan Electr Power. 2017;45:17–9.

Sequentially Distributed Detection and Data Fusion with Two Sensors

LI Cheng[(⊠)]

College of Underwater Acoustic Engineering, Harbin Engineering University,
Harbin, China
lichengong1@163.com

Abstract. The relationship of decision rule of sensor for each other is relevant to data fusion, so different topological networks of sensors usually results in different performances. This paper considers the sequential network fusion with two sensors in some detail and compares its performance with that of single detection and fusion. In this paper, the detection model is specified for binary hypotheses testing problem. In particular, this paper supposes that Bayesian risk cost of different decisions and the prior probability distribution of two hypotheses are known. Finally, this paper simulates the probabilities of error and Bayesian risk by these fusion rules with corresponding to different values of prior probabilities of two hypotheses by these fusion methods. And compared to single detection and fusion, the performance of sequential detection and fusion is better.

Keywords: Distributed detection · Single detection and fusion · Optimal fusion · Sequential network fusion

1 Introduction

Multi-sensor distributed detection and data fusion is an emerging technology that is applied in a wide range of areas such as automated target recognition, field surveillance, and so on [1, 2]. Contrast to the multi-sensor centralized detection which needs to collect the whole data received from all sensors to make decision, multi-sensor distributed detection refers to that each sensor could make temporary decision to be referred to the fusion center to make final decision, so multi-sensor data fusion refers to synergistic combination of information inferred by sensors for a better decision usually corresponding to a less probabilities of error [3].

2 Single Detection Fusion and Sequential Fusion

The relationship of decision rule of sensor for each other is relevant to data fusion, so different topological networks of sensors usually results in different performance. This paper considers sequential topological data fusion for two sensors on binary hypothesis testing problem in some detail. And then, this paper discusses its difference from the fusion of single detection and sequential fusion.

© Springer Nature Singapore Pte Ltd. 2020
Q. Liang et al. (eds.), *Communications, Signal Processing, and Systems*, Lecture Notes in Electrical Engineering 516,
https://doi.org/10.1007/978-981-13-6504-1_127

2.1   Distributed Detection with Two Sensors

In a binary hypothesis testing problem, each hypothesis represents either the absence of a target or the presence of a target, and it can also represent symbol zero or one. Usually, the two hypotheses are respectively denoted by H_0 or H_1. For Bayesian detection method, the prior probabilities of the two hypotheses are known and denoted by P_0 or P_1. And what will be declared yields to a certain conditional probability under each hypothesis. This paper denotes the average cost or Bayesian risk function by R.

$$R = \sum_{i=0}^{1} \sum_{j=0}^{1} C_{ji} P_i P\left(\text{declaring } H_j | H_i \text{ is present}\right) \tag{1}$$

When it comes to two sensors, the conditional probability is related to the two declarations u_1 and u_2, respectively, from two sensors. u_1 is 1 if sensor 1 declares H_1 or else is 0, likewise for u_2. So, the conditional probability above can be expressed by

$$P\left(H_j | H_i\right) = \sum_{u_2=0}^{1} \sum_{u_1=0}^{1} P\left(\text{declaring } H_j | u_1, u_2\right) P\left(u_1, u_2 | H_i \text{ is present}\right) \tag{2}$$

And the conditional probability $P(u_2, u_1 | H_i \text{ is present})$ is determined by observations of sensors, y_2 and y_1. For analog observations, y is stochastically sampled in continuous space and yields to a certain conditional probability distribution. And hence,

$$P(u_2, u_1 | H_i \text{ is present}) = \int_{Z_1(u_1)} \int_{Z_2(u_2)} p(u_2, u_1, y_2, y_1 | H_1) dy_2 dy_1 \tag{3}$$

where $Z_1(u_1)$ expresses the region corresponding to declaration u_1 for observation y_1, likewise, it is the same with $Z_2(u_2)$.

In above equation, the factor $P(u_2, u_1 | H_i \text{ is present})$ usually means distributed detection process and factor $P\left(\text{declaring } H_j | u_2, u_1\right)$ does data fusion process. However, not in all distributed detection and data fusion architecture, the two processes can be clearly separated. As to sequential network, the two processes of distributed detection and data fusion are combined.

2.2   Single Detection and Fusion with Two Sensors

For single detection, performances of sensors are conditionally independent,

$$P(u_1, u_2 | H_i) = P(u_1 | H_i) P(u_2 | H_i) \tag{4}$$

And in the case of two sensors' detection, there are 2^{2^2} fusion rules, however, only six of them are monotonic fusion rules [4]. The two of the six are simply all-one and all-zero, and their probabilities of error are the prior probabilities. And another two are the decision of each sensor selves, totally disregarding the other. So, the probabilities of error of fusion are that of corresponding sensor. And the other two fusion rules are AND rule and OR rule, which comply with logical functions as follow,

$$\text{AND rule}: u_0 = \begin{cases} 1, & \text{if } u_1 = 1 \text{ and } u_2 = 1 \\ 0, & \text{otherwise} \end{cases} \tag{5}$$

$$\text{OR rule}: u_0 = \begin{cases} 1, & \text{if } u_1 = 1 \text{ or } u_2 = 1 \\ 0, & \text{otherwise} \end{cases} \tag{6}$$

According to logical function and conditional independence, by any OR rule or AND rule, we can obtain the probability of detection, false alarm and missing which are denoted by P_D, P_F and P_M, respectively.

$$\text{AND rule: } \begin{aligned} P_D &= P_{D1}P_{D2} \\ P_F &= P_{F1}P_{F2} \\ P_M &= 1 - P_D \end{aligned} \tag{7}$$

$$\text{OR rule: } \begin{aligned} P_M &= (1 - P_{D1})(1 - P_{D2}) \\ P_F &= 1 - (1 - P_{F1})(1 - P_{F2}) \\ P_D &= 1 - P_M \end{aligned} \tag{8}$$

And, the probability of error is given by

$$P(\text{error}) = P(H_0)P_F + P(H_1)P_M \overset{\text{def}}{=} P_E \tag{9}$$

2.3   Sequential Network Detection and Fusion

The architecture of sequential network as followed picture, two sensors observe the same source, and their observations are conditionally independent. The first sensor will transmit its decision to the second. And the second sensor will integrate the decision of the first sensor and its observation. The combination mode of decision is serial, that is to say, the second has two different thresholds corresponding to the decision of sensor 1. And the decision of second is not only the second decision but also the fusion result (Fig. 1).

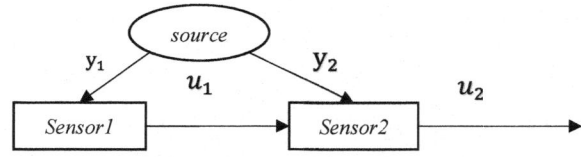

Fig. 1. Sequentially distributed detection and data fusion structure

The Bayesian risk function for this architecture is given by

$$R = \sum_{i,j,k} \iint_{y_1,y_2} C_{jk} P_k p(u_2|u_1,y_1,y_2,H_k) p(u_1,y_1,y_2|H_k) dy_1 dy_2 \qquad (10)$$

where $i,j,k = 0,1$, and i is the decision u_1 and j is decision u_2. And the decision u_2 does not depend on y_1, therefore,

$$R = \sum_{i,j,k} \iint_{y_1,y_2} C_{jk} P_k p(u_2|u_1,y_2) p(u_1,y_1|H_k) p(y_2|H_k) dy_1 dy_2 \qquad (11)$$

Explicitly summing over u_2 and considering $p(u_2 = 1|u_1,y_2) + p(u_2 = 0|u_1,y_2) = 1$.

$$R = \sum_{i,k} \iint_{y_1,y_2} C_{jk} P_k p(u_1,y_1|H_k) p(y_2|H_k) dy_1 dy_2$$
$$+ \sum_i \int_{y_2} p(u_2 = 0|u_1,y_2) \sum_k \int_{y_1} (C_{0k} - C_{1k}) \qquad (12)$$
$$\times P_k p(u_1,y_1|H_k) p(y_2|H_k) dy_1 dy_2$$

In order to minimize the Bayesian risk, the decision $u_2 = 1$ is under the condition that $\sum_k (C_{0k} - C_{1k}) P_k p(u_1|H_k) p(y_2|H_k) > 0$, then the decision rule of sensor 2 is,

$$\frac{p(y_2|H_1)}{p(y_2|H_0)} \begin{array}{c} u_2 = 1 \\ > \\ < \\ u_2 = 0 \end{array} \frac{P_0(C_{10} - C_{00}) p(u_1|H_0)}{P_1(C_{01} - C_{11}) p(u_1|H_1)} \qquad (13)$$

Therefore, u_1 is 0 or 1, representing different values of $\frac{p(u_1|H_0)}{p(u_1|H_1)}$. And, we denote the two thresholds by t_2^1 and t_2^0. They are expressed by

$$t_2^1 = \frac{P_0(C_{10} - C_{00}) p(u_1 = 1|H_0)}{P_1(C_{01} - C_{11}) p(u_1 = 1|H_1)}, \qquad t_2^0 = \frac{P_0(C_{10} - C_{00}) p(u_1 = 0|H_0)}{P_1(C_{01} - C_{11}) p(u_1 = 0|H_1)} \qquad (14)$$

Next, we come back to risk function Eq. (11), and rewrite it as follows

$$R = \sum_{i,j,k} \iint_{y_1,y_2} C_{jk} P_k p(u_2|u_1,y_2) p(u_1,y_1|H_k) p(y_2|H_k) dy_1 dy_2 \qquad (15)$$

Expanding over u_1 and also considering $p(u_1 = 1|y_1) + p(u_1 = 0|y_1) = 1$

$$R = \sum_{j,k} \iint_{y_1,y_2} C_{jk}P_k p(u_2|u_1 = 1, y_2)p(y_1|H_k)p(y_2|H_k)dy_1dy_2 + \int_{y_1} p(u_1 = 0|y_1)$$

$$\sum_{j,k} \int_{y_2} C_{jk}P_k p(y_1|H_k)p(y_2|H_k)[p(u_2|u_1 = 0, y_2) - p(u_2|u_1 = 1, y_2)]dy_1dy_2$$

$$(16)$$

In order to minimize the Bayesian risk, the decision $u_1 = 1$ is under the condition that $\sum_{j,k} C_{jk}P_k p(y_1|H_k)[p(u_2|u_1 = 0, H_k) - p(u_2|u_1 = 1, H_k)] > 0$ [5]. Expanding over j and k, then considering $C_{10} > C_{00}$ and $C_{01} - C_{11}$, we can obtain the decision rule of sensor 1 with considering t_2^1 and t_2^0,

$$\frac{p(y_1|H_1)}{p(y_1|H_0)} \mathop{\gtrless}_{u_1 = 0}^{u_1 = 1} \frac{P_0(C_{10} - C_{00})\left[P_{F2}\left(t_2^1\right) - P_{F2}\left(t_2^0\right)\right]}{P_1(C_{01} - C_{11})\left[P_{D2}\left(t_2^1\right) - P_{D2}\left(t_2^0\right)\right]} \overset{\text{def}}{=} t_1 \qquad (17)$$

And, the probability of false alarm and missing are, respectively, following

$$P_F = P_{F1}P_{F2}\left(t_2^1\right) + (1 - P_{F1})P_{F2}\left(t_2^0\right) \qquad (18)$$

$$P_M = P_{D1} \cdot \left(1 - P_{D2}\left(t_2^1\right)\right) + (1 - P_{D1}) \cdot \left(1 - P_{D2}\left(t_2^0\right)\right) \qquad (19)$$

The probability of error is the same as Eq. (9).

3 Simulation Result

3.1 Description of Two Hypotheses

Suppose the noise and target observations yield to unit Gaussian distribution for both two sensors. The distribution of observation and Bayesian cost factors is shown in Tables 1 and 2. In the simulation, the average cost is expressed by minus number, so the lower is the average cost, the bigger is Bayesian risk.

Model

Table 1. Likelihood function at two sensors

Sensor 1	$p(y_1|H_0) \sim N(0,1)$ $p(y_1|H_1) \sim N(2,9)$ '	Sensor 2	$p(y_2|H_0) \sim N(0,1)$ $p(y_2|H_1) \sim N(1,1)$

Table 2. Bayesian cost factors

Cost factors	Declaring H_0	Declaring H_1
Presenting H_0	0	-1
Presenting H_1	-1	0

3.2   Single Sensor Detection and Fusion

First, according to Bayesian rules for single sensor detection, we check the probability of error and Bayesian risk of both two sensors. And the performances of each sensor are shown in Fig. 2. We find that the performance of sensor 1 is better than that of sensor 2, because the difference of observation means under both hypotheses is larger at sensor 1 than that at sensor 2. And, the maximal probability of error of sensor 1 is 0.24, while that of sensor 2 is 0.31. And then, we check the performances by the fusion of AND rule and OR rule as shown in Fig. 3, the maximal probability of error by the fusion of OR rule is 0.25, lower than that by the fusion of AND rule, 0.29, and a bit higher than that of single detection at sensor 1, 0.24. That is because, in this case, the probability of missing is reduced by fusion of OR rule and the probability of false alarm, however, is increased a bit.

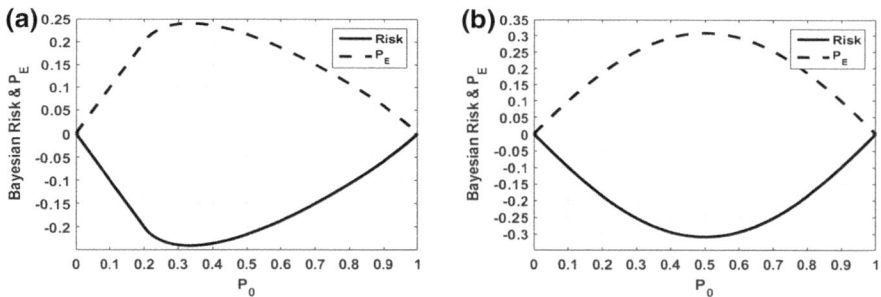

Fig. 2. a Performance of sensor 1 **b** Performance of sensor 2

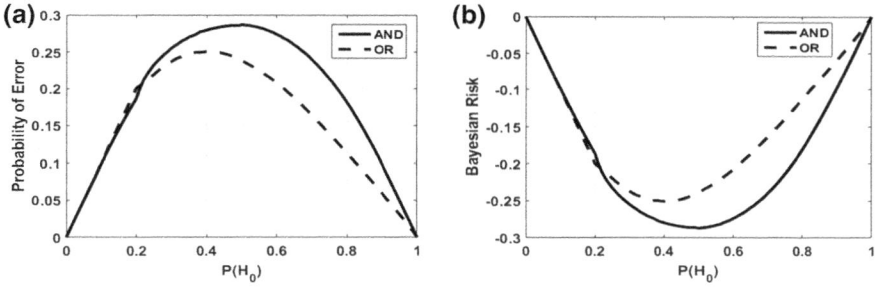

Fig. 3. a Probability of error by the fusion of AND rule and OR rule **b** Bayesian risk by the fusion of AND rule and OR rule

3.3   Sequential Fusion Result

For two sensors' detection, the sequential fusion has two states corresponding different sequences of sensors. And according to Eqs. (14) and (17), the decision of the latter sensor depends on the former. And we know by Fig. 2 that sensor 1 is better than sensor 2. So, the thresholds and performances in two states may be different. We first simulate the sequence in which sensor 2 is placed latter to fuse decisions, and then the other sequence. And the simulation result of sequential fusion network is showed in Fig. 4a, and the worst Bayesian Risk is -0.21, better than that of both OR rule fusion, -0.25, and single detection at sensor 1, -0.24. Because we adapt numerical methods to solve the thresholds. Figure 4b shows the logarithm of three thresholds of two sensors versus the prior probability of hypothesis H_0. When sensor 1 is placed latter to fuse decisions, the simulation result is showed in Fig. 5a. And its worst Bayesian risk is -0.21, similar to that of above sequence. But by comparing Bayesian risks of both sequences respect to P_0 as 0.7, 0.8, and 0.9, the Bayesian Risks of current sequence are better (Fig. 4).

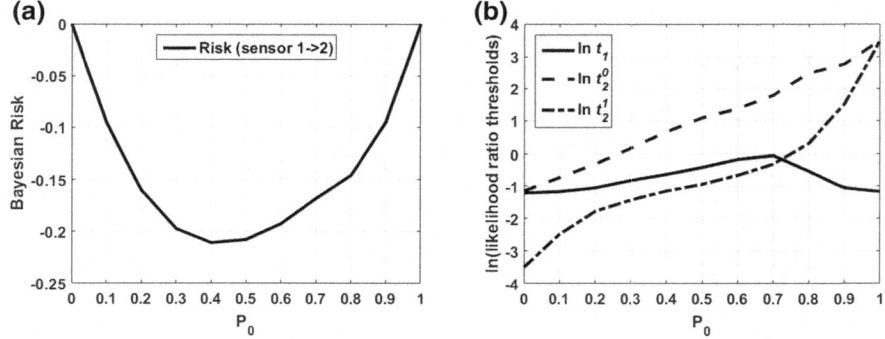

Fig. 4. a Performance of sequential network with sensor 2 as the latter to fuse decisions. **b** Logarithm of three thresholds of two sensors versus prior probability of hypothesis H_0

Figures 4b and 5b show the three logarithmic thresholds versus the prior probability of H_0. When we consider the thresholds rationally at the point that prior probability of H_0 is approaching zero, the logarithmic thresholds of sensor 2 should tend to be negative infinite so that the probability of detection is 1, and they should tend to be infinite when the prior probability of H_0 is approaching one. In fact, in Fig. 6a and b, the logarithmic thresholds lnt_2^1 and lnt_2^0 haven't converged yet, but the Bayesian risk and prob-ability of error has been convergent to zero along with iterations.

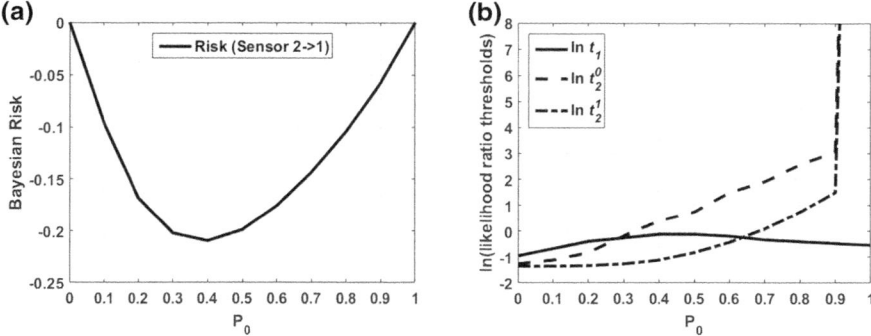

Fig. 5. a Performance of sequential network with sensor 1 as the latter to fuse decisions. **b** Logarithm of three thresholds of two sensors versus prior probability of hypothesis H_0

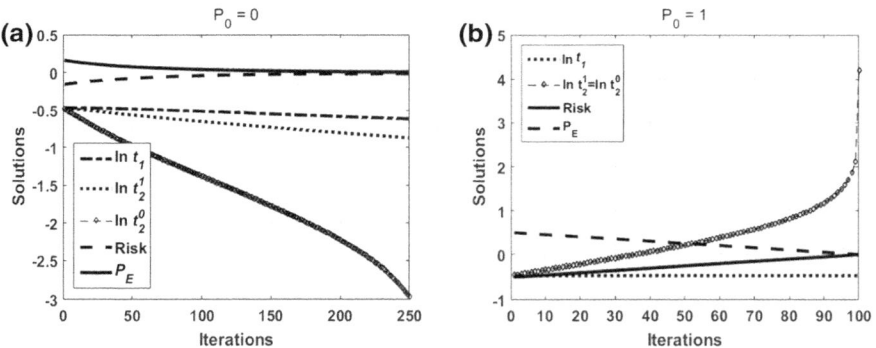

Fig. 6. a Iterative solutions of performance and thresholds when prior probability of H_0 is 0. **b** Iterative solutions of performance and thresholds when prior probability of H_0 is 1

4 Conclusion and Discussion

This paper discusses the fusion rule of two sensors' detection, and the observations at two sensors are conditionally independent. For single detection and fusion with two sensors, the two decisions are not conditionally independent, and there are six monotonic fusion rules. Two of the six fusion rules are all-one and all-zero which probability of error are probability of hypotheses selves, so maybe not referred in most occasions. This paper, particularly, compares the other four fusion rules with sequentially distributed detection data fusion. For sequential detection, the decisions of two sensors are not conditionally independent any more in sequential fusion, and they are coupled serially. And the thresholds of the latter sensor are dependent on the declaration of the former sensor, so the sequencing is relevant. The latter the better sensor is placed, the better is the fusion performance.

References

1. Chair Z, Varshney PK. Optimal data fusion in multiple sensor detection systems. IEEE Trans Aerosp Elect Syst. 1986;AES-22(1):98–101.
2. Llinas J, Hall DL. Introduction to multi-sensor data fusion. Proc. IEEE. 1998;(6). 537–540 vol.6. 10.1109/ISCAS.1998.705329.
3. Blum RS. Quantization in multisensor random signal detection. IEEE Trans On Info Theory. 1995;41(1):204–215.
4. Chair Z, Varshney PK. Distributed bayesian hypothesis testing with distributed data fusion. IEEE Trans Syst Man Cybern. 1988;SMC-18(5):695–9.
5. Varshney PK, Burrus CS. Distributed detection and data fusion. New York: Springer; 1997.

Part V
Localization and Navigation

Particle Filter with Correction of Initial State for Direction of Arrival Tracking

Huang Wang, Qiyun Xuan, Yulong Gao$^{(\boxtimes)}$, and Xu Bai

Harbin Institute of Technology, Harbin 150028, China
16S105138@stu.hit.edu.cn, hitxuanqiyun@126.com,
{ylgao,x_bai}@hit.edu.cn

Abstract. Generally, particle filter is used in the single snapshot situation and the initial state is assumed to be known. To make the measurement interval be small enough, we construct a multiple measurement vectors model for DOA tracking since it usually outperforms the single measurement vector model. And we take the initial state into consideration. The initial tracking error of the particle filter becomes very large when the initial state is unknown. Thus, we modify the initial state according to the likelihood of the generated random samples. The method is numerically evaluated using a uniform linear array in simulations. The results show that the proposed algorithm has higher tracking accuracy.

Keywords: DOA tracking · Particle filter · Multiple measurement vectors

1 Introduction

Decades of research have given rise to many algorithms that solve the direction of arrival (DOA) estimation problem which has important applications in fields like radar, sensor networks, and communications. The high-resolution DOA estimation is realized by algorithms like multiple signal classifications (MUSIC) [1] and estimation of signal parameters via rotational invariance techniques (ESPRIT) [2]. But these methods fail or suffer performance degradations when the DOA is variant during the observation period. However, in many scenarios, the sources are moving. Therefore, it is necessary to study the DOA tracking algorithm.

Some DOA tracking algorithms are based on subspace tracking [3, 4]. The subspace tracking algorithm, such as projection approximation subspace tracking (PAST) [5], updates the subspace by minimizing the objective function and avoids the heavy eigenvalue decomposition operation. In general, the performance of subspace-based algorithms degrades with signal correlation. Another DOA tracking algorithms are based on state filtering, including extended Kalman filter (EKF) [6], unscented Kalman filter (UKF) [7], and particle filter (PF) [8–11]. EKF transforms the nonlinear model into a linear model by using the Taylor series expansion and ignoring the higher

This work is supported by National Natural Science Foundation of China (NSFC) (Grant No. 61671176).

order terms. UKF uses unscented transform to approximate the nonlinear distribution by sampling. Although these two methods are used to deal with nonlinear systems, they can not be guaranteed to be convergent. Particle filter is a recursive algorithm which is widely used in nonlinear/non-Gaussian system. A filtering idea based on Bayesian sequential importance sampling is first put forward by Gordon, Salmond, and Smith, which establishes the theoretical foundation of particle filter algorithm. Then, a variety of improved algorithms has been developed. In [9], PF is introduced to DOA tracking problem. In [10, 11], DOA tracking algorithms based on PF are proposed to deal with two-dimensional DOA and impulsive noise, respectively.

In this paper, we implement DOA tracking in the framework of sampling importance resampling (SIR) filter and improve the algorithm. First, we construct a dynamic state-space model and use multiple measurement vectors in the observation equation. We make the measurement interval very small so that the source is regarded to be stationary. Through the use of multiple snapshot data, the estimation accuracy is improved. Then, DOA tracking is carried out by sequential importance sampling and resampling. In addition, we consider the case that the initial state is unknown. In this case, the initial tracking accuracy is very low. To solve this problem, we modify the initial state using the observation data. It follows that higher accuracy is obtained with the correction of initial state.

2   System Mode

Suppose that K narrowband far-field signals impinge onto an array of M omnidirectional sensors from unknown time-varying directions $\boldsymbol{\theta}(t) = [\theta_1(t), \theta_2(t), \ldots, \theta_K(t)]^T$. The wavelength of signals is λ. The space of the adjacent antennas is d. The sequential sampling approach we adopt admits a first-order state-space hidden Markov model. The time-varying direction vector is modeled as

$$\boldsymbol{\theta}(t) = \boldsymbol{\theta}(t-1) + \mathbf{v}(t) \tag{1}$$

where $\mathbf{v}(t)$ is normally distributed with zero mean and the covariance matrix $\sigma_v^2 \mathbf{I}$. Here, σ_v^2 represents the noise spectral parameter, and $\mathbf{I}$ is the identity matrix.

For a collection of observed output date of M sensors in the array, $\mathbf{x}(t) = [x_1(t), \ldots, x_M(t)]^T$, the matrix formulation of outputs of the sensors at time t is obtained as follows

$$\mathbf{x}(t) = \mathbf{A}(\boldsymbol{\theta}(t))\mathbf{s}(t) + \mathbf{n}(t) \tag{2}$$

where $\mathbf{s}(t) = [s_1(t), \ldots, s_K(t)]^T$ is the signal waveform, $\mathbf{n}(t) = [n_1(t), n_2(t), \ldots, n_M(t)]^T$ is the complex independent Gaussian noise with zero mean and variance σ_n^2, $\mathbf{A}(\boldsymbol{\theta}(t)) = [\mathbf{a}(\theta_1(t)), \ldots, \mathbf{a}(\theta_K(t))]$ is the array manifold matrix, and $\mathbf{a}(\theta_k(t)) = [1, e^{-j2\pi d \, \sin \theta_k(t)/\lambda}, \ldots, e^{-j2\pi(M-1)d \, \sin \theta_k(t)/\lambda}]^T$ is the steering vector.

We assume that the signal is observed very T seconds. L measurements are taken for each increment T, and the time interval over which these are gathered is sufficiently small proportion of T for us to approximate the DOA as stationary over the measurement interval. That is, $\boldsymbol{\theta}(t) = \boldsymbol{\theta}(t+\tau) = \cdots = \boldsymbol{\theta}(t+(L-1)\tau)$, where τ is the time between measurements and $L\tau \ll T$. Then, the observation model with multiple measurement vectors is expressed as

$$\mathbf{X}(t) = \mathbf{A}(\boldsymbol{\theta}(t))\mathbf{S}(t) + \mathbf{N}(t) \tag{3}$$

where $\mathbf{X}(t) = [\mathbf{x}(t), \mathbf{x}(t+\tau), \ldots, \mathbf{x}(t+(L-1)\tau)]$, $\mathbf{S}(t)$ and $\mathbf{N}(t)$ are defined similarly.

3 Description of the Proposed Method

3.1 Particle Filter

From a Bayesian perspective, the tracking problem is to recursively calculate some degree of belief in the state $\boldsymbol{\theta}_t$ at time t, taking different values, given the data $\mathbf{X}_{1:t}$ up to time t. In principle, the posterior probability density function $p(\boldsymbol{\theta}_t|\mathbf{X}_{1:t})$ may be obtained recursively in two stages: prediction and update. Suppose that $p(\boldsymbol{\theta}_{t-1}|\mathbf{X}_{1:t-1})$ at time t-1 is available, the prediction stage is as follows

$$p(\boldsymbol{\theta}_t|\mathbf{X}_{1:t-1}) = \int p(\boldsymbol{\theta}_t|\boldsymbol{\theta}_{t-1})p(\boldsymbol{\theta}_{t-1}|\mathbf{X}_{1:t-1})\mathrm{d}\boldsymbol{\theta}_{t-1} \tag{4}$$

At time step t, the measurement $\mathbf{X}_t$ becomes available, the update stage is implemented via Bayes rule

$$p(\boldsymbol{\theta}_t|\mathbf{X}_{1:t}) \propto p(\mathbf{X}_t|\boldsymbol{\theta}_t)p(\boldsymbol{\theta}_t|\mathbf{X}_{1:t-1}) \tag{5}$$

The basic idea behind the particle filter is that we represent the posterior distribution of interest with a set of weighted particles, each of which forms an independent hypothesis of the state at a given time. The general particle filter consists of three stages. First, we use the set of particles from the previous time step to propose a new set for the new time step, according to the importance function $q(\boldsymbol{\theta}_t|\boldsymbol{\theta}_{t-1}, \mathbf{X}_t)$. Then, the weights of particles are calculated by

$$w_t^{(i)} \propto \tilde{w}_{t-1}^{(i)} \frac{p\left(\mathbf{X}_t|\boldsymbol{\theta}_t^{(i)}\right)p\left(\boldsymbol{\theta}_t^{(i)}|\boldsymbol{\theta}_{t-1}^{(i)}\right)}{q\left(\boldsymbol{\theta}_t^{(i)}|\boldsymbol{\theta}_{t-1}^{(i)}, \mathbf{X}_t\right)} \tag{6}$$

where $i = 1, \ldots, Ns$, Ns is the number of particles. And, the normalized weights are given by

$$\tilde{w}_t^{(i)} = \frac{w_t^{(i)}}{\sum_{i=1}^{Ns} w_t^{(i)}} \tag{7}$$

Finally, the weighted particles may be resampled to convert them into an unweighted set without changing the distribution they represent. We use SIR algorithm to implement DOA tracking which chooses the prior $p(\boldsymbol{\theta}_t|\boldsymbol{\theta}_{t-1})$ as the importance function so that the particle weights are then given by the likelihood only.

3.2   Correction of the Initial State

Considering the case that the initial state is unknown, we improve the algorithm under the framework of SIR. First, we produce a set of random samples $\left\{\boldsymbol{\theta}_0^{(i)}\right\}_{i=1}^{Ns}$ which obey the distribution $U[-90°, 90°]$, where $U[\cdot]$ denotes a uniform distribution. Then, the likelihood function of each particle is calculated according to the observation data at the initial time, which is followed as

$$p\left(\mathbf{X}_0\middle|\boldsymbol{\theta}_0^{(i)}\right) = \frac{\exp\left(-tr\left(\left(\mathbf{X}_0 - \mathbf{A}\left(\boldsymbol{\theta}_0^{(i)}\right)\hat{\mathbf{S}}_0\right)^H \left(\sigma_n^2\mathbf{I}_M\right)^{-1}\left(\mathbf{X}_0 - \mathbf{A}\left(\boldsymbol{\theta}_0^{(i)}\right)\hat{\mathbf{S}}_0\right)\right)\right)}{\left(\pi^M \det\left(\sigma_n^2\mathbf{I}_M\right)\right)^L} \tag{8}$$

where $tr(\cdot)$ denotes the trace, $\det(\cdot)$ denotes the determinant. $\hat{\mathbf{S}}_0$ is the least squares estimate of $\mathbf{S}_0$, which is given by

$$\hat{\mathbf{S}}_0 = \left(\mathbf{A}\left(\boldsymbol{\theta}_0^{(i)}\right)^H \mathbf{A}\left(\boldsymbol{\theta}_0^{(i)}\right)\right)^{-1} \mathbf{A}\left(\boldsymbol{\theta}_0^{(i)}\right)^H \mathbf{X}_0 \tag{9}$$

Each particle is weighted by the likelihood function, that is $w_0^{(i)} = p\left(\mathbf{X}_0\middle|\boldsymbol{\theta}_0^{(i)}\right)$. Then, we get the normalized weights $\tilde{w}_0^{(i)}$ and the corrected initial state can be obtained by

$$\tilde{\boldsymbol{\theta}}_0 = \sum_{i=1}^{Ns} \tilde{w}_0^{(i)}\boldsymbol{\theta}_0^{(i)} \tag{10}$$

To eliminate particles that have small weights and to concentrate particles with large weights, we generate a new set $\left\{\boldsymbol{\theta}_0^{(i*)}\right\}_{i=1}^{Ns}$ by resampling which satisfies

$$P\left(\boldsymbol{\theta}_0^{(i*)} = \boldsymbol{\theta}_0^{(j)}\right) = \tilde{w}_0^{(j)} \tag{11}$$

Then, the weights are reset to $\tilde{w}_0^{(i)} = \frac{1}{Ns}$. Thus, we can treat this new set as the initial particle set in the SIR algorithm and perform SIR to realize DOA tracking in the following time.

A summary of the proposed algorithm is as follows

- Correct the initial state
  - Produce a set of random samples $\left\{\boldsymbol{\theta}_0^{(i)}\right\}_{i=1}^{Ns}$ from $U[-90°, 90°]$

- – Calculate the likelihood function of each sample by (8)
- – Obtain the corrected initial state by (10)
- – Determine the initial particle set by resampling
- • Carry out SIR filter
  - – Draw $\boldsymbol{\theta}_t^{(i)} \sim p\left(\boldsymbol{\theta}_t \middle| \boldsymbol{\theta}_{t-1}^{(i)}\right)$, $i = 1, \ldots, Ns$
  - – Calculate the particle weights $w_t^{(i)} = p\left(\mathbf{X}_t \middle| \boldsymbol{\theta}_t^{(i)}\right)$ and normalize them by (7)
  - – Resampling
- • Output the state estimation

4   Numerical Simulations

Several simulations are performed to verify the feasibility of the proposed method. In the simulations, $K = 1$, $M = 10$, $d = \lambda/2$, $\sigma_v^2 = 1$, $\sigma_n^2 = 1$, $T = 1$, $Ns = 100$ and SNR = 5 dB.

We first analyze the DOA tracking error of SIR for the various numbers of snapshots and the root mean square error (RMSE) is given by $\mathrm{RMSE}(t) = \frac{1}{N}\sum_{i=1}^{N}\left(\hat{\theta}^i(t) - \theta^i(t)\right)^2$, where N is the times of Monte Carlo trials. In this simulation, the times of Monte Carlo are 100, and the initial state is assumed to be known. The results are shown in Fig. 1. It can be seen that the tracking error reduces with the increasing of the number of snapshots, which proves that the multiple measurement vectors model has better performance than the single measurement vector model.

Then, we compare the DOA tracking results of SIR algorithm with the known initial state and the unknown initial state. In this simulation, we utilize the multiple

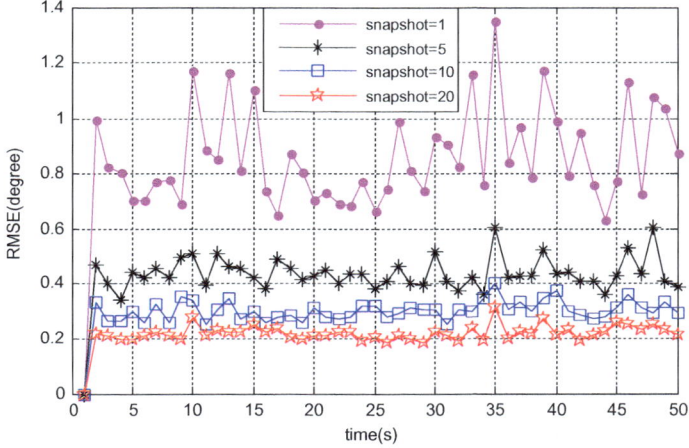

Fig. 1. Tracking error of SIR with different numbers of snapshots

measurement vectors model and set $L = 5$. The simulation results are illustrated in Fig. 2. It is indicated that the information of the initial state will affect the performance of the algorithm. The performance is obviously degraded when the initial state is unknown. As shown in the figure, the tracking error is stayed small in the known initial state case while it is very large in the first few seconds in the unknown initial state case. At time $t = 5$, the estimated DOA with unknown initial state starts to get close to the true DOA

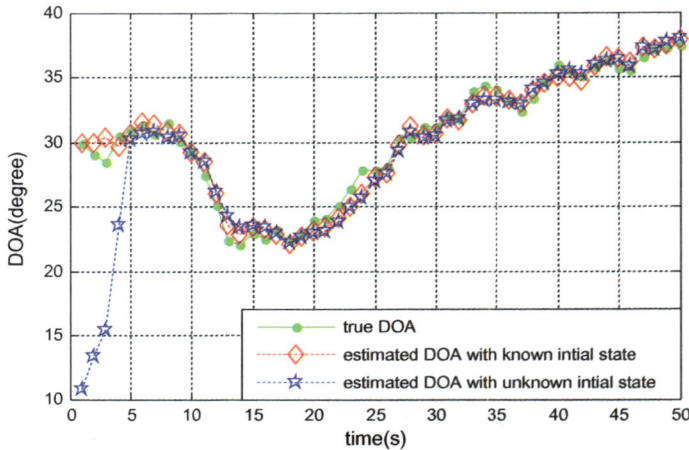

Fig. 2. DOA tracking with known initial state and unknown initial state

To reduce the initial tracking error in the case of the unknown initial state, we add the correction step. In the following, we evaluate the performance of the proposed method. In this simulation, we set $L = 5$ and carry out 1000 Monte Carlo trials. The results are shown in Fig. 3. The RMSE of SIR with unknown initial state is higher than 30 degree while that of the proposed method is lower than 3 degree. It is illustrated that the proposed method can improve the initial tracking performance.

5   Conclusions

In this paper, a particle filter algorithm with correction of the initial state for DOA tracking is proposed. A dynamic state-space model is constructed to describe the state variation tendency. In this model, we utilize multiple measurement vectors in the observation equation in order to take advantage of more data information. To solve the problem that the initial tracking performance is degraded under the situation of unknown initial state, we correct the initial state by weighting the random particles

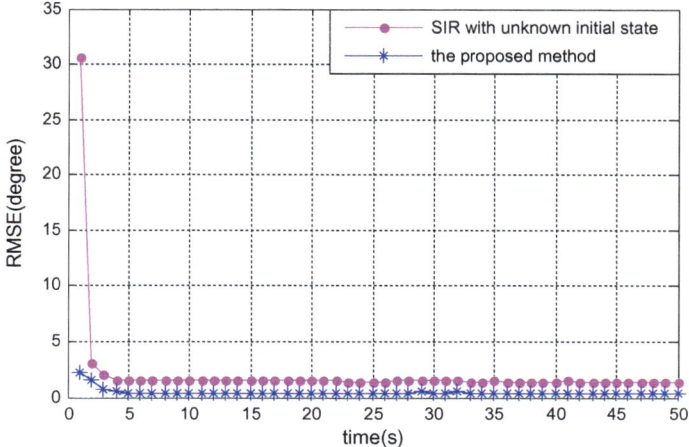

Fig. 3. Performance comparison between the proposed method and SIR with unknown initial state

based on likelihood function. The proposed algorithm performs better than SIR filter algorithm.

References

1. Schmidt RO. Multiple emitter location and signal parameter estimation. IEEE Trans Antennas Propag. 1986;34(3):276–80.
2. Roy R, Kailath T. ESPRIT-estimation of signal parameters via rotational invariance techniques. IEEE Trans Acoust Speech Signal Process. 1989;37(7):984–95.
3. Badeau R, David B, Richard G. Fast approximated power iteration subspace tracking. IEEE Trans Signal Process. 2005;53(8):2931–41.
4. Liao B, Zhang ZG, Chan SC. DOA estimation and tracking of ULAs with mutual coupling. IEEE Trans Aerosp Electron Syst. 2012;48(1):891–905.
5. Yang B. Projection approximation subspace tracking. IEEE Trans Signal Process. 1995;43 (1):95–107.
6. Kong D, Chun J. A fast DOA tracking algorithm based on the extended Kalman filter. In: National aerospace and electronics conference, 2000. Naecon 2000. Proceedings of the IEEE. IEEE Xplore; 2000. p. 235–8.
7. Kumar CV, Rajagopal R, Kiran R. An optimal integrated tracking (ITS) for passive DOA tracking using unscented Kalman filter. In: Information, decision and control, 2002. Final Program and. IEEE; 2002. p. 253–8.
8. Arulampalam MS, Maskell S, Gordon N, et al. A tutorial on particle filters for online nonlinear/non-Gaussian Bayesian tracking[J]. IEEE Trans Signal Process. 2002;50(2):174–88.
9. Orton M, Fitzgerald W. A Bayesian approach to tracking multiple targets using sensor arrays and particle filters. IEEE Trans Signal Process. 2002;50(2):216–23.
10. Zhong X, Premkumar AB, Madhukumar AS. Particle filtering and posterior cramér-rao bound for 2-D direction of arrival tracking using an acoustic vector sensor. IEEE Sens J. 2012;12(2):363–77.
11. Zhong X, Premkumar AB, Madhukumar AS. Particle filtering for acoustic source tracking in impulsive noise with alpha-stable process. IEEE Sens J. 2013;13(2):589–600.

Localization of a Mobile Node Using Fingerprinting in an Indoor Environment

Sohaib Bin Altaf Khattak[1], Min Jia[1(✉)], Mir Yasir Umair[2], and Attiq Ahmed[2]

[1] School of Electronics and Information Engineering, Harbin Institute of Technology, Harbin 150001, China
jiamin@hit.edu.cn
[2] National University of Sciences and Technology, Islamabad, Pakistan

Abstract. Localization is an important requirement in today's world, and numerous modern applications require location tracking. An indoor localization of a mobile node using the range-free fingerprinting technique in WLAN environment is presented. The work focuses on improvement in the accuracy of localization using some additional parameters in the fingerprint, along with the conventional received signal strength (RSS). ToA has been used to enrich the fingerprint data for more unique fingerprints. The impact of AP placement on localization accuracy is also addressed. In this paper, a technique is proposed that is not complex to implement using existing infrastructure and is also easy to understand. Significant improvement has been achieved from about 20 to 40% in different scenarios including line of sight and non-light of sight scenarios, small and large areas.

Keywords: Indoor localization · WLAN AP placement · LBS for IoT

1 Introduction

Localization is the process of finding the position of a specific target, based on some observable phenomenon. It has numerous applications like robot movement, navigating self-driving cars, automatic object location detection. Recently, location-based services (LBS) have a great impact on our everyday lives, it is also related to IoT. The poor performance of GPS indoors and the popularity of WLANs have opened up many opportunities for LBS. Numerous approaches have been proposed that find user locations without the aid of GPS [1]. Among these approaches, Wi-Fi-based localization has gained immense attention, it saves extra infrastructural costs, because the Wi-Fi access points are widely deployed in every building nowadays. Indoor localization method using Wi-Fi can generally be divided into two categories: range based (an approach based on triangulation/ trilateration) and range free (fingerprinting approach).

Location fingerprinting is a technique that matches the fingerprint of certain characteristics of a signal as a function of the position. It has two phases: offline

Q. Liang et al. (eds.), *Communications, Signal Processing, and Systems*, Lecture Notes in Electrical Engineering 516,
https://doi.org/10.1007/978-981-13-6504-1_129

and online. In the offline stage, the location coordinates and RSS from APs are collected. In the online phase, a positioning technique uses the currently observed RSS and previously collected information to figure out an online query by the user.

RADAR [2] and Horus [6] systems will be used for comparison with the proposed technique, as these are accepted and famous algorithms in the area, also widely used for comparison. RADAR system is a deterministic algorithm based on nearest neighbor algorithm, and Horus is probabilistic based on maximum likelihood and Bayes' rule. We propose a system which uses a combination of Wi-Fi RSSI values and ToA values for localization based on fingerprinting framework. In the literature, multiple techniques have been used along with location fingerprinting [3,4]. TOA is previously used along with RSS but as range-based technique [3], whereas in this paper, it is taken in range-free method and combined it with the RSS in the fingerprint. AP placement techniques [5] are also used to further improve the accuracy of the system. The paper is divided into the following sections. Section 1 is the introduction , Sect. 2 is the proposed technique, and Sect. 3 is simulation and results. The paper is concluded in Sect. 4. At last, bibliography is given.

2   Proposed Technique

This section describes the system model. The system is simulated, and the proposed technique is tested along both deterministic and probabilistic for comparison in different scenarios.

2.1   System Model

A 2D indoor environment is considered covered by a certain number of WLAN APs, which are visible throughout the area of interest. To build a radio map, area of interest is divided into grids of 2 m by 2 m. In the offline phase, the readings from Wi-Fi APs are taken at the RPs and stored in the FP database. The point can be represented as (x, y) 2D coordinates on the floor. As shown in Fig. 1, this figure will be further used to test localization algorithm. The notations used are given as follows, and the parameters used in simulation are given in Table 1.

χ = two-dimensional space $\qquad\qquad$ (x, y) = coordinates in 2D

Γ = vector having RSS reading $\qquad\quad$ τ = vector having ToA readings

N_{ap} = number of APs $\qquad\qquad\qquad\qquad$ $Pixelvalue = 2$

ψ = samples of RSS from S $\qquad\qquad$ ξ = samples of ToA from T

L_x = dimensions of area along x-axis $\quad$ L_y = dimensions of area along y-axis

$Dif_x = [0.5, 2.5, \ldots, L_x]$ $\qquad\qquad\qquad$ $Dif_y = [0.5, 2.5, \ldots, L_y]$

2.2   Offline Phase

RSSI can be used to estimate distance because the power of electromagnetic waves is inversely proportional to the distance [7]. Practically, the RSS along with distance depends on other environmental factors also. So the following pathloss model is used:

$$P(d) = P(d_0) + 10.n.\log(d/d_0) + \zeta + W \tag{1}$$

Here, P(d) is the RSS at distance d, $P(d_0)$ is the RSS at the reference distance d_0 . 'd' is the distance of the RP from the AP. 'n' is the path loss coefficient, ζ is the random noise element, W is the wall attenuation factor. Multiple number of samples for RSS had to be taken, as values of RSS change at every instant. All the values of RSS for a certain RP from a certain AP are averaged and are stored in the fingerprint database.

$$RSS_{ij}(avg) = \frac{\sum_{n=1}^{N} RSS_{ij}}{N} \tag{2}$$

Time of arrival (TOA) of each AP is also calculated at all RPs in the proposed technique and stored in the database along with the RSS values to enrich the fingerprint data. The values for TOA in the simulation are found by using the formula $s = v.t$, where s is the distance, v is the velocity, and t is the time. This relation has been used as $t = d/c$, here d is the distance between the AP and RP, c is the propagation velocity of the signal, and t is the TOA for that specific RP. Due to the scattering and multipath effect 'd' is not accurate, so a random noise is added as $d = d_i + d_n$. Using the values of RSS and TOA together in fingerprint, certain weight(W_n) has to be assigned to TOA values because there is a large difference among the values of these parameters which can lead to wrong estimation during matching. So, the ToA calculation can be mathematically represented by the following equation:

$$ToA = \frac{\sum_{i=1}^{N} ToA(n)}{N} * W_n \tag{3}$$

The system detects the walls in the channel between the AP and RP and includes the effect in the propagation model for RSS calculation.

The training data in the offline phase is stored and then database of radio map will have the form:

$$\lambda = \begin{bmatrix} x_1, y_1 & \Gamma_{1,1} = (\psi_1, .., \psi_{Nap}) & \tau_{1,1} = (\xi_1, .., \xi_{Nap}) \\ \vdots & \cdots & \vdots \\ x_{L_x}, y_{L_y} & \Gamma_{Lx,Ly} = (\psi_1, \ldots, \psi_{Nap}) & \tau_{Lx,Ly} = (\xi_1, \ldots, \xi_{Nap}) \end{bmatrix}$$

Order of the matrix is $\alpha * \beta$, where α is the number of data points and β is dimensionality.

2.3   Online Phase

A query is put to the system having the values (RSS and TOA), and the system provides the estimated location by using matching technique, the most similar match to the query is given as the estimated location and coordinate points are returned. One thousand random test points are created in the simulation, where the matrix (R, S) has x, y coordinates of the generated test points. The query matrix is also saved in the form:

$$Q = [(R_i S_i), \Gamma_i = (\psi_1, \ldots, \psi_{Nap}), \tau_i = (\xi_1, \ldots, \xi_{Nap})] \tag{4}$$

The two matrices λ and Q will be matched for location determination. For the matching purpose, the K-nearest neighbor (KNN) algorithm is used. The KNN algorithm basically calculates the distance of the query from the fingerprinting database, selects the K-nearest neighbors, and gives the averaged value.

Both RSS and ToA are used in the proposed technique, so the combined vectors will be used,

$$\sqrt{(Query_t - DataBase_f)^2} \tag{5}$$

So, $(Query_t)$ is the query, by the user that wants to find its location, and $(DataBase_f)$ is the vector from fingerprint database. K-nearest neighbors are selected on the basis of minimum distance, their coordinate points are then averaged, and the position is estimated.

$$(x, y) = \frac{1}{k} \sum_{i=1}^{K} (x_i, y_i) \tag{6}$$

(x, y) is the estimated position, K is number of nearest neighbors, (x_i, y_i) are coordinate points of the K-nearest neighbors. The root mean square error (RMSE) is calculated to check the overall performance of the system,

$$RMSE = \sqrt{\sum (P_R - P_{Es})^2} \tag{7}$$

P_R is the real position coordinates, and P_{Es} are the estimated position coordinates.

In the simulation for KNN implementation and location determination, the following procedure was adopted.

KNN (λ , Q)

$$Q - (R_i S_i) = [\Gamma_i = (\psi_1, \ldots, \psi_{Nap}), \tau_i = (\xi_1, \ldots, \xi_{Nap})] = \omega \tag{8}$$

$$(R_i, S_i) = Actual test locations = L \tag{9}$$

$$D = \sqrt{\sum (\omega_i - \lambda_i)^2} \tag{10}$$

$$D_{sort} = argmin \sum_{i=1}^{1000} \sqrt{\sum (\omega_i - \lambda_i)^2} \tag{11}$$

$$D_{sort} = [d_{min}, \ldots, d_{max}] \tag{12}$$

$$Sort = Indexes(D_{sort}) \tag{13}$$

$$EstimatedPosition = \frac{\sum P(x), P(y)}{K} \tag{14}$$

$$EstimatedPositions(i = 1 : 1000) = E = P(x_i), P(y_i) \tag{15}$$

$$Error = \varepsilon = \sqrt{\sum (E_i - L_i)^2} \tag{16}$$

After the results from KNN and the position estimation tests performed for 1000 random queries, their cumulative density function (CDF) of Error was taken as a metric for precision of the system.

The values of parameters used in simulation are shown in Table 1.

Table 1. Parameters used in simultaion

Parameter	Value
Path loss exponent n	3
Wall attenuation factor WAF	2
Signal propagation speed c	299792458
Reference distance d_0	1
Power at d0 $P(d_0)$	−30
Transmission power Ptx	10
K in KNN	5
Grid size	2*2
No. of position queries	1000

3   Simulations and Results

In this section, the simulation scenario is explained and discussed with the obtained results. The RSS-based localization techniques and the proposed technique are tested in different indoor scenarios, and results are compared. The number and placement of AP are also addressed.

3.1   Case A and Case B

The first scenario is mapped upon an area of 50×50 (m) in which there is a LOS distance between the APs and the RPs and there are four APs placed at the four corners of the simulation area. See Fig. 1. The results were very encouraging, the proposed technique achieved 21% improvement. The CDF plot is shown in Fig. 2 and the detailed results in Table 2.

Second scenario considers an area of 100×100 m. The number of APs is 5. See Fig. 3. The proposed technique gave better results as compared to the simple RSS-based localization techniques. CDF plot is shown in Fig. 4 and results in Table. 2.

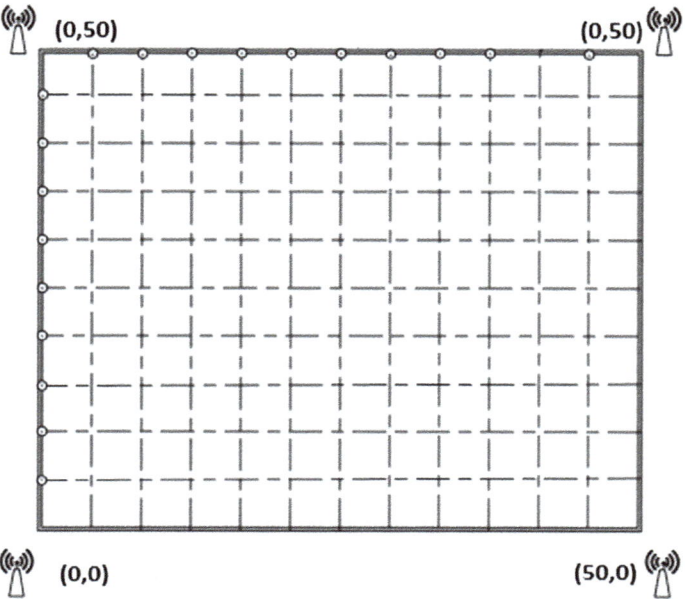

Fig. 1. Case A: 4 APs 50 m by 50 m area

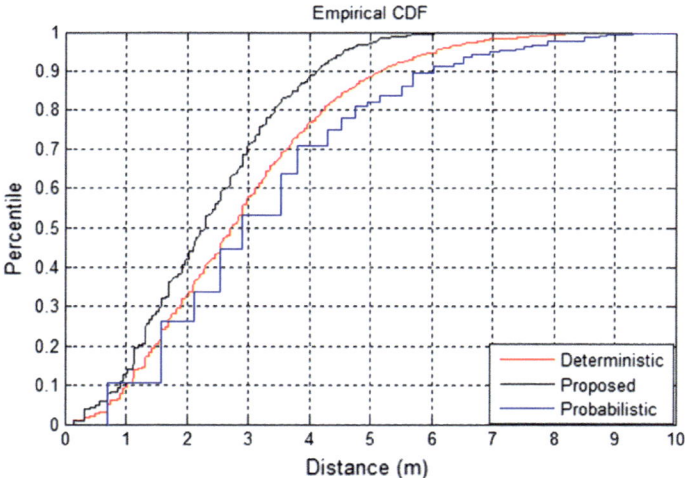

Fig. 2. Simulation results for case A

Table 2. CDF plot case A and B

Case	Technique	CDF (error) at 0.7 (m)
A	Probabilistic	3.8
A	Determinisric	3.6
A	Proposed	2.9
B	Probabilistic	6
B	Deterministic	5.4
B	Proposed	3.3

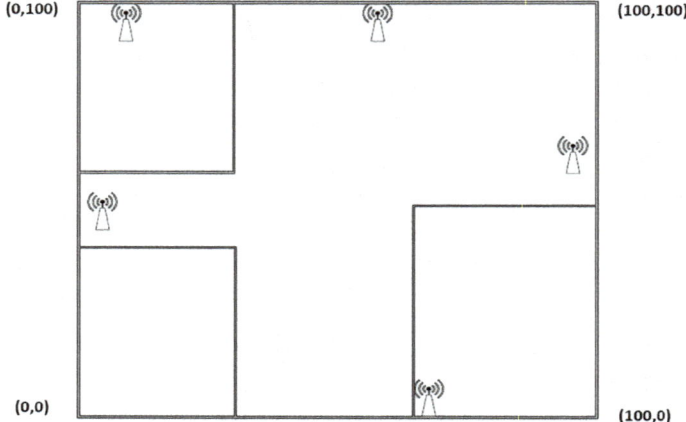

Fig. 3. Coordinates of 5 APs deployed (Case B 5 APs 100 m by 100 m Area)

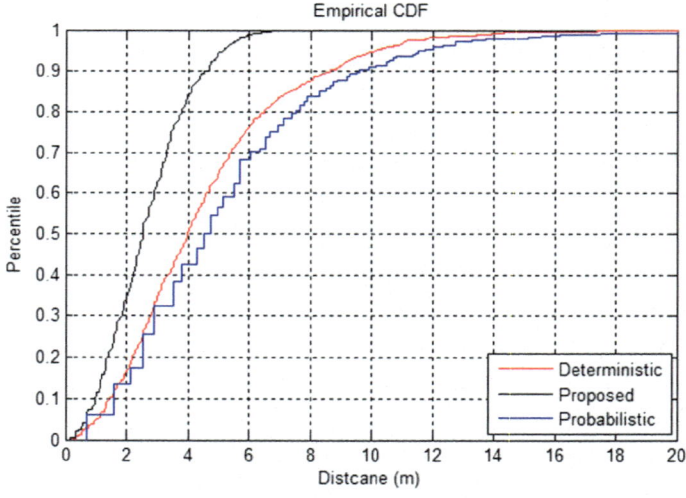

Fig. 4. Simulation results of case B (5 APs 100 m by 100 m area)

3.2 Impact of Number of APs and Placement

Increasing the number of APs to 8, the impact of number of APs and their placement on accuracy will be checked. Eight APs are deployed in the same model used in Case B as shown in Fig. 3. The APs are deployed first randomly, then two times in different symmetries. The coordinates of the APs can be seen in Table 3, and detailed results in Table 4. The CDF plots of the three AP deployment configurations are shown in Figs. 5, 6, and 7, respectively.

It can be seen from the results that placement of APs has an impact on the accuracy of the system, and it can also be observed that increasing the number of APs can have a positive impact on the accuracy. But one thing has to be noted that the proposed technique is not much prone to changes due to APs placements as compared to the RSS-based system and the results are consistent over all scenarios.

Table 3. APs coordinates

AP number	Random	Symmetry 1	Symmetry 2
1	0, 0	0, 0	13, 25
2	0, 50	0, 100	13, 75
3	31, 100	100, 100	37, 25
4	100, 0	100, 0	37, 75
5	50, 50	35, 20	63, 25
6	75, 80	35, 80	63, 75
7	100, 20	75, 20	87, 25
8	50, 0	75, 80	87, 75

Table 4. Error comparison and results

AP placement strategy	Technique	CDF (error) at 0.7
Random APs placement	Probabilisitc	6.3
	Deterministic	5.3
	Proposed	3.6
APs placement symmetry 1	Probabilisitc	5.7
	Deterministic	5.1
	Proposed	3.2
APs placement symmetry 2	Probabilisitc	3.8
	Deterministic	3.5
	Proposed	2.8

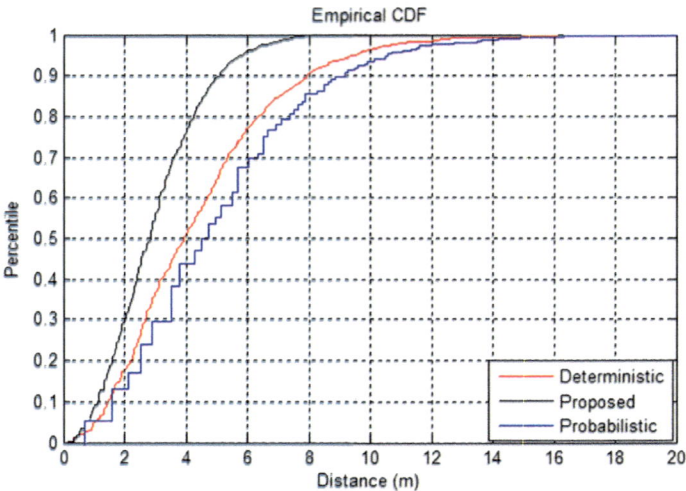

Fig. 5. CDF plot (8 APs randomly deployed)

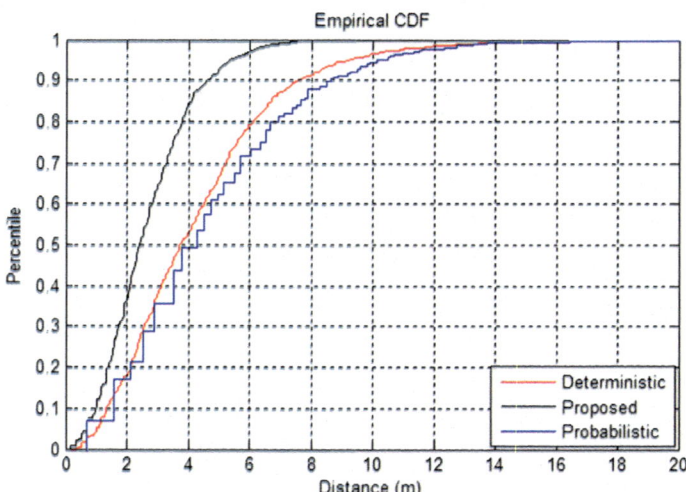

Fig. 6. CDF plot (8 APs deployed in symmetry 1)

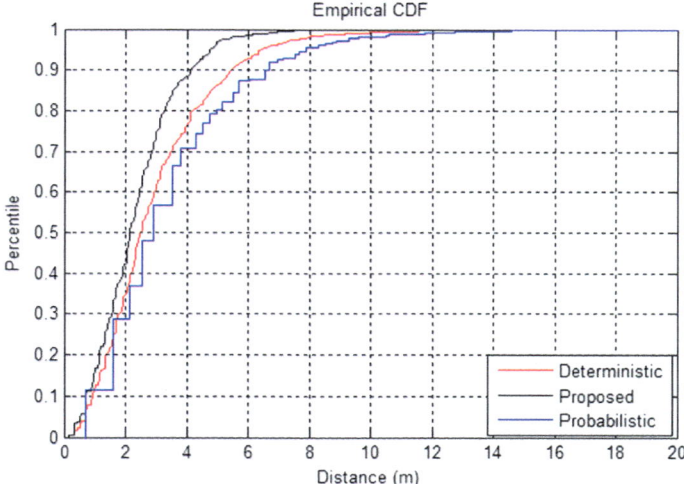

Fig. 7. CDF plot (8 APs deployed in symmetry 2)

4   Conclusion

The proposed technique that uses an additional parameter of ToA in finger-print shows significant improvement in overall results. It can be concluded that utilizing other location dependent parameters in WLAN or any other wireless network can improve the localization accuracy. The impact of AP placement is also addressed and the results show changes due to placement. The proposed technique is consistent overall and provides much better results. The idea can be taken further by investigating other location dependent parameters for fin-gerprinting and AP placement can be also investigated, relations depending on the characteristics of the signal and geometry of the area can provide optimal AP locations.

Acknowledgments. This work was supported by the National Science Foundations of China (No.61671183 and 61771163).

References

1. Liu H, Darabi H, Banerjee P, Liu J. Survey of wireless indoor positioning techniques and systems. Ieee Trans Syst Man Cybern Part C: Appl Rev. 2007;37(6):6.
2. Bahl P, Padmanabhan VN. RADAR: an in-building RF-based user location and tracking system. In: IEEE INFOCOM 2000, 2000.
3. Torteeka P, Chundi X, Dongkai Y. Hybrid Technique for indoor positioning system based on wi-fi received signal strength indication. In: 2014 international conference on indoor positioning and indoor navigation, 27th–30th October, 2014.
4. Machaj J, Brida P. Using of GSM and wi-fi signals for indoor positioning based on fingerprinting algorithms. Inf Commun Technol Serv. 2015;13(3).

5. Chen Y, Francisco J-A, Trappe W, Martin RP. A Practical approach to landmark deployment for indoor localization IEEE SECON 2006 proceedings.
6. Youssef M, Agrawala A. The horus WLAN location determination system. In: MobiSys05: proceedings of the 3rd international conference on Mobile systems, applications and servicesl, USA; 2005. p. 205–18.
7. Rappaport TS. Wireless communications: principles and practice. New Delhi: Prentice-Hall of India; 2003.

An Improved State Coherence Transform Algorithm for the Location of Dual Microphone with Multiple Sources

Shan Qin and Ting Jiang[✉]

Beijing University of Posts and Telecommunications, Haidian District,
Beijing 100000, China
tjiang@bupt.edu.cn

Abstract. This paper proposes a new kernel function in state coherence transform to perform multiple time difference of arrival estimation in order to increase the resolution of location in frequency-domain blind source separation. The state coherence transform associated with each source generalizes the GCC for multiple sources and generates envelopes with clear peaks corresponding to the maximum-likelihood TDOAs. However, the weight allocation of the kernel function is unreasonable for small spacing microphones. We propose an improved kernel function to enhance the resolution of small values, which means that a larger weight allocated to smaller values. Experimental results show that the proposed approach allows to separate four speakers, using very short utterances, in highly reverberant environment even with small-spaced microphones of 2 cm.

Keywords: Nonlinear weighting compensation · State coherence transformation · Blind source separation

1 Introduction

During the last thirty years, a huge number of methods of blind source separation have been proposed but separation in real life with small microphone pairs is still a challenging problem [1, 2]. The estimation of time difference of arrival (TDOA) is an essential step in several approaches for blind source separation [3]. In the indoor reverberation environment, frequency-domain approaches outperform the time-domain methods in computing complexity performance and convergence property, which is the most investigated like independent component analysis [4] and nonnegative matrix factorization [5]. However, the removal performance of internal permutation ambiguity directly affects the separation effect. The separation perform on small-spaced microphone pair is still lag, since the accuracy of time delay estimation is limited for microphone space, resulting in permutation problem. Among the most promising ones, a robust way to solve the permutation problem is to apply state coherence transform to increase the inter-source TDOA resolution [6]. However, this approach has a poor performance for the case of separation with small-spaced microphone pairs when reverberation and spatial aliasing existing.

© Springer Nature Singapore Pte Ltd. 2020
Q. Liang et al. (eds.), *Communications, Signal Processing,
and Systems*, Lecture Notes in Electrical Engineering 516,
https://doi.org/10.1007/978-981-13-6504-1_130

In this work, we present a new kernel function to improve SCT solving the permutation problem on small-spaced microphone pairs, which is robust both to spatial aliasing and to reverberation even for high T_{60}. The principle of time delay estimation is analyzed in Sect. 2, and the real data are used for experimental verification. Finally, it confirms that the proposed improved method performs effectively allowing the separation with microphone spaced at 2 cm in highly reverberant environments.

2   Physical Interpretation of SCT

To simplify the understanding of the proposed improved approach, we first give a simple physical interpretation of the TDOA and SCT. For simplicity, we analyze a number of source N of 1 and the number of the microphones M of 2. A schematic diagram is shown in Fig. 1.

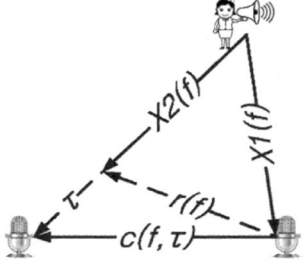

Fig. 1.   Wave propagation model of source

It is known that X_1 and X_2 are signals received by microphones, and the microphone spacing is represented by $c(f, \tau)$. The TDOA of the source relative to the microphones is represented by τ. $r(f)$ represents the phase difference of two observed acoustic propagation from the source to the microphone pair (a, b). Under ideal conditions, the model of the inter-microphone delay for a given source can be represented by $c(f, \tau)$. According to the mathematical relation in the graph, TDOA is expressed as

$$\vec{\tau} = \overrightarrow{c(f, \tau)} - \overrightarrow{r(f)} \tag{2.1}$$

Furthermore, as long as the acoustic waves related to the direct propagation paths are relatively strong compared with the reflection paths, a likelihood measure of the TDOAs for multiple sources can be estimated. TDOAs estimation associated with each source can be performed by minimizing the following quantity [7]:

$$\vec{\tau_k} = \underset{N \ M}{\operatorname{argmin}} \sum \sum \|c(f, \tau) - r(f)\| \tag{2.2}$$

SCT will be maximized for values of t that for each frequency minimize the sum of the above distances. In practice, given the coherence of the states across the frequency, the state coherence transform is formulated as follows

$$\text{SCT}(\tau) = \max \sum_N \sum_M \left[1 - g\left(\frac{\|c(f,\tau) - r(f)\|_2}{2} \right) \right] \qquad (2.3)$$

Since $\vec{\tau} \leq \overrightarrow{c(f,\tau)}$, if we do not perform any nonlinear transformation of the Euclidean distance with the decreasing of microphone space, the inter-source TDOAs will be too small to be distinguished with decreasing resolution, which makes the peak selection more sensitive to errors. $g(\cdot)$ is a generic nonlinear function which operates as a weighting function and allocates more weight to the small value τ that corresponds to the maximum likelihood of propagation parameters of each source. The common nonlinear monotonic function has been described in [8] as

$$g(\cdot) = \tan h(\alpha x) \qquad (2.4)$$

Although the SCT is effective in normal situations, it lacks reliability if speech separation is analyzed with small microphone spacing at 2 cm or smaller. $g(\cdot)$ function has not enough resolution to distinguish multiple peaks in small space condition, which may cause the separation performance of small microphone array decline.

3   Proposed Kernel Function

Although the SCT in estimation of TDOAs is more accurate than GCC, the decreasing resolution reduces the accuracy of estimation [7]. When the distance between microphones decreases, the reduction in resolution may be occurred. Improving the ability of $g(\cdot)$ in separating multiple local peaks may enhance the resolution.

In this work, we propose an improved function with reasonable performance by using a locally confined kernel function that enhances the resolution on $t \approx 0$, which implicitly resolves the permutation problem and implements the estimation of TDOAs of multiple sources with small spacing microphone pair.

$$g(\cdot) = 1 - \tan h(\alpha x) \qquad (3.1)$$

Figure 2 shows the graph of different kernel functions with $\alpha = 8$. Abscissa represents time delay, and ordinate represents the weight assigned to distance. Compared from the diagram, it can be inferred that SCT does have a resolution magnification for the distance, but the allocation of weights under the smaller size is not reasonable. The smaller space requires greater weights to enhance the resolution of the delay estimation. The higher weights are allocated to τ when τ has small values when the kernel function $g(\cdot) = 1 - \tan h(\alpha x)$ with better performance than $g(\cdot) = \tan h(\alpha x)$, and vice versa. The removal performance of internal permutation ambiguity outperforms the SCT. When $N = 2$, the improved SCT formula is shown as follows:

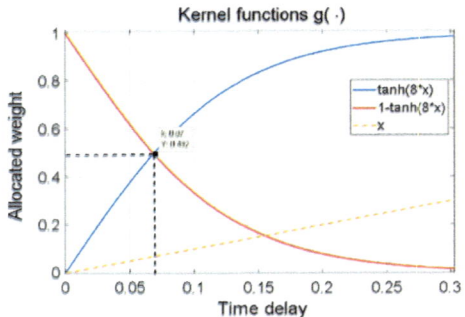

Fig. 2. Values of different kernel functions with $\alpha = 8$

$$\text{ISCT} = \left(1 - \left(1 - \tan h\left(\frac{\alpha\|c(f,\tau) - r_1(f)\|_2}{2}\right)\right)\right)$$
$$+ \left(1 - \left(1 - \tan h\left(\frac{\alpha\|c(f,\tau) - r_2(f)\|_2}{2}\right)\right)\right) \tag{3.2}$$

And as known that $\|c(f,\tau) - r(f)\|_2 = \sqrt{[c(f,\tau) - r_2(f)][c(f,\tau) - r_2(f)^*]}$, so the above formulator can be expressed as:

$$\text{ISCT} = \max \sum_N \sum_M \tan h\left(\alpha\sqrt{1 - \text{Re}[c(f,\tau) * r(f)]}\right) \tag{3.3}$$

In previous research, it was proved that $\text{GCCPHAT}(\tau) = \text{Re}[c(f,\tau) * r(f)]$ [6]. The (3.4) describes the relationship between ISCT and GCCPHAT, which indicates that the essence of SCT is to assign different weights to the value of GCC and enhance the resolution of location to estimate the TDOA more accurately:

$$\max \text{ISCT} = \tan h\left(\alpha\sqrt{1 - GCC}\right) \tag{3.4}$$

Regarding the values of parameter α, it is decided by doing experiments or experiences of researchers. In this paper, we propose a constraint function to restrict α:

$$\alpha(d) = \begin{cases} \frac{1}{5d} & (0 < d \leq 0.2) \\ 1 & (0.2 < d \leq 1) \end{cases} \tag{3.5}$$

where d is the microphone space in meters.

Inverse proportional sequence has a good convergence property. Monotonic bounded sequence must converge in the defined domain. Therefore, the (3.4) can obtain the local extreme values which represent the estimation of the TDOAs of multiple sources.

4   Experiments and Results

In this work, we solve the removal permutation problem according to the estimated TDOA and the optimization in formula (3.4). The parameter α is set according to the (3.5). The nonnegative matrix factorization algorithm has been applied to separate three sources with distances as shown in Fig. 3. Speakers place about 1.2 meter away from the array with the microphones spacing of $d = 2$ cm and the room impulse response with $T_{60} = 600$ ms. The separation performance is evaluated by waveform. The STFT was implemented with hamming window at the length of 1024 samples and 50% overlap. The speech collected by the microphone array is considered as experimental data. The voice acquisition module is shown in Fig. 4.

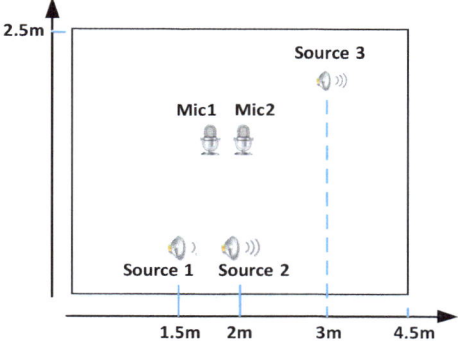

Fig. 3. Recording scene of the speech data

Fig. 4. Circuit board diagram of the microphone array

In order to demonstrate the ability of the proposed ISCT methods, we used the GCC-NMF algorithm to obtain the estimation of the TDOAs and separated sources. The voice acquisition module consists of six microphones spaced of 2 cm to each one. Here, we only keep the voice data of two adjacent microphones a and b. We compare our proposed algorithm ISCT with SCT.

Figure 5 describes the results estimation of TDOA of SCT, and Fig. 6 describes the result of ISCT. The abscissa is the angular spectrum, and the ordinate is the sum of the SCT in all frequencies. The solid line indicates the value of SCT and the dotted lines are the local maximum value, that is, the time difference we estimate. Compared with the two figures, the location resolution of ISCT algorithm is higher than SCT algorithm.

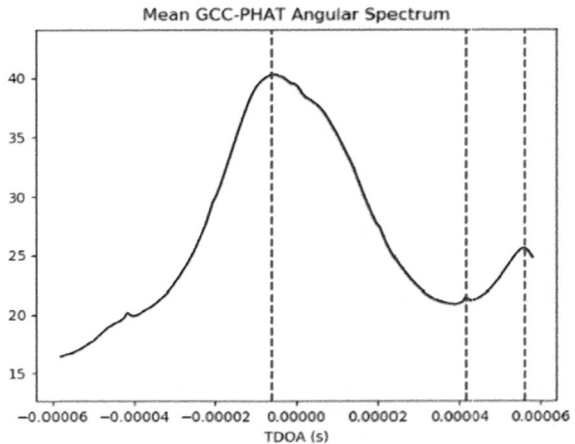

Fig. 5. SCT angular spectrum

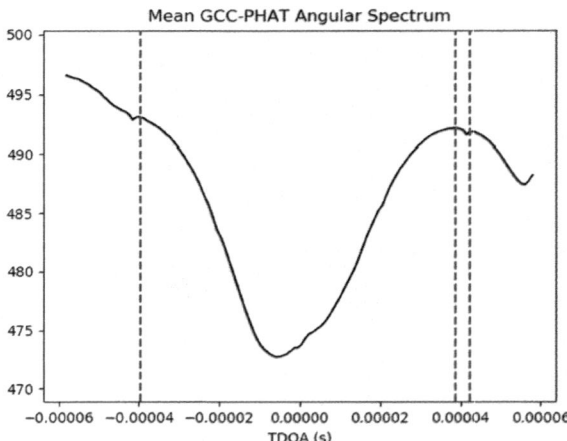

Fig. 6. ISCT angular spectrum

Since our experimental data are actually measured data, there is no pure speech contrast to calculate SIR. Comparing the waveform of the actual speech with the waveform of the separated speech is selected to evaluate the performance. The performance of the two algorithms is shown in the following waveform diagrams.

Figure 7 is the waveform of the separation results of ISCT. Figure 8 is the waveform of the separation results of SCT. The separation effect based on ISCT

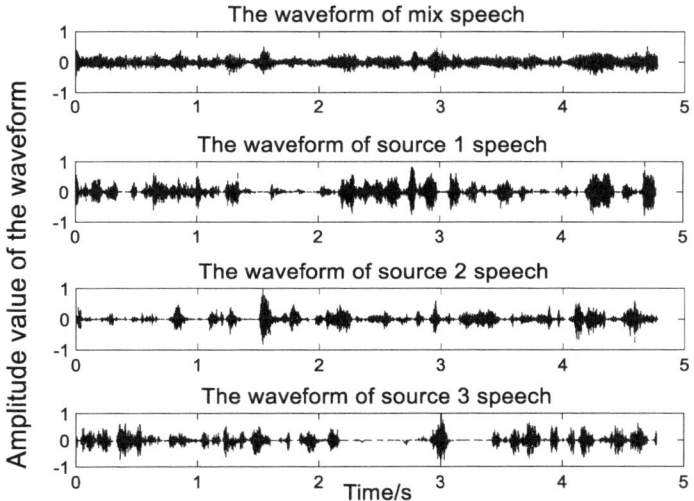

Fig. 7. Waveform of separated speech of ISCT

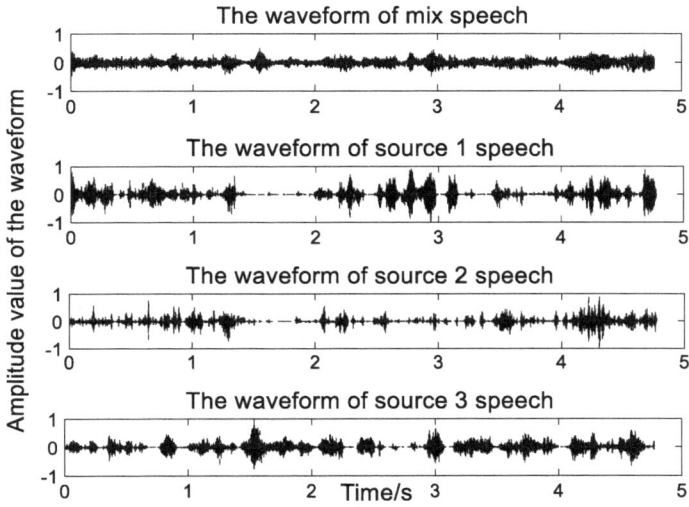

Fig. 8. Waveform of separated speech of SCT

outperforms the SCT. Compared with the separation results in Figs. 7 and 8 shows only two separation sources, which indicates that the separation effect based on ISCT outperforms the SCT.

ISCT performs better when the size of the microphone array is very small about 2 cm or smaller. If the size is changed to 5 cm or bigger, the effect of SCT is obvious. Therefore, different nonlinear functions can be selected for compensation according to the size of different microphone arrays as follows:

$$\mathrm{maxSCT}(\tau) = \begin{cases} \tan h\left[\alpha(d)\sqrt{1 - GCC(\tau)}\right] & 0 < d \le 0.04 \\ 1 - \tan h\left[\alpha(d)\sqrt{1 - GCC(\tau)}\right] & 0.05 \le d < 1 \end{cases} \tag{4.1}$$

5   Conclusion

In this paper, we analyze the principle of estimation of TDOA. Spatial resolution decreases with decreasing size of microphone array if no nonlinear compensation is made. Compared with the nonlinear function of SCT, the new kernel function proposed in this paper will allocate a larger weight to smaller values. Therefore, there will be a higher spatial resolution in the case of small microphone spacing to obtain more accurate positioning accuracy, solve the problem of arrangement, and get the purpose of blind source separation. We also propose constraint functions for the parameters α according to the microphone space. Experimental results show that the proposed approach allows to separate four speakers, using very short utterances, in highly reverberant environment even with small-spaced microphones of 2 cm.

Acknowledgments. This work was supported by National Natural Science Foundation of China (NSFC) (No.61671075) and Major Program of National Natural Science Foundation of China (No.61631003).

References

1. Hosseini MS, Rezaie A, Zanjireh Y. Time difference of arrival estimation of sound source using cross correlation and modified maximum likelihood weighting function. Sci Iran. 2017;24(6).
2. Jia RS, Gong Y, Peng YJ, Sun HM, Zhang XL, Lu XM. Time difference of arrival estimation of microseismic signals based on alpha-stable distribution. 2017; p. 1–17.
3. Zhu H, Li Z, Cheng Q. Sound source localization through optimal peak association in reverberant environments. 2017; p. 1–6).
4. Mirzal A. NMF versus ICA for blind source separation. Adv Data Anal Classif. 2017;11 (1):25–48.
5. Wood SU, Rouat J. Real-time speech enhancement with GCC-NMF. In: INTERSPEECH. 2017
6. Nesta F, Omologo M. Generalized state coherence transform for multidimensional TDOA estimation of multiple sources. IEEE Trans Audio Speech Lang Process. 2012;20(1):246–60.

7. Azadi M, Abutalebi HR. Modified state coherence transform to reduce spatial aliasing in TDOA estimation of multiple sound sources. In: International symposium on telecommunications. IEEE; 2015. p. 492–6
8. Nesta F, Omologo M. Generalized state coherence transform for multidimensional localization of multiple sources. In: Applications of signal processing to audio and acoustics, 2009. WASPAA '09. IEEE workshop on Vol.4. IEEE; 2009. p. 2360–71

Route Navigation System with A-Star Algorithm in Underground Garage Based on Visible Light Communication

Ying Yu, Jinpeng Wang$^{(\boxtimes)}$, Xinpeng Xue, and Nianyu Zou

School of Information Science and Engineering, Dalian Polytechnic University,
No. 1st Qinggongyuan, Ganjingzi, Dalian 116034, Liaoning, China
wangjp@dlpu.edu.cn

Abstract. In order to solve the problem of parking lot difficulty, communication security and low efficiency of garage, with the garage using LED lighting, designed an underground garage navigation system based on visible light communication, which uses the lighting system in the garage to realize the real-time monitoring of the parking space and the navigation of the vehicle. This paper designed the navigation system of underground garage based on the principle of visible light communication in the garage with LED lighting. Result shows that the system achieves signal transmission in the range of 4 m, which can meet the needs of vehicle navigation.

Keywords: Visible light communication · Underground garage ·
A* algorithm · LED

1 Introduction

In big cities, particularly in megacities, serials of problems, including the city traffic jam, environment pollution, and easy-angry drivers, etc., are caused by the parking problem. In the traditional way of parking, most people choose the parking space blindly, and some parking lots have to use manual guidance to enter the parking space aiming at the problem of parking [1]. However, the manual guidance has caused the waste of human resources. Some underground parking lots use the local sound control [2] method to guide the vehicle to stop, which is considering that the garage cannot receive the outside signal, and the ordinary navigation cannot be used in the underground garage. However, for the open environment of underground garage, when multiple vehicles enter the parking garage, the way of voice control can easily cause confusion for the garage [3]. For an underground garage, it is difficult for the owner to quickly find the best parking space without good guidance, thus reducing the efficiency of the garage.

This paper uses visible light communication technology (VLC) [4, 5], sensing technology, and the A*(A-Star) [6–8] search algorithm, to give a new underground garage parking mode. The photoelectric detector, consisted of LED lamps above the parking lots and photosensitive resistors on the ground, is used to detect the condition of parking lots; when vehicles enter the garage, the A* algorithm is selected to

© Springer Nature Singapore Pte Ltd. 2020
Q. Liang et al. (eds.), *Communications, Signal Processing,
and Systems*, Lecture Notes in Electrical Engineering 516,
https://doi.org/10.1007/978-981-13-6504-1_131

calculate the optimal path according to the vacant condition of the parking space. The system will use A* algorithm to find optimal path giving to the drivers according to the condition of vacant parking space. The path, which can be used to guide, could be expressed as square waves with different frequency generated via single chip. When the system works, a modem demodulates signals to LED, then receivers transform optical signals to different frequency electrical signals, and demodulates the electrical signal to the LCD screen through the modem [9]. After processing, LCD displays the path information, which can prompt the driver's driving direction and improve the efficiency of the garage.

2 The Plane Model of the Underground Parking Lot

The system has to be abstracted the entity appearance of the garage into a plane model, so it can determine the vehicle's running path. The A* algorithm, most effective direct search algorithm, is suitable for path planning. This paper used the A* search algorithm to find the optimal path, and the plane model of the garage is shown in Fig. 1.

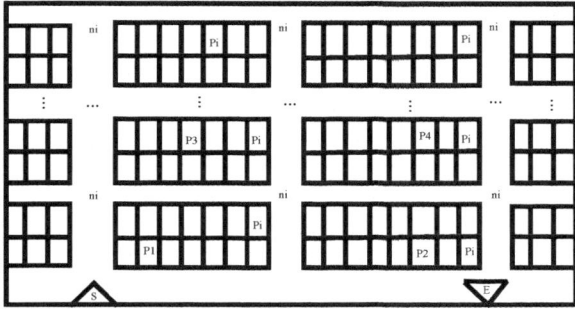

Fig. 1. Underground garage plane model

The principle of A* search algorithm is to design an evaluation function:
$f(n) = h(n) + g(n)$ In this function:

$f(n)$ the evaluation function from the initial point s to the target point P through the node n;

$h(n)$ the real cost of from the starting node 's' to the current node 'n' in space;

$g(n)$ the estimated cost from the current node 'n' to the destined node 'p'. The evaluation function is used to evaluate each point that can be reached at the next step of the current position, and the system can find the smallest point of the value $f(n)$ as the path node by searching each step. The algorithm must restart when the path is forwarded one step.

The specific steps of the A* algorithm to calculate the path:

1. Create a 'start' table and a 'route' table. The two tables are initiated to empty, and the starting point 's' is placed in the 'start' table;
2. Search for nodes in 'start' table. If 'start' table is empty, it means that no path is found, and fail to search;
3. If the 'start' table is not empty, this system will select a node with the minimum value 'f' to be the optimal node, which is denoted as 'm' and put into the 'route' table;
4. Determine whether node 'm' is the target node 'e,' and if node 'm' is the target node, a path is found successfully;
5. If the node 'm' is not the target node, then it is extended to generate the sub-node m_1, m_2 ... For each child node, the following procedure is performed (with the example of child node 'm_1'):

   - If 'm_1' is already in the 'start' table, calculate '$g(m_1)$,' and the original node 'm_1' in the 'start' table is called node 'oa,' compared with '$g(m_1)$' and '$g(oa)$.' If $g(m_1) < g(oa)$, the parent pointer of 'oa' is modified to 'm,' and the '$g(oa)$' value is corrected, and the smaller value '$g(m_1)$' is assigned to '$g(oa)$,' and the corresponding update '$f(oa)$' value; If '$g(m_1)$' is greater than '$g(oa)$', then the extension node is stopped;
   - If 'm_1' is already in the 'route' table, then this system skips this node and continues to extend the other nodes;
   - If 'm_1' is not in the 'start' table and not in the 'route' table at the same time, then put it in the 'start' table, and add a pointer to 'm^1' to its parent node 'm,' calculate $g(m_1)$;

6. Turn to step (2) and continue circulation until the solution is found or no solution is left.

The block diagram of the algorithm is shown in Fig. 2.

3 Experimental Model of Communication System

The VLC system [10] needs the following steps to accomplish: Firstly, the information that needs to be transferred is converted into an electrical signal. Then, the signal is processed by the transmitter circuit, the signal is loaded to the LED through the drive circuit, and the LED sends out the light wave with information after the modulation. Finally, the modulated optical signal is converted to electrical signal through photo-electric detector. After receiving the terminal circuit, the electrical signal is restored to the data information. The principle block diagram of the system work is shown in Fig. 3.

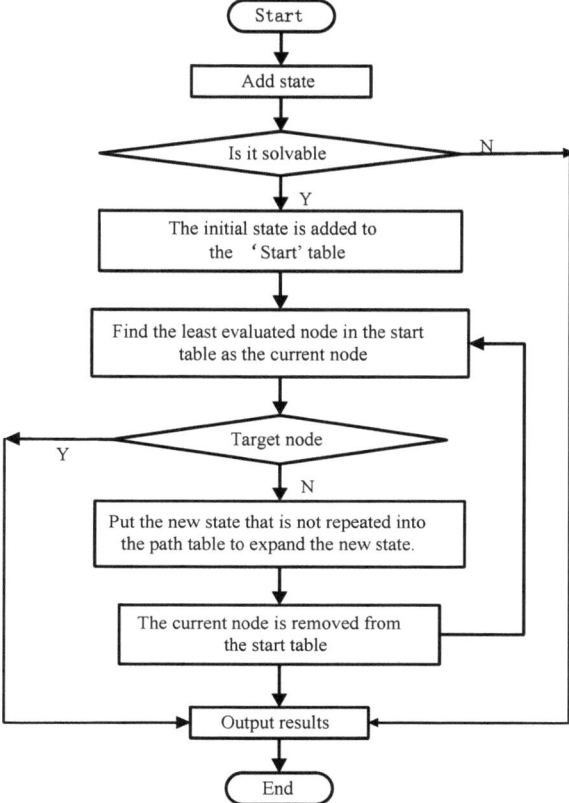

Fig. 2. Algorithm flowchart

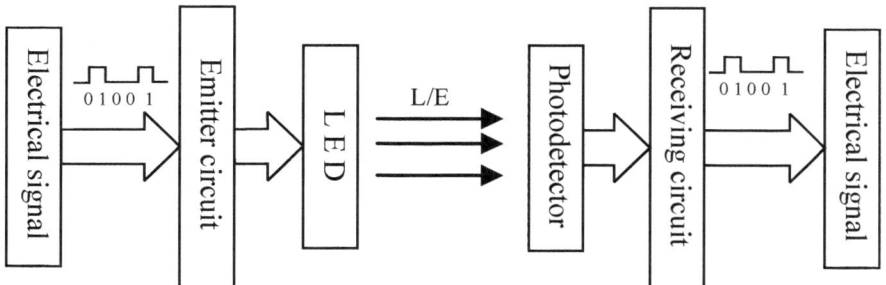

Fig. 3. System block diagram

3.1 Parking Information Monitoring

MSP430 control chip is used for the acquisition of parking information, and the monitoring of the parking lot is realized by the circuit of LED light source, photosensitive resistor, power supply, and voltage comparator. The photosensitive resistor is placed in the center of the parking space. The LED light source is placed above the photosensitive resistance, when the car is parked in the parking space, the light signal is blocked, the resistance value of the photosensitive resistor increases, and the voltage on both ends will increase. The signals produced by the amplifier are transferred to the voltage comparator and converted into high and low-level signals that can be identified by the single-chip microcomputer. Display on LCD screen after processing by single-chip microcomputer, the red block indicates that the car is parking, and the green block indicates no parking, in order to monitor the parking space.

3.2   The Signal Transmitter

After calculating by the A* algorithm, the system is used in expressing the path of vehicle planning by means of different frequency square wave signals produced by MSP430 and modulates the generated square wave signal to the LED drive circuit. Using the modulation characteristics of the high-frequency flicker of the LED light source, the direction information of the vehicle is loaded into the light wave and radiated in the free space, in which the structure diagram of the transmitter is shown in Fig. 4.

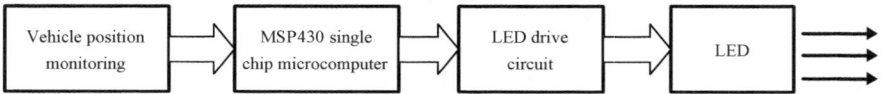

Fig. 4. Structure of the transmitter

The LED drive circuit is the main part of the transmitter. Combined with the actual garage situation, the power of LED lighting lamps is very large. In order to ensure the voltage constant at both ends of the lamp, the current drive LED method is adopted to ensure the normal work of the LED. The driving module takes the LM324 operational amplifier as the core driver and converts the voltage change of the front-stage circuit into the current change of the LED through the V/I conversion, in order to provide the normal work of the LED light source, and the LED light source drive circuit is shown in Fig. 5. At the same time, the regulation of variable resistance Rf can change output, and the circuit response speed is fast.

3.3   The Signal Receiver

The signal receiver is mainly composed of photodetector, second-order active preamplifier, filter circuit, and MSP430 single-chip microcomputer. In the optical receiving terminal changes the optical signals into electrical signals, which is amplified by the preamplifier circuit and sent to the single chip by serial port. Finally, the driving information is displayed on the LCD screen, and Fig. 6 shows the structure of the receiver.

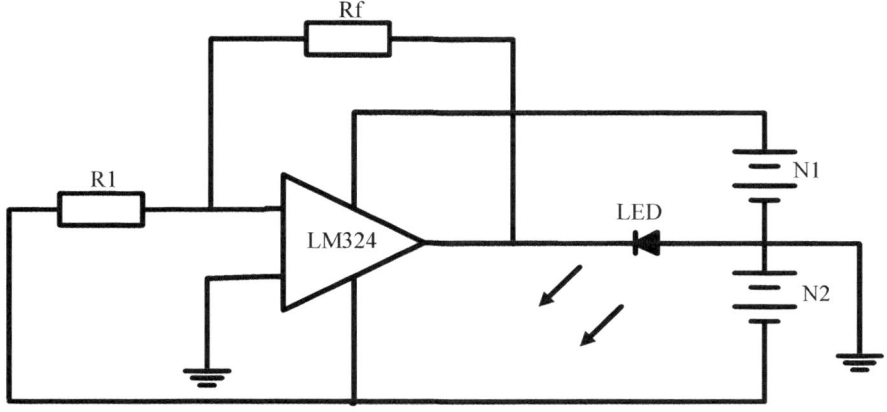

Fig. 5. LED light source drive circuit

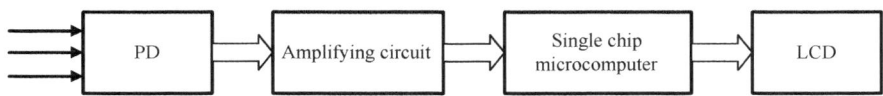

Fig. 6. Structure of the receiver

The preamplifier circuit of the receiver is the key part of the recovery of the optical signal and the sensitivity of the butt end has a direct influence, and it will also affect the transmission quality of the signal. The preamplifier circuit is based on the OP37 device, which has the advantages of extremely low input bias current and voltage noise. It is an ideal choice for high-speed trans-impedance amplifier and high impedance sensing amplifier. Figure 7 shows the trans-impedance preamplifier circuit and the two-order active filter circuit. When the receiver receives light at different frequencies, there will be a current Ir in the silicon photocell. Through the preamplifier circuit, the weak current signal will be amplified and converted to the voltage signal. The signal generated by the filter circuit is processed into a standard voltage signal and input to the single-chip microcomputer.

3.4 Modulation and Coding

In visible light communication system, there are mainly OOK modulation, PPM modulation, DMT modulation, OFDM modulation, and so on. In terms of the complexity of circuit design, the advantages of OOK modulation and PPM modulation are obvious, while DMT modulation and OFDM modulation have better spectrum utilization and overcoming multipath performance. In this system, because the circuit design is not complex, the OOK modulation mode is adopted. Through the '0' and '1' high and low pulse coding, we can control the brightness and darkness of LED lights and send data.

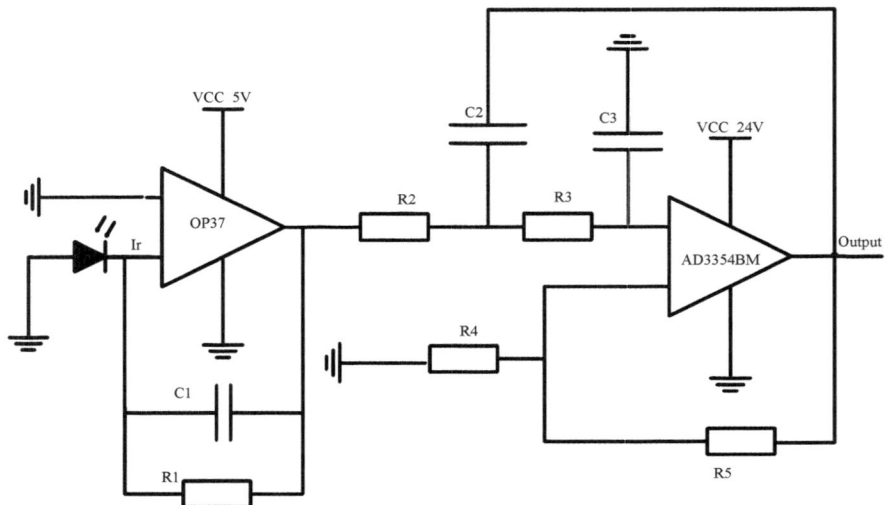

Fig. 7. Trans-impedance preamplifier circuit and active filter circuit

As the transmitter of the visible light communication, the LED as the transmission carrier can transmit a long series of 0 and 1 sequence signals with the brightness of LED lights which will not interfere with the radio frequency signal. Usually, when the frequency of LED in 50–60 Hz, because of the persistence of vision of the human eye, does not pay attention to the LED high-speed flash lamp; however, if the signal '0' and '1' are longer, it is possible that people will perceive the brightness change of LED. In order to meet the lighting requirements of the parking lot, the LED can still be used as a lighting appliance after transmitting the corresponding digital signal, so it is necessary to make the LED work at a high level after the signal transmission. In this paper, the non-return-to-zero (NRZ) coding of the signal can avoid the obvious brightness change of the LED to a certain extent, and it will not affect the normal work of the LED. The encoding of NRZ is shown in Fig. 8.

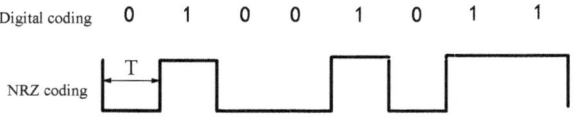

Fig. 8. NRZ encoding

4   Path Navigation

Motion sensors are used to determine the location of the moving vehicle, and the system acquirement of vehicle's path by algorithm, the LED at each node in the path indicates the driving direction of the vehicle in that position by emitting different '0' and '1' combination signals. When the motion sensor [11] detects the movement of the vehicle, the LED on the node emits the frequency signal, and the light detector on the car receives the signal and displays the signal on the LCD, and converts the path information that the driver can identify. When the vehicle leaves the node, the point can continue executing the operation instructions of the system for subsequent vehicles; similarly, when the vehicle is away from the parking location, the sensor detects the change of the vehicle, and the system plans the outgoing route for the vehicle. The program schematic diagram for path navigation is shown in Fig. 9.

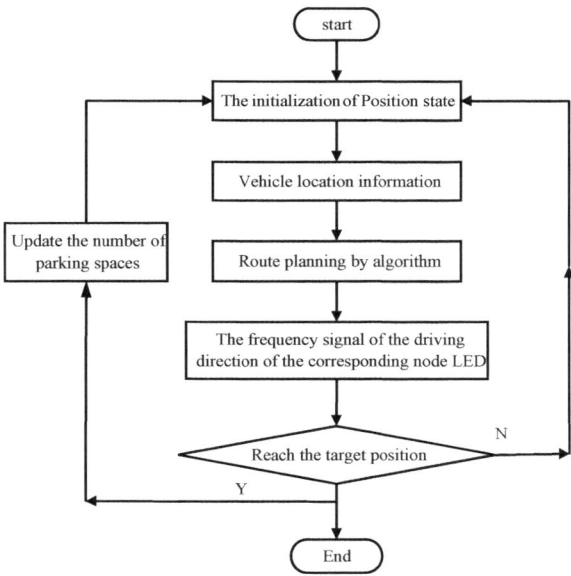

Fig. 9. Path navigation flowchart

5   Experiments and Discussions

Refer to the design standard of JGJ100-98 'specifications for building design of garage,' as shown in Table 1. In order to ensure that the system can be used for minivans, small cars, light cars, buses, and truck garages, the distance between the photodetectors of the LED light source vehicle is set to 4 m for testing.

Table 1. Minimum height of the car garage

Models	Minimum height (m)
Small cars	2.20
Light vehicle	2.80
Medium, large, coach	3.40
Medium, large freight car	4.20

The signal transmitting terminal of LED and the signal receiving terminal of photodetector are connected to the oscilloscope. We observe whether there are two similar waveforms on the oscilloscope. According to the design of each module circuit, the physical experiment model generated is shown in Fig. 10.

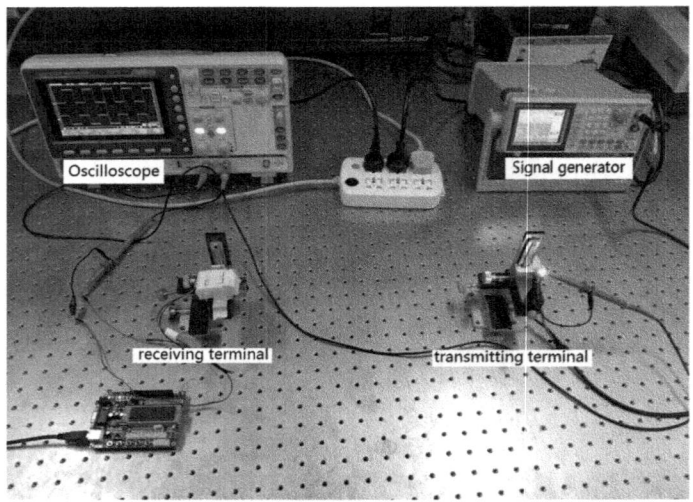

Fig. 10. Physical experiment model

Figure 11 shows the transmitting terminal and the receiving terminal signals detected by oscilloscope during signal transmission. The 4 m distance set in the experiment is acceptable to the signal, and the communication performance is good,

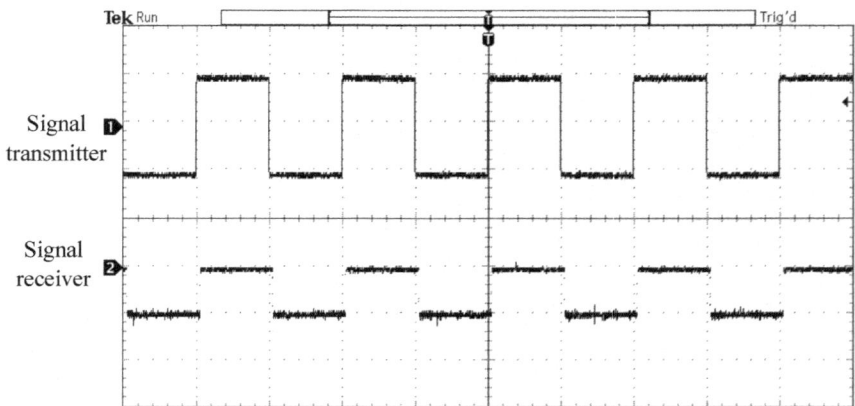

Fig. 11. Intercepted send and receive waveforms

and the received signal waveform is almost identical with the emitted signal waveform. Therefore, it can be seen that the system has certain practical application significance in underground garage.

6 Conclusion

This paper designed the navigation system of underground garage based on the principle of visible light communication in the garage with LED lighting. The system effectively reduces time to parking, avoids the blindness of people to find parking, by the LED lighting of the garage and implements the underground garage navigation based on visible light communication, is again for energy utilization. The system should be further studied and developed, which can be combined with the secondary searching system, and the influence of the moving speed of the vehicle on the reception signal should be considered in the practical application. The system has great market development value.

Acknowledgments. This research was financially supported by Project of the National Natural Science Foundation of China (61402069), 'the Fundamental Research Funds for the Central Universities' (3132016317), 2017 Project of the Natural Science Foundation of Liaoning province (20170540059), and General Project of Liaoning Education Department in 2016 (2016J205).

References

1. Bao X, Dai J, Zhu X. Visible light communications heterogeneous network (VLC-HetNet): new model and protocols for mobile scenario. Wirel. Netw. 2017;23(1):299–309.
2. Beijing Institute of construction and technology, editor in chief. JGJ100–98 garage design code Design Code for Garage. Beijing: China Construction Industry Press. August 2002.
3. Wang J, Zou N, Zhang Y, Li P. Study on downlink performance of multiple access algorithm based on antenna diversity. ICICExpress Lett. 2015;9(4):1221–5.
4. Zhao J, Li Y, Zhang Y, Zou N, Wang J. A supplementary lighting system for plant growth with lighting-emitting diode based on DT TS&IC. 2016;61(7):548–51.
5. Pan G, Ye J, Ding Z. Secure hybrid VLC-RF systems with light energy harvesting. IEEE Trans. Commun. 2017;65(10):4348–59.
6. Li L. A* algorithm analysis and research. Sci. Wealth. 2016;(9):3–3. https://doi.org/10.3969/j.issn.1671-2226.2016.09.003.
7. Iskander M, Aboumoussa W, Gouvin P. Instrumentation and monitoring of a distressed multistory underground parking garage. J. Perform. Constr. Facil. 2001;15(3):115–23.
8. Onan A, Bulut H, Korukoglu S. An improved ant algorithm with LDA-based representation for text document clustering. J Inf Sci. 2017;43(2):275–292.
9. Yao Q, Mou X, Jia Y, Zhao G, Zhang H. Underground garage intelligent lighting control system. Sci. Technol. Eng. 2014;(14):239–243.

10. Rahman MS, Kim B-Y, Bang M-S, Park, Y-I, Kim, K-D. Color space mapping and medium access control techniques in visible light communication. J Inst Internet Broadcast. Commun. 2009;9(4):99–107.
11. Rabadan J, Guerra V, Rodríguez R, et al. Hybrid visible light and ultrasound-based sensor for distance estimation. Sensors (Basel). 2017;17(2):330.

The Research of Fast Acquisition Algorithms in GNSS

Xizheng Song[✉]

Department of Electronic Engineering, Dalian Neusoft University of Information,
Dalian 116023, Liaoning, China
songxizheng@neusoft.edu.cn

Abstract. Recently, GNSS has been applied in various domains deeply and widely. In some of applications such as carbon canyon, GNSS signals degrade severely. The conventional receivers have no ability to deal with such weak signals. The sensitivity performance has already been one of the most important features in modern receivers. Consider the advantages of easy implement and high efficiency, we choose it with coherent integration and differential coherent integration to acquire signals which of power is −145 dBm.

Keywords: High-Sensitivity · Weak signal · PMF

1 Introduction

Signal acquisition, also known as coarse synchronization, is roughly obtaining the code phase of the satellite signal and carrier frequency, which is the first step for the GPS receiver to process the baseband signal. However, the signal strength of GPS signals in an indoor environment or a sheltered environment is 15–30 dB lower than that in outdoor open areas. Therefore, how to capture such a weak signal has become the first hard problem that needs to be solved for high-sensitivity receivers.

The traditional pseudo-code acquisition is achieved through correlation and energy detection. When the correlator outputs an energy peak and that exceeds the threshold, it indicates that the input signal pseudo-code phase is consistent with the local pseudo-code phase. However, this algorithm search is very slow, which is more obvious especially during the high-speed movement of the carrier.

Furthermore, a faster speed of acquisitions is required to improve the capturing sensitivity; therefore, it has become a necessity to take the fast capturing technology.

In this paper, the structure of the matched filter +FFT is adopted, and according to the design parameters, the signal with the signal strength of −145 dBm can be captured.

Q. Liang et al. (eds.), *Communications, Signal Processing, and Systems*, Lecture Notes in Electrical Engineering 516,
https://doi.org/10.1007/978-981-13-6504-1_132

2   The Partial Matching Filtering and the FFT Fast Acquisition Algorithm

2.1   The Integral Forms

Integration is necessary to improve the acquisition sensitivity. When processing the communication information, there are three integral forms, namely, coherent integration, non-coherent integration and differential coherent integration. Since the integrator is a low-pass filter, the high-frequency signal components and noise in the signal are filtered out, and therefore, the output signal has a high signal-to-noise ratio. The gain obtained is called the coherent integration gain G_{ci} [1]. Coherent integration is very effective for improving receiver sensitivity. However, the length of integration time is affected by various factors, which limits the further improvement of the coherent integration gain. Therefore, in order to have the ability to capture lower energy signals, the receiver generally performs N times of incoherent integration or differential coherent integration on the basis of coherent integration. Assuming that the total time for capturing points is T_{acq}, the coherent integration time is T_{COH}, then the incoherence times are:

$$N_{NC} = \frac{T_{acq}}{T_{COH}} \qquad (1)$$

Since a differential coherency requires the use of the two before and after coherent integration sampling points, the number of differential coherence then becomes:

$$N_{DC} = \frac{T_{acq}}{T_{COH}} - 1 \qquad (2)$$

For the same capture time T_{acq}, there are different combinations like T_{COH} and N_{NC}, and T_{COH} and N_{DC}. That is, if the coherent integration time increases, the number of incoherent coherence or differential coherence decreases; if the coherent integration time decreases, incoherent coherence or differential coherence will increase. Theoretically, the coherent integration has the best gain, while the non-coherent integration and the differential coherence integration have different degrees of loss. Therefore, the coherent integration time should be maximized to reduce the number of the non-coherent integration or the differential coherence integration. To compare the signal capturing capability, Monte Carlo experiments were performed on the acquisition of 1000 different 2 ms input signals with SNR = −20 dB, and the peak-to-noise gain was measured using the peak-to-average ratio. In the case of short bit-hopping, coherent integration is the best, and the coherence integration is better than non-coherent integration (as shown in Fig. 1). That is, in the same capture time, coherent integration time should be increased as much as possible. However, in the hardware receiver, the acquisition module is implemented by hardware, and longer coherent integration time means more sampling points and that more storage units are required.

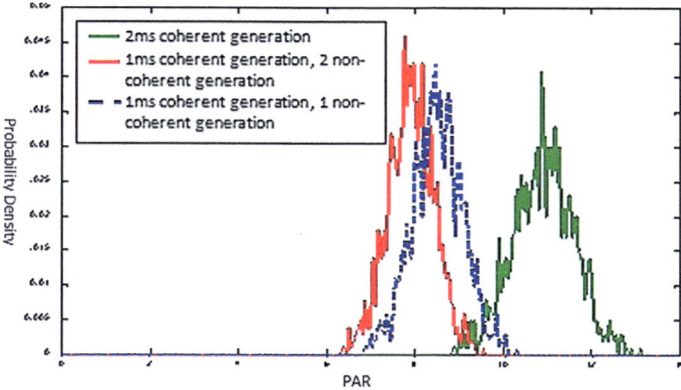

Fig. 1. Gain comparison of three kinds of integral form

2.2   The Basic Principle of Partially Matched Filter and FFT Fast Acquisition Algorithm

The fast acquisition algorithm with the combination of partially matched filter and Fourierism transform is a combination of time-frequency domain method [2], in which the matched filter to complete the serial search of code dimension, and the Fourierism transform to complete Doppler compensation to achieve the parallel Doppler dimensional search algorithm. Compared with the parallel frequency-domain-based processing technique, this algorithm is simpler, and more flexible, and it reduces the number of FFT kernels greatly, and saves resources, and it is more suitable for capturing under low signal-noise ratio and high dynamic environment. The matching filter is implemented in the form of piecewise matched filtering. The specific schematic diagram is shown in Fig. 2.

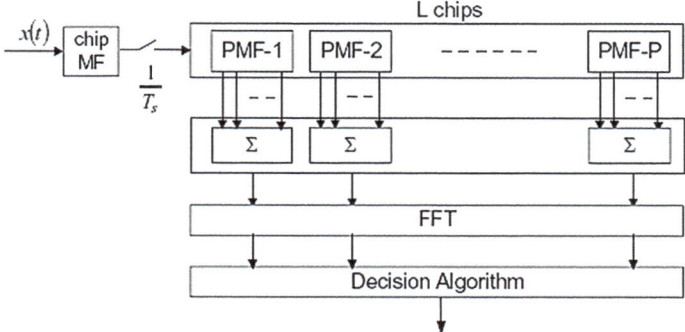

Fig. 2. Diagram of the piecewise matched filtering and FFT algorithm

Matching filter implementation is the core of the algorithm. For a communication system, the detection probability of a signal is directly related to the signal-to-noise ratio: The greater the signal-to-noise ratio, the greater the detection probability and the smaller the error probability; on the contrary, the smaller the signal-to-noise ratio, the smaller the probability of detection and the greater the probability of error. Therefore, in order to improve the detection probability of the communication system or to suppress the error probability, a preprocessing of the received signal should be performed so that we can get a greater output signal-to-noise ratio after the preprocessing. This principle is called the maximum signal-to-noise ratio criterion.

2.3   The Impact of the Partial Correlation Length and FFT Points on the Acquisition Performance

Under normal circumstances, when the spread spectrum gain is selected (i.e., the number of coherently added points is a fixed value), the number of segment points, the number of chips per segment and the number of FFT points will lead to a decrease of the accumulated peak attenuation coefficient [3]. Figure 3 shows the normalized curve of the system peak when different segments M and FFT value P, when there is no noise involved.

As can be seen from Fig. 3, the larger the correlation length N, the smaller the normalized correlation peak of the FFT output. Therefore, the selection of the partial correlation length M and the FFT value P is determined by the bandwidth of the frequency difference, the frequency resolution and the system attenuation function, and depending on the different types of application, it can be selected flexibly.

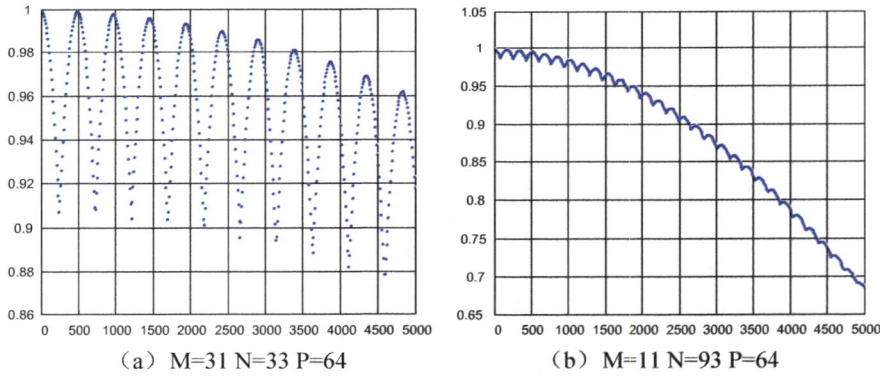

(a) M=31 N=33 P=64   (b) M=11 N=93 P=64

Fig. 3. Impact of the partial correlation length and FFT points on the acquisition performance

2.4   The Structure of Matching Filter

Matched filter is the core of the whole design of the fast acquisition module, which is equivalent to a FIR filter [4], shown in Fig. 4. R stands for the register, and assuming

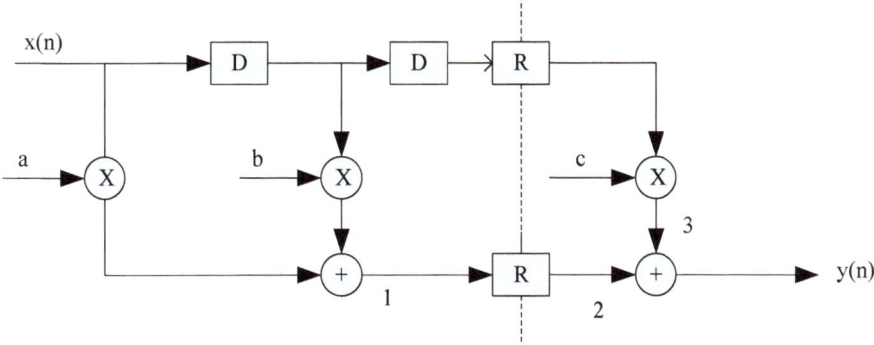

Fig. 4. diagram of the pipeline matching filter structure

there is no register case, T_M is the multiplier delay and T_A is the adder delay, then the critical path delay is $T_M + 2T_A$, so the data cycle of the filter output T_{out} needs to meet

$$T_{out} = T_M + 2T_A \tag{3}$$

The pipelined approach is used in the actual design. Two registers were introduced in the figure, where x(n) is equivalent to the local C/A code sequence, and a, b, c is equivalent to the received data after AD.

By using the pipeline, the critical path delay is reduced from $T_M + 2T_A$ to $T_M + T_A$, which has effectively improved the data processing speed. Table 1 describes the processing timing of the water matched filter.

Table 1. Processing timing of the pipeline matching filter structure signal

Clk	Input	Node 1	Node 2	Node 3	Output
0	x(0)	ax(0) + bx(−1)	−	−	−
1	x(1)	ax(1) + bx(0)	ax(0) + bx(−1)	cx(−2)	y(0) = ax(0) + bx(−1) + cx(−2)
2	x(2)	ax(2) + bx(1)	ax(1) + bx(0)	cx(−1)	y(1) = ax(1) + bx(0) + cx(−1)
3	x(3)	ax(3) + bx(2)	ax(2) + bx(1)	cx(0)	y(2) = ax(2) + bx(1) + cx(0)

As can be seen from the table, in the case when there is a large amount of data, theoretically, each clock will produce an output, and because of the match filtering hierarchy, the logic delay of each stage decreases, and the clock speed increases accordingly.

3   Partially Matched Filter and FFT Fast Acquisition Scheme

According to the actual requirements, a compromise was made between the acquisition time and the sensitivity. Therefore, in this scheme, the coherent integration length of the matched filter is 10 ms, which is divided into 128 segments with integral time of (10/128) ms, and then 128 points of FFT were used to estimate Doppler. Through the simulation analysis of MATLAB, the acquisition sensitivity gets to 141 dBm after using 10 ms coherent integration. Considering the quantification loss, noise correlation and various control loss (2 dB) of the limited data bits, the acquisition sensitivity can reach −139 dBm. For weak signals, non-coherent integration capture is started by software setting. Non-coherent acquisition is performed seven times, and in order to solve the message bit reversal, two consecutive 10 ms data segments are captured, of which a bit must not be bit-flipped. The specific capture strategy is to pre-store 140 ms of data, and then for every 10 ms process a fast acquisition, and then do the accumulation of the odd times of 10 ms and even times of 10 ms processed FFT matrix, with a total of seven cumulative to complete the non-coherent processing. The capture results are shown in Fig. 5. As can be seen from the figure, the capture peak is obvious compared with other clutter peaks, with a capture sensitivity of −145 dBm (remove the loss of 2 dB).

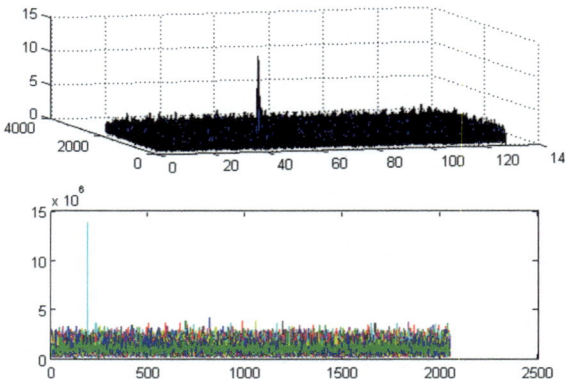

Fig. 5. Simulation results when SNR = −36 dB (−147 dBm), fd = 3800 Hz

According to the theoretical analysis, when using the above parameters, the theoretical Doppler search range reaches −12.8 to 12.8 kHz, but when the Gaussian white noise is added in the simulation, the Doppler search range is significantly reduced. In addition, since at this time the Doppler accuracy is 100 Hz, when the Doppler frequency is exactly the valley of the scallop loss, the acquisition peak is no longer noticeable. Figure 6 illustrates the impact that both cases have on the acquisition results.

As can be seen from Fig. 6, when Doppler reaches 4.8 kHz, the acquisition peak is nearly submerged in the noise, while for a typical receiver, −4.8 to 4.8 kHz basically can cover the normal Doppler range, but for the receivers that have higher dynamically demands, the search range cannot cover all Doppler frequencies. Therefore, the local carrier frequency must be changed to search for other Doppler frequency ranges.

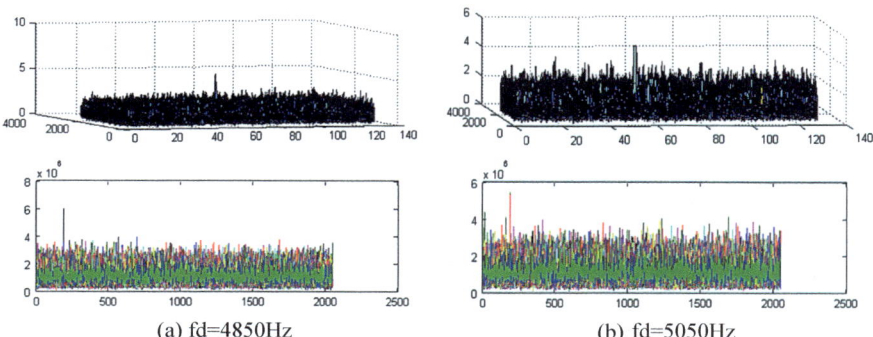

(a) fd=4850Hz   (b) fd=5050Hz

Fig. 6. Impact of different doppler on the acquisition results when SNR = −36 dB

4   Conclusion

The first step of processing baseband signal in the acquisition of satellite navigation signals, the improvement of the acquisition sensitivity and the improvement of the acquisition speed has always been the primary goals of satellite navigation signal acquisition. In this paper, by analyzing the matched filter and several integral forms, by using the method of the combination of the partial matched filter and FFT fast acquisition, the acquisition speed is increased, and at the same time the acquisition sensitivity is improved. Moreover, under the general dynamic conditions, this method generally covers the general Doppler range, and the experimental simulation and actual hardware implementation tests both show that the method has a great effect of the signal acquisition.

References

1. yi J, Shufang Z, Qing H, Xiaowen S. A new FFT-based acquisition algorithm for GPS signals. In: 2008 education technology training and 2008 international workshop geoscience remote sensing. ETT and GRS 2008 international workshop on, vol. 2, 2008.
2. Tian S, Pi Y. Research of weak GPS signal acquisition algorithm. In: International conference on communications circuits and systems 2008 (ICCCAS 2008). 2008; p. 793–6.
3. Akopian D, Fast FFT based GPS satellite acquisition methods. IEE Proceeding-Radar Sonar Navig. 2005;152(4).
4. Spillard CL, Spangenberg SM, Povey GJR. A serial-parallel FFT correlator for PN code acquisition from LEO Satellites. In: Proceedings of ISSSTA'98, 1998.

Research on BDS/GPS Combined Positioning Algorithm

Hong-Fang He$^{(\boxtimes)}$, Xin-Yue Fan, and Kang-Ning An

Key Laboratory of Optical Communication and Networks, School of Communication
and Information Engineering, Chongqing University of Posts and
Telecommunications, Chongqing 400065, China
hehfang@hotmail.com, fanxy@cqupt.edu.cn, ankangning@163.com

Abstract. When single satellite navigation system is used for position-ing, there exist the following problems: accuracy of positioning is low and reliability of positioning is also low. This paper investigates the com-bined positioning algorithm of BDS and GPS for static and dynamic observation point. For static observation point positioning, the weight coefficient integrated positioning algorithm of BDS and GPS is proposed. For dynamic observation point positioning, the Kalman filter can achieve smoothing of the movement trajectory which is the once combined obser-vation point. The experimental results show that the combination of BDS and GPS is more accurate and reliable than any single system for static observation point positioning. At the same time, for the consideration of the weight coefficient, the system has good adjustability and practi-cality, and the Kalman filter can better modify the dynamic combined observation point.

Keywords: Combined positioning · Static point positioning ·
Dynamic point positioning · Weight coefficient

1 Introduction

Beidou Navigation Satellite System (BDS) is one of the four satellite navigation systems independently developed by China. Since the laboratory system was established in 2000, 15 orbital satellites have been developed to cover the entire Asia-Pacific region by 2012, and it will provide services around the world by 2020. It includes geostationary (GEO) satellites, inclined geosynchronous satellite orbit (IGSO) satellites, and medium earth orbit (MEO) satellites distributed over 21,000 km and over 35,000 km [1]. Global Positioning System (GPS) is developed by the US Land and Air Force. It was researched quite early and had achieved a global coverage of 98% in 1994. There are MEO satellites distributed over 20,200 km [2]. Under the research of all parties, the accuracy of single point positioning of BDS and GPS can reach a level of 10 m and an elevation of 10 m [3,4].

© Springer Nature Singapore Pte Ltd. 2020
Q. Liang et al. (eds.), *Communications, Signal Processing,*
and Systems, Lecture Notes in Electrical Engineering 516,
https://doi.org/10.1007/978-981-13-6504-1_133

With the gradual improvement of the four major satellite navigation systems, the future air constellation satellites will have more than 120 satellites, providing more stable, reliable, and accurate services of navigation and positioning for global users [5]. However, facing many satellite navigation information resources, its body of the application has gradually evolved from single navigation positioning to a new era of multi-information integration, such as collective positioning and the Internet. Global navigation satellite system (GNSS) has also entered the era of multi-system compatible and cooperative development [6]. The development time of BDS is short, and the technology is not yet perfect in this filed. Therefore, the combination of mature GPS and BDS can further improve accuracy and reliability of navigation positioning [7,8].

The satellites of BDS and GPS have different orbital altitudes, which has created some obstacles for the study of tight combinations [9]. Currently, BDS and GPS are capable of doing independently navigation positioning, the accuracy of single point positioning is not accurate enough, and researching the combined positioning of observation points under single point positioning is rare [10]. This paper designs the combined observation point positioning algorithm of BDS and GPS. For once positioning and multiple positioning of static observation points, a corresponding weight coefficient integrated positioning algorithm is put forward. At the same time, based on once positioning of static observation points, the Kalman filter can smooth the movement trajectory for the dynamic observation point. This design not only has adjustability, but also has strong practicality and confidentiality.

2   Static Point Positioning

When the static observation point is located, once and multiple positioning can be performed on the same observation point by using BDS and GPS simultaneously. Therefore, we can get different positioning coordinates. A more accurate coordinate can be obtained by combined method ultimately.

2.1   Once Positioning

At a fixed point, once positioning is used to do the same static observation point by using BDS and GPS simultaneously. Thus, two positioning coordinates are obtained to perform the combined positioning. In the longitude and latitude direction of the observation point, assume that (x_B, y_B) and (x_G, y_G), respectively, are predictive coordinates of BDS and GPS, and (x, y) is the combined coordinate of BDS and GPS. The positioning accuracy of specific receiver determines observed gain errors of BDS and GPS. $\varepsilon_B = 5\,\mathrm{m}$ and $\varepsilon_G = 2\,\mathrm{m}$, respectively, are the gain error counted from the receiver in the paper. When the positioning coordinates of two-systems are the same, $(x, y) = (x_B, y_B) = (x_G, y_G)$ is the combined coordinate of BDS and GPS. However, the positioning accuracy of two systems is different in practical operation. Assume that the combined coordinate and range, respectively, are calculated as follows:

$$(x, y) = \frac{1}{2}(x_B, y_B) + \frac{1}{2}(x_G, y_G)$$

$$r_i = \sqrt{(x_i - x)^2 + (y_i - y)^2} \quad (i = B, G) \tag{1}$$

When $r_i \leq 2\varepsilon_i$ is established, the weight coefficient as follows:

$$\delta_i = 1 - \frac{r_i}{r_B + r_G} \quad (i = B, G) \tag{2}$$

When $r_i > 2\varepsilon_i$ is established, if the range is large, the weight coefficient will be 0, and if the range is small, the weight coefficient will be 1, which means that if one system suffers from strong interference, it will not participate in positioning. Therefore, the combined coordinate of BDS and GPS is calculated as follows:

$$(x, y) = \delta_B(x_B, y_B) + \delta_G(x_G, y_G) \tag{3}$$

2.2   Multiple Positioning

At a fixed point, multiple positioning is used to do the same static observation point by using BDS and GPS simultaneously. Thus, multiple discrete coordinates are obtained to perform combined positioning. The idea of combined positioning is to leave dense coordinates and remove discrete coordinates. When there are enough sample points, it is evenly distributed among a certain neighborhood of the precise coordinates, and offset errors can cancel each other out.

In the longitude and latitude direction of the observation point, assume that (x_i^B, y_i^B) and (x_i^G, y_i^G), respectively, are predictive multiple discrete coordinates of BDS and GPS, and (x, y) is accurate combined coordinate of BDS and GPS. Let us take BDS as an example, the specific method is as follows:

Step 1: Calculating the desired point coordinate in all discrete coordinates is as follows:

$$\left(x_0^B, y_0^B\right) = E\left\{\left(x_1^B, y_1^B\right), \left(x_2^B, y_2^B\right), \ldots, \left(x_n^B, y_n^B\right)\right\} \tag{4}$$

Step 2: Respectively, taking the discrete point (x_i^B, y_i^B) $(i = 0, 1, \ldots, n)$ as the center of circle O_{Bi} and the error gain ε_B as the radius. The number of all coordinates contained in the circle O_{Bi} is n_{Bi}. Therefore, their probability is $p_i = \frac{n_{Bi}}{n}$.

Step 3: When $p_i \geq p_0$ is established, corresponding positioning coordinates are remarked as $\left(x_j^B, y_j^B\right)$ $(j = 0, 1, 2, \ldots, m_B)$. If the number of observed effective positioning coordinates is $m_B \geq 1$, the collection of available positioning positions is $\theta_B = \bigcap_{j=1}^{m_B} O_{Bj}$, and if the number of observed effective coordinates is $m_B = 0$, the collection of available positioning positions is $\theta_B = O_{B0}$.

As above, we can derive the number of effective coordinates m_G and the collection of available positioning positions θ_G observed by GPS. Then, (x_B, y_B)

and (x_G, y_G), respectively, are barycentric coordinates of the collection of available positioning positions. Finally, the precise combined coordinate of BDS and GPS is shown in formula (3), where the weight coefficient as follows:

$$\delta_i = \frac{m_i}{m_B + m_G} \ (i = B, G) \tag{5}$$

3   Dynamic Point Positioning

When the dynamic observation point is located, data is gathered using BDS and GPS simultaneously in a very short period of time, in which combined positioning of observation points is carried out. The specific method is the same as the static observation point. In fact, the movement trajectory of the BDS/GPS combined observation point is determined, and two ways can be used generally. The first is to collect many discrete coordinates as combined positioning coordinates in very short time intervals, which fit it directly to the movement trajectory. The second is to collect once coordinates as combined positioning coordinates in a very short time interval, which smooth the movement trajectory using Kalman filter. The first operation is less difficult, and the second effect is good.

This paper determines the movement trajectory of BDS/GPS combined observation point by the second method. The Kalman filter has a correlation with each epoch. The observations of the previous moment have an effect on the latter moment, and it has good smoothness. The study, which is about the movement trajectory of combined observation points of BDS and GPS based on Kalman filter, is assumed to be performed under linear conditions. Now only considering the uniform tracking position of the observation point in the two-dimensional plane, the selected state vector of the observation point is as follows:

$$x_k = \begin{bmatrix} x_t\,(k) \ \dot{x}_t\,(k) \ y_t\,(k) \ \dot{y}_t\,(k) \end{bmatrix}^T \tag{6}$$

where the coordinate of observation point under each epoch is $(x_t\,(k)\,, y_t\,(k))$, and the speed of observation point under each epoch is $(\dot{x}_t\,(k)\,, \dot{y}_t\,(k))$.

Owing to lack of control variable, the available system model and measurement model of the observation point are as follows:

$$\begin{aligned} x_{k+1} &= F x_k + B_w w_k \\ z_{k+1} &= H x_k + v_k \end{aligned} \tag{7}$$

where the state transition matrix F, B_w and the measurement matrix H are as follows:

$$F = \begin{bmatrix} 1 & T & 0 & 0 \\ 0 & 1 & 0 & 0 \\ 0 & 0 & 1 & T \\ 0 & 0 & 0 & 1 \end{bmatrix} \quad B_w = \begin{bmatrix} \frac{T^2}{2} & 0 \\ T & 0 \\ 0 & \frac{T^2}{2} \\ 0 & T \end{bmatrix} \quad H = \begin{bmatrix} 1 & 0 & 0 & 0 \\ 0 & 0 & 1 & 0 \end{bmatrix} \tag{8}$$

The sampling interval is set to T. w_k and v_k, respectively, are system state noise and measurement noise of Gaussian white noise sequence. D_{w_k} and D_{v_k}, respectively, are its covariance matrix. The equations are as follows:

$$\begin{cases} E(w_k) = 0 \\ E(v_k) = 0 \\ \text{cov}(w_k, w_j) = D_w(k)\delta_{kj} \\ \text{cov}(v_k, v_j) = D_v(k)\delta_{kj} \\ \text{cov}(w_k, v_j) = 0 \end{cases} \tag{9}$$

δ_{kj} is the Kronecker function, and $E(x_0) = u_x(0)$ var$(x_0) = D_x(0)$ cov$(x_k, v_k) = 0$, cov$(x_k, w_k) = 0$ denotes the initial state of the system respectively.

4   Experiment Analysis

We choose the outdoor playground in Chongqing University of Posts and Telecommunications as our experiment area. Experimental data were collected by GPS/BDS two-system compatible receiver, which respectively collected 600 static data and 100 dynamic data by GPS and BDS. The measurement interval was set to 1s. The following is a test of static positioning error using multiple positioning, and the dynamic positioning error using the Kalman filter. Respectively, we analysis the positioning error in the longitude direction of the x-axis and latitude direction of the y-axis.

4.1   Static Point Positioning Analysis

To evaluate the combined positioning performance of static observation point, we compare the positioning error of single-system and two-system combined observation point after multiple positioning. In the two-dimensional coordinate system of longitude and latitude, standard deviation and root mean square error are defined as follows:

$$\sigma_{2D} = \sqrt{\frac{\sum_{i=1}^{n} \left((x_i - u_x)^2 + (y_i - u_y)^2 \right)}{n}} \tag{10}$$

$$\delta_{2D} = \sqrt{\frac{\sum_{i=1}^{n} \left((x_i - x_0)^2 + (y_i - y_0)^2 \right)}{n}}$$

where n is the number of epochs, and (x_i, y_i) is the positioning coordinates of the epoch data. u_x and u_y, respectively, are the expected values of their corresponding positioning point. The true coordinate of the test location is (x_0, y_0).

Table 1 shows the STD and RMS values obtained from the statistics of static observation points after multiple positioning.

As can be seen from Table 1, in the longitude and latitude direction of the observation point, the positioning effect of static observation point by GPS is better in the single system, and the positioning effect of BDS is relatively poor. At the same time, compared with the positioning of single system, the positioning error of the combined observation point of BDS and GPS is the smallest, and the positioning error of the combined observation point is relatively minimal in the two-dimensional coordinate system of longitude and latitude. Therefore, the combined positioning of static observation point of BDS and GPS is more accurate after multiple positioning.

Table 1. STDs and RMSs of single-system and two-system combined observation point

Scheme	STD			RMS		
	σ_x	σ_y	σ_{2D}	δ_x	δ_y	δ_{2D}
BDS	1.4441	2.0834	2.9590	1.5391	2.3390	3.0588
GPS	0.8524	1.4384	2.0506	1.3493	1.8708	2.5992
BDS/GPS	0.7614	0.9062	1.7318	0.8063	0.9329	1.8331

4.2   Dynamic Movement Trajectory Analysis

To evaluate the combined positioning performance of dynamic observation point, we compare the movement trajectory of single-system and two-system combined observation point after once positioning. At the same time, we compare the positioning error of directly fitting combined observation points and the Kalman filter smoothing observation points. Figure 1a shows the movement trajectory of single-system and two-system combined observation point and the Kalman filter smoothing observation point. Figure 1b shows the positioning errors of BDS/GPS combined and the Kalman filter smoothing observation points.

It can be seen from Fig. 1a that the corrected movement trajectory using the Kalman filter is closer to the true movement trajectory. Tracking points of

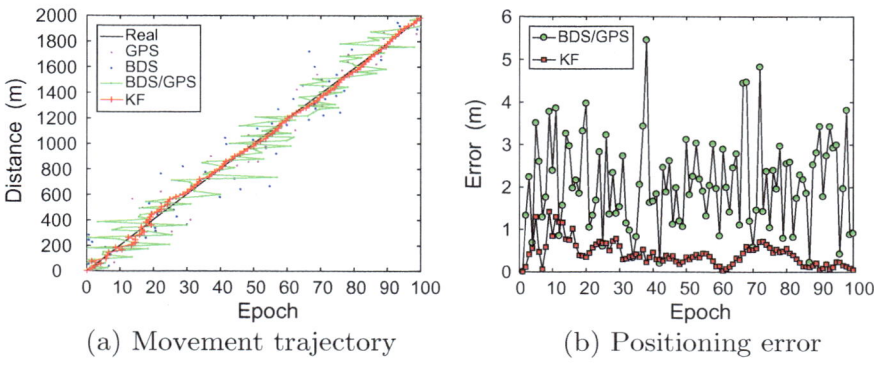

(a) Movement trajectory   (b) Positioning error

Fig. 1. Movement trajectory and positioning error

BDS and GPS are distributed on both sides for real movement trajectory. The fitted movement trajectory did not achieve very good effect after the combined observation points of BDS and GPS, but the movement trajectory is smoother using the Kalman filter.

It can be seen from Fig. 1b that the positioning error using the Kalman filter is less than the combination of BDS and GPS, and the positioning error is stable within 1.5 m, which means that the Kalman filter has a good corrected effect on the processing of dynamic observation points.

5   Conclusion

In this paper, we investigate the combined positioning of static and dynamic observation points. Once positioning and multiple positioning of static observation point are separately analyzed, two situations that can be used for the movement trajectory of the dynamic combined observation points have been introduced. Finally, for static combined observation point positioning, multiple positioning is chosen to determine a more accurate combined positioning coordinate. For the movement trajectory of the dynamic combined observation points, the Kalman filter has a good corrected effect after once combined positioning.

At the same time, the proposed weight coefficient integrated algorithm is adjustable, which means that not only an accuracy positioning can get by BDS and GPS, but also the positioning does not depend on BDS or GPS. The algorithm can be used in the military field, because they are practical and confidential.

Acknowledgments. This work was supported by the National Natural Science Foundation of China (61471077).

References

1. Wang B, Lou Y, Liu J, Zhao Q, Su X. Analysis of BDS satellite clocks in orbit. GPS Solut. 2015;20(4):1–12.
2. Yang D, Yang J, Li G, Zhou Y, Tang CP. Globalization highlight: orbit determination using BeiDou inter-satellite ranging measure-ments. GPS Solut. 2017;21(3):1395–404.
3. Chen H, Xu C, Gao J, Song X, Yuan L. Precision analysis of pseudorange single point positioning by BDS. GPS and combined BDS/GPS: J Shandong Univ Sci Technol; 2015.
4. Man X, Sun F, Liu S, Li H, Ding H. Analysis of positioning performance on combined BDS/GPS/GLONASS. Berlin, Heidelberg: Springer; 2015.
5. Wei E, Liu X, Liu J. Accuracy evaluation and analysis of single point positioning with BeiDou and GPS. Bulletion Surv Mapp. 2017.
6. Yuanxi Y. Progress, contribution and challenges of Compass/Beidou satellite navigation system. Acta Geod Cartogr Sin. 2010;39(1):1–6.
7. Dai F, Mao X. BDS/GPS dual systems positioning based on kalman filter in urban canyon environments, In: IEEE International conference on intelligent transportation systems; 2014. p. 1882–3.

8. Zhang K, Hao J. Research on BDS/GPS combined single-epoch attitude determination performance (2017).
9. Zeng A, Yang Y, Ming F, Jing Y. BDS-GPS inter-system bias of code observation and its preliminary analysis. GPS Solut. 2017;21(2):1–9.
10. Liu R, Gao X. The integrated positioning algorithm of "Beidou". GPS and GLONASS Satellite navigation system: Basic Sci J Text Univ; 2017.

Indoor Positioning with Sensors in a Smartphone and a Fabricated High-Precision Gyroscope

Dianzhong Chen(✉), Wenbin Zhang, and Zhongzhao Zhang

Communication Research Center, Harbin Institute of Technology,
Harbin 150001, China
dc2e12@163.com, zwbgxy1973@hit.edu.cn,
zzzhang@hope.hit.edu.cn

Abstract. In the paper, an indoor positioning scheme combining pedestrian dead reckoning (PDR) and magnetic strength matching (MSM) is proposed. PDR is conducted by sensing acceleration and angular speed through the 3-axis accelerometer in iphone7 and a fabricated high-precision rotational gyroscope. Low bias stability (0.5°/h) of the gyroscope contributes to a small accumulative error in heading angle estimation. Through data analysis to outputs of the accelerometer and the gyroscope, human motion, such as walking a step, walking upstairs or downstairs, turning left or right, is recognized and walking path is reckoned with motion information. Magnetic strength is measured by the magnetometer in iphone7 and MSM positioning result is used to reduce error of reckoned heading angle. The error rate of downstairs/upstairs step count is low and after heading angle correction by MSM, a satisfactory indoor positioning result is obtained.

Keywords: Indoor positioning · Pedestrian dead reckoning (PDR) · Magnetic strength matching (MSM) · Modified dynamic time warping (DTW) algorithm

1 Introduction

Service requirements of indoor positioning have been growing rapidly over time. Though global navigation satellite system (GNSS) for outdoor navigation has been mature and widely utilized, indoor positioning is still a task for the facts that: (1) GNSS signal in indoor environment is unavailable or seriously attenuated [1]; (2) indoor environment is complex with the phenomenon of not line of sight (NLoS) [2] and interference from humans [3]; (3) human path is not restricted to specific route as vehicles [4], however, with unpredictable turns [5]. Existing indoor positioning technology includes Wi-Fi positioning, Bluetooth Low Energy (BLE) positioning, radio frequency identification (RFID) positioning, dead reckoning (DR), MSM, and so on. Wi-Fi positioning, with accuracy of 10–20 m, is widely used for mature IEEE 802.11 standard and universality of Wi-Fi signal receiver, such as smartphones [6]. However, this technology has problems of high reliability on availability and distribution of Wi-Fi signals [7], attenuation [8], and multipath effect [9]. BLE positioning depends on dense arrangement of signal sources and RFID positioning requires special signal

© Springer Nature Singapore Pte Ltd. 2020
Q. Liang et al. (eds.), *Communications, Signal Processing,
and Systems*, Lecture Notes in Electrical Engineering 516,
https://doi.org/10.1007/978-981-13-6504-1_134

receivers, which restrict their widespread application. DR, based on measured linear acceleration signal and angular speed signal from inertial sensors (accelerometer, gyroscope) to calculate motion trail without receiving any external measured signals, is a positioning method different from others. Thus, DR is immune to outside interference or change of environment. However, for frequently used inertial MEMS sensors in watches or smartphones, inherent problems such as temperature sensitivity [10, 11], non-ideal repeatability of input–output characteristic for different measurements, low bias stability, cause measurement errors. Moreover, heading angles and linear speeds are the accumulation of measured angular speeds and linear accelerations, which will accumulate measurement errors [12]. DR with MEMS sensors is often used in short-distance path reckoning, and the other method is needed to modify the path. MSM can be a supplementary method to modify positioning error by PDR, as introduced in [13, 14]. Positioning performance of MSM relies highly on distribution condition of magnetic strength [15, 16]. Higher stability in time and larger variations in space of magnetic strength distribution contribute to better positioning precision by MSM.

In this paper, an indoor positioning scheme combining PDR and MSM is proposed. The three-axis accelerometer in iphone7 and a fabricated high-precision gyroscope (with the bias stability of 0.5°/h) with a ball-disk shaped rotor [17, 18] compose the inertial measurement unit (IMU). Construction of IMU is introduced in Sect. 2 with the structure of the fabricated gyroscope illustrated. In Sect. 3, PDR based on human motions through analysis to IMU signals is described in detail. In Sect. 4, the principle of MSM positioning and procedure of modified DTW algorithm are introduced and an example of MSM positioning with the modified DTW algorithm is given. An experiment of indoor positioning is conducted and the precision of the proposed scheme is analyzed in Sect. 5. Section 6 concludes the significance of the proposed indoor positioning scheme.

2 IMU Construction

The bias stability of the MEMS rate-grade gyroscope in smartphones is in the range of 10–1000°/h, which will lead to a large accumulative error in indoor positioning. A tactical-grade gyroscope with the bias stability of 0.5°/h is fabricated, constructing the IMU with the 3-axis accelerometer in iphone7. The fabricated gyroscope is a rotational gyroscope with a ball-disk shaped rotor supported by a water-film bearing, based on magnetic self-restoring effect to balance Coriolis torque, as shown in Fig. 1. Measurement range of the fabricated gyroscope is −30°/s to 30°/s, which is restricted by the linearity of differential capacitance detection [17, 18]. Nevertheless, for positioning application with memory module and processing module, it is feasible to construct a lookup table with corresponding differential capacitances and input angular speeds. To further increase the measurement range, the distance between the detection electrode and the rotor disk is increased from 100 μm to 500 μm. And the rotor deflection angle range increases from 1° (restricted by linearity) to 6°. And correspondingly, maximum measurable input angular speed increases from 30°/s to 180°/s (3.14 rad/s), enough for walking action recognition. But as a trade-off, sensitivity decreases, however, enough for PDR application. Low bias stability value of the

(a)

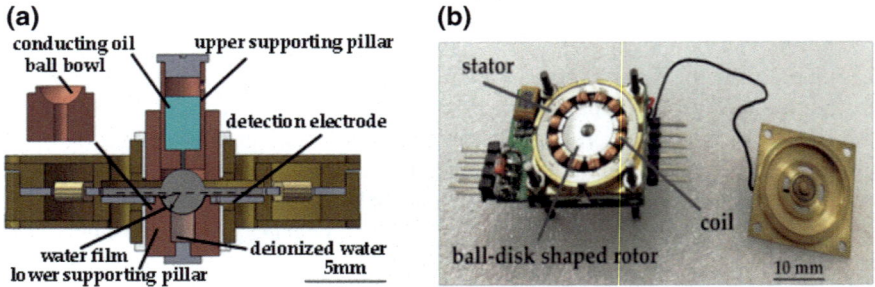

(b)

Fig. 1. **a** Engineering diagram of the gyroscope. **b** Photograph of the gyroscope

fabricated gyroscope (0.5°/h) contributes to a small accumulative error of heading angle estimation. The gyroscope is fixed on its side upon the iphone7 screen and put in a box, named as positioning box. There is a belt on the box to be fastened on the waist of the human for motion tests.

3   PDR Based on Human Motion

PDR is conducted by detecting typical human motions of walking a step on the flat floor, walking upstairs/downstairs, turning right/left through inertial sensors. I wear the positioning box on the waist, collect data from sensors during these actions, analyze their features, select those can be utilized for motion detection. In order to express it more clearly, the forward direction (Y-axis direction of the accelerometer) and the vertical direction (Z-axis direction of the accelerometer) during walking are defined as shown in Fig. 2. Motion of walking on the flat floor is divided into two stages. In stage one, from the time when heel of the left (right) foot touches the floor to the right time before tiptoes of the right (left) foot leave the floor, acceleration in forward direction is increased, as curve A–B in Fig. 3a. In stage two, from the time when tiptoes of the right (left) foot leave the floor to the time when heel of the right (left) foot touches the floor, acceleration in forward direction is decreased, as curve B–C in Fig. 3a. Thus, motion of walking with two stages can be detected by change of acceleration in forward direction.

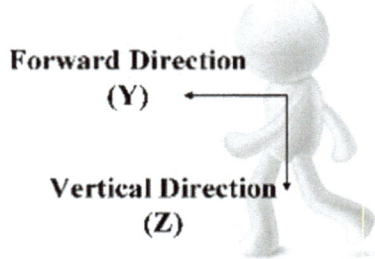

Fig. 2. Schematic diagram of directions during walking motion

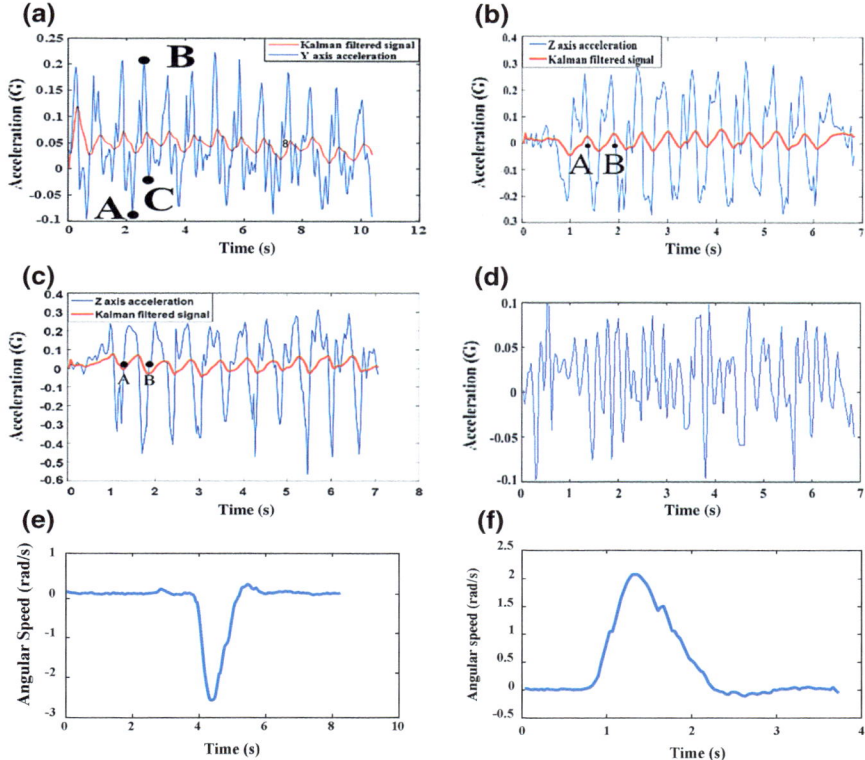

Fig. 3. **a** *Y*-axis acceleration while walking on the flat floor. **b** Z-axis acceleration of walking upstairs of 10 stairs. **c** Z-axis acceleration of walking downstairs of 10 stairs. **d** Z-axis acceleration of walking on the flat floor. **e** Angular speed of turn right. **f** Angular speed of turn left

Kalman filtering an algorithm jointing newly measurement and prediction based on last output to decrease measurement error is adopted to filter *Y*-axis acceleration. Filtered signal is noted as the red line as shown in Fig. 3a and peak number of the filtered signal is equal to sum of positive and negative peaks of the blue curve. Thus, through Kalman filtering, walking steps can be obtained by the function findpeaks in Matlab. Motion of walking upstairs and downstairs of 10 steps is detected by Z-axis acceleration (Fig. 3b, c), and the curve for motion of one step walk is marked between points A, B. As with walking on the flat floor, curves are Kalman filtered and the number of steps is equal to positive (negative) peaks of the filtered red curve. Figure 3d represents the curve of Z-axis acceleration in motion of walking on the flat floor. Compare Fig. 3b–d, curve difference of Z-axis accelerations is obvious, which can be a distinguish standard to differentiate three motions before step counting. Heading angle change is sensed by the fabricated gyroscope in IMU. Gyroscope can sense the angular speed and the turning angle is the integral of angular speed to time. Gyroscope outputs for experiments of turning right and turning left are as shown in Fig. 3e, f. Turning direction can be easily distinguished by angular speed direction. Turning angle is calculated by $\theta = \sum_i \frac{1}{f} \omega_i$, where f is the sampling frequency (30 Hz) and ω_i is the angular speed at the sampling point. Calculated turning angle of turning right and turning left is -1.62 and 1.54 rad, respectively, absolute value of which are approaching to $\pi/2$ rad.

4   Magnetic Strength Matching (MSM) Indoor Positioning

Magnetic strength matching works under the precondition of stable magnetic strength distribution in time and existence of magnetic strength variation in space. Indoor positioning by MSM is conducted through two stages: database construction by mobile matching and positioning. This scheme increases the dimension of information from 3 to the length of magnetic strength sequence. Database stores magnetic strength sequences of continuous RPs and positions of the first and the last RP as the form of:
$R_k = \{m_{k1}, m_{k2}, m_{k3}, \dots m_{kn}, \text{pos}_{k1}, \text{pos}_{k2}\}$, where $m_{k1}, m_{k2}, m_{k3}, \dots m_{kn}$ represent magnetic field strength of n RPs in route R_k. When the measured continuous magnetic strength data sequence matches R_k in shape, the walking path is positioned to be between pos_{k1} and pos_{k2}.

For a different height and walking speed of humans, curves of magnetic strength sequences through the same path will have similar overall component shapes but with different average amplitudes and sequence lengths as graph of Fig. 4a [19]. To identify similarity of signal 1 and signal 2, a time domain signal processing method of modified DTW is adopted. DTW algorithm stretches or compresses signal curves for dissimilarity analysis as shown in Fig. 4b. After signal 2 stretches to the same length as signal 1, signal similarity is evaluated by sum of squared euclidean metric of corresponding points. Modification to DTW algorithm is to eliminate influence of difference in average amplitude of magnetic strength. Procedure of modified DTW is as below: let data sequences $A(a_1, a_2, \dots a_n)$, $B(b_1, b_2, \dots b_m)$ be the reference data sequence and measured data sequence, respectively. For magnetic strength change trend matching, measured data sequence B is modified $(B'(b_1 + d, b_2 + d, \dots, b_m + d))$ to be with the same average as the reference data sequence A, initially. Then, a $n \times m$ matrix grid is constructed with one element at each grid (i, j) representing squared euclidean distance between a_i and b_j. A route from $(0, 0)$ to (m, n) is searched for with the smallest sum of elements in grids. A chosen grid (g, h) means a_g, b_h are corresponding data in

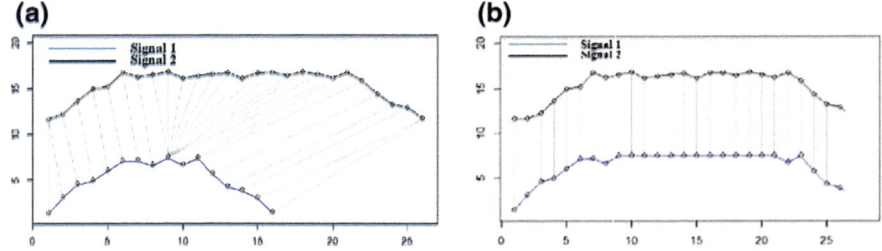

Fig. 4. Example of signal alignment by DTW

sequences A, B and the smallest sum is the similarity expressed in value between A and B, noted as parameter SUM. The smaller the SUM is, the better sequence A matches sequence B. The algorithm is implemented in Matlab.

To evaluate the functionality of MSM with modified DTW algorithm, a path is walked through twice with measured data sequence 1 (D1), 2 (D2) and a path nearby is walked through with measured data sequence 3 (D3). D2 is set as the reference data sequence and data pair of (D1, D2), (D3, D2) are input to the Matlab algorithm of modified DTW. (D1, D2), (D3, D2) with modified sequences MD1, MD3 are shown as Fig. 5a, b. SUM of (D1, D2), (MD1, D2), (MD3, D2) after DTW processing is 59.4, 30.1, 498.1, respectively. The result reveals that average modification can improve similarity between data sequences and discriminability of magnetic matching positioning with modified DTW algorithm is satisfactory.

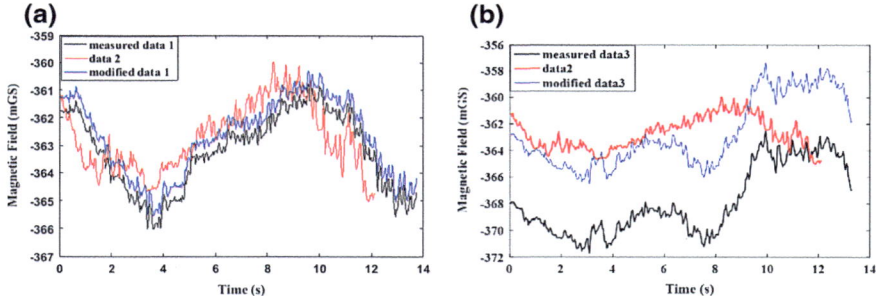

Fig. 5. Magnetic field strength signal alignment by modified DTW of **a** two walks through the same path **b** two walks through two nearby paths

5   Experiment and Discussion

To examine the functionality and the precision of the proposed indoor positioning scheme with the positioning box worn, I walk from the first floor to the second floor of a building in my campus. Layout of two floors is approximately the same, as shown in Fig. 6, with position of stairs marked. After constructing database of MSM by mobile mapping, I walked as the path of the red line in Floor 1, up two flights of stairs (20 steps), and as the path of green line of Floor 2 (Fig. 6). An additional experiment of walking downstairs is conducted, afterward. During the walking, a number of steps in Floor 1 and Floor 2 are counted. Magnetic matching corrects obvious route error by PDR which crosses areas with large magnetic field strength change. PDR path and magnetic matching correcting (MMC) path are marked in blue, yellow lines (Fig. 6). PDR path can roughly indicate the walking path and MMC path decreases errors, especially those caused by heading angle estimation (accumulative error by the gyroscope). Curves of PDR error, MMC error, and error of heading angle at corners (Fig. 7) reflect precision of positioning. It can be seen that maximum PDR error and PDR error at ending point are 3.5 m and 3 m, respectively. Heading angle error is

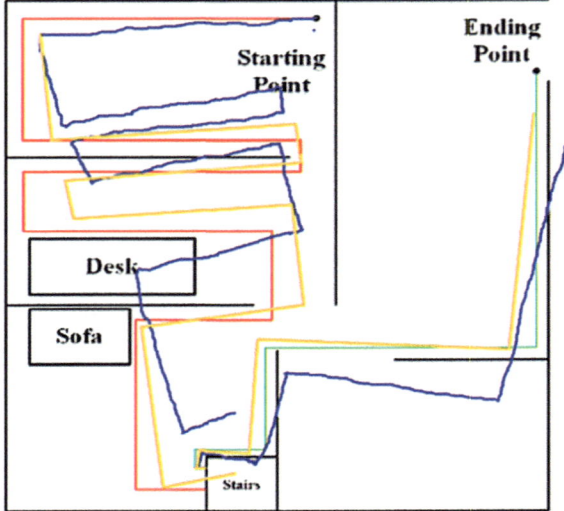

Fig. 6. Similar layout of Floor 1 and Floor 2

within 20°. After correcting of heading angle error (MMC path), positioning error at ending point is decreased to 1.5 m. Heading angle correction can effectively find out PDR route errors of crossing walls or other barriers. Figure 7b reveals that heading angle error accumulates with successive turn of the same direction, which derives from scale factor error of the gyroscope. Counted upstairs and downstairs numbers are 19, 20 respectively, with 1 step upstairs count error. The accuracy rate of stair count is high enough.

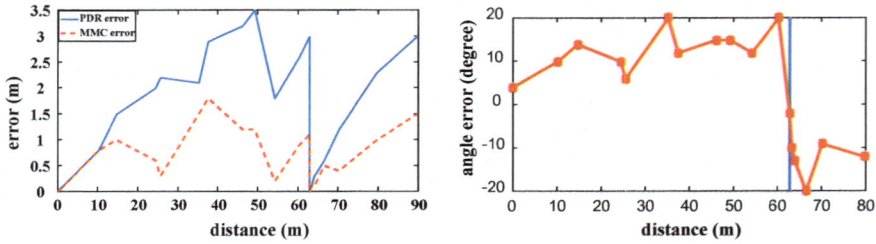

Fig. 7. **a** PDR and MMC error **b** heading angle error (blue line represents position of stairs)

6   Conclusion

An indoor positioning scheme combining PDR and MSM is proposed in the paper. The 3-axis accelerometer and the magnetometer in iphone7 with a fabricated high-precision gyroscope construct the hardware system of the positioning box. Through Kalman filtering to outputs from accelerator and gyroscope, typical human motions during walking are recognized with low error rate. MSM positioning as a supplementary method to decrease accumulative errors is conducted through data matching between measured magnetic strength sequence and sequences in database by modified DTW algorithm. The proposed positioning scheme is applied in an experiment of walking path tracking from floor 1 to floor 2 in a building of my campus. PDR error caused by heading angle estimation error is obvious. After MSM correction, positioning error is lowered. Further work will focus on developing a real-time data processing module to process MSM data as the walking action and correct heading angle error in real time to decrease accumulative effect of PDR positioning by inertial sensors.

References

1. He Z. High-sensitivity GNSS Doppler and velocity estimation for indoor navigation. Engineering Engineering–AerospaceEngineering–Electronics and Electrical, 2013.
2. Xiao Z, Wen H, Markham A, Trigoni N, Blunsom P, Frolik J. Non-line-of-sight identification and mitigation using received signal strength. IEEE Trans Wirel Commun. 2015;14:1689–702.
3. Schmitt S, Adler S, Kyas M. The effects of human body shadowing in RF-based indoor localization. In: 30th international conference on indoor positioning and indoor navigation, 2014.
4. Saeedi S. Context-aware personal navigation services using multi-level sensor fusion algorithm. Ph.D., University of Calgary, 2013.
5. Morrison A, Renaudin V, Bancroft JB, Lachapelle G. Design and testing of a multi-sensor pedestrian location and navigation platform. Sensors. 2012;12:3720–38.
6. Talvite J, Renfors M, Lohan ES. Distance-based interpolation and extrapolation methods for RSS-based localization with indoor wireless signals. IEEE Trans Veh Technol. 2015;64 (4):1340–53.
7. Cheng Y, Wang X, Morelande M, Moran B. Information geometry of target tracking sensor networks. Inf Fusion. 2013;14:311–26.
8. Torres-solis J, Falk TH, Chau T. A review of indoor localization technologies: toward navigational assistance for topographical disorientation. Ambient Intell. 2010;51–84.
9. Bose A, Foh CH. A practical path loss model for indoor WiFi positioning enhancement. In: 6th international conference on information, communication & signal processing. IEEE; 2007. p. 1–5.
10. Niu X, Li Y, Zhang H, Wang Q, Ban Y. Fast thermal calibration of low-grade inertial sensors and inertial measurement units. Sensors. 2013;13:12192–217.
11. Wang Q, Li Y, Niu X. Thermal calibration procedure and thermal characterisation of low-cost inertial measurement units. J Navig. 2015;1–18.
12. Akeila E, Salcic Z, Swain A. Reducing low-cost INS error accumulation in distance estimation using self-resetting. Trans Instrum Meas IEEE. 2014;63:177–84.
13. Xie H, Gu T, Tao X, Ye H, Lv J. Maloc. A practical magnetic fingerprinting approach to indoor localization using smartphones. In: Proceedings of the 2014 ACM international joint conference on pervasive and ubiquitous computing. ACM; 2014. p. 243–53.
14. Zhang C, Subbu K, Luo J, Wu J. GROPING: geomagnetism and crowdsensing powered indoor navigation. IEEE Trans Mob Comput. 2015;14(2):387–400.
15. Pritt N. Indoor navigation with use of geomagnetic anomalies. In: Geoscience and Remote Sensing Symposium (IGARSS), 2014 IEEE International; IEEE. 2014. p. 1859–62.

16. Li Y, Zhuang Y, Lan H, Zhang P, Niu X, El-sheimy N. WiFi-aided magnetic matching for indoor navigation with consumer portable devices. Micromachines. 2015;6:747–64.
17. Chen D, Liu X, Zhang H, Li H, Weng R, Li L, Rong W, Zhang Z. A rotational gyroscope with a water-film bearing based on magnetic self-restoring effect. Sensors. 2018;18(2).
18. Chen D, Liu X, Zhang H, Li H, Weng R, Li L, Rong W, Zhang Z. Friction reduction for a rotational gyroscope with mechanical support by fabrication of a biomimetic superhydrophobic surface on a ball-disk shaped rotor and the fabrication of a water film bearing. Micromachines. 2017;8(7):223.
19. Zhen D, Zhao H, Gu F, Ball A. Phase-compensation-based dynamic time warping for fault diagnosis using the motor current signal. Meas Sci Technol. 2012;23:055601.

Design and Verification of Anti-radiation SPI Interface in Dual Mode Satellite Navigation Receiver Baseband Chip

Yi Ran Yin[✉] and Xiao Lin Zhang

Electronic and Information Engineering, BeiHang University,
XueYuan Road no. 37, Haidian District, Beijing, China
728660626@qq.com

Abstract. This paper designs the SPI interface module in the baseband chip of the dual mode satellite navigation receiver based on the AMBA bus, and this paper is based on the first edition of the baseband chip accepted by the Science and Industry Corp. This design provides the IP core of SPI for the SoC using the LEON series processor and uses the TMR on register transfer level and evaluates the final results after the reinforcement. The results of verification and evaluation show that the SPI master designed by this paper can communicate normally based on the AMBA bus, and the SPI master after reinforcement can resist SEU to a certain extent and improve its stability in the space radiation environment.

Keywords: Anti-radiation · Baseband chip · AMBA bus

1 Introduction

The digital baseband chip is an important part of the navigation receiver compatible with the GPS and BD-2 systems and is the core unit of the dual system compatible receiver. This paper designs the SPI module as a part of a major special project. From a global point of view, there are many kinds of on-chip buses, and many semiconductor manufacturers have developed their own bus standards. The literature [1] shows that the wishbone bus has been widely used in the market with its diverse topological structures and most of the IP cores of SPI are currently designed based on the wishbone bus. The literature [2] shows that the AMBA bus designed by ARM has been widely recognized in the market with excellent architecture. The literature [3] shows that the failure of the space radiation in the satellite accounts for 55% of the total number of faults, so the stability of the integrated circuit is very important. So this paper describes the top-down design process for SPI based on the AMBA bus in detail and verifies the design with the LEON3 platform.

© Springer Nature Singapore Pte Ltd. 2020
Q. Liang et al. (eds.), *Communications, Signal Processing, and Systems*, Lecture Notes in Electrical Engineering 516,
https://doi.org/10.1007/978-981-13-6504-1_135

2   Design of SPI Bus Control Module

There are two main methods for designing IP cores. One is top-down, and the other is from bottom to top. Because the structure of SPI master is relatively simple and there are less modules, it is relatively easy to grasp from the whole. Therefore, the top-down approach is used to design.

The main function of the SPI master is to define and describe the related registers, so the APB bus can control the SPI master to select the SPI slave and to receive data from the SPI slave and to send data to SPI slave. The architecture of the SPI master section is shown in Fig. 1.

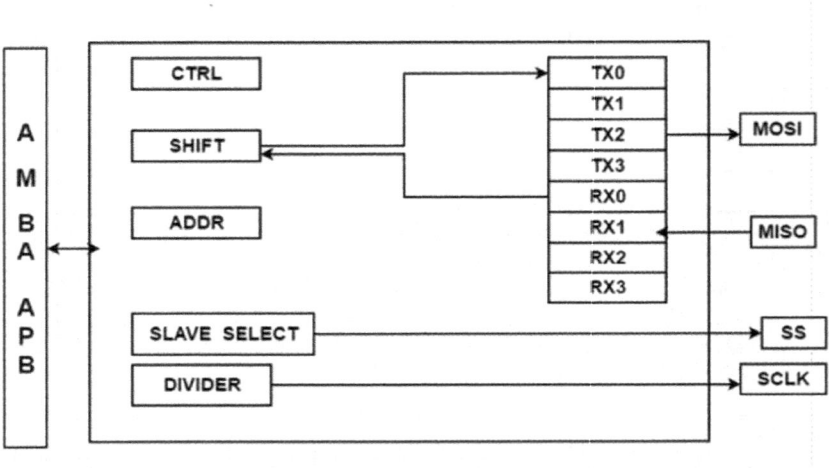

Fig. 1. Architecture of SPI master

The modules involved in Fig. 1 include: control and state register module, clock generation module, shift register module, address decoding module, etc.

2.1   Clock Generation Module

In order to match the high-speed clock of the APB bus and the low-speed clock of the external device, the clock division module 'spi_clgen' is needed in the design of the SPI master. The main function of the module is to divide the clock frequency PCLK from the APB bus and generate the clock signal SCLK to control the sending and receiving data.

This module divides the frequency from the APB bus and generates the internal clock signal. The frequency division function is realized by the reciprocal counter (CNT). The initial value of the counter (CNT) is stored in the register named 'divider.' The idle state of the 'clk_out' is low, and the value of the counter (CNT) is reduced by 1 at the coming of the rising edge of each PCLK. When the value of counter (CNT) is reduced to 0, the value of the signal named 'clk_out' is turned to high level, and the high level lasts for a PCLK period, and then, it returns to low level. The signal passes through the transceiver logic module and generates a signal which is half of the signal named 'clk_out.' The signal generated is the main clock signal of the SPI communication.

2.2   Address Decoding Module

When it needs to write data or read data, the module will compare the address from the APB bus with the defined address constants one by one. If the APB bus address is consistent with the defined address, it performs the operations of reading and writing. In this judgment, the conditions should be met; that is, only when the PENBALE is enabled, PSELx chooses the corresponding SPI slave. The address signal named PADDR decodes the address and configures corresponding registers when it receives data.

2.3   Shift Register Module

The main function of this module is to transform parallel data from APB bus to serial data and deliver the serial data to the SPI master, transform serial data from SPI master to parallel data and deliver the parallel data to APB bus.

The main function of this module is completing the conversion of serial and parallel data through the shift register. The specific implementation process is as follows: The serial data named 's_out' from SPI slave stores in the receiving register RX in a certain order and then the data in the RX is passed to the register named 'data.' When the APB bus sends the corresponding task to the SPI master, the SPI master sends the data stored in the register named 'data' in parallel to the APB bus. In the same way, when the APB bus sends the corresponding command to the SPI master, SPI stores the parallel data named 'p_out' from the APB bus in the shift register named 'data,' and the data in the shift register is then transferred to the register TX in a certain order. Then, the data transfers to serial data and delivers to the SPI slave one by one.

2.4   Control and State Register Module

The main function of the control and state register is to control the transmission state and transmission mode of the SPI master. The transfer mode can be controlled by setting the related bits of the register. The control and state register has 32 bits, and the main internal signals are as follows: ASS, this bit is the twelfth bit of the register. If this bit is set, SS signals are generated automatically. IE, this bit is the eleventh bit of the register. If this bit is set, the interrupt output is set active after a transfer is finished. The interrupt signal is deasserted after reading from any registers or writing to any registers. TX_NEG, this bit is the second bit of the register. If this bit is set, the MOSI signal is changed on the falling edge of a SCLK clock signal, or otherwise, the MOSI signal is changed on the rising edge of SCLK.

3   Simulation and Verification of SPI Master Interface

This paper verifies the designed SPI master in two aspects: One part is writing testbench to do the preliminary verification, and the other is linking the design which has passed the initial verification to the LEON3 platform provided by ESA.

3.1   Write a Testbench for Verification

Before connecting the designed SPI master to the LEON3 SoC platform, we first need to write a testbench to verify the designed code preliminarily, and test whether the function of the designed circuit is consistent with the expected target. The structure diagram of testbench is shown in Fig. 2.

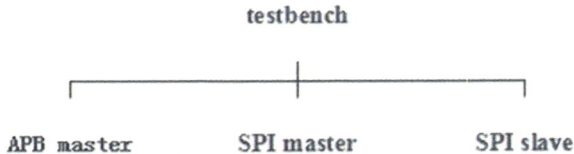

Fig. 2. Architecture of testbench

This paper uses Verilog to design the basic functions of the APB bus to verify the normal work of the designed SPI master. In this section, the writing task, reading task, and data comparison task of the APB bus are defined. In the testbench, this paper describes the corresponding WADDR and the corresponding WDATA, and the corresponding RADDR and the corresponding RDATA in detail. Through the COMPARE TASK this paper defined, we can judge whether the APB bus can write data to the corresponding address and read data from the corresponding address. The result of the test is shown in Fig. 3.

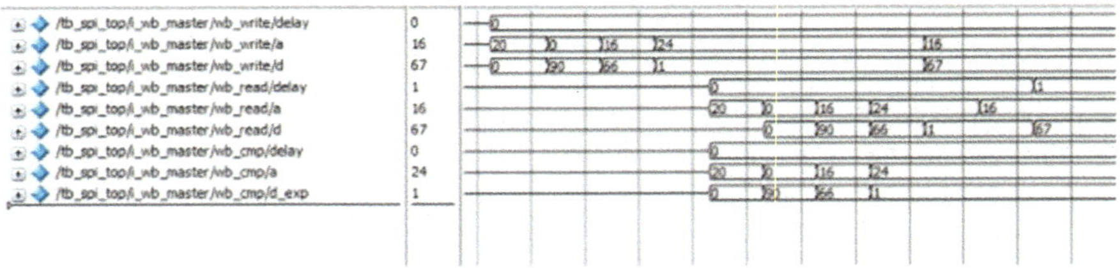

Fig. 3. Test of communication between APB and SPI master

In order to improve the reliability of the design results, this paper verifies the function of the designed SPI master fully. The SPI master also reads and writes 1000 times randomly to the related registers and compares the correctness of the written data and the data which is read from the related registers. The result of the test is shown in Fig. 4.

```
# the result is right, rand is 00000000000000000000011010000101,prdata is 00000000000000000000011010000101
# the result is right, rand is 11111111111111111111010000101101,prdata is 11111111111111111111010000101101
# the result is right, rand is 00000000000000000000010000101011,prdata is 00000000000000000000010000101011
# the result is right, rand is 00000000000000000001110011010101,prdata is 00000000000000000001110011010101
# the result is right, rand is 00000000000000000001111100111101,prdata is 00000000000000000001111100111101
# the result is right, rand is 11111111111111111101101101110111,prdata is 11111111111111111101101101110111
# the result is right, rand is 00000000000000000001011010100100,prdata is 00000000000000000001011010100100
# the result is right, rand is 11111111111111111111010000000111,prdata is 11111111111111111111010000000111
# the result is right, rand is 11111111111111111111110010110111,prdata is 11111111111111111111110010110111
# the result is right, rand is 11111111111111111101101001101001,prdata is 11111111111111111101101001101001
# the result is right, rand is 11111111111111111110110010111001,prdata is 11111111111111111110110010111001
# the result is right, rand is 00000000000000000000100101001011,prdata is 00000000000000000000100101001011
# the result is right, rand is 00000000000000000001010011011001,prdata is 00000000000000000001010011011001
# the result is right, rand is 00000000000000000010010000100001,prdata is 00000000000000000010010000100001
# the result is right, rand is 11111111111111111111101010100001,prdata is 11111111111111111111101010100001
# the result is right, rand is 00000000000000000001011000101010,prdata is 00000000000000000001011000101010
# the result is right, rand is 00000000000000000000100101111010,prdata is 00000000000000000000100101111010
# the result is right, rand is 11111111111111111110010110010101,prdata is 11111111111111111110010110010101
# the result is right, rand is 00000000000000000000011011010011,prdata is 00000000000000000000011011010011
# the result is right, rand is 11111111111111111111101000001100,prdata is 11111111111111111111101000001100
# the result is right, rand is 11111111111111111101101010010010,prdata is 11111111111111111101101010010010
# the result is right, rand is 11111111111111111110111101011110,prdata is 11111111111111111110111101011110
# the result is right, rand is 11111111111111111110100101011000,prdata is 11111111111111111110100101011000
# the result is right, rand is 11111111111111111101111001011111,prdata is 11111111111111111101111001011111
# the result is right, rand is 00000000000000000001000101100110,prdata is 00000000000000000001000101100110
# the result is right, rand is 00000000000000000010000010110000,prdata is 00000000000000000010000010110000
# the result is right, rand is 00000000000000000001100010111000,prdata is 00000000000000000001100010111000
# the result is right, rand is 00000000000000000000010101000000,prdata is 00000000000000000000010101000000
```

Fig. 4. Final test of communication between APB and SPI master

3.2 Connect to the LEON3 Platform for Verification

LEON3 uses the 7-level pipeline structure, and it connects the INSTRUCTION REGISTER with the MEMORY CONTROL REGISTER through the high-speed AMBA-AHB bus. The DATA CACHE REGISTER is connected with the high-speed external interface to transmit data. Low-speed AMBA-APB bus connects with low-power peripherals, such as timers, serial ports, and network interfaces [4].

To connect to the LEON3 platform, it is needed to understand the plug-and-play mechanism of APB bus, configure the designed SPI address, and connect the SPI master to the LEON3 platform. This paper uses the Verilog language for the design of SPI, while the open source code of LEON3 uses the VHDL language, so this paper needs to write a wrapper to complete the conversion of two languages.

For this paper, the function of verifying the designed SPI is to write to the SPI master and read data from the SPI master through the APB bus. The APB bus writes data to the related registers to transmit the related data and configure the related registers.

The basic function of this code is to write data 0 × 5a to the TX register of the SPI master. By comparing the data read out with the written data, we can judge whether the SPI master can be used to transmit data with the APB bus.

This code is written in/grlib/software/leon3/apbuart.c. This paper writes the test function int spitest(int addr) and invokes the function in systest.c. When this paper invokes the function, the APB bus must know the address of the related registers to be written.

According to the above method, the APB bus can get the address of the related registers. So, the APB bus can write data into the related registers and read data from the related registers. The final results are shown in Fig. 5.

As is shown in Fig. 5, first, 000005A is written to the DIVIDER REGISTER of the SPI master. The address of the DIVIDER REGISTER is 0000084, and then, the APB

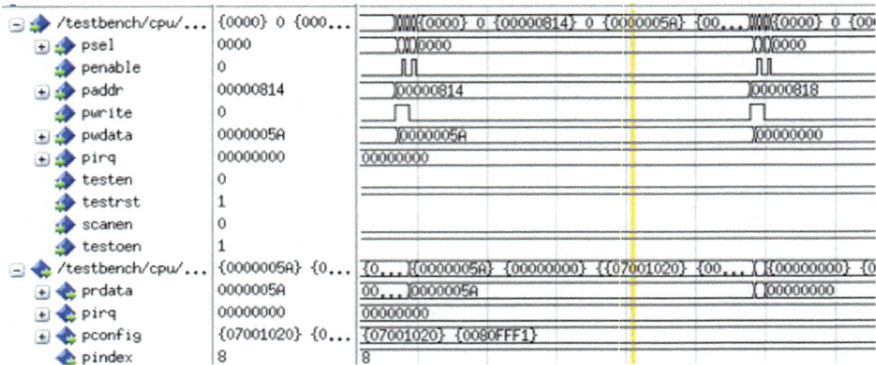

Fig. 5. Final result of test

bus can read 000005A from the DIVIDER REGISTER. In the same way, 000000 is written to the SS REGISTER and the APB bus can read the same data from the SS REGISTER. The address of the SS REGISTER is 00000818.

It can be concluded that the SPI master has been successfully connected to the APB bus and can be written and read out through the APB bus.

4   TMR of the SPI Master

In the actual aerospace circuit, to meet the reliability requirements of SPI, we usually use the fault-tolerant technology to ensure the function of the SPI master. The most common fault-tolerant technology is TMR. The main parts of TMR are three identical modules and a voter [5]. The output of the voter is in accordance with the majority of the three inputs. The precondition of the voter is that there are two or more than two modules working normally in the circuit, and the overall circuit function is normal, so that the fault of the single circuit module is shielded.

TMR technology is considered the most reliable anti-radiation method. To achieve the best effect of anti-SEU, we can instantiate all the modules into three identical modules. While this method will bring about an increase in resources and cost, we should consider the importance of the module and sensitivity to SEU. This paper only reinforces the main registers of the SPI master. There are the data receive and transmit registers named RX0, RX1, RX2, RX3; slave select register named SS; divider register named DIVIDER; control and state register named CTRL.

The main implementation process is that this paper instantiates three identical modules and assumes that one of the three registers has SEU that means the data in this register has changed. Then, this paper observes the output of the register whether has changed; if there is no change in the output, we can consider that the SPI master designed in this paper has a certain ability of anti-radiation. Because the probability of happening SEU at one time is very low and the probability of happening SEU is lower, it can be regarded as impossible. In this paper, we only discuss the situation of only one register has SEU [6].

To judge whether the design achieves the purpose of reinforcement, this paper verifies the final results after TMR. The results of TMR are shown in Fig. 6.

This paper also reinforces other registers of the SPI master in the same way. The total test result after TMR is shown in Fig. 7.

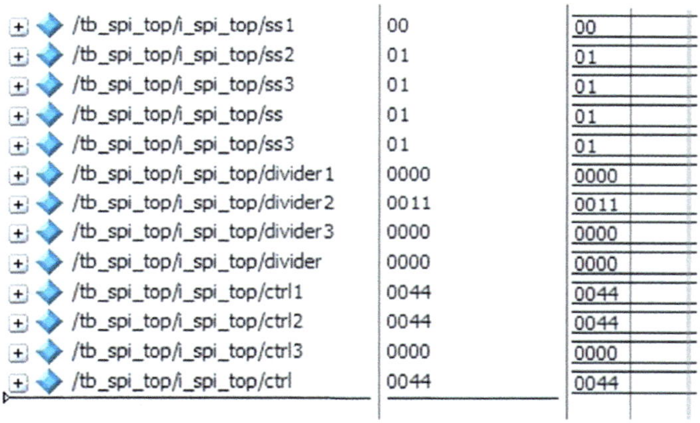

/tb_spi_top/i_spi_top/ss1	00	00
/tb_spi_top/i_spi_top/ss2	01	01
/tb_spi_top/i_spi_top/ss3	01	01
/tb_spi_top/i_spi_top/ss	01	01
/tb_spi_top/i_spi_top/ss3	01	01
/tb_spi_top/i_spi_top/divider1	0000	0000
/tb_spi_top/i_spi_top/divider2	0011	0011
/tb_spi_top/i_spi_top/divider3	0000	0000
/tb_spi_top/i_spi_top/divider	0000	0000
/tb_spi_top/i_spi_top/ctrl1	0044	0044
/tb_spi_top/i_spi_top/ctrl2	0044	0044
/tb_spi_top/i_spi_top/ctrl3	0000	0000
/tb_spi_top/i_spi_top/ctrl	0044	0044

Fig. 6. Test result of TMR

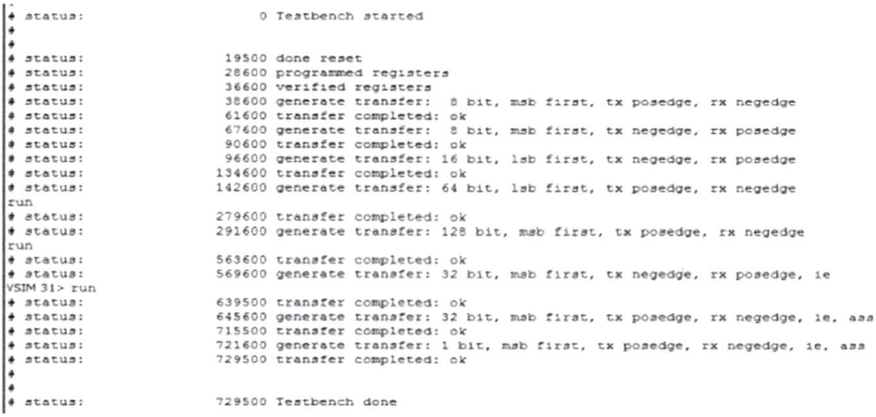

```
 status:                    0 Testbench started


 status:                19500 done reset
 status:                28600 programmed registers
 status:                36600 verified registers
 status:                38600 generate transfer:  8 bit, msb first, tx posedge, rx negedge
 status:                61600 transfer completed: ok
 status:                67600 generate transfer:  8 bit, msb first, tx negedge, rx posedge
 status:                90600 transfer completed: ok
 status:                96600 generate transfer: 16 bit, lsb first, tx negedge, rx posedge
 status:               134600 transfer completed: ok
 status:               142600 generate transfer: 64 bit, lsb first, tx posedge, rx negedge
run
 status:               279600 transfer completed: ok
 status:               291600 generate transfer: 128 bit, msb first, tx posedge, rx negedge
run
 status:               563600 transfer completed: ok
 status:               569600 generate transfer: 32 bit, msb first, tx negedge, rx posedge, ie
VSIM 31> run
 status:               639500 transfer completed: ok
 status:               645600 generate transfer: 32 bit, msb first, tx posedge, rx negedge, ie, ass
 status:               715500 transfer completed: ok
 status:               721600 generate transfer: 1 bit, msb first, tx posedge, rx negedge, ie, ass
 status:               729500 transfer completed: ok


 status:               729500 Testbench done
```

Fig. 7. Final test result of TMR

5   Concluding Remarks

This paper completes the design and verification of the SPI interface module in the baseband chip of the dual mode satellite navigation receiver, reinforces the SPI interface modules through the TMR method, and completes simple evaluation after reinforcement. In summary, this paper improves the radiation resistance and improves the stability of its work.

Acknowledgments. This work was partially supported by the National Natural Science Foundation of China under Grant No.61601295.

This work was financially supported by Xi'an Aisheng innovation and Development Foundation (ASN-IF2015-1405).

References

1. Tian Ze, Zhang YH, Yu DS. The Summary of SoC OCB. Semicond. T Echnology. 2003;11:11–5.
2. ARM Ltd. AMBA Specification Revision 2. 1999.
3. Hui X, Wei T, Yuan HY, Liang Z. Research of radiation harden based on triple modular redundancy for ASIC. Microprocessors. 2015;10:1–4.
4. Lin L, Lin ZX, Xi Y. Building and testing of SoC platform based on LEON open source soft core. Microcontrollers Embed. Syst. 2007;01:32–5.
5. Hua GJ, Run X, Lin Z. The research of radiation harden for ASIC. J. CAEIT. 2013;12: 644–645.
6. Xing K, Yang J. Study on radiation harden method for SRAM based FPGA in space. Chin. J. Electron Devices. 2007;30(1).

Indoor Localization Algorithm Based on Particle Filter Optimization in NLOS Environment

Weiwei Liu$^{(\boxtimes)}$, Tingting Liu, and Lei Tang

Nanjing Institute of Technology, Hongjing Road 1, Nanjing, China
466346830@qq.com, lwwlll7@njit.edu.cn

Abstract. The performance of indoor localization algorithm is limited by non-line-of-sight (NLOS) error, a positioning system includes Bluetooth module, Bluetooth gateway and cloud monitoring center based on particle filter is presented to enhance positioning accuracy. Our experimental results indicate that the proposed localization scheme leads to higher localization accuracy and lower power consumption.

Keywords: Localization · Non-line-of-sight · Particle filter

1 Introduction

Over the past decades, localization service has been applied to a variety of different applications, especially in civil class positioning navigation. While Global Positioning System (GPS) localization does not work well in certain indoor regions. To tackle the problems with GPS, many researchers have proposed a series of alternative indoor localization schemes. Ultrasound positioning has high accuracy and simple structure, but it is greatly affected by multipath effect and non-line-of-sight propagation. Wi-Fi is applied to small-scale indoor positioning with low cost; meanwhile, its accuracy is easy to be affected by other signals. ZigBee location determines the location of the object by calculating the distance between the object and the reference nodes with known locations, which is not common in practical applications. Based on Bluetooth positioning network, it can span a large physical space, increases the coverage of monitoring area, reduces blind area, and be suitable for remote communication process. That is to say, the Bluetooth LAN with intelligent terminals and several Bluetooth beacons, which transmits data between Bluetooth gateway and cloud monitoring center [1–9].

At present, the Bluetooth location method mainly is based on triangular location algorithm. Other auxiliary methods such as Bayesian-based [10] sequential Monte Carlo [11–13] particle filter (PF), Kalman filter algorithm, and Gauss filter algorithm.

Many improved methods have been proposed [14–16]. Xie et al. [14] came up with an unscented particle filter (UPF). The main idea of UPF is to use unscented Kalman filter method integrating the latest observation information to get a good proposal distribution. Gaussian particle filter (GPF) was also proposed to solve degeneracy problem [16] with a low error when the environment is highly non-Gaussian, especially in line-of-sight (NLOS) environment.

© Springer Nature Singapore Pte Ltd. 2020
Q. Liang et al. (eds.), *Communications, Signal Processing, and Systems*, Lecture Notes in Electrical Engineering 516,
https://doi.org/10.1007/978-981-13-6504-1_136

In our research, we focus on the outdoor car localization schemes that can be adopted by smartphones. In this paper, we present the original of car location detection algorithm in the database of cloud data center maps the one-to-one mapping relationship between car location and intelligent LED positioning lamp. The mobile phone submits the query request service through the beacon protocol communication. The nearest Bluetooth lamp receives the request information and broadcasts the request information to the LED wireless ad hoc MESH network for broadcast traversal to search of car bits information, which reduces the cost of hardware compared to traditional video parking detection methods.

The rest of this paper is organized as follows. In Sect. 1, the related work on existing localization schemes is presented. The problem formulation and describes the proposed hybrid localization scheme are included in Sect. 2. Section 3 includes the details of our experimental results. Finally, our conclusions are presented in Sect. 4.

2 The Problem Formulation and Describes the Proposed Hybrid Localization Scheme

Particle filter is to generate a set of random samples (particles). The weights and positions of the particles are adjusted according to the measurement information, and the initial empirical condition distribution is modified by the adjusted particle information. Observation model is given as

$$z_k = o_k(x_k, v_k, u_k) \tag{1}$$

where z_k is observation position at time k, o_k is Observation function, v_k and u_k are observation noise and test noise, respectively.

2.1 The Model Structure

The model structure of the localization in NLOS environment, which is shown in Fig. 1. In order to realize the stochastic characteristic analysis of the NLOS environment in the region, the dynamic process of observation noise and test noise should be modeled first in the region.

2.2 The Proposed Hybrid Localization Scheme

The position determines its position based on the LED data received with non-line-of-sight (NLOS) serving for this investigation. The optical power received from a single LED without overlap is given as

$$x_k^i \sim q(x_k | x_{1:k-1}^i, z_{1:k}) \tag{2}$$

where k is the time indicator, $x_k = \{x_k^1, x_k^2, \cdots x_k^N\}$ is the system state vector consisting of all states at the time k, $z_{1:k} = \{z_1, z_2, \cdots z_k\}$ is a vector set of all measurement, $q(\bullet)$ is the prior probability function, $i \in [1, N]$, N is the number of generated samples.

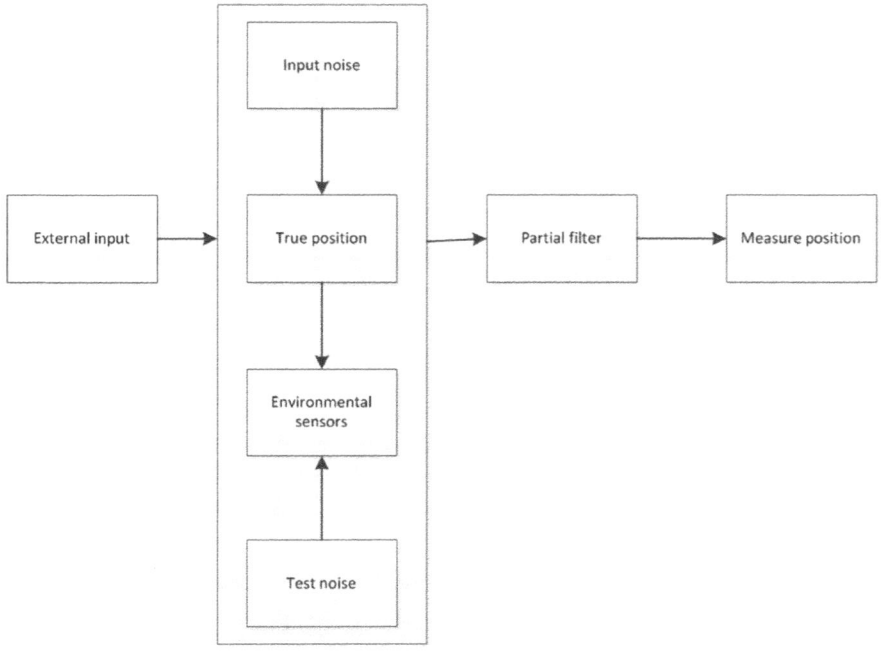

Fig. 1. Model structure

Suppose

$$w_k^i = w_{k-1}^i \frac{p(z_k|x_k^i)p(x_k^i|x_{k-1}^i)}{q(x_k^i|x_{0:k-1}^i, z_{1:k})} \tag{3}$$

Denote their corresponding weights.
Thus, we have

$$p(x_k|z_{1:k}) = \sum_{1}^{N} w_k^i \delta(x_k - x_k^i) \tag{4}$$

as random quantity to represent the posteriori probability density function. Here, the weight satisfies the regular condition. Where δ is the Dirac delta function.

After updating the state of particles, we weighted and estimated the location of particles.

$$\hat{x}_k = \sum_{1}^{N} w_k^i x_k^i \tag{5}$$

3   Simulation Result

In this paper, we propose a position mechanism, and the simulation results of particle filtering are shown as shown. Then, we discuss the effects of car position data.

3.1   Localization Accuracy

In our research, we implemented a testbed around the Academic Building at Nanjing Institute of Technology, China. In this testbed, an android system running the proposed localization scheme was used to achieve localization.

Figure 2 shows the evolution of the entire system. On the one hand, by using the original measurement equations and the random disturbances, measurements are obtained in the forward process. On the other hand, by using the partial filter process of the system, measurement function and the partial observation of the system state, we are able to realize a dynamic backtracking from the measurement data to the system state. We compare the effect of the improved state evolution process with the process without the direct observation for partial system states. And this dimensionality of the system is tested to have the best performance.

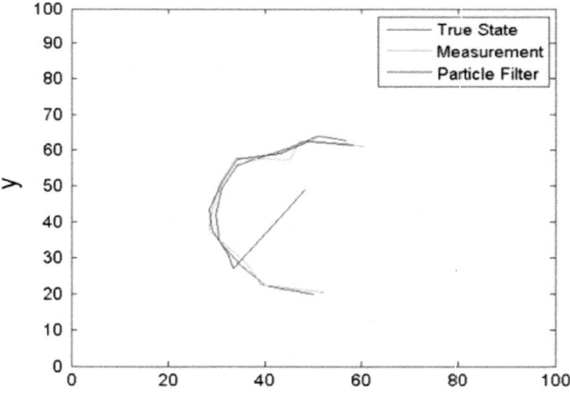

Fig. 2. Evolution of the entire system

In the following section, we present a numerical case to the effect of the indoor navigation observation for partial system.

3.2   The Effective of the Proposed System

We plotted these results into indoor car position as Fig. 3 shows.

The car owners can match the real-time parking demand and parking lot of accurate data through the mobile phone APP intelligent parking. The intelligent parking owners completely and efficiently realize the parking lot to maximize the use of resources in Fig. 3a. The parking users who are needed to parking can get parking navigation

(a) (b)

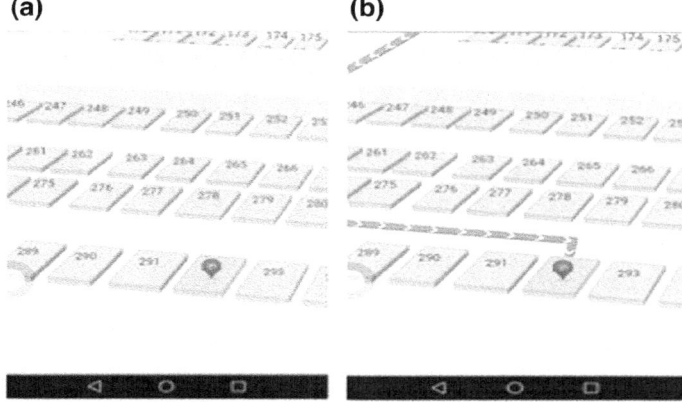

Fig. 3. **a** Three-dimensional map distribution of empty car position. **b** Empty car position navigation

services in Fig. 3b. The parking lot operators can master the parking data in real time through the system. The analysis of parking big data can provide strong decision-making basis for operators to improve the operation efficiency of parking lots, and support the unified management of multiple parking lots and the unified management of multiple massive devices.

4 Conclusions

In this paper, we present an indoor car localization scheme, which utilizes smartphones. With the partial filter mechanism, this scheme is expected to be more feasible than the existing indoor localization methods. Our experimental results indicate that the proposed localization scheme leads to higher localization accuracy and lower power consumption.

Acknowledgments. This work is supported in part by the National Natural Science Foundation of China under Grant 61702258, in part by the China Postdoctoral Science Foundation under grant 2016M591852, in part by Postdoctoral research funding program of Jiangsu Province under grant 1601257C, in part by the China Scholarship Council Grant 201708320001 and the NJIT Foundation(Grant No. YKJ201419).

References

1. Chen R, Guinness R. Geospatial computing in mobile devices. Artech House. 2014.
2. Zhou B, Li Q, Mao Q. Activity sequence-based indoor pedestrian localization using smartphones. IEEE Trans Hum-Mach Syst. 2015;45(5):562–74.
3. Groves PD. principles of GNSS inertial, and multisensor integrated navigation systems, 2nd edn. Aerosp. Electron. Syst. Mag. IEEE. 2015;30(2):26–27.

4. Shi K, Ma Z, Zhang R. Support vector regression based indoor location in IEEE 802.11 Environments. Mob Inf Syst. 2015.
5. Harle R. A survey of indoor inertial positioning systems for pedestrians. Commun Surv Tutor, IEEE. 2013;15(3):1281–93.
6. Cheng J, Yang L, Li Y. Seamless outdoor/indoor navigation with WIFI/GPS aided low cost inertial navigation system. Phys Commun. 2014;13:31–43.
7. Chen Z, Zou H, Jiang H. Fusion of WiFi, smartphone sensors and landmarks using the Kalman filter for indoor localization. Sensors. 2015;15(1):715–32.
8. Jiao J, Courtade TA, Venkat K. Justification of logarithmic loss via the benefit of side information. Inf Theory IEEE Trans On. 2015;61(10):5357–65.
9. Han B, Joo SW, Davis LS. Adaptive resource management for sensor fusion in visual tracking. Theory and applications of smart cameras. Springer Netherlands; 2016. p. 187–213.
10. Srkk S. Bayesian filtering and smoothing. Inst Math Stat Textb. 2013;40(4):S407.
11. Lu X,Dong Y, Wang XA. Monte carlo localization algorithm for 2-D indoor self-localization based on magnetic field. International ICST conference on communications and networking in China. IEEE; 2013. p. 563–8.
12. Smith A. Sequential monte carlo methods in practice. Springer Science & Business Media; 2013.
13. Klaas M, Freitas ND, Doucet A. Toward practical N2 monte carlo: the marginal particle filter. J Neurophysiol. 2012;94(3):2045–52.
14. Xie H, Gu T, Tao X. A reliability-augmented particle filter for magnetic fingerprinting based indoor localization on smartphone. IEEE Trans Mob Comput. 2016;15(8):1877–92.
15. Yin S, Zhu X. Intelligent particle filter and its application to fault detection of nonlinear system. Ind Electron, IEEE Trans On. 2015;62(6):3852–61.
16. Ades M, Van Leeuwen PJ. An exploration of the equivalent weights particle filter. Q J R Meteorol Soc. 2013;139(672):820–40.

An Improved Sensor Selection
for TDOA-Based Localization
with Correlated Measurement Noise

Yue Zhao[1(✉)], Zan Li[1], Feifei Gao[2], Jia Shi[1], Benjian Hao[1,3], and Chenxi Li[1]

[1] State Key Laboratory of Integrated Services Networks,
Xidian University, Xi'an, China
{yuezhao, chenxili}@stu.xidian.edu.cn
{zanli, jiashi, bjhao}@xidian.edu.cn
[2] National Laboratory for Information Science and Technology,
Tsinghua University, Beijing, China
feifeigao@ieee.org
[3] Collaborative Innovation Center of Information Sensing and Understanding,
Xi'an, China

Abstract. This paper focuses on the problem of sensor selection in time-difference-of-arrival (TDOA) localization scenario with correlated measurement noise. The challenge lies in how to select the reference sensor and ordinary sensors simultaneously when the TDOA measurement noises are correlated. Specifically, the optimal sensor subset is found by introducing two independent Boolean selection vectors and formulating a nonconvex optimization problem, which motivates to minimize the localization error in the presence of correlated noise and energy constraints. Upon transforming the original nonconvex problem to the semidefinite program (SDP), the randomization method is leveraged to tackle the problem, and thereby proposing the novel algorithm for sensor selection. Simulations are included to validate the performance of proposed algorithm by comparing with the exhaustive search method.

Keywords: Sensor selection · Time-difference-of-arrival · Source localization · Convex optimization

1 Introduction

Source localization has been of considerable interest for many applications in radar [1], wireless communications [2], and spectrum monitoring [3]. In the wireless sensor network, source localization can be boosted by time-of-arrival (TOA), time-difference-of-arrival (TDOA), frequency-difference-of-arrival (FDOA), angle-of-arrival (AOA) measurements, or a combination of them [4,5]. This paper focuses on the TDOA localization for a stationary source.

© Springer Nature Singapore Pte Ltd. 2020
Q. Liang et al. (eds.), *Communications, Signal Processing, and Systems*, Lecture Notes in Electrical Engineering 516,
https://doi.org/10.1007/978-981-13-6504-1_137

In practice, sensors are always deployed in the harsh conditions and powered by batteries [6]. Although localization with a great number of sensors can improve the accuracy [7], the system energy consumption and complexity are also enhanced [8]. To balance the energy consumption, the number of sensors, with the localization accuracy [7], the sensor selection problem arises and aims at invoking only a subset of sensors used for localization.

The straightforward approach to obtain the optimal sensor subset is enumerating all the possible sensor combinations and selecting one which has the optimal performance. However, this solver is computationally intensive and cannot be applied to the actual system. A heuristic approach has been proposed in [9] to solve the sensor selection problem by convex optimization. The authors of [10] have proposed a method that selects the minimum number of sensors in the premise of guaranteeing a certain estimation performance. All the above literatures deal with sensor selection problems by convex relaxation that relaxes the original nonconvex problem as a semidefinite program (SDP). In addition, several papers addressed the sensor selection problem by greedy approaches [11,12].

In the existing literatures, whether the measurement model is linear or nonlinear, only single-sensor selection vector is used to determine whether a sensor is selected. However, in TDOA localization scenario, both reference sensor and other ordinary sensors have a great influence on localization performance [13]. Besides, ubiquitous TDOA correlated noise makes the sensor selection problem more complicated since the contribution of each sensor to the Fisher information matrix (FIM) is no longer in an additive manner [14]. Therefore, development of sensor selection scheme in TDOA localization scenario in the presence of correlated noise, which can select the optimal reference sensor and ordinary sensors simultaneously, is crucial.

This paper develops an optimization framework for sensor selection in TDOA localization based on two Boolean vectors, which motivates to minimize the localization error in the presence of correlated noise and energy constraints. Though not focusing on a specific estimate algorithm, the Cramer-Rao Lower Bound (CRLB) is used as the performance metric [10], which results in the lowest possible variance that an unbiased linear estimator can achieve [2] and the trace being the minimum achievable localization mean square error (MSE) [15].

The rest of this paper is organized as follows. Section 2 introduces the measurement model and CRLB. Section 3 presents the problem formulation. Sensor selection approach is derived in Sect. 4. Section 5 presents the simulation results. Finally, conclusions will be drawn in Sect. 6.

2 Measurement Model and CRLB

We consider a wireless radio-based sensor localization network scenario where M sensors are used to determine the source position $\boldsymbol{u}^o$ employing TDOAs. Let us denote the true position of the source as $\boldsymbol{u}^o = [x^o, y^o, z^o]^T$ and the position of sensor i as $\boldsymbol{s}_i = [x_i, y_i, z_i]^T$, $i = 1, 2, \ldots, M$, where $(*)^T$ stands for transpose operation. The true distance between the source and sensor i is

$$r_i^o = \|\boldsymbol{u}^o - \boldsymbol{s}_i\| = \sqrt{(\boldsymbol{u}^o - \boldsymbol{s}_i)^T(\boldsymbol{u}^o - \boldsymbol{s}_i)}, \tag{1}$$

where $\| * \|$ denotes the Euclidean distance. Assume that each sensor in the network can act as the reference sensor, and sensor $\boldsymbol{s}_1$ is regarded as the reference sensor in our analysis without loss of generality. The true TDOA of a signal received by the sensor pair i and 1 is

$$t_{i1}^o = \frac{1}{c}r_{i1}^o = \frac{1}{c}(r_i^o - r_1^o), \tag{2}$$

where c is the signal propagation speed.

In practice, the observed TDOA measurements are

$$\boldsymbol{t} = \boldsymbol{t}^o + \Delta\boldsymbol{t}, \tag{3}$$

where $\boldsymbol{t} = [t_{21}, \ldots, t_{(M-1)1}]^T$ is the collection of the TDOA measurements, $\Delta\boldsymbol{t} = [\Delta t_{21}, \ldots, \Delta t_{(M-1)1}]^T$ is the additive zero mean Gaussian noise with covariance matrix $\boldsymbol{Q}_t$. Δt_{i1} is the difference of TOA noise Δt_i and Δt_1, where Δt_i is zero mean Gaussian noise with standard deviation σ_i with defining $\boldsymbol{\sigma} = (\sigma_1, \ldots, \sigma_M)$. In this paper, we assume that TDOA noises between different sensor pairs are correlated due to the common reference, so the covariance matrix $\boldsymbol{Q}_t$ is not diagonal [16].

The CRLB of $\boldsymbol{u}$ is equal to the inverse of the Fisher information matrix [17]

$$CRLB(\boldsymbol{u}) = (\boldsymbol{J}_u)^{-1}, \tag{4}$$

where FIM is defined as

$$\boldsymbol{J}_u = \boldsymbol{H}\boldsymbol{Q}_t^{-1}\boldsymbol{H}^T = \left(\frac{\partial \boldsymbol{t}}{\partial \boldsymbol{u}^T}\right)^T \boldsymbol{Q}_t^{-1} \frac{\partial \boldsymbol{t}}{\partial \boldsymbol{u}^T}, \tag{5}$$

$$\frac{\partial \boldsymbol{t}}{\partial \boldsymbol{u}^T} = \frac{1}{c} \begin{bmatrix} \frac{(\boldsymbol{u}-\boldsymbol{s}_2)^T}{r_2} - \frac{(\boldsymbol{u}-\boldsymbol{s}_1)^T}{r_1} \\ \vdots \\ \frac{(\boldsymbol{u}-\boldsymbol{s}_M)^T}{r_M} - \frac{(\boldsymbol{u}-\boldsymbol{s}_1)^T}{r_1} \end{bmatrix}, \boldsymbol{Q}_t = \begin{pmatrix} \sigma_2^2 + \sigma_1^2 & \cdots & \sigma_1^2 \\ \vdots & \ddots & \vdots \\ \sigma_1^2 & \cdots & \sigma_M^2 + \sigma_1^2 \end{pmatrix}. \tag{6}$$

3   Problem Formulation

In this section, it discusses the general theory for the sensor selection problem in TDOA localization scenario when the noises are correlated, and the optimization problem is formulated and analyzed. In particular, we are motivated to select the best sensor subset with K sensors of $M(M > K)$ available sensors that minimizing the localization error, which is subject to a constraint on the number of active sensors.

We introduce two Boolean vectors

$$\boldsymbol{q} = [q_1, q_2, \ldots, q_M]^T, \ q_i \in \{0, 1\} \tag{7}$$

$$\boldsymbol{p} = [p_1, p_2, \ldots, p_M]^T, \ p_j \in \{0, 1\} \tag{8}$$

to select the reference sensor and other ordinary sensors, respectively. The ith element of $\boldsymbol{q}$ describes whether or not the ith sensor is selected as the reference sensor and the jth element of $\boldsymbol{p}$ indicates whether or not the jth sensor is selected into the ordinary sensor subset.

We define two matrices $\boldsymbol{\Phi}_p$ and $\boldsymbol{\Phi}_q$, which are related to Boolean selection vectors and can formulate the FIM matrix and TDOA noise covariance matrix for selected sensors by picking out some columns or rows from information matrix. $\boldsymbol{\Phi}_p$ is a submatrix of $diag(\boldsymbol{p})$ after all columns corresponding to the unselected sensors are removed and $\boldsymbol{\Phi}_q$ is expanded by $\boldsymbol{q}$ as $\boldsymbol{\Phi}_q = [\boldsymbol{q}, \boldsymbol{q}, \ldots, \boldsymbol{q}]$.

Since two Boolean vectors are orthogonal, some properties are presented:

1. The products of two matrices:

$$\boldsymbol{\Phi}_p^T \boldsymbol{\Phi}_q = \boldsymbol{0}_{(K-1)}, \quad \boldsymbol{\Phi}_p \boldsymbol{\Phi}_q^T = \boldsymbol{p}\boldsymbol{q}^T, \quad \boldsymbol{\Phi}_p^T \boldsymbol{\Phi}_p = \boldsymbol{I}_{(K-1)}, \quad \boldsymbol{\Phi}_p \boldsymbol{\Phi}_p^T = diag\{\boldsymbol{p}\},$$
$$\boldsymbol{\Phi}_q \boldsymbol{\Phi}_q^T = (K-1) diag\{\boldsymbol{q}\}, \quad \boldsymbol{\Phi}_q^T \boldsymbol{\Phi}_q = \boldsymbol{1}_{(K-1) \times 1} \boldsymbol{1}_{1 \times (K-1)},$$

where $\boldsymbol{I}_{(K-1)}$ is an identity matrix with size $(K-1)$, $\boldsymbol{0}_{(K-1)}$ is a zero matrix with size $(K-1)$, $\boldsymbol{1}_{(K-1) \times 1}$ is a vector composed of '1' with size $(K-1) \times 1$.
2. We have

$$(\boldsymbol{\Phi}_p + \boldsymbol{\Phi}_q)^T \boldsymbol{E} (\boldsymbol{\Phi}_p + \boldsymbol{\Phi}_q) = (\boldsymbol{\Phi}_p - \boldsymbol{\Phi}_q)^T \boldsymbol{E} (\boldsymbol{\Phi}_p - \boldsymbol{\Phi}_q),$$

where $\boldsymbol{E}$ can be any diagonal matrix. This property plays an important role in the derivation of optimization problem.

In our analysis, since the source position $\boldsymbol{u}$ is the only unknown parameter, the FIM for selected sensors can be expressed as

$$\boldsymbol{J}_u^{sel} = (\boldsymbol{\Gamma}_t \boldsymbol{\Phi}_p - \boldsymbol{\Gamma}_t \boldsymbol{\Phi}_q) \left(\boldsymbol{\Phi}_p^T \boldsymbol{\Gamma}_q \boldsymbol{\Phi}_p + \boldsymbol{\Phi}_q^T \boldsymbol{\Gamma}_q \boldsymbol{\Phi}_q \right)^{-1} (\boldsymbol{\Gamma}_t \boldsymbol{\Phi}_p - \boldsymbol{\Gamma}_t \boldsymbol{\Phi}_q)^T \tag{9}$$

where

$$\boldsymbol{\Gamma}_t = \frac{1}{c} \left[\frac{(\boldsymbol{u} - \boldsymbol{s}_1)}{r_1}, \ \frac{(\boldsymbol{u} - \boldsymbol{s}_2)}{r_2}, \ \ldots, \ \frac{(\boldsymbol{u} - \boldsymbol{s}_M)}{r_M} \right], \tag{10}$$

$$\boldsymbol{\Gamma}_q = diag\{\Delta t_1^2, \Delta t_2^2, \ldots, \Delta t_M^2\}. \tag{11}$$

Theorem 1. *The closed form of the FIM for selected sensors $\boldsymbol{J}_u^{sel}$ with respect to $\boldsymbol{p}$ and $\boldsymbol{q}$ is*

$$\boldsymbol{J}_u^{sel} = \boldsymbol{\Gamma}_t \left(\boldsymbol{\Gamma}_o^{-1} - \boldsymbol{\Gamma}_o^{-1} \left(\boldsymbol{\Gamma}_o^{-1} + \boldsymbol{C} \right)^{-1} \boldsymbol{\Gamma}_o^{-1} \right) \boldsymbol{\Gamma}_t^T, \tag{12}$$

where

$$\boldsymbol{C} = \alpha^{-1} \left(diag\{\boldsymbol{p}\} + diag\{\boldsymbol{q}\} - \frac{(\boldsymbol{p} + \boldsymbol{q})(\boldsymbol{p} + \boldsymbol{q})^T}{K} \right)$$

is the dependence between these two Boolean vectors and FIM, positive definite matrix $\boldsymbol{\Gamma}_o$ and scalar α are related to the TOA noise covariance matrix $\boldsymbol{\Gamma}_q$.

Proof. To present the closed form of the FIM with respect to p and q, the TOA noise covariance matrix $\boldsymbol{\Gamma}_q$ is decomposed and the TDOA noise covariance matrix for selected sensors is simplified after some algebraic manipulations.

The TOA noise covariance matrix $\boldsymbol{\Gamma}_q$ can be decomposed as

$$\boldsymbol{\Gamma}_q = \boldsymbol{\Gamma}_o + \alpha \boldsymbol{I} \tag{13}$$

where α is a positive scalar, $\boldsymbol{I}$ is the identity matrix, and $\boldsymbol{\Gamma}_o$ is a positive definite matrix [14]. This decomposition can be obtained by setting α to be half of the any eigenvalue, and it can help simplify the inverse operation in (9).

The TDOA noise covariance matrix for selected sensors is expressed as

$$\boldsymbol{\Gamma}_q^{sel} = \boldsymbol{\Phi}_p^T \boldsymbol{\Gamma}_q \boldsymbol{\Phi}_p + \boldsymbol{\Phi}_q^T \boldsymbol{\Gamma}_q \boldsymbol{\Phi}_q.$$

According to the properties of $\boldsymbol{\Phi}_p$ and $\boldsymbol{\Phi}_q$, it derives

$$\begin{aligned}
\boldsymbol{\Phi}_p^T \boldsymbol{\Gamma}_q \boldsymbol{\Phi}_p + \boldsymbol{\Phi}_q^T \boldsymbol{\Gamma}_q \boldsymbol{\Phi}_q &= (\boldsymbol{\Phi}_p + \boldsymbol{\Phi}_q)^T \boldsymbol{\Gamma}_q (\boldsymbol{\Phi}_p + \boldsymbol{\Phi}_q) \\
&= (\boldsymbol{\Phi}_p - \boldsymbol{\Phi}_q)^T \boldsymbol{\Gamma}_q (\boldsymbol{\Phi}_p - \boldsymbol{\Phi}_q) \\
&= \boldsymbol{\Lambda} + (\boldsymbol{\Phi}_p - \boldsymbol{\Phi}_q)^T \boldsymbol{\Gamma}_o (\boldsymbol{\Phi}_p - \boldsymbol{\Phi}_q) \tag{14}
\end{aligned}$$

where $\boldsymbol{\Lambda} = \alpha(\boldsymbol{\Phi}_p - \boldsymbol{\Phi}_q)^T \boldsymbol{I} (\boldsymbol{\Phi}_p - \boldsymbol{\Phi}_q)$.

Substituting (14) into $(\boldsymbol{\Phi}_p - \boldsymbol{\Phi}_q)\left(\boldsymbol{\Gamma}_q^{sel}\right)^{-1}(\boldsymbol{\Phi}_p - \boldsymbol{\Phi}_q)^T$ and employing the matrix inversion lemma [14], it derives

$$(\boldsymbol{\Phi}_p - \boldsymbol{\Phi}_q)\left(\boldsymbol{\Gamma}_q^{sel}\right)^{-1}(\boldsymbol{\Phi}_p - \boldsymbol{\Phi}_q)^T = \boldsymbol{\Gamma}_o^{-1} -$$
$$\boldsymbol{\Gamma}_o^{-1}\left(\boldsymbol{\Gamma}_o^{-1} + (\boldsymbol{\Phi}_p - \boldsymbol{\Phi}_q)(\alpha(\boldsymbol{\Phi}_p - \boldsymbol{\Phi}_q)^T \boldsymbol{I}(\boldsymbol{\Phi}_p - \boldsymbol{\Phi}_q))^{-1}(\boldsymbol{\Phi}_p - \boldsymbol{\Phi}_q)^T\right)^{-1}\boldsymbol{\Gamma}_o^{-1} \tag{15}$$

where $\boldsymbol{\Gamma}_o$ is determined by the matrix decomposition in (13) and has no effect on the analysis.

Hence, the major challenge is to simplify the element in the inverse operation

$$\begin{aligned}
\boldsymbol{C} &= (\boldsymbol{\Phi}_p - \boldsymbol{\Phi}_q)\left(\alpha(\boldsymbol{\Phi}_p - \boldsymbol{\Phi}_q)^T \boldsymbol{I}(\boldsymbol{\Phi}_p - \boldsymbol{\Phi}_q)\right)^{-1}(\boldsymbol{\Phi}_p - \boldsymbol{\Phi}_q)^T \\
&= \alpha^{-1}(\boldsymbol{\Phi}_p - \boldsymbol{\Phi}_q)\left(\boldsymbol{I}_{K-1} + \boldsymbol{1}_{(K-1)\times 1}\boldsymbol{1}_{(K-1)\times 1}^T\right)^{-1}(\boldsymbol{\Phi}_p - \boldsymbol{\Phi}_q)^T. \tag{16}
\end{aligned}$$

Considering the relationship

$$\left(\boldsymbol{I}_{K-1} + \boldsymbol{1}_{(K-1)\times 1}\boldsymbol{1}_{(K-1)\times 1}^T\right)^{-1} = \left(\boldsymbol{I}_{K-1} - \frac{1}{K}\boldsymbol{1}_{(K-1)\times 1}\boldsymbol{1}_{(K-1)\times 1}^T\right),$$

we have

$$\boldsymbol{C} = \alpha^{-1}\left(diag\{\boldsymbol{p}\} + diag\{\boldsymbol{q}\} - \frac{(\boldsymbol{p}+\boldsymbol{q})(\boldsymbol{p}+\boldsymbol{q})^T}{K}\right).$$

Substituting (15) into (9), it derives

$$\boldsymbol{J}_u^{sel} = \boldsymbol{\Gamma}_t\left(\boldsymbol{\Gamma}_o^{-1} - \boldsymbol{\Gamma}_o^{-1}\left(\boldsymbol{\Gamma}_o^{-1} + \boldsymbol{C}\right)^{-1}\boldsymbol{\Gamma}_o^{-1}\right)\boldsymbol{\Gamma}_t^T.$$

$\square$

Hence, the optimization problem can be formulated as

$$\underset{p,q}{\text{minimize trace}} \ (J_u^{sel})^{-1} \tag{P1}$$

$$\text{subject to} \quad \mathbf{1}^T (p + q) = K$$
$$p^T q = 0$$
$$p, q \in \{0, 1\}^M.$$

4   Sensor Selection Approach

Since the presence of Boolean constraint and the inner product of vectors, the optimization problem (P1) is nonconvex and is not easy to address [10]. We present the sensor selection method based on convex relaxation techniques and the major operation is relaxing these nonconvex constraints.

Substituting (12) into (P1), it derives

$$\underset{p,q}{\text{min tr}} \ \left(B - D \left(\Gamma_o^{-1} + C \right)^{-1} D^T \right)^{-1}$$

$$s.t. \quad \mathbf{1}^T (p + q) = K$$
$$p^T q = 0 \tag{17}$$
$$p, q \in \{0, 1\}^M$$

where $B = \Gamma_t \Gamma_o^{-1} \Gamma_t^T$ and $D = \Gamma_t \Gamma_o^{-1}$.

Two auxiliary matrices Z and V are introduced to successively satisfy

$$B - D \left(\Gamma_o^{-1} + C \right)^{-1} D^T \succeq Z^{-1}, \tag{18}$$

$$V - D \left(\Gamma_o^{-1} + C \right)^{-1} D^T \succeq 0, \tag{19}$$

where notation '$\succeq$' stands for an operation that $A \succeq B$ means that $A - B$ is positive semidefinite. Besides, defining a vector $\iota = p + q$ and a matrix $R_\iota = \iota \iota^T$, C can be rewritten as

$$C = \alpha^{-1} \left(diag\{p\} + diag\{q\} - \frac{1}{K} R_\iota \right). \tag{20}$$

We can use the Schur complement to transform (18)–(20) into the following linear matrix inequalities (LMIs)

$$\begin{bmatrix} B - V & I \\ I & Z \end{bmatrix} \succeq 0, \quad \begin{bmatrix} V & D^T \\ D & \Gamma_o^{-1} + C \end{bmatrix} \succeq 0, \quad \begin{bmatrix} R_\iota & \iota \\ \iota^T & 1 \end{bmatrix} \succeq 0. \tag{21}$$

As for (17), $p \in [0, 1]^M$ and $q \in [0, 1]^M$ are used to relax the last constraint and the optimal solution may be fractional. The equation $p^T q = 0$ means that the values of corresponding elements in p and q cannot be '1' at the same time. This constraint is abandoned in the algorithm and is considered when the fractional vector is judged to be a Boolean vector.

Substituting (21) into (17) and considering the above relaxation operations, the problem can be transformed to a SDP

$$\min_{p,q,Z,V,R_{\iota}} \quad \text{tr}\,(Z)$$

$$\begin{aligned} s.t. \quad &\text{LMIs\ \ in}\ (21)\\ &\mathbf{1}^T(p+q) = K\\ &0 \leq p_i \leq 1,\ \ i=1,2,\ldots,M\\ &0 \leq q_i \leq 1,\ \ i=1,2,\ldots,M. \end{aligned} \tag{22}$$

The SDP problem can be solved efficiently by using interior-point approach [9,18]. However, since the relaxation operation, the optimal selection vectors for (22) are fractional and must be transformed to the Boolean vectors by judgement to select the suboptimal sensor subset from all the sensors. The simplest method is to select the maximal $(K-1)$ weight values from p and one maximal value from q [19]. Besides, we can also use the randomization method [14,20], generating a set of random fractional vectors related to the SDP solution and giving a set of random Boolean selection vectors, to select the minimum value of objective function when each random Boolean selection vector pair is substituted.

5   Simulation Results

This section presents the simulation results of the proposed sensor selection method and compares its performance with the exhaustive search and closest selection method. The performance metric is CRLB when the true position of source is substituting.

In the simulation, the sensor network contains 30 sensors in which 8 of them are selected. Their positions are randomly obtained and the coordinate components are Gaussian parameters when standard deviation is 3000. The position of source is also generated randomly in a large cube at $[0,0,0]^T$ with a length of 1000. TOA noise covariance matrix $\varGamma_q$ is a diagonal matrix with diagonal elements as σ^2, where σ is a random vector. In the SDP with randomization method, we set 50 random sensor combinations to calculate. Figure 1 shows the localization accuracy comparison when the range of arrival (ROA) noise strength varies from -5 to $10\,\text{dB}$. The circular symbol denotes the SDP method, the cross symbol represents the SDP with randomization method, and the inverted triangle represents the exhaustive search method. It is evident from the figure that the performance of the proposed SDP with randomization method is able to reach the exhaustive search method and is close to the scene when all sensors participate in localization. In particular, when five sensors of network are selected and ROA noise strength is fixed at 0.1 m, the selected reference sensor and ordinary sensors indices comparison of different methods are shown in Fig. 2. The pink and blue lines denote the estimation fractional vector of SDP and SDP with randomization method, and the solid symbols represent selected sensors.

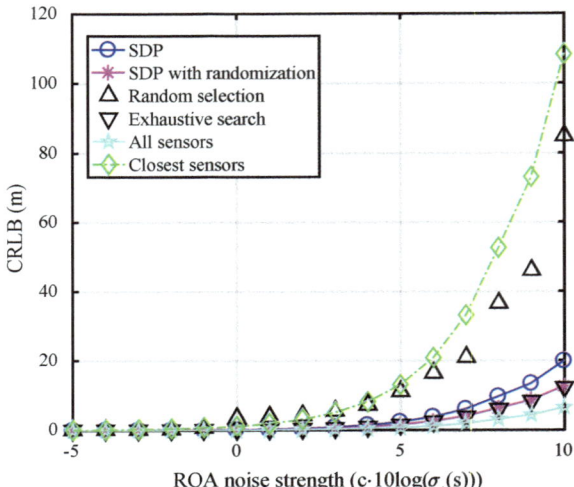

Fig. 1. CRLB versus ROA noise strength

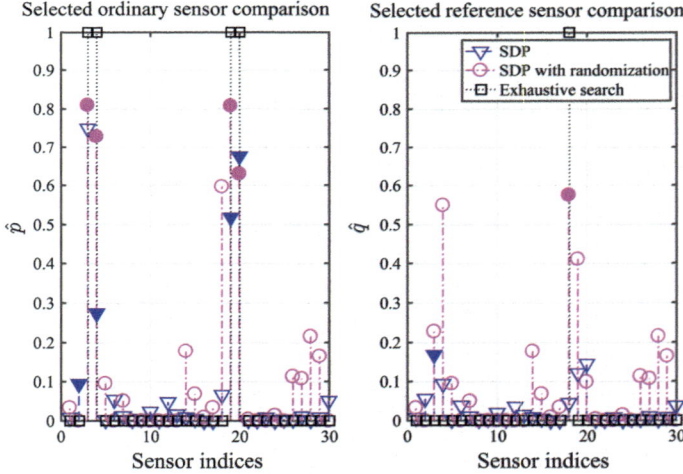

Fig. 2. Selected sensor indices comparison

Besides, for ease of comparison, we set the selection vector of exhaustive search method to Boolean. It is seen that the selected sensors of proposed SDP with randomization method are similar to the exhaustive search method including the reference sensor and other ordinary sensors.

6   Conclusion

This paper has proposed a sensor selection method to select the reference sensor and ordinary sensors simultaneously in TDOA localization scenario when the measurement noises are correlated. Two Boolean vectors are introduced to formulate the nonconvex optimization problem, which motivates to minimize the localization accuracy in the constraint of energy constraint. The problem has been solved by convex relaxation technique and randomization method. Simulation results have validated that the localization performance and selected sensor indices of proposed SDP with randomization method are similar to the exhaustive search method.

Acknowledgments. This work was supported by the National Natural Science Foundation of China under Grant 61631015, 61401323, 61471395 and 61501356, by the Key Scientific and Technological Innovation Team Plan (2016KCT-01), the Fundamental Research Funds of the Ministry of Education (7215433803 and XJS16063).

References

1. Yin J, Wan Q, Yang S, Ho KC. A simple and accurate TDOA-AOA localization method using two stations. IEEE Signal Process Lett. 2016;23:144–8.
2. Ho KC, Xu W. An accurate algebraic solution for moving source location using TDOA and FDOA measurements. IEEE Trans Signal Process. 2004;52:2453–63.
3. Li Z, Guan L, Li C, Radwan A. A secure intelligent spectrum control strategy for future THz mobile heterogeneous networks. IEEE Commun Mag. 2018.
4. Wang Y, Ho KC. An asymptotically efficient estimator in closed-form for 3-D AOA localization using a sensor network. IEEE Trans Wireless Commun. 2015;14:6524–35.
5. Ho KC, Lu X, Kovavisaruch L. Source localization Using TDOA and FDOA measurements in the presence of receiver location errors: analysis and solution. IEEE Trans Signal Process. 2007;55:684–96.
6. Calwo-Fullana M, Matamoros J, Anton-Haro C. Sensor selection and power allocation strategies for energy harvesting wireless sensor networks. IEEE J Sel Areas Commun. 2016;34:3685–95.
7. Kaplan LM. Global node selection for localization in a distributed sensor network. IEEE Trans Aerosp Electron Syst. 2006;42:113–35.
8. Hao B, Zhao Y, Li Z, Wan P. A sensor selection method for TDOA and AOA localization in the presence of sensor errors. In: Proceedings of IEEE/CIC International Conference on Communications in China, Qingdao; 2017. p. 1–6
9. Boyd S, Vandenberghe L. Convex optimization. New York, NY, USA: Cambridge Univ. Press; 2004.
10. Chepuri SP, Leus G. Sparsity-promoting sensor selection for non-linear measurement models. IEEE Trans Signal Process. 2015;63:684–98.
11. Rao S, Chepuri S P, Leus G. Greedy sensor selection for non-linear models. In: Proceedings of IEEE International Workshop on Computational Advances in Multi-Sensor Adaptive Processing. 2015. p. 241–4
12. Shamaiah M, Banerjee S, Vikalo H. Greedy sensor selection: Leveraging submodularity. In: Proceedings of IEEE Conference on Decision and Control. 2010. 2572–7

13. Rene J E, Ortiz D, Venegas P, Vidal J. Selection of the reference anchor node by using SNR in TDOA-based positioning. In: Proceedings of IEEE Ecuador Technical Chapters Meeting. 2016. p. 1-4

14. Liu S, et al. Sensor selection for estimation with correlated measurement noise. IEEE Trans Signal Process. 2016;64:3509–22.

15. Hao B, Li Z, Ren Y, Yin W. On the cramer-rao bound of multiple sources localization using RDOAs and GROAs in the presence of sensor location uncertainties. In: Proceedings of IEEE Wireless Communications and Networking Conference. 2012. 3117–22

16. Qi Y, Kobayashi H, Suda H. Analysis of wireless geolocation in a non-line-of-sight environment. IEEE Trans Wireless Commun. 2006;5:672–81.

17. Kay SM. Fundamentals of statistical signal processing, estimation theory. Englewood Cliffs, NJ: Prentice-Hall; 1993.

18. Shen X, Liu S, Varshney PK. Sensor selection for nonlinear systems in large sensor networks. IEEE Trans Aerosp Electron Syst. 2014;50:2664–78.

19. Joshi S, Boyd S. Sensor selection via convex optimization. IEEE Trans Signal Process. 2009;57:451–62.

20. Luo Z, et al. Semidefinite relaxation of quadratic optimization problems. IEEE Signal Process Mag. 2010;27:20–34.

Location Precision Analysis of Constellation Drift Influence in TDOA Location System

Liu Shuai[1], Song Yang[1], Guo Pei[2], Meng Jing[1],
and Wu Mingxuan[1(✉)]

[1] China Academy of Space Technology, NO.104, You Yi Road,
Haidian District, Beijing, China
lsshr@163.com, wumingxuan@sina.com
[2] Institute of Manned Space System Engineering, NO.104, You Yi Road,
Haidian District, Beijing, China
stillif@sina.com

Abstract. In the TDOA location system,the constellation drift can affect the location precision. Aiming at this problem, based on the analysis of the three satellites constellation drifting in a period of time, we proposed the location precision results of constellation drift influence in TDOA location system. Considering the constraint of location accuracy and satellite resources, a feasible recommendation for location maintenance of three satellites system is put forward, which provides the theoretical support for the location maintenance strategy of three satellites TDOA location.

Keywords: Three satellites constellation · Constellation drift influence · Location precision

1 Introduction

Three satellites TDOA(time difference of arrival) location system is a kind of passive electronic reconnaissance location mode, which uses three formation flying satellites to determine two equal time difference surface by measuring the time arrival difference of radar emitter signal. Combined with the surface constraint equations, real-time and high precise location on electronic signal is achieved [1].

However, when the satellite is in orbit, its position will drift, which can affect the configuration of the satellite constellation than can affect the location precision of TDOA location system [2, 3]. In the paper [4], based on the best configuration mathematical expressions of the minimum positioning error variance, Li Wenhua demonstrated the effect of three satellites location precision with design of the constellation configuration, but did not explain the effect of the constellation changes on the location precision. In the paper [5], based on the TDOA location error model under the constraint of the earth, Gu Liming et al. demonstrated the effect of location performance by satellite constellation configuration, but did not make the relationship between constellation configuration and time changes.

© Springer Nature Singapore Pte Ltd. 2020
Q. Liang et al. (eds.), *Communications, Signal Processing, and Systems*, Lecture Notes in Electrical Engineering 516,
https://doi.org/10.1007/978-981-13-6504-1_138

To solve the above problems, based on the three satellites TDOA system, we analyze the location precision results of constellation drift influence in a period of time, put forward the recommendation of location maintenance and improve the location precision.

2   Location Mechanism of TDOA System

By measuring the arrival time of the electronic signal pulse, the TDOA location system realizes the high precision location of the electronic signal radiation source. The location mechanism is shown in Fig. 1, $S0$, $S1$ and $S2$ are the obtained two independent pulse arrival times, which can determine the location points of a hyperboloid. Three satellites can make two hyperboloids as $A1$ and $A2$, which can form two curves $L1$ and $L2$ by intersection with the earth's surface respectively, and the intersection point of two curves is the location of radiation source.

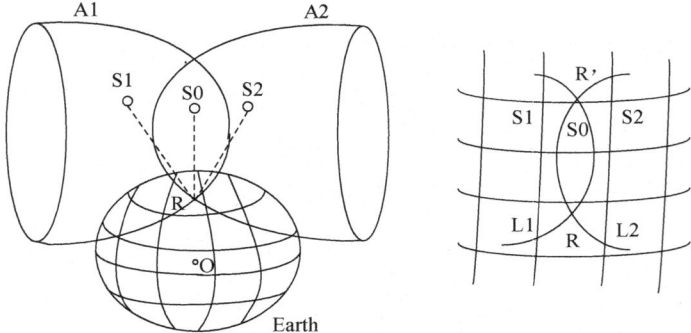

Fig. 1. Mechanism of three satellites TDOA location

Generally, there is only one location point R of the intersection between $A1$ and $A2$ and the earth's surface. When the distribution of three satellites is unreasonable, there will be two location points R and R'. The true radiation source is R and the false radiation source location is R'. With the motion of the satellites, R' is divergent, R is stable convergent [6, 7].

3   Analysis of the Constellation Drift

3.1   Analysis of Constellation Configuration Drift

Select a time as the starting point, the satellite constellation position of the current time is the initial position [8] and then the latitude and longitude location of the three satellites constellation is simulated at an interval of one month. In Fig. 2, the far right is

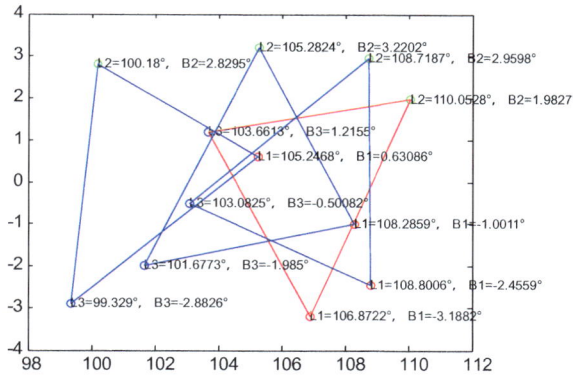

Fig. 2. Three months constellation change

the three satellites position at the starting point, and the left constellation lines are three satellites position after one month, two months and three months.

Table 1 is the latitude and longitude positions of the three satellites with three months change.

Table 1. Longitude and latitude drift with an interval of three months

	Initial position (°)	Three months later (°)	Drifting degree (°)
Satellite longitude (A)	106.8722	105.2468	−1.62533
Satellite latitude (A)	−3.18817	0.630862	3.819034
Satellite longitude (B)	110.0528	100.18	−9.87278
Satellite latitude (B)	1.98275	2.829539	0.84679
Satellite longitude (C)	103.6613	99.32903	−4.33232
Satellite latitude (C)	1.215527	−2.88258	−4.09811

As we can see, the maximum constellation drifting degree of the east and west direction (longitude) is −9.87°, and the maximum constellation drifting degree of the north and south direction (latitude) is −4.10°.

3.2   Analysis of Constellation Angle

The changed angle of constellation lines also illustrated the changes of constellation. In the initial design, the three satellites are an equilateral triangle and the angles are 60°. In the observation of a month, the baseline angle changes within the range of 5°, which will infect the location precision. The angles of the constellation are changed as shown in Fig. 3.

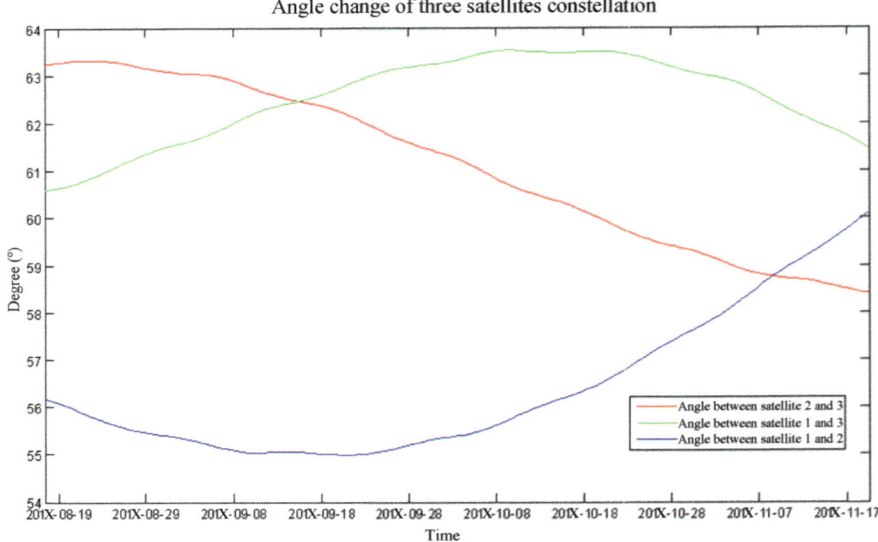

Fig. 3. Baseline angles change of satellite constellation

3.3   Analysis of the Coverage Area of Constellation

The coverage area of the three satellites constellation is also the main factor affecting the location precision. In order to compare the area changes with the initial constellation during different periods visually, we simulated the constellation changes each month, the results are shown in Fig. 4.

From Fig. 4, we can see that there is no overlap between the changed constellation and the initial constellation since October 16. As time goes on, the distance between the constellations is bigger and bigger, and the satellite coverage target area decreases. Table 2 listed the coverage area and coverage changing trends per ten days from August 17 to December 16.

As shown in Table 2, in the period of three months from August 17, the coverage area of three satellites is decreased. The coverage rate is decreased to 96.60% one month later and decreased to 57.01% three months later, which will affect the location precision of three satellites largely.

4   Simulation Analysis of Location Precision Influence

Choosing key parameters such as the constellation position error, time error and satellites velocity measurement error [9], then we can obtain the system location precision results, as shown in Fig. 5.

Select the specific single point in the coverage area, then we can analyze the influence of location precision of the three satellites in the drift state. Choosing the ground electronic signal emitter with the distances from the location below satellites

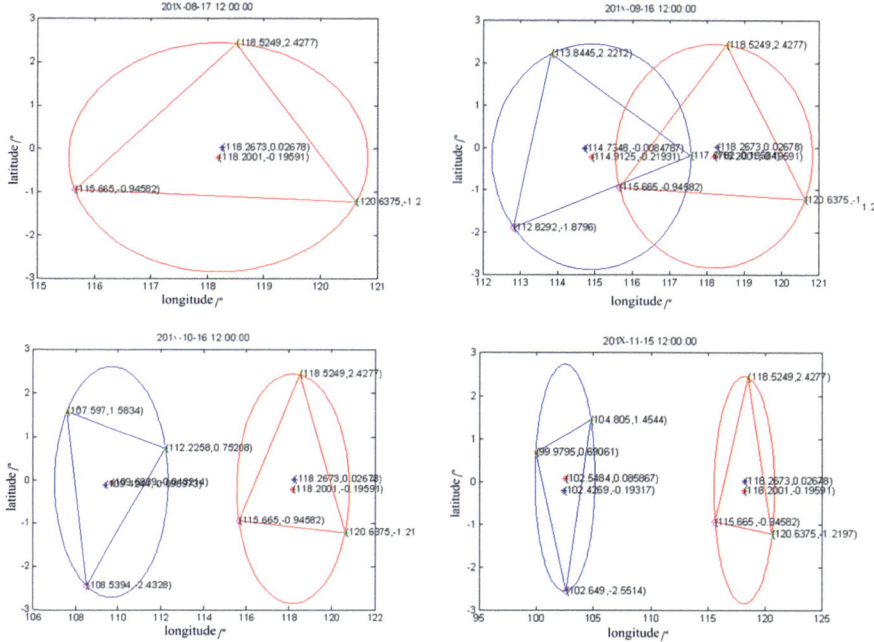

Fig. 4. Constellation changes each month

Table 2. Changing trends of coverage area

Time (12 h)	Initial area (km²)	Coverage area (km²)	Coverage rate (%)
8.17	19,595,000	19,595,000	100
8.27		19,576,000	99.90
9.6		19,356,000	98.78
9.16		18,929,000	96.60
9.26		18,250,000	93.14
10.6		17,113,000	87.34
10.16		15,787,000	80.57
10.26		14,341,000	73.19
11.5		12,804,000	65.34
11.15		11,171,000	57.01

center is 2000, 3000, 4000 and 5000 km respectively, then we can obtain the location precision of the three satellites(GDOP)with the period of three months, as shown in Fig. 6. In Fig. 6, select the worst value (the maximum value of GDOP) of the location precision in the all equal distance target points at any time.

Simulation results show that the farther of the distance below the satellites center point, the longer of the three satellites constellation drift time, the worse of the location

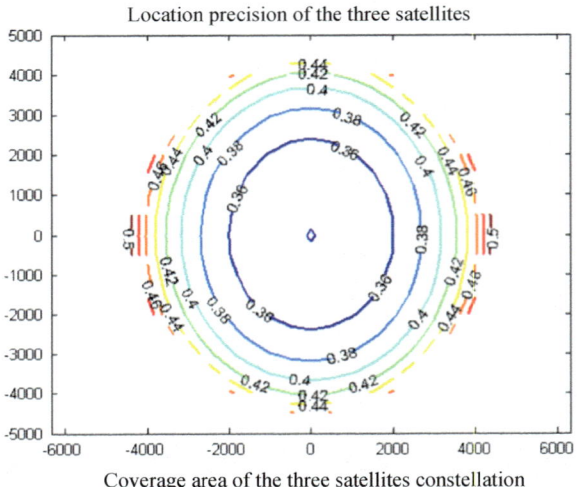

Fig. 5. Location precision of the three satellites

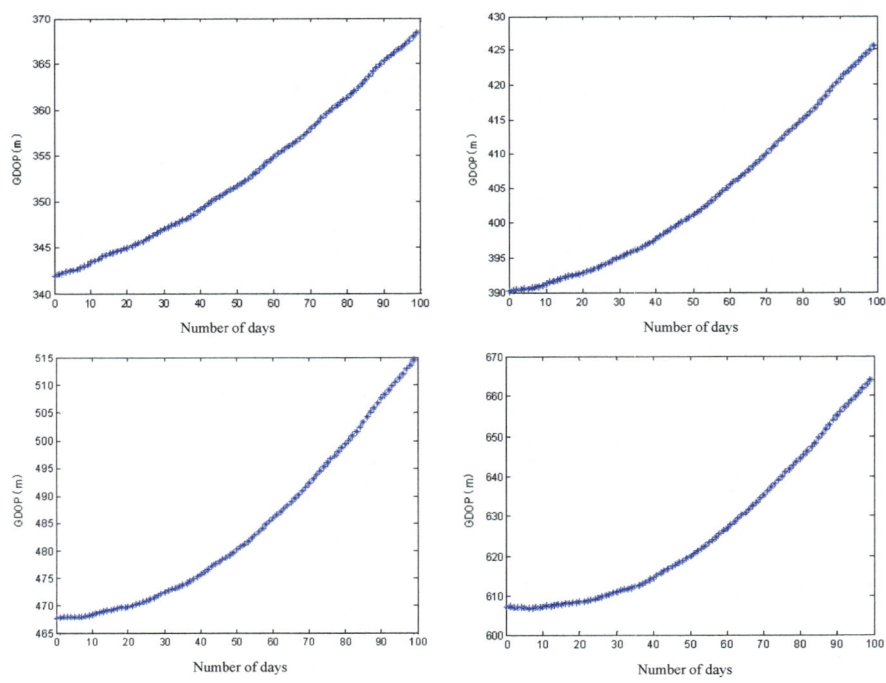

Fig. 6. Location precision for distances of 2000, 3000, 4000 and 5000 km

precision, as shown in Table 3. Generally, considering the location precision results and the satellite resources, we recommend that the three satellites TDOA system carried out a location maintenance for about 100 days.

Table 3. Summary of location precision at different distances

Distance below the center point of satellites (km)	GDOP (m)
2000	342–368
3000	390–426
4000	468–515
5000	608–665

5 Summary

In the three satellites TDOA system, the three satellites will drift, which affects the location precision of system. In this paper, the constellation change of the three satellites over a period of time is demonstrated, and the influence of constellation drift on location precision is also analyzed. Considering the constraints of location precision and satellite resources, we proposed feasible location maintenance recommendations for the three satellites, which provided a theoretical support for the location protection strategy of three satellites TDOA constellation.

References

1. Zhu W, Huang P, Ma Q,Lu X. Emitter location with multi-station using TDOA/FDoA measurements. J Data Acquis Process. 2010; 05:307–12.
2. Yu M, Xu J, Xia Y, Lu BK. A Study of optimal configuration of three-satellite constellation for localizing the object on earth. ACTA Astron Sin. 2003;08:302–9.
3. Chen Y, Li C, Li X. A precision analyzing&reckoning model in tri-station TDOA location. ACTA Electron Sin..2004;32(9):1452–5.
4. Li W. Research on configuration of tri-satellites and location precision of TDOA. J Astronaut. 2010;03:701–6.
5. Gu L, Zhao Y, Zhu J, Miao S. Constellation configuration and its performance based on TDOA with four satellites. Telecommun Eng. 2017; 01:33–8.
6. Fu W. DTOA Location precision of target on the ground.Electron Inf Warf Technol. 2008;23 (6):17–20.
7. Rao H. Research on key technologies of TDOA with multi-satellites. Nan jing: Nanjing University of Science and Technology; 2009.
8. Du L, Zhang Z, Li X, Wang R, Liu L, Guo R. Station-keeping maneuver monitoring and moving-window ground track fitting of GEO satellites. Act Geod Cartogr Sin. 2014;03: 233–9.

9. Ping Z, Zhen H, Jianhua L. Passive stationary target positioning using adaptive particle filter with TDOA and FDOA Measurements, vol. 4. Issue 12. 9th Asia-peck Conference about Communications. 2010. p. 224–235.

A Weighted and Improved Indoor Positioning Algorithm Based on Wi-Fi Signal Intensity

Guanghua Zhang and Xue Sun$^{(\boxtimes)}$

Northeast Petroleum University Youth Fund: XN2014111, Daqing, China
`1054351628@qq.com`

Abstract. In order to solve the problem of the influence of signal strength fluctuation on indoor positioning accuracy, an improved indoor positioning algorithm based on WiFi signal strength is proposed in the paper. Based on the *K*-nearest neighbor location algorithm, the weight of the signal strength is further increased, the characteristics of the received signal and the fingerprint database are optimized, the interference of the weak signal on the positioning is reduced, and the accurate indoor positioning is achieved. The calculation in this work suggests that the positioning error can be reduced on the original basis, the accuracy of the algorithm is improved. On the basis of the original algorithm, the error range is narrowed and the positioning accuracy is improved. The average error of the improved algorithm is controlled at about 1.87 m.

Keywords: Wi-Fi · Indoor positioning · Weighted signal strength

1 Introduction

Indoor positioning technology refers to determining the position of a receiving terminal in a certain reference coordinate system at a certain time in an indoor environment. The "indoor" here not only refers to the general interior of a typical building, but also includes underground mines and dense high-rise buildings. Some fully enclosed or half enclosed space are also called indoors, such as buildings or woods [1]. These enclosed or semi-enclosed spaces are collectively referred to as interior spaces. With the development of science and technology, there are more and more indoor positioning technologies, such as wireless network technology, Bluetooth, infrared, ZigBee technology, and RFID. The principle of different positioning methods is also different; there are positioning methods based on signal arrival time or angle of arrival, or signal strength, long time, etc. For different indoor environments, the most suitable positioning method should be used to reduce the positioning error as much as possible. At present, several commonly used positioning software, such as Baidu Maps and Gaode maps, have a good positioning effect in the outdoors and also bring great convenience to people's lives. However, the positioning in the room is severely degraded due to the blockage of the GPS signal by the obstacles and cannot be applied indoors. The massive use of mobile phones has led to an increasing demand for indoor positioning. In addition, the coverage of Wi-Fi has been continuously expanding, which has led to

© Springer Nature Singapore Pte Ltd. 2020
Q. Liang et al. (eds.), *Communications, Signal Processing, and Systems*, Lecture Notes in Electrical Engineering 516,
https://doi.org/10.1007/978-981-13-6504-1_139

the emergence of indoor Wi-Fi positioning technology. This article mainly introduces the commonly used indoor positioning algorithms. Based on this, it proposes a method to increase the weights and improves the positioning accuracy through database filtering and error analysis.

2   Traditional Indoor Positioning Algorithm

2.1   KNN Algorithm

KNN algorithm is the K-nearest neighbor classification algorithm. The core principle of the algorithm is that if the most neighboring samples of a sample in the feature space mostly belong to a certain category, then the sample is also said to belong to this category, and has the basic characteristics of the sample of this category [2, 3]. In the KNN algorithm, the most important affair is to correctly select the value of K. When the value of K is small, the number of nearby points to be selected is small, and the category of the reference point cannot be accurately determined, thereby increasing the error. When the K value is large, the farther point will be selected, resulting in inaccurate results. Therefore, multiple measurements are required determining the value of parameter K during the experiment. Then selecting a certain number of collection points in the measurement area, and collecting the position information and signal intensity values of each point. When the user issues a positioning request, the coordinate information of the point is fed back to the user through the RSS of the signal.

2.2   Triangle Positioning Algorithm

Triangle principle of distance measurement: First we must determine the coordinate information of at least 3 Wi-Fi stores, by calculating the signal strength received by the mobile receiver, according to the formula to calculate the distance from the measuring point to each AP $P_r(d) =$

$$\frac{P_t G_t G_r \lambda^2}{4\pi d^2 l} \tag{1}$$

P_t is the transmission power of the AP, G_t is the gain of the signal transmitting end, G_r is the gain of the mobile receiving end, λ is the wavelength of the electric wave, $P_r(d)$ is the signal power received by the mobile receiving end, d is the distance between AP and the mobile terminal, l is the loss factor [3]. Only d in the formula is unknown. Solve the distance from each AP to the point to be measured. Examples are A, B, C APs with known coordinates. D is the point to be measured, the three transmitting ends are centered, the distance between A and D is d1, the distance between B and D is d_2, and the distance between C and D is d_3. The distance d_3 is a radius, and the intersection of the three circles is the point to be measured. Let D coordinate be (x, y) A, B, and C coordinates (x_1, y_1), (x_2, y_2), (x_3, y_3), and calculate the known Wi-Fi hot spot to the point D to be determined by formula For distances d_1, d_2, and d_3, $(x - x_i)^2 + (y - y_i)^2 = d_i^2$, where $i = 1, 2$, and 3. The values of x and y can be obtained by solving a system of linear equations.

2.3 Disadvantages and Demerits of Traditional Location Algorithms

The K-nearest neighbor method determines that the user equipment is located near the sensor position when the signal of the user equipment is detected by a sensor of a fixed position. In this way, the location of the sensor can be used to characterize the spatial location of the user which is a relative location data. And the initial investment is not suitable for a wide range of applications. Triangulation, in the data acquisition process, has a high requirement for signal stability and is affected by complex indoor environments and equipment. Signals can fluctuate and distort due to occlusion of buildings or the flow of people.

3 Improved Weight Measurement

The main idea based on the weight measurement method is to compare the RSSI collected in real time with the feature database, and select the centroid of the nearest point as the target's estimated position. In order to complete the matching, a feature database should be established first, including: location identification, location coordinates, signal strength, and intensity change range. The location identifier is an identifier for each AP in the collected data in the indoor area. Positioning is a constant parameter used [4, 5].

3.1 Based on Weight Measurement

Position coordinates refer to the X and Y coordinates of a signal strength set, which is used to calculate and determine the target position. The signal strength refers to a list of signal strengths $(AP_n, RSSI_n)$ after a part of the location-specific reference AP has averaged the signal strengths several times, which is the number of selected reference points in the target area.

The signal strength and intensity variation range in the feature database are the features used to determine the position. The current position is determined by matching the feature library with the signal strength acquired in real time at the current position. Current Wi-Fi signal strength $(AP_1, RSSI_1)$ $(AP_2, RSSI_2)$ ... $(AP_n, RSSI_n)$, n is the number of currently scanned APs. The first name from the list starts with matching in the signature database, finds the AP with the same name, and then judges its signal strength. When the current signal strength is within the variation range of the signal strength of the gather, the weight of the position is increased by 1, so repeated will get a certain number of weight lists, in different positions, the maximum weight q_{max} is selected as a reference, and the weight and the position of the phase difference within e are used as reference points, where the value of e varies with the environment, and $q_{max} - q_i < e$. q_i is the first i weights. The weights of the reference points are arranged from the largest to the smallest and are recorded as $Q_1 = (q_1, x_1, y_1)$ to $Q_n = (q_n, x_n, y_n)$ according to the formula as below.

$$x = \frac{\sum_{i=1}^{n} q_i x_i}{\sum_{i=2}^{n} q_i} \tag{2}$$

$$y = \frac{\sum_{i=1}^{n} q_i y_i}{\sum_{i=2}^{n} q_i} \tag{3}$$

3.2   The Main Influencing Factors of Indoor Positioning

In the indoor positioning process, due to the different environments, the influence on the positioning accuracy is also different. The main influencing factors on the positioning accuracy are: the impact of obstacles on the strength of Wi-Fi signals, different obstacles such as walls, furniture, and people. These obstacles reflect or absorb the Wi-Fi signal. If these influencing factors are brought into the calculation, it will increase the positioning error. In order to reduce the error, an improvement factor is introduced [6]. We denote the correction of the influence of obstacles on the signal as a. The transmission distance will also affect the signal strength. As the distance increases, the signal strength will continue to attenuate. When the distance exceeds a certain distance, the signal will reach the receiving end and the intensity will be weak or even absent. In this paper, the distance will affect the signal. Correction is marked b. There are also multipath effects in the propagation of Wi-Fi signals. Because the signal is transmitted to the receiving process, it will be blocked by different interference and obstructions, which will cause multiple paths during the propagation process. In this way, the arrival time of the signal will be different, and the positioning longitude will be attenuated. In this paper, the correction of the effect of multipath effects on the signal is denoted as c.

3.3   The Realization of Improved Algorithm

In this article, improvements are made based on signal strength acquisition and impact factors. Through a large number of data acquisitions, it is found that the intensity advantage of Wi-Fi signals collected at two locations with a certain distance will be greatly different. If an AP's signal is lost several times within the same time, it cannot be scanned. This is not the case at another point that is a certain distance away from this point. Through comparison, when the signal strength at a certain point is lower than a certain value, the AP signal has a large fluctuation at this point. When the characteristic database is established, stocks can be excluded from this type of point and can be to a certain extent. Improve positioning accuracy.

After positioning based on the weights, the values of x and y can be obtained, and then a judgment is made on the result of the last accurate positioning based on the coordinates. Then, the difference between the two sets of coordinates is calculated, and if the difference is within the required range of the positioning accuracy, the positioning is considered to be true. Otherwise, relocate. This method is used in the positioning stage. After the results are obtained by the positioning algorithm, firstly, the position coordinates obtained by the positioning algorithm are obtained, the distance d_1 between the coordinates and the coordinates of the last correct positioning is calculated, and the accelerometer and timing on the communication device are used. Calculate the speed and time from the first coordinate to the second coordinate by using the application of the acceleration sensor and the timer on the communication device, and then calculate the distance d_2 that the target moves during this time, and record the positioning accuracy of the system. For d, when $|d_1 - d_2| < d$, it means the positioning result is correct, otherwise the result is wrong.

However, the influencing factors in the signal propagation process cannot be ignored. To improve the positioning accuracy, the corresponding x and y values can be multiplied by the correction factor of the corresponding influencing factor. Then the formulas for calculating x and y are as follows: $x' = x * a * b * c, y' = y * a * b * c$, where x' and y' are the results of the last positioning. The value of a, b, c, is obtained from the review of the reference or the measurement of a large amount of data. For different positioning environments, the three values are not the same.

4   The Main Performance of the Positioning System

4.1   Data Collection and Filtering

In the initial period of database establishment, a large amount of data must be collected, and the strength of the Wi-Fi signal has great volatility. If it can be filtered based on changeable range of the intensity, the positioning accuracy will be greatly improved. In this paper, based on the RSS signal strength filtering method, the propagation of wireless signals in space is more complex, and the RSS values measured at the same point are also not the same, and its measured values are affected in many ways [7, 8], for example, the multipath of signal propagation, non-line-of-sight propagation, interference of other signals and obstacles. This result in different RSS worth cases, and because of the non-linear time-varying nature of the signal, the single-measured RSS value will also be different. Provided that only one RSS value is used to measure, the error is large, so the processing of the RSS value is necessary. In this paper, the average value filtering method is used to process the RSS value. The main idea of the mean filter is to perform multiple measurements at the same location to obtain different RSS values and then average these values so that the effect of a single intensity value can be avoided. The formula is as follows.

$$\text{RSS}' = \frac{1}{N} \sum_{i=1}^{N} \text{RSS}_i \tag{4}$$

RSS_i represents the value obtained by the i-th measurement, to avoid the error caused by a single value.

4.2   Positioning Function

The positioning function is the main function of the positioning system. The main workflow of the positioning system is as follows. First, the positioning module needs to complete the data acquisition in real time. Then, the acquired data is stored, and the obtained data and data in the database are completed through data calculation. Matching and calculating the current position coordinates. Since there is a certain error in the result of the positioning, an inaccurate judgment is made on this result. If the error is within the required positioning accuracy, the result is correct; otherwise, the result is discarded and repositioned [9, 10].

5   Experimental Results and Error Analysis

After the data is collected, the system diagram is simulated with MATLAB to obtain the positioning result. According to the positioning results, the positioning error is calculated, a chart depicting the distance of the error of the x-axis and y-axis is plotted to calculate the cumulative probability. From Fig. 1, the cumulative probability that the error distance of the x-axis is within 0.5 m is approximately 50% within 1 m. The cumulative probability is about 72%, and the cumulative probability within 2 m can reach more than 90%.

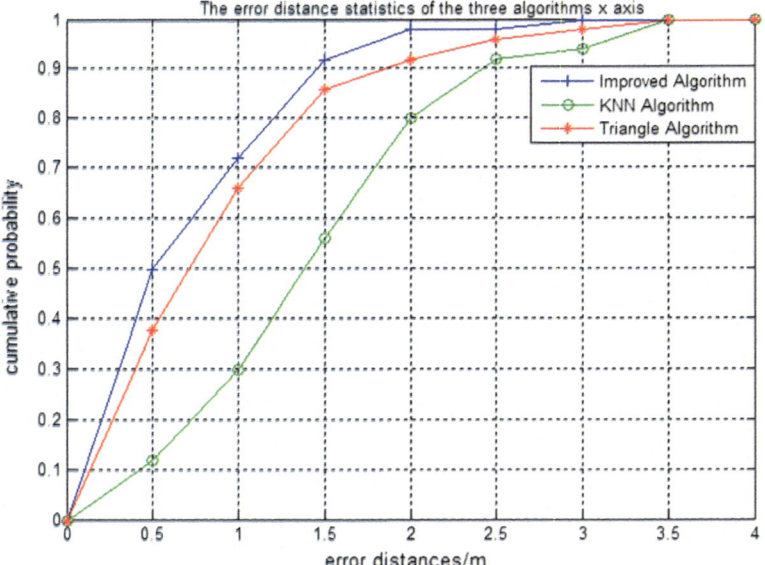

Fig. 1. Three algorithms x-axis error distance chart

From Fig. 2 we can see that the cumulative probability of *y*-axis error within 0.5 m is 38%, the cumulative probability of error within 1 m is about 70%, and the cumulative probability within 2 m is 98%.

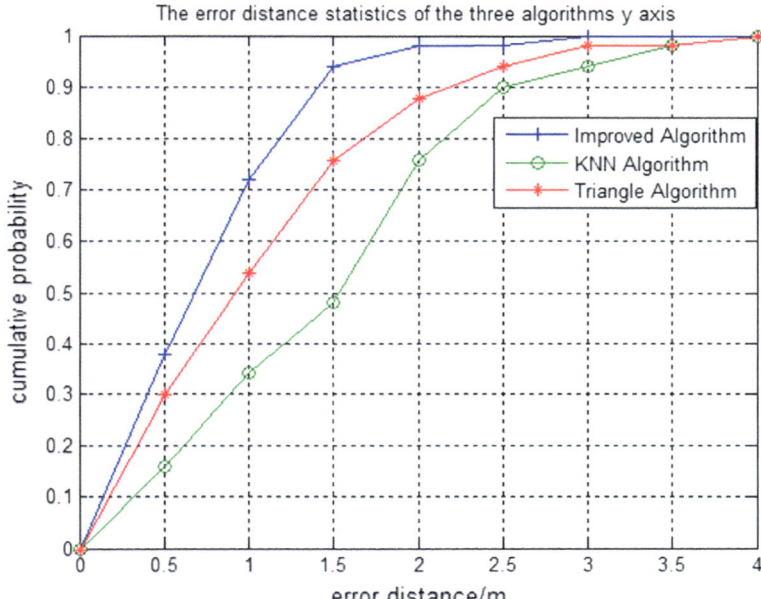

Fig. 2. Three algorithms *y*-axis error distance chart

According to the Fig. 3, the error histogram and average error histogram are plotted and compared from the error maximum, minimum, and average error. It can be seen that the improved positioning algorithm is smaller than the other two algorithms in numerical value. The positioning error can be shortened to within 2 m.

According to the Fig. 4, the paper uses CEP to calculate accuracy. It can be calculated that CEP = 1.87 m. From the figure, we can see that just 50% of the points fall within the circle.

6   Conclusion

This paper improves the positioning algorithm based on the weights and improves the positioning accuracy through database filtering, error analysis, and correction of impact factors. By comparing the advantages and disadvantages of the KNN algorithm and the triangle positioning algorithm, an improved positioning method based on weights is proposed to lay a foundation for the future development of indoor positioning technology. Wi-Fi-based positioning technology can be used in vehicle positioning systems, large shopping malls, libraries, etc. to ensure the accuracy of positioning through

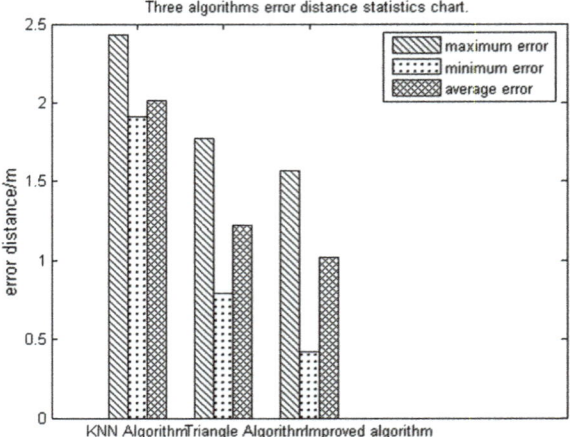

Fig. 3. Three kinds of algorithm error comparison chart

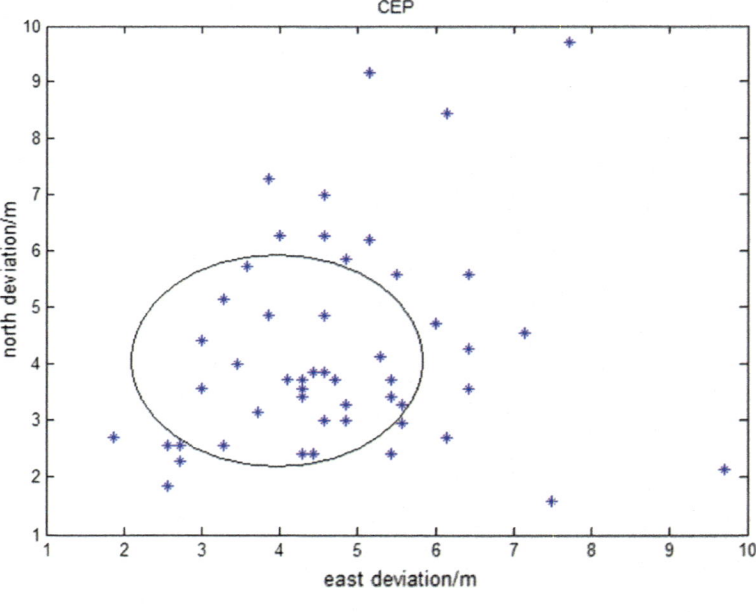

Fig. 4. CEP

real-time updates to the database. In the past two years, the application and research in China have become more and more widespread. Many hot spots have been installed in tourist attractions, restaurants, or subways. You can use mobile devices such as mobile phones and tablet computers to connect to the Internet, and various Wi-Fi-based applications are also used. The research on Wi-Fi positioning technology has become more and more in-depth, and there are more and more positioning algorithms. People hope to solve the limitations of the indoor environment through the advantages of this technology and achieve indoor precise positioning. With the deepening of research, it is constantly improving. The accuracy of positioning brings greater convenience to people's lives.

References

1. Qing W. Design and implementation of WiFi indoor positioning system. Beijing: Beijing Jiao tong University; 2014.
2. Bi-Chao Y. Research on indoor location technology based on WiFi. Chengdu: University of Electronic Science and Technology; 2017.
3. Yang P. A Weighted value selection and weighted localization algorithm based on RSSI. Inf Electron Eng. 2012;148–151.
4. Jin C, Qiu D. Research on indoor positioning technology based on WiFi signal. Bull Surv Mapp. 2017;21–25.
5. Lin H. Location-fingerprint indoor positioning algorithm based on Wi-Fi. Shanghai, Nanjing: East China Normal University; 2016.
6. Yan J. Research on indoor localization technology based on Wi-Fi. Guangzhou: South China University of Technology; 2013.
7. Rui M, Qiang G. An improved WiFi indoor positioning algorithm by Weighted Fusion. Sensors. 2015;21824–21843.
8. Hung-Huan L, Wei-Hsiang L. A WiFi-based weighted screening method for indoor positioning systems. Wireless Pers Commun. 2014;611–627.
9. Yang C, Shao H-R. WiFi-based indoor positioning. IEEE Commun Mag. 2015;150–157.
10. Huang H. WiFi Indoor positioning system design. J Guangxi Aademc Sci. 2016;59–61.

Evaluation Distance Metrics for Pedestrian Retrieval

Zhong Zhang[1,2]($\boxtimes$), Meiyan Huang[1,2], Shuang Liu[1,2], and Tariq S. Durrani[3]

[1] Tianjin Key Laboratory of Wireless Mobile Communications and Power
Transmission, Tianjin Normal University, Tianjin, China
{zhong.zhang8848,meiyanhuang7295}@gmail.com
[2] College of Electronic and Communication Engineering, Tianjin Normal University,
Tianjin, China
shuangliu.tjnu@gmail.com
[3] Department of Electronic and Electrical Engineering, University of Strathclyde,
Glasgow Scotland, UK
t.durrani@strath.ac.uk

Abstract. Pedestrian retrieval is an important technique of searching
for a specific pedestrian from a large gallery. In this paper, we introduce three types of distance metrics for pedestrian retrieval, including
learning-free distance metric methods, metric learning methods, and convolution neural network (CNN) methods, and evaluate the performance
of different distance metrics using the Market-1501 database. The experiment shows that the CNN methods achieve the best results.

Keywords: Pedestrian retrieval · Learning-free distance metric
methods · Metric learning methods · CNN methods

1 Introduction

Pedestrian retrieval mainly refers to a task of matching persons from different
cameras. It is widely applied to pedestrian tracking, image retrieval, human
behavior analysis, and so on [1–5]. However, the challenges of low resolution,
viewpoint change, occlusion and illumination difference make pedestrian retrieval
a non-trivial problem.

In addition to extracting robust feature representation to deal with the
above-mentioned challenges, distance metric plays an essential role in pedestrian retrieval. An effective distance metric could improve the matching accuracy for pedestrian retrieval. The existing distance metrics are mainly divided
into three categories, i.e., learning-free distance metric methods, metric learning methods, and convolution neural network (CNN) methods. The learning-
free distance metric methods exploit pre-defined distance functions, including
Euclidean distance [6], Cosine distance [7], Bhattacharyya distance [8], and so

© Springer Nature Singapore Pte Ltd. 2020
Q. Liang et al. (eds.), *Communications, Signal Processing,
and Systems*, Lecture Notes in Electrical Engineering 516,
https://doi.org/10.1007/978-981-13-6504-1_140

on, to calculate the similarity between two pedestrian images in the Euclidean space as shown in Fig. 1a. The metric learning methods are based on learning a transformation matrix M using the positive and negative sample pairs. The learned matrix M can transfer the feature representation from the Euclidean space to a new space as shown in Fig. 1b, where the positive samples are close to each other and the negative samples are relatively far away. The CNN methods employ the end-to-end CNN architecture to obtain the classification results directly as shown in Fig. 2.

In this paper, we review the Euclidean distance [6] and Cosine distance [7] to analyze the learning-free distance metric methods and enumerate several typical algorithms to describe the metric learning methods concretely. The CNN methods are introduced based on two types of CNNs that are the verification network and the identification network, respectively. We aim to evaluate the impact of different distance metrics on pedestrian retrieval.

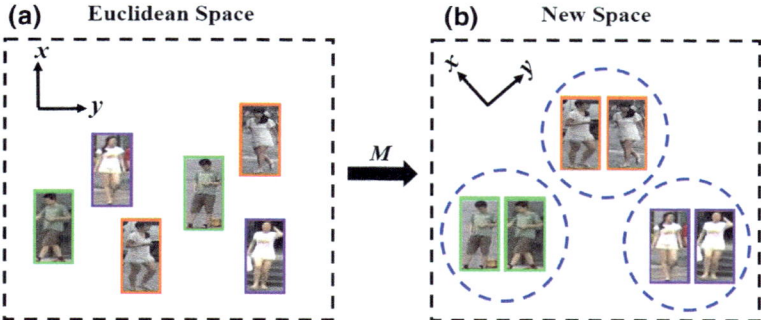

Fig. 1. Pedestrian images in the Euclidean space are mapped into a new space by the transformation matrix M.

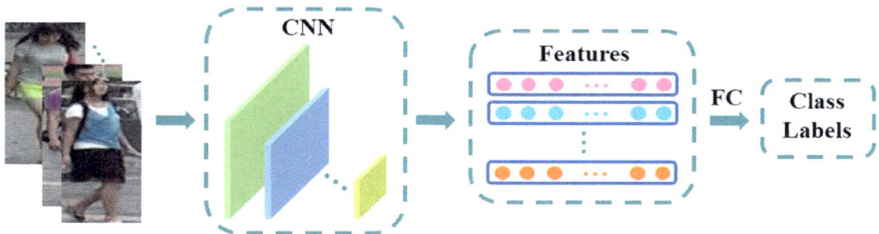

Fig. 2. End-to-end CNN architecture.

2   Distance Metrics for Pedestrian Retrieval

In this section, we introduce three types of distance metrics, i.e., the learning-free distance metric methods, metric learning methods, and CNN methods.

2.1   Learning-Free Distance Metric Methods

The learning-free distance metric methods utilize the ready-made distance functions, for example, Euclidean distance, Cosine distance, and Bhattacharyya distance, to compute the distance between two pedestrian images in the Euclidean space. We take the Euclidean distance and the Cosine distance as examples to describe the learning-free distance metric methods, and the smaller the distance is, the bigger the similarity score between two pedestrian images is. Given two feature vectors corresponding to two pedestrian images $P = \{p_1, p_2, \ldots, p_n\}$ and $Q = \{q_1, q_2, \ldots, q_n\}$, the Euclidean distance between two pedestrian images is expressed as

$$dist_E(P, Q) = (\sum_{i=1}^{n}(p_i - q_i)^2)^{1/2} \tag{1}$$

Different from the Euclidean distance that takes the absolute distance between two pedestrian images as the evaluation criterion of similarity score, the Cosine distance, also known as cosine similarity, exploits the cosine value to obtain the similarity score. If the two images are the same pedestrian, then the cosine value is close to 1, otherwise 0. It is formulated as:

$$dist_C(P, Q) = \frac{\sum_{i=1}^{n} p_i q_i}{(\sum_{i=1}^{n} p_i{}^2)^{1/2}(\sum_{i=1}^{n} q_i{}^2)^{1/2}} \tag{2}$$

The learning-free distance metric methods compute the similarity score between two pedestrian images in a fixed way, which neglects the properties of samples.

2.2   Metric Learning Methods

We first define two kinds of sample pairs. A pair of pedestrian images with the same class label is named positive sample pair, and a pair of pedestrian images with different class labels is called negative sample pair. Metric learning methods focus on learning a transformation matrix M by positive and negative sample pairs. The transformation matrix M maps the feature vectors of pedestrian images to a new space, which helps decrease the distance between positive sample pairs and enlarge the distance between negative sample pairs, simultaneously. The Mahalanobis distance proposed by Xing et al. [10] is one of the classical metric learning methods, and it is formulated as

$$dist_M(z_i, z_j) = [(z_i - z_j)^T M(z_i - z_j)]^{1/2} \tag{3}$$

where $z_i \in \mathbb{R}^n$ and $z_j \in \mathbb{R}^n$ represent the feature vectors of the ith and jth pedestrian images, respectively. $M = YY^T$ is the transformation matrix, and $Y \in \mathbb{R}^{n \times c}$ is the projection direction. c is the dimension of feature vector in the new space.

Many metric learning methods based on the Mahalanobis distance are proposed to improve the performance of pedestrian retrieval. Weinberger et al. [11] describe a new kind of metric learning method termed large margin nearest neighbor (LMNN), and the goal of this method is that k-nearest neighbors usually possess the same identity, and meanwhile, pedestrian images with different identities are partitioned by a large margin. An information-theoretic approach [12] is presented to learn the Mahalanobis distance, which regards the optimal solution as searching the minimum of the LogDet divergence under linear constraints. Guillaumin et al. [13] propose the logistic discriminant metric learning (LDML), a logistic discriminant method, to force the positive sample pairs having smaller distance than that of negative sample pairs. Koestinger et al. [14] come up with the KISS metric learning method, which learns the distance function built on the statistical inference perspective. Liao et al. [15] present the metric learning method called cross-view quadratic discriminant analysis (XQDA) that maps the feature vectors of pedestrian images into a low-dimensional subspace. Specifically, they build two covariance matrices that are intra-class covariance matrix computed by positive sample pairs and extra-class covariance matrix calculated by the negative sample pairs. They then introduce the generalized Rayleigh quotient to solve the transformation matrix (Fig. 3).

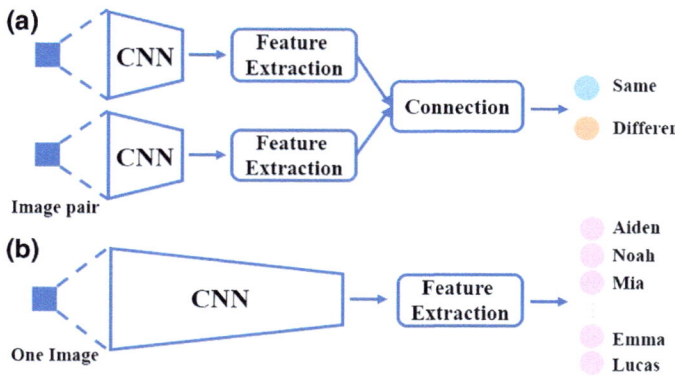

Fig. 3. Verification network and identification network.

2.3   CNN Methods

Two types of CNNs are illustrated in Fig. 2, i.e., verification network and identification network. Figure 2a shows the verification network. Its input is a pair of pedestrian images, and its output tells the two pedestrian images belonging to the same class or not. The verification network does not require explicit image labels. Figure 2b shows the identification network, which considers one pedestrian image as input and output the class labels. Both the verification network and the identification network adopt the end-to-end CNN architecture to directly obtain the classification results. The CNN methods mainly rely on two kinds of CNNs and employ different loss functions. Here, we introduce the triplet loss and the cross-entropy loss corresponding to the verification network and the identification network, respectively.

The triplet loss [17] is an efficient CNN loss function which optimizes the embedding space so that samples with the same class are closer to each other than those with different classes. The triplet loss is defined as

$$Loss_{Tri} = \sum_{\substack{x,b,e \\ L_x=L_b \neq L_e}} [s + dist_{x,b} - dist_{x,e}] \tag{4}$$

$$s.t. \ [s + dist_{x,b} - dist_{x,e}] > 0$$

where s is the value of margin, x is the target image, b denotes the pedestrian image having the same class label with the target image, e represents the pedestrian image having the different class labels with the target image, and L_x, L_b, and L_e are the class labels of three pedestrian images, respectively. Here, $dist_{x,b}$ and $dist_{x,e}$ express the distance between two pedestrian images in an embedding space, and the constraint term means that the distance between the same pedestrian is less than the distance between different pedestrians, by at least a margin s. The cross-entropy loss [16] is simple and easy to calculate and accelerates the speed of training network. The cross-entropy loss is represented as

$$Loss_{Cross-E} = \sum_{n=1}^{N} -x_n log\hat{x}_n \tag{5}$$

where x_n is the true identity, and $\hat{x}_n$ indicates the prediction probability belonging to the nth pedestrian. N is the number of identities.

3   Experimental Results

In this section, we evaluate the performances of three kinds of distance metrics on the Market-1501 database [9]. The Market-1501 database is composed of 32,668 pedestrian images of 1501 identities collected with six cameras, and it is split into the training set and test set. The training set contains 12,936 pedestrian images of 751 identities, and the test set includes 19,732 pedestrian images of 750 identities. There are still 3368 query images, and we adopt a single query setting. We find the same identity for each query in the test set. The Local Maximal Occurrence (LOMO) feature representation proposed by Liao et al. [15] is utilized for learning-free distance metric methods and metric learning methods, and the ResNet-50 network [18,19] is taken as the pre-trained CNN model for CNN methods. Table 1 lists the rank-1 accuracies of the different distance metrics.

From Table 1, the metric learning methods outperform the learning-free distance metric methods because the new space is more discriminative than the Euclidean space. Furthermore, the CNN methods achieve the best accuracy and surpass the first two kinds of methods by a large margin, especially, combining the verification network and the triplet loss. Experimental results show that the CNN methods have great potential to improve performance for pedestrian retrieval.

Table 1. Rank-1 accuracies (%) of different distance metrics on the Market-1501 database [9]. V represents the verification network, and I denotes the identification network.

Methods	Rank-1 (%)
Euclidean distance	7.83
Cosine distance	9.07
LMNN	17.88
LDML	22.43
KISSME	29.59
XQDA	45.31
Cross-entropy loss + I	73.69
Triplet loss + V $(s = 0.1)$	**75.50**

4   Conclusion

In this paper, three kinds of distance metrics are described which are learning-free distance metric methods, metric learning methods, and CNN methods. The learning-free distance metric methods ignore the properties of samples with the fix distance functions. The metric learning methods learn the similarity between two pedestrian images in a new space, and the CNN methods employ the end-to-end CNN architecture to obtain the classification results. Based on the results of the Market 1501 database, we conclude that the CNN methods are effective for pedestrian retrieval.

Acknowledgments. This work was supported by the National Natural Science Foundation of China under Grant No. 61711530240 and No. 61501327, Natural Science Foundation of Tianjin under Grant No. 17JCZDJC30600 and No. 15JCQNJC01700, the Fund of Tianjin Normal University under Grant No.135202RC1703, the Open Projects Program of National Laboratory of Pattern Recognition under Grant No. 201700001 and No. 201800002, the China Scholarship Council No. 201708120039 and No. 201708120040, the NSFC-Royal Society grant, and the Tianjin Higher Education Creative Team Funds Program.

References

1. Zheng F, Shao L. Learning cross-view binary identities for fast person re-identification. In: International joint conference on artificial intelligence. New York: USA; 2016. p. 2399–406.
2. Chen J, Wang Y, Qin J, Liu L, Shao L. Fast person re-identification via cross-camera semantic binary transformation. In: IEEE conference on computer vision and pattern recognition. Honolulu, HI, USA; 2017. p. 5330–9.
3. Zhang Z, Wang C, Xiao B, Zhou W, Liu S, Shi C. Cross-view action recognition via A continuous virtual path. In: IEEE conference on computer vision and pattern recognition. Portland, OR, USA; 2013. p. 2690–7.
4. Zhang Z, Wang C, Xiao B, Zhou W, Liu S. Action recognition using context-constrained linear coding. IEEE Signal Proc Let. 2012;19(7):439–42.
5. Zhang Z, Wang C, Xiao B, Zhou W, Liu S. Attribute regularization based human action recognition. IEEE T Inf Foren Sec. 2013;8(10):1600–9.
6. Farenzena M, Bazzani L, Perina A, Murino V, Cristani M. Person re-identification by symmetry-driven accumulation of local features. In: IEEE conference on computer vision and pattern recognition. San Francisco, CA, USA; 2010. p. 2360–7.
7. Zhang D, Lu G. Evaluation of similarity measurement for image retrieval. In: International conference on neural networks and signal processing. Nanjing, China; 2003. p. 928–31.
8. Cheng D, Cristani M, Stoppa M, Bazzani L, Murino V. Custom pictorial structures for re-identification. In: British machine vision conference. Dundee, UK; 2011. p. 6.
9. Zheng L, Shen L, Tian L, Wang S, Wang J, Tian Q. Scalable person re-identification: a benchmark. In: IEEE international conference on computer vision. Santiago, Chile; 2015. p. 1116–24.
10. Xing E, Jordan M, Russell S, Ng A. Distance metric learning with application to clustering with side-information. In: Advances in neural information processing systems. Vancouver, British Columbia, Canada; 2003. p. 521–8.
11. Weinberger K, Blitzer J, Saul K. Distance metric Learning for large margin nearest neighbor classification. In: Advances in neural information processing systems. Vancouver, British Columbia, Canada; 2006. p. 1473–80.
12. Davis J, Kulis B, Jain P, Sra S, Dhillon I. Information-theoretic metric learning. In: International conference on machine learning. Cincinnati, Ohio, USA; 2007. p. 209–16.
13. Guillaumin M, Verbeek J, Schmid C. Is that you? Metric learning approaches for face identification. In: IEEE international conference on computer vision. Berthold K.P. Horn; 2009. p. 498–505.
14. Koestinger M, Hirzer M, Wohlhart P, Roth P, Bischof H. Large scale metric learning from equivalence constraints. In: IEEE conference on computer vision and pattern recognition. Providence, RI, USA; 2012. p. 2288–95.
15. Liao S, Hu Y, Zhu X, Li S. Person re-identification by local maximal occurrence representation and metric learning. In: IEEE Conference on computer vision and pattern recognition. Boston, Massachusetts; 2015. p. 2197–206.
16. Zheng Z, Zheng L, Yang Y. A Discriminatively learned CNN embedding for person re-identification. ACM T Multim Comput. 2017;14(1):13.

17. Hermans A, Beyer L, Leibe B. In defense of the triplet loss for person re-identificationn. 2017. arXiv:1703.07737.
18. He K, Zhang X, Ren S, Sun J. Deep residual learning for image recognition. In: IEEE conference on computer vision and pattern recognition. Las Vegas, Nevada; 2016. p. 770–8.
19. Zhang Z, Huang M. Learning local embedding deep features for person re-identification in camera networks. Eurasip J Wirel Comm. 2018;1–9.

Indoor Visible Light Positioning and Tracking Method Using Kalman Filter

Xudong Wang, Wenjie Dong[(✉)], and Nan Wu

Information Science Technology College, Dalian Maritime University,
Dalian 116026, Liaoning, China
{wxd, dwenjie}@dlmu.edu.cn

Abstract. In order to improve the accuracy and tracking performance of the indoor positioning system based on visible light communication (VLC), an indoor positioning and tracking method is proposed in this paper. This method utilizes time difference of arrival (TDOA) solved by nonlinear least squares (NLLS) method to realize indoor positioning and uses Kalman filter to obtain the tracking capability. The performance of the proposed positioning method is evaluated in the room measuring 5 m × 5 m × 3 m. The simulation results show that the average location errors by adopting the NLLS method can reach to 2.99 cm and the accuracy of positioning can be promoted to 1.33 cm by using Kalman filter, the positioning accuracy increased by 55.52%.

Keywords: Indoor positioning · Visible light communication ·
Time difference of arrival · Kalman filter

1 Introduction

In recent years, visible light communication technology based on white light LED has gained more and more attention in academia and industry [1, 2]. White light LED with high-speed modulation and short response time as well as other characteristics makes the application of LED extended from illumination to communication. The dual function of illumination and communication can be achieved simultaneously. Visible light communication (VLC) as a new means of communication has obvious advantages in terms of electromagnetic radiation, application environment, and safety compared with the RF wireless communication [3]. For these reasons, visible light is an effective choice for indoor positioning. Recently, VLC-based positioning is of particular interest hotspot to be further investigated [4–6]. Visible light positioning technique including received signal strength (RSS), time of arrival (TOA), angle of arrival (AOA), and time difference of arrival (TDOA) can be used currently [7–9]. The technique of AOA can achieve high accuracy, but require deploying an array of image sensors which is expensive at the receiver side. The technique of RSS is relatively easier to realize, but it is hard to acquire a high accuracy. In indoor positioning, the traveling time of signal is so short due to the short distance between transmitters and receiver. This makes TOA method need additional hardware equipment because it requires a precision of the clock at the receiver side. But for TDOA technique, the synchronization between transmitters and the receiver is not necessary. As long as the

© Springer Nature Singapore Pte Ltd. 2020
Q. Liang et al. (eds.), *Communications, Signal Processing,
and Systems*, Lecture Notes in Electrical Engineering 516,
https://doi.org/10.1007/978-981-13-6504-1_141

information of time difference of arrival can be achieved, we can achieve high estimation accuracy.

In this paper, we use the TDOA method to realize indoor positioning. By considering the transmission time differences between each LED and positioning node are obtained to construct an objective function of positioning, and then the target location is computed through nonlinear least squares method. Furthermore, the Kalman filter is applied to modify the positioning results and promote the accuracy of mobile users tracking [10]. Simulation results show that a high positioning accuracy has been achieved when the Kalman filter is adopted.

The remainder of this paper is organized as follows. The principle of VLC positioning is presented in Sect. 2. In Sect. 3, a method of indoor positioning using NLLS and Kalman filter is proposed. Numerous simulation results and performance analysis are given in Sect. 4. Finally, the main conclusions of this paper are summarized in Sect. 5.

2 Principle of VLC Positioning

2.1 Positioning System Structure

VLC as the most promising technology of indoor wireless communication can provide various information transmission scheme. A VLC indoor location system structure diagram is shown in Fig. 1. Our system is modeled in the room measuring 5 m × 5 m × 3 m. Assuming the coordinate of LED, Tx_i $(i = 1, 2, 3, 4)$ to be Tx_i (x_i, y_i, H), where H is the height of the room. The four LEDs located in $(1 \text{ m}, 1 \text{ m}, H)$, $(1 \text{ m}, 4 \text{ m}, H)$, $(4 \text{ m}, 1 \text{ m}, H)$, and $(4 \text{ m}, 4 \text{ m}, H)$, respectively, where $H = 3$ m. When VLC is applied to indoor positioning, lighting LED can transmit reference signal, and the target position can be obtained by different positioning algorithm. This layout ensures that the mobile user in the radiation zone can receive information transmitted from different LEDs at the same time. The receiver can be placed at any point in the room.

2.2 VLC Channel Model

LED is used as the reference signal emission source in this positioning system based on VLC. Only direct signal is considered in this paper as direct signal strength is far greater than the reflection and diffraction signal strength. Lambertian radiation pattern is applied. In optical wireless channel, if the transmission distance d is much larger than the detector size A_r, then the received irradiance is approximately constant over the surface of the detector and all of the signals arrive at the same time. Thus, the impulse response of VLC wireless channel can be expressed as

$$h(t) = \frac{A_r(m+1)}{2\pi d^2}\cos^m(\varphi)T_s(\phi)g(\phi)\delta\left(t - \frac{d}{c}\right)$$
$$0 \le \phi \le \text{FOV}$$

(1)

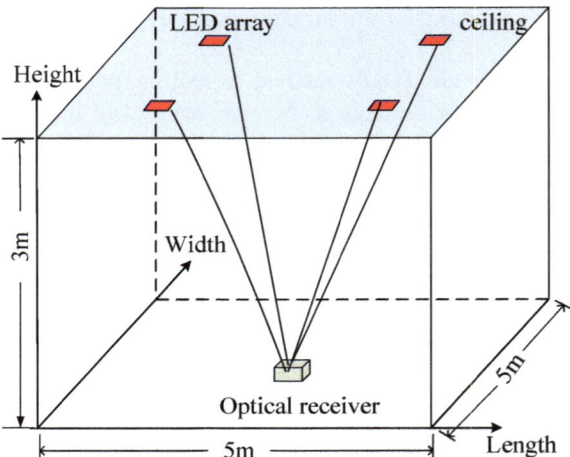

Fig. 1. System model of visible light positioning

where FOV is field of view of the receiver, c is the speed of the light in free space (approximately 3.00×10^8 m/s), φ is the light angle of incidence, ϕ is the angle of reception of the light at the receiver, $Ts(\phi)$ is the optical filter gain at the receiver, $g(\phi)$ is the gain of the optical concentrator, and m is the mode number of the radiation lobe given by $m = -\ln 2 / \ln \left(\cos \varphi_{1/2} \right)$, where $\varphi_{1/2}$ is the LED semi-angle at half power. Thus, in VLC system, receiver signal $y(t)$ can be approximated as

$$y(t) = H(0) \cdot s(t) + n(t) \tag{2}$$

where $s(t)$ represents the incident optical signal, $H(0)$ is the channel DC gain given by $H(0) = \int_{-\infty}^{+\infty} h(t)\mathrm{d}t$, $n(t)$ denotes shot noise and thermal noise in the measurement and is modeled as AWGN with mean of zero and variance σ^2 given as $\sigma^2 = \sigma_{\text{shot}}^2 + \sigma_{\text{thermal}}^2$. The variance for shot noise and thermal noise is σ_{shot}^2 and $\sigma_{\text{thermal}}^2$, respectively. The electrical SNR can be expressed in terms of the photodetector responsivity R, received optical power P_r, and noise variance as

$$\text{SNR} = \frac{(RP_r)^2}{\sigma^2} \tag{3}$$

3   Positioning Method

3.1   Time Delay Estimation

Each LED transmits a single pulse signal after every guard time period τ. All signals have the same pulse width determined by the size of positioning room and width of guard time. According to the simulation environment, the pulse width is chosen as 30 ns. The length of guard time period is set to be 100 ns so that the previous signal causes no or very little interference to the next signal. In the VLC channel model, the amplitude of the receiver signal is changed and arrives with different delay time. Assuming four transmitters is used in this model and the received four signals are $s_1(t)$, $s_2(t)$, $s_3(t)$, and $s_4(t)$, the waveforms with noise interference are shown in Fig. 2. The cross-correlation is used to detect each signal. Then the information of peak time of correlation signal is obtained. If correlation peak time is τ_1, τ_2, τ_3, and τ_4, respectively, the delay difference of arrival time between various signals can be given as $\Delta t_{21} = \tau_2 - \tau_1 - \tau$, $\Delta t_{31} = \tau_3 - \tau_1 - 2\tau$, $\Delta t_{41} = \tau_4 - \tau_1 - 3\tau$. Using this value, we can obtain the transmission distance difference $R_{21} = \Delta t_{21} \times c$, $R_{31} = \Delta t_{31} \times c$, $R_{41} = \Delta t_{41} \times c$. Figure 3 shows how the time delay difference is obtained using cross-correlation detection.

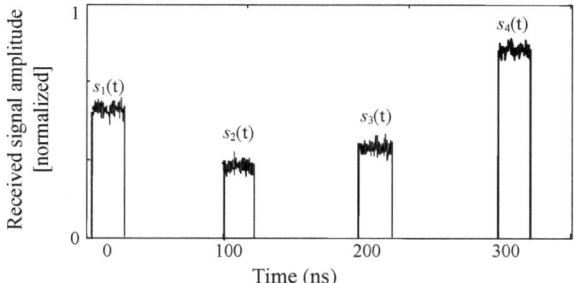

Fig. 2. Received signal waveform

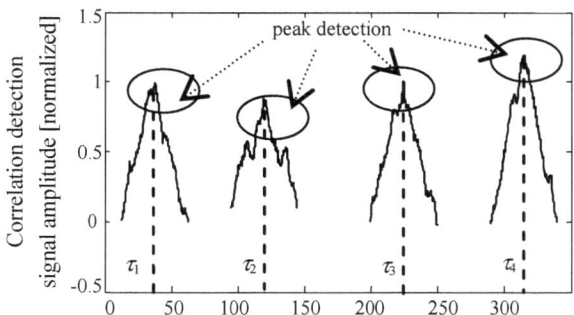

Fig. 3. Correlation detection signal

3.2   NLLS Method

As we all know, a number of methods can be applied to realize the TDOA positioning in the indoor scenario, such as the Chan method, the Fang method, and Taylor series expansion method. Due to the short transmission distance in indoor environment, even very small detection error of time would lead to a large deviation of the eventual positioning results. Thus, they are not suitable for high precision indoor positioning problem in terms of VLC. In this paper, we convert position estimation into the problem of nonlinear optimization with constraints and then use nonlinear least squares (NLLS) algorithm to achieve high precision indoor positioning. We assume that the coordinates of mobile user and beacons used in location are set to be (x, y) and (x_i, y_i, z_i), $i = 1, 2, 3 \ldots M$, where M is the number of beacons. We can get the time difference τ_{s1} between the sth signal and the first signal (as reference) to the target receiver through correlation calculation. So, the distance difference can be expressed as $R_{s1} = R_s - R_1$ ($s = 2, 3 \ldots M$). Considering the case of $M = 4$, for TDOA positioning estimation, the objective function can be formulated as follows

$$E(x, y) = (d_{21} - R_{21})^2 + (d_{31} - R_{31})^2 + (d_{41} - R_{41})^2$$
$$\text{s.t} \begin{cases} 0 \le x \le d_{\mathrm{L}} \\ 0 \le y \le d_{\mathrm{W}} \end{cases} \tag{4}$$

where $d_i = \sqrt{(x - x_i)^2 + (y - y_i)^2 + H^2}$, $d_{i1} = d_i - d_1$ ($i = 2, 3 \ldots M$), d_{L} and d_{W} are the length and width of the room, respectively. The coordinate (x, y) of mobile user can be selected when $E(x, y)$ is minimized making use of nonlinear least squares algorithm to solve the above problem.

3.3   Kalman Filter

To realize indoor tracking for the mobile user, the Kalman filter (KF), which is a recursive state estimator of a linear dynamic system, is introduced into the positioning method mentioned above. The Kalman filter addresses the problem of estimating the state of a noisy system that can be described by a linear system and a linear measurement model. It generally describes the behavior of dynamical systems. The dynamic model can be described by

$$X_k = AX_{k-1} + BU_{k-1} + W_{k-1} \tag{5}$$

$$Z_k = HX_k + V_k \tag{6}$$

where X_k is the vector of system state at time k, Z_k is the vector of system measurements, U_k is the control vector, A is the state transition matrix, B is the system control matrix, and H is the measurement matrix. The state noise is $W_k \sim N(0, Q_k)$, zero mean, and Gaussian distributed noise with covariance Q_k, and $V_k \sim N(0, R_k)$ is the measurement noise with covariance R_k.

The Kalman filter can be divided into two steps as follow:

(1) The prediction process:

$$X_{k,k-1} = AX_{k-1} + BU_{k-1} \tag{7}$$

$$P_{k,k-1} = AP_{k-1}A^T + Q \tag{8}$$

where $X_k,k-1$ is the priori state estimate at time k given knowledge of the process prior to time $k-1$, and $P_k,k-1$ is the covariance matrix of the state-prediction error.

(2) The updated process:

$$
\begin{aligned}
K_k &= P_{k,k-1}H^T(HP_{k,k-1}H^T + R)^{-1} \\
X_k &= X_{k,k-1} + K_k(Z_k - HX_{k,k-1}) \\
P_k &= (I - K_kH)P_{k,k-1}
\end{aligned} \tag{9}
$$

where X_k is the posteriori estimate at time k, K denotes the Kalman gain, and Z_k is the measurement matrix.

4   Simulation Results

Numerical simulations are performed to evaluate the performance of the considered TDOA-based positioning and Kalman filter-based tracking algorithm. We assume that the speed of mobile target is constant, and the average speed of moving on the x-axis and y-axis is 0.1 and 0.05 m/s, respectively. And the initial position of mobile user is set at (1, 1.5). The sampling interval is 1 s. In order to evaluate the performance of positioning, the position error is defined as

$$ER_i = \sqrt{(x_{ei} - x_{ci})^2 + (y_{ei} - y_{ci})^2} \tag{10}$$

where ER_i is the position error of the node numbered i, and (x_{ei}, y_{ei}) and (x_{ci}, y_{ci}) is the estimation coordinate and real coordinate of the detection node, respectively.

The positioning error of proposed positioning method with and without KF is shown in Fig. 4. It can be seen that the average positioning error with and without KF is 1.33 and 2.99 cm, respectively. It is recorded that KF generates a 55.52% higher accuracy than without KF. The error of positioning is remarkably decreased throughout KF. Figure 5 illustrates the cumulative distribution of positioning error, which indicates that the error with and without KF is guaranteed to be within 2.78 and 6.03 cm 95% confidence interval. It proved that KF method has achieved higher performance.

It is generally acknowledged that SNR is lower around the room edge, and lower SNR will result in lower accuracy of positioning. We assume the initial position of mobile user is set at (0.5, 0.5) with SNR = −2 dB, and the user moves on along the

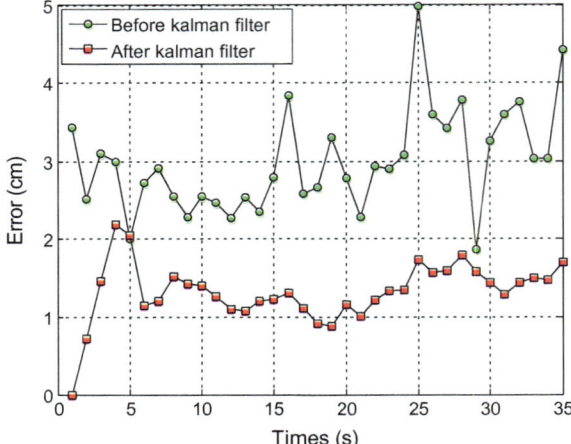

Fig. 4. Positioning error with and without KF

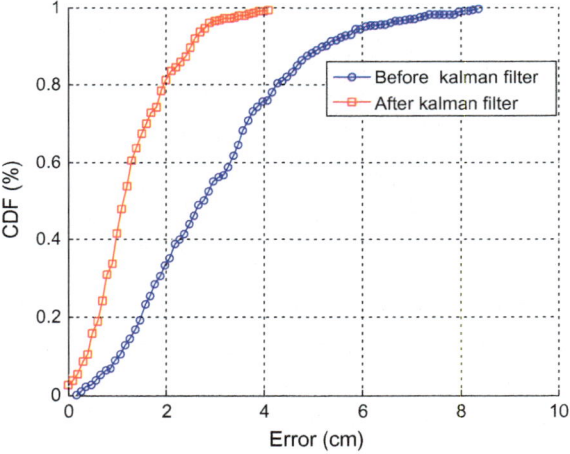

Fig. 5. CDF of positioning error

room edge. The positioning trajectory is shown in Fig. 6. The average positioning error with and without KF are respectively 1.79 and 3.62 cm. The positioning accuracy significantly increases by 50.55%. It proves that accuracy of positioning has been promoted when the KF is applied. Figure 7 shows the trajectory of mobile user move on the *x*-axis. As we can see, the filtering path is closer to actual path. This proves that the proposed algorithm is still valid when target moves to indoor corner areas.

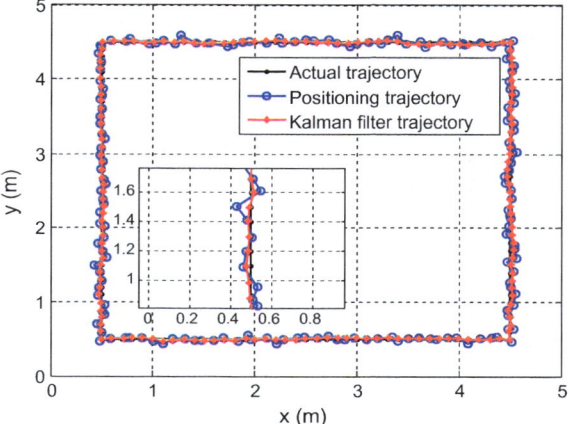

Fig. 6. Trajectory of users around the room edge

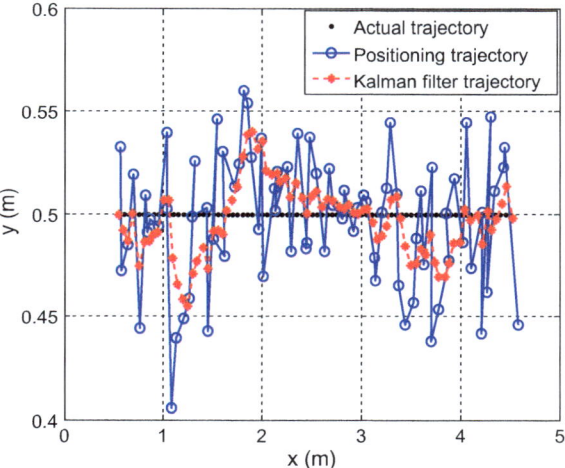

Fig. 7. Trajectory of users move on the x-axis

5 Conclusion

In this paper, we proposed an indoor positioning and tracking method for indoor positioning system based on VLC. In order to realize indoor positioning, a NLLS method was adopted. Main objective of this paper is then promoting the accuracy of proposed positioning method by using KF method. Evaluation by simulation proves that KF has achieved higher accuracy than the NLLS method, and the performance of positioning and tracking has been improved significantly.

References

1. Pant K, Armstrong J. Indoor localization using white LEDs. Electron Lett. 2012;48(4):228–30.
2. Iturralde D, Azurdia-Meza C, Krommenacker N, et al. A new location system for an underground mining environment using visible light communications. In: International symposium on communication systems, networks and digital signal processing. IEEE, Manchester, UK; 2014. p. 1165–9.
3. Yan K, Zhou H, Xiao H, et al. Current status of indoor positioning system based on visible light. In: International conference on control, automation and systems. IEEE, Busan, South Korea; 2015. p. 565–9.
4. Dardari D, Closas P, Djuri PM. Indoor tracking: theory, methods, and technologies. IEEE Trans Veh Technol. 2015;64(4):1263–78.
5. Zhang W, Chowdhury M, Kavehrad M. Asynchronous indoor positioning system based on visible light communications. Opt Eng. 2014;53(4):045105.
6. Zhang W, Kavehrad M. A 2-D indoor localization system based on visible light. Seattle, USA: IEEE Photonics Society Summer Topical Meeting Series; 2012. p. 80–1.
7. Hassan NU, Naeem A, Pasha MA, et al. Indoor positioning using visible LED lights: a survey. ACM Comput Surv. 2015;48(2):20.
8. Do TH, Yoo M, An in-depth survey of visible light communication based positioning Systems. Sensors. 2016;16(5):678.
9. Ghassemlooy Z, Popoola W, Rajbhandari S. Optical wireless communications. Boca Raton: CRC Press, inc; 2012.
10. Eroglu YS, Guvenc I, Pala N, et al. AOA-based localization and tracking in multi-element VLC systems. In: Wireless and microwave technology conference. IEEE; 2015. p. 1–5.

A Novel Method for 2D DOA Estimation Based on URA in Massive MIMO Systems

Bo Wang, Deliang Liu$^{(\boxtimes)}$, Dong Han, and Zhuanghe Zhang

Army Engineering University, Shijiazhuang, China
{liudeliang82, han58228}@sina.com

Abstract. Massive MIMO is one of the enabling technologies to cope with exponential data growth. It is very crucial for downlink precoding to accurately estimate direction-of-arrival. A lot of work has been done for 2D DOA estimation based on uniform rectangular array. However, in cases that snapshots are severely limited, such as extremely complex communication environment, conventional 2D DOA estimation method cannot work properly. Iterative adaptive approach (IAA) is one of the sparse algorithms which can handle heavy snapshot limitations. In this paper, we propose a novel 2D DOA estimation algorithm based on IAA for massive MIMO systems. Unlike conventional methods, an estimator in this algorithm is updated by previous iteration instead of the snapshots. The iteration ends until convergence. Simulation results demonstrate that the proposed algorithm is superior to conventional methods in low snapshot cases.

Keywords: 2D DOA estimation · Massive MIMO · Few snapshots

1 Introduction

It is a great challenge for future wireless systems to handle exponential traffic data growth. To meet such a challenge, the massive MIMO technology has received substantial attention for its higher data rates, enhanced link reliability, and potential power savings [1]. Accurate 2D DOA estimation is very important for massive MIMO systems to achieve reliable downlink communication.

In recent years, a lot of literatures have been addressed for 2D DOA estimation in massive MIMO. Classical multiple signal classification (MUSIC) is expanded into 2D version in [2] by using a unique steering matrix structure. This algorithm still requires spectral peak search and has high computational complexity. A low-complexity 2D DOA estimation algorithm is developed in [3] by utilizing the rotational invariance of two pairs of shifted URAs, in which a large scale of URA is needed. 2D DOA estimation based on ESPRIT without spectrum peak search is investigated in [4], at which DOA and the channel impulse responses are jointly estimated. 2D unitary ESPRIT is applied to massive MIMO systems in [5], which reduces computational complexity compared with 2D MUSIC. However, these methods are not suitable for coherent multipath signals.

The above-mentioned methods assume that snapshot numbers are large. However, in many practical application scenarios, such as underwater or indoor signal processing,

© Springer Nature Singapore Pte Ltd. 2020
Q. Liang et al. (eds.), *Communications, Signal Processing,*
and Systems, Lecture Notes in Electrical Engineering 516,
https://doi.org/10.1007/978-981-13-6504-1_142

signals are easily interfered or even interrupted, which result in snapshot numbers for 2D DOA estimation are very low. In this case, the performance of these algorithms deteriorates dramatically.

Sparsity-based algorithms can deal with snapshots limitation. In [6], Yardibi. T, etc. propose a user-parameter-free, weighted-least-square-(WLS)-based iterative adaptive approach for amplitude and phase estimation (IAA-APES), which has excellent estimation performance with few snapshots.

In this paper, we propose a novel 2D DOA estimation method based on IAA-APES for 2D DOA estimation in massive MIMO systems. In the first step, the optimal weighted least square estimator is obtained. In the second step, utilizing the estimator to update the energy of each scanning angle point. Step 1 and step 2 are iteratively calculated until convergence. Finally, the azimuth and elevation angles can be got from the spectrum peak without pairing process.

Notation: $(\cdot)^*, (\cdot)^T, (\cdot)^H$ denote the complex conjugate, transpose, Hermitian transpose, respectively. Symbol '$\otimes$' denotes Kronecker product, '$\odot$' stands for Khatri-Rao product.

2   Problem Formulation

In this paper, we consider a uniform rectangular array (URA) with $M \times N$ antenna elements is equipped with the base station. Assume that URA locates in the xoz plane. The positive direction of y-axis is determined by the right-hand rule from x-axis and z-axis. The azimuth angle with respect to x-axis, $\varphi_k \in (0°, 180°)$, and the elevation angle with respect to z-axis, $\theta_k \in (0°, 90°)$ (Fig. 1).

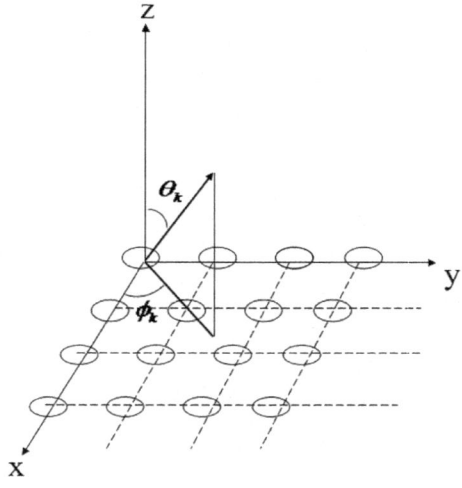

Fig. 1. URA for massive MIMO systems

Without loss of generality, the reference point is placed at the antenna element in origin o. The (m, n)th antenna element locates in the xoz plane with coordinate, $(\frac{\lambda}{2}m, 0, \frac{\lambda}{2}n)$ where λ is the wavelength of carrier frequency, antenna element spacing $d = \lambda/2$, and $m = 0, 1, \ldots, M - 1$; $n = 0, 1, \ldots, N - 1$. We assume that there are K resolvable far-field narrowband signals, each of which has its own angle pair θ_k and φ_k, $k = 1, 2, \ldots, K$.

The received signal at qth snapshot on (m, n)th antenna element can be written as [7]

$$y_{m,n}(q) = \sum_{k=1}^{K} \gamma_k(q) e^{j(m-1)\mu_k} e^{j(n-1)\upsilon_k} + \omega_{m,n}(q) \tag{1}$$

where

$$\mu_k = \frac{2\pi d \cos \theta_k}{\lambda}, \ \upsilon_k = \frac{2\pi d \cos \phi_k}{\lambda} \tag{2}$$

$\gamma_k(q)$ represents the unknown complex amplitude of the kth source at qth snapshot with power

$$E\{|\gamma_k(q)|^2\} = \sigma_k^2, \ k = 1, 2, \ldots, K \tag{3}$$

$\omega_{m,n}(q)$ is the Gaussian white noise with variance σ^2. The total number of snapshots is Q.

Let $\mathbf{a}(\mu_k)$ and $\mathbf{b}(\upsilon_k)$ be the z-axis and x-axis steering vector, respectively. Among them,

$$\mathbf{a}(\mu_k) = [1, e^{j\pi\mu_k}, e^{j2\pi\mu_k}, \ldots, e^{j\pi(M-1)\mu_k}]^{\mathrm{T}} \tag{4}$$

$$\mathbf{b}(\upsilon_k) = [1, e^{j\pi\upsilon_k}, e^{j2\pi\upsilon_k}, \ldots, e^{j\pi(M-1)\upsilon_k}]^{\mathrm{T}} \tag{5}$$

Assume that

$$\mathbf{y}(q) = [y_{1,1}(q), y_{2,1}(q), \ldots, y_{M,1}(q), \ldots, y_{M,N}(q)]^{\mathrm{T}} \tag{6}$$

be the data matrix at qth snapshot, then it has

$$\mathbf{y}(q) = [\mathbf{b}(\upsilon_1) \otimes \mathbf{a}(\mu_1), \ldots, \mathbf{b}(\upsilon_k) \otimes \mathbf{a}(\mu_k)]\boldsymbol{\gamma}(q) + \boldsymbol{\omega}(q) \tag{7}$$

where vector $\boldsymbol{\gamma}(q) = [\gamma_1(q), \gamma_2(q), \ldots, \gamma_K(q)]^{\mathrm{T}}$, and (m, n) $\boldsymbol{\omega}(q)$ has the same data stacking form as $\mathbf{y}(q)$. Furthermore, defining $\mathbf{A} = [\mathbf{a}(\mu_1), \mathbf{a}(\mu_2), \ldots, \mathbf{a}(\mu_K)]^{\mathrm{T}}$, $\mathbf{B} = [\mathbf{b}(\upsilon_1), \mathbf{b}(\upsilon_2), \ldots, \mathbf{b}(\upsilon_k)]^{\mathrm{T}}$, and considering total Q snapshots, (7) can be rewritten as a compact form

$$\mathbf{Y} = (\mathbf{B} \odot \mathbf{A})\mathbf{S} + \mathbf{Z} \tag{8}$$

where $\mathbf{Y} = [y(1), y(2), \ldots, y(Q)]$, $\mathbf{S} = [\gamma(1), \gamma(2), \ldots, \gamma(Q)]$ and $\mathbf{Z} = [\omega(1), \omega(2), \ldots, \omega(Q)]$.

3 Proposed Algorithm

First of all, we refine the interested region into a grid with predefined increment. Each point of the grid is assumed as a potential signal with an azimuth and elevation angle. Actual signals are much smaller than potential ones so that sparse representation algorithm can be used in 2D DOA estimation.

We make an assumption that $\mathbf{P}$ is a $(m, n)K \times K$ diagonal matrix, whose diagonal is the power estimation of each scanning point. $\mathbf{P}$ is given by

$$P_k = \frac{1}{N} \sum_{n=1}^{N} |\gamma_k(q)|^2, \, k = 1, 2, \ldots, K \tag{9}$$

In addition, the noise covariance matrix is written as

$$\mathbf{Q}(\theta_k, \phi_k) = \mathbf{R} - P_k \mathbf{h}(\theta_k, \phi_k)\mathbf{h}(\theta_k, \phi_k)^H \tag{10}$$

in which

$$(m, n)\, \mathbf{h}(\theta_k, \phi_k) = \mathbf{b}(\upsilon_k) \otimes \mathbf{a}(\mu_k) \tag{11}$$

$\mathbf{R}$ is given by $\mathbf{R} \triangleq (\mathbf{B} \odot \mathbf{A})\mathbf{P}(\mathbf{B} \odot \mathbf{A})^H$, which is dependent on the unknown signal power.

Then, the weighted least square (WLS) cost function is expressed as

$$\sum_{t=1}^{Q} \|\mathbf{y}(t) - \gamma_k(t)\mathbf{h}(\theta_k, \phi_k)\|_{\mathbf{Q}^{-1}}^2 \tag{12}$$

where $\|\mathbf{x}\|_{\mathbf{Q}^{-1}}^2 = \mathbf{x}^H \mathbf{Q}\mathbf{x}$. Minimizing (12) with respect to $\gamma_k(q)$, $t = 1, 2, \ldots, Q$, yields

$$\hat{\gamma}_k(t) = \frac{\mathbf{h}(\theta_k, \phi_k)^H \mathbf{Q}(\theta_k, \phi_k)^{-1}\mathbf{y}(t)}{\mathbf{h}(\theta_k, \phi_k)^H \mathbf{Q}(\theta_k, \phi_k)^{-1}\mathbf{h}(\theta_k, \phi_k)} \tag{13}$$

among them $\mathbf{Q}(\theta_k, \phi_k)$ and $\mathbf{h}(\theta_k, \phi_k)$ are known before. According to (10) and the matrix inversion lemma, (13) can be expressed as

$$\hat{\gamma}_k(t) = \frac{\mathbf{h}(\theta_k, \phi_k)^H \mathbf{R}^{-1}\mathbf{y}(t)}{\mathbf{h}(\theta_k, \phi_k)^H \mathbf{R}^{-1}\mathbf{h}(\theta_k, \phi_k)} \tag{14}$$

Unlike conventional DOA estimation methods, IAA-APES [6] gets $\mathbf{Q}_k$ from iteration.

Table 1 shows the calculation steps. The initialization of $\mathbf{P}_k$ is got from traditional DAS method

$$\hat{P}_k = \mathbf{h}(\theta_k, \phi_k)^H \mathbf{\Gamma} \, \mathbf{h}(\theta_k, \phi_k), k = 1, \ldots, K \tag{15}$$

$$\mathbf{\Gamma} = \frac{1}{Q} \sum_{t=1}^{Q} \mathbf{y}(q) \mathbf{y}^H(q) \tag{16}$$

The azimuth and elevation angles' estimation is obtained from spectrum peak after the above steps.

Table 1. Proposed algorithm

Initialization
calulate by $\hat{\mathbf{P}}_k^{(0)}$ (15)
Iteration
$\mathbf{R}^{(i)} = (\mathbf{B} \mathbf{e} \ \mathbf{A}) \hat{\mathbf{P}}^{(i-1)} (\mathbf{B} \mathbf{e} \ \mathbf{A})^H$
for k=1,2,...,K
calculate $\hat{\gamma}_k^{(i)}(q)$ by (14), q=1,2,...,Q
calculate $\hat{\mathbf{P}}_k^{(i)}$ by (9)
end for
Termination
$\|\hat{\gamma}_k^{(i)}(q) - \hat{\gamma}_k^{(i-1)}(q)\|^2$ is less than a specified tolerance

4   Numerical Examples

In this section, we demonstrate the performance and advantages of proposed method in comparison with the 2D MUSIC algorithm. Simulation is based on a URA with $M = 11$ and $N = 11$, and the element space along the X-axis and Y-axis is $\lambda/2$. The number of signals is $K = 3$, and the search step for 2D MUSIC is $0.2°$.

Assume one mobile terminal with three coherent signals impinging on the antenna array from three directions $\theta \in \{66°, 62°, 61°\}$ and $\phi \in \{62°, 66°, 68°\}$. The SNR and the number of snapshots are set to 12 dB and 5 respectively. The spatial spectrum obtained from the proposed algorithm and MUSIC-based algorithm is shown in Figs. 2 and 3. From the results, we can see that under the case that snapshots are only five and three DOAs are closely separated, the 2D MUSIC algorithm fails to estimate the three directions while the proposed method can accurately resolve the three directions with three sharp spectrum peaks. The estimated DOAs are $(66°, 62°)$, $(62°, 66°)$ and $(68°, 61°)$, respectively. Figures 2 and 3 show that the proposed method is much superior to the 2D MUSIC algorithm in conditions of low snapshot numbers.

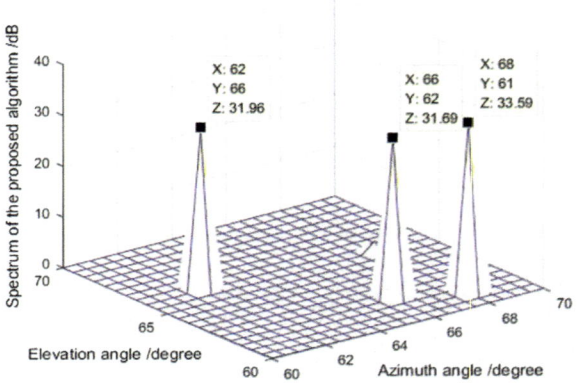

Fig. 2. Spectrum of the proposed method

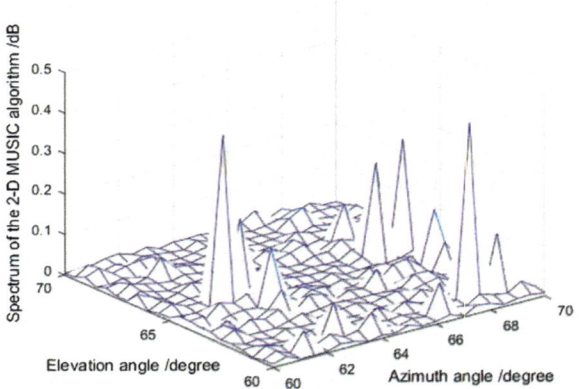

Fig. 3. Spectrum of a MUSIC-based algorithm

5 Conclusions

We present a new two-dimensional DOA estimation algorithm based on URA for massive MIMO systems. There are numerous practical applications that a large number of snapshots are unavailable, and this algorithm can effectively deal with this case. Moreover, it provides super-resolution. Simulation results verify the proposed algorithm has superior estimation performance with few snapshots.

Acknowledgments. This work is funded by the National Natural Science Foundation of China under Grant 61601494.

References

1. Larsson E, Edfors O, Tufvesson F, Marzetta T. Massive MIMO for next generation wireless systems. IEEE Commun Mag. 2014;52(2):186–95.
2. Yang, K-Y, Wu J-Y, Li W-H. A low-complexity direction-of-arrival estimation algorithm for full-dimension massive MIMO systems. In: 2014 IEEE International Conference on Communication Systems (ICCS). IEEE; 2014. p. 472–6.
3. Meng H, Zheng Z, Yang Y, et al. A Low-complexity 2-D DOA estimation algorithm for massive MIMO systems. In: 2016 IEEE/CIC international conference on communications in China (ICCC). IEEE; 2016. p. 1–5.
4. Kuang J, Zhou Y, FEI Z. Joint DOA and channel estimation with data detection based on 2D unitary ESPRIT in massive MIMO systems. Front Inform Technol Electron Eng. 2017;18(6): 841–9.
5. Wang T, Ai B, He RS, Zhong ZD. Two-dimension direction-of-arrival estimation for massive MIMO system. IEEE Access. 2015;3:2122–8.
6. Yardibi T, Li J, Stoica P, et al. Source localization and sensing: a nonparametric iterative adaptive approach based on weighted least squares. IEEE Trans Aerosp Electron Syst. 2010;46(1):425–43.
7. Nion D, Sidiropoulos ND. Tensor algebra and multidimensional harmonic retrieval in signal processing for MIMO radar. IEEE Trans Signal Process. 2010;58(11):5693–705.

Two-Dimensional DOA Estimation for 5G Networks

Zhuanghe Zhang$^{(\boxtimes)}$, Dong Han, Deliang Liu, and Bo Wang

Shijiazhuang Campus of Army Engineering University, Shijiazhuang, China
zhangzhuanghe1995@163.com

Abstract. Mobile communication is coming to the fifth generation (5G) networks. In the age of 5G, the three-dimensional (3D) beamforming is one of the highlighted technologies. 3D beamforming increases vertical dimension in terms of space domain, but also has brought a challenge to beamforming, especially the problem of direction-of-arrival (DOA) estimation. In this paper, a DOA estimation method based on multiple signal classification (MUSIC) of uniform circular array (UCA) with mutual coupling compensation is presented for the future 5G networks. 2D MUSIC does well in estimating DOA without an accurate number of signals. The MATLAB software is employed in order to conduct the modeling and 2D MUSIC algorithm simulations.

Keywords: 2D MUSIC · DOA · UCA · 5G networks

1 Introduction

In 5G networks, different from conventional beamforming techniques, the 3D beamforming shows us the possibility of combining the vertical dimension with horizontal dimension in order to enhance system performance [1]. So, the method for jointly estimating azimuth and elevation is important.

In [2], we can know the design of antenna array architecture, in which the antenna elements can be designed in the shapes of a hexagon, circle, or cross, in addition to the conventional rectangle. The simulation results indicate that while there always exists a non-trivial gain fluctuation in other common antenna arrays, the circular antenna array has a flat gain in the main lobe of the radiation pattern with varying angles. This makes the circular antenna array more robust to angle variations that frequently occur due to antenna vibration in a complex environment. So, in array signal processing, UCA with its unique array structure and widely spread, but in practical application, the mutual coupling effect is also very obvious, especially in the future 5G networks, which use the narrow millimeter wave beam technology and usually at a very high frequency of work.

DOA estimating techniques are of particular interest in communication systems, especially with antenna technology. There are many algorithms such as multiple signal classification (MUSIC), estimation of signal parameters via rotation invariance techniques (ESPRIT), but most of them have been proposed to estimating the one-dimensional direction of arrival with uniform linear array (ULA) due to their simplicity [3]. Recent years, more and more 2D DOA estimation algorithms are proposed,

© Springer Nature Singapore Pte Ltd. 2020
Q. Liang et al. (eds.), *Communications, Signal Processing,
and Systems*, Lecture Notes in Electrical Engineering 516,
https://doi.org/10.1007/978-981-13-6504-1_143

in [4], a 2D DOA based on iterative adaptive approach (IAA) algorithm has been introduced. As for [5], it presented a 2D DOA estimation with MUSIC algorithm using the
modulus constraint, but most of the high-resolution DOA estimation algorithm for array
mutual coupling effect caused by flow pattern error is very sensitive.

Aiming at the above problems, this article expounds how uniform circular array is
established, and a DOA estimator is introduced to jointly estimate both azimuth and
elevation based on MUSIC method under the condition of UCA mutual coupling effect,
the computer simulation results by MATLAB verify the validity of it.

2   Data Model

In this section, consider an antenna with uniform circular array, in which N isotropic
elements with radius R as shown in Fig. 1. The center of the circle is at the origin of the
frame, let K narrowband far-field sources at arbitrary locations whose angles are $\mathbf{a}(\theta,$
$\varphi) = [(\theta_1, \varphi_1), (\theta_2, \varphi_2),\ldots, (\theta_K, \varphi_K)]$, respectively, impinge on this array, and wavelength of sources is denoted as λ [6, 7].

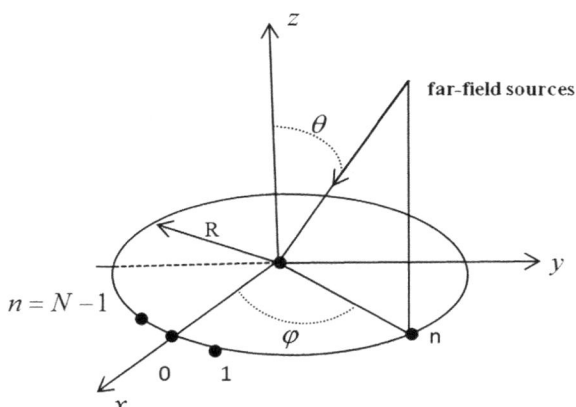

Fig. 1. Uniform circular array

In the ideal case, the received signals can be expressed as

$$X(t) = AS(t) + N(t) \tag{1}$$

And

$$A = [a(\theta_1, \varphi_1), a(\theta_2, \varphi_2), \ldots, a(\theta_K, \varphi_K)] \tag{2}$$

Matrix A is a matrix of the K steering vectors, which represents the possible value
set of DOA.

In which

$$a(\theta_i, \varphi_i) = [e^{j2\pi R \sin\theta_i \cos\varphi_i/\lambda}, \ldots, e^{j2\pi R \sin\theta_i \cos(\varphi_i - \gamma_n)/\lambda}]^{\mathrm{T}} \tag{3}$$

where

$$\gamma_n = \frac{2\pi n}{N} \qquad \begin{matrix} i = 1, 2, \ldots, K \\ n = 1, 2, \ldots, N \end{matrix} \tag{4}$$

And

$$S(t) = [s_1(t), s_2(t), \ldots, s_K(t)]^{\mathrm{T}} \tag{5}$$

$$N(t) = [n_1(t), n_2(t), \ldots, n_K(t)]^{\mathrm{T}} \tag{6}$$

where $S(t)$ is signal source vector of size $(K \times 1)$ and $N(t)$ is the received additive white Gaussian noise whose variance is σ^2. Here, it is assumed that there is no correlation between $S(t)$ and $N(t)$, based on (1), the correlation matrix of received vector can be computed as

$$R_x = E[XX^H] = AVA^H + \sigma^2 I = R_s + \sigma^2 I \tag{7}$$

V is covariance matrix of signal vector (S) which is a full rank matrix of order $(K \times K)$ given by

$$V = \begin{bmatrix} E\left[|S_1|^2\right] & \cdots & \cdots & 0 \\ 0 & E\left[|S_2|^2\right] & \cdots & 0 \\ \vdots & \ddots & \cdots & \vdots \\ 0 & 0 & \cdots & E\left[|S_K|^2\right] \end{bmatrix} \tag{8}$$

R_s is a signal covariance matrix of order $N \times N$ with rank K given by

$$R_s = \begin{bmatrix} E\left[|S_1|^2\right] & \cdots & \cdots & 0 & \cdots & 0 \\ 0 & E\left[|S_2|^2\right] & \cdots & 0 & \cdots & 0 \\ \vdots & \ddots & \cdots & \vdots & \cdots & 0 \\ 0 & 0 & \cdots & E\left[|S_K|^2\right] & \cdots & 0 \\ 0 & 0 & \cdots & 0 & \cdots & 0 \end{bmatrix} \tag{9}$$

3   UCA-MUSIC Algorithm

Given the decoupling effect, steering vector can be expressed as [8]

$$\tilde{a}(\theta, \varphi) = Z(\theta)a(\theta, \varphi) \tag{10}$$

where $Z(\theta)$ is mutual coupling matrix (MCM) of UCA, the elevation dependence of the mutual coupling effect is taken into account so that the MCM will vary with elevation angle for a fixed elevation angle, it is well known that a complex symmetric circular matrix provides a satisfactory model for the MCM of a UCA. If only consider about the mutual coupling between adjacent arrays of three elements, the cycle of cyclic matrix-vector can be defined as

$$z = [z(0), z(1), z(2), 0 \cdots, 0, z(2), z(1)] \tag{11}$$

According to (9), R_S has $N-K$ eigenvectors corresponding to zero eigenvalues. Steering vector $\mathbf{a}(\theta, \varphi)$ in the signal subspace is orthogonal to noise subspace. Let Q_n be such an eigenvector

$$R_s Q_n = AVA^H Q_n = 0 \tag{12}$$

Since V is a positive definite matrix

$$a^H(\theta_i, \varphi_i)Q_n = 0 \tag{13}$$

This implies that signal steering vectors are orthogonal to eigenvectors corresponding to noise subspace. So, the MUSIC algorithm searches through all angles and plots the spatial spectrum

$$P_{\text{MUSIC}}(\theta, \varphi) = \frac{1}{(a^H(\theta, \varphi)Q_n Q_n^H a(\theta, \varphi))} \tag{14}$$

Therefore, when there is mutual coupling, the mth snapshot of the array can be expressed as

$$\tilde{X}(t) = \tilde{A}s(t) + n(t) = Z(\theta)As(t) + n(t) \tag{15}$$

Because of the existence of array mutual coupling error, the guiding vector of array becomes

$$\hat{A} = Z(\theta)A \tag{16}$$

So, the actual direction vector is

$$\hat{a}(\theta, \varphi) = Z(\theta)a(\theta, \varphi) \tag{17}$$

Consider the effect of array mutual coupling error on array performance, the actual covariance matrix is

$$\hat{R} = ZR_x Z^H \qquad\qquad (18)$$

Mutual coupling effect mainly affects the direction vector of array, thus to influence the array receiving signal matrix.

4 Simulation Results

In the first simulation, consider three sources which are uncorrelated with each other incident on a UCA consists of $N = 8$ elements, sources located at angles $(88°, 80°)$, $(25°, 40°)$, and $(245°, 5°)$. The number of snapshots is set to 500 and the signal-to-noise ratio (SNR) is 30 dB.

Figure 2 shows that three sources make three corresponding peaks in the plot when considering mutual coupling. However, it makes some pseudo peaks when decoupling error is not considered.

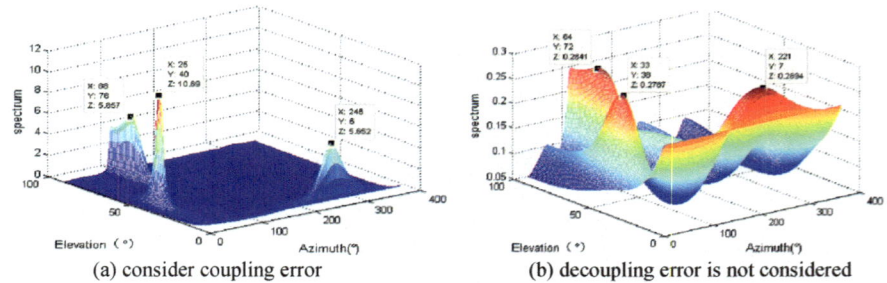

(a) consider coupling error   (b) decoupling error is not considered

Fig. 2. Influence of the mutual coupling error

In the second experiment, we present 200 Monte Carlo simulations to illustrate the performance of proposed methods. Define root mean squared error (RMSE). The array structure is a uniform circular array (UCA) composed of $N = 8$ elements, and there are two noncoherent sources located at angles $(10°, 20°)$ and $(30°, 60°)$, the signal-to-noise ratio (SNR) is 15 dB in Fig. 3b. We demonstrate the performance of UCA-MUSIC algorithm with respect to SNR and the number of snapshots. Comparisons are performed with CRB.

From Fig. 3, RMES of the algorithm with mutual coupling compensation is in a downtrend with the increase of SNR, but elevation has no obvious changes. Also the performance of estimation gets better with the increase of snapshot.

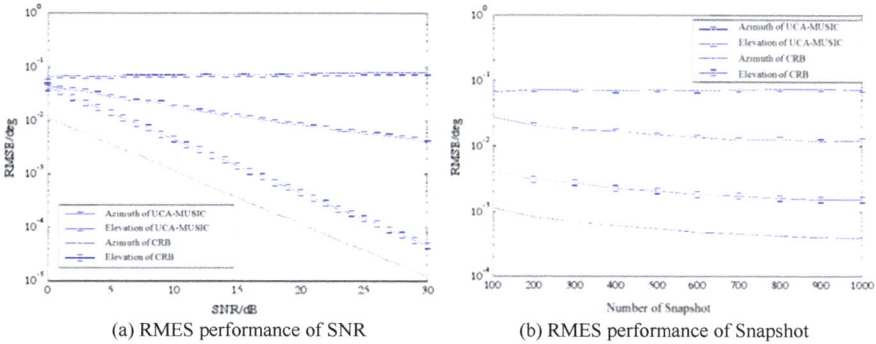

(a) RMES performance of SNR   (b) RMES performance of Snapshot

Fig. 3. Estimation performance of RMES

5   Conclusion

In this letter, An 2D DOA estimation algorithm based on MUSIC is proposed for UCA in the presence of mutual coupling. Simulations are presented to show the effectiveness of the proposed algorithm on SNR and snapshots. It can be concluded that the proposed algorithm improves the performance for both azimuth estimates and elevation estimates.

Acknowledgments. This work is supported by the National Natural Science Foundation of China under Grant 61601494.

References

1. Kelif J-M, Coupechoux M, Mansanarez M. A 3D Beamforming analytical model for 5G wireless networks. In: 14th International Symposium on 2016. Modeling and optimization in mobile, Ad Hoc, and wireless networks. WiOPT 2016.
2. Zhang J, Ge X, Li Q, Guizani M, Zhang Y. 5G millimeter-wave antenna array: design and challenge. IEEE Wirel Commun. April 2017;106–12.
3. Yilmazer N, Koh J, Sarkar TK. Utilizaion of a unitary transform for efficient computation in the matrix pencil method find the direction of arrival. IEEE Trans Antenna Propag. 2006;54 (1):175–81.
4. Barcelo M, Vicario JL, Seco-Granados G. A reduced complexity approach to IAA beamforming for efficient DOA estimation of coherent sources. EURASIP J Adv Signal Process. 2011;1:1–1:16.
5. Cai JJ, Qin GD, Li P. Tow- dimensional DOA estimation with reduced- dimensional MUSIC algorithm using the modulus constraint. Syst Eng Electron. 2014;9:1681–6.
6. Mathews CP, Zoltowski MD. Performance analysis of the UCA-ESPRIT algorithm for circular ring arrays. IEEE Trans Signal Process. 1995;2535–9.
7. Yang L, Zhang H, Yang X, DOA estimation for wideband sources based on UCA. J. Electron. (China). 2006;128–31.
8. Hui HT. Improved compensation for the mutual coupling effect in a dipole array for direction finding, Antennas and Propagation. IEEE Trans. On. 2003;51:2498–503.

A Grid-Map-Oriented UAV Flight Path Planning Algorithm Based on ACO Algorithm

Wei Tian and Zhihua Yang[(✉)]

Communications Engineering Research Center, Shenzhen Graduate School,
Harbin Institute of Technology, Shenzhen, China
tianwei0323@foxmail.com, yangzhihua@hit.edu.cn

Abstract. With the extensive applications of unmanned aerial vehicle (UAV), typical algorithm for path planning is usually restricted for its low efficiency and easy failure, especially for the complex obstacle environments. Therefore, in this paper, a new UAV path planning algorithm is proposed based on ant colony optimization (ACO) for such complex obstacle environment. In particular, the proposed algorithm optimizes the distribution of pheromones and modifies the transfer probability by considering the regional security factors. As a result, it can increase search speed and avoid local optimum and deadlock. Simulation results verify the feasibility and effectiveness of the proposed method.

Keywords: UAV · Path planning · Grid map · ACO

1 Introduction

In these years, UAV takes more important roles in military missions and industrial applications, such as reconnaissance and tracking [1]. Generally, path planning enables an UAV with the capabilities of automatically deciding and executing a sequence of collision-free and safety motions in order to achieve certain tasks in a given environment [2]. Path planning is the basis and prerequisite for a series of autonomous control activities of UAV assignment planning system, such as formation control and multi-UAV coordination.

Currently, there are various types of path planning algorithms, such as A* algorithm, Dijkstra algorithm, artificial potential field algorithm, and genetic algorithm [3]. Among them, the ACO algorithm is widely used for its advantages on simple coding, strong robustness, and positive feedback. ACO algorithm was firstly used to solve traveling salesman problem (TSP) [4], by using a distributed feedback parallel computer system. Moreover, it is easy to merge with other algorithms, with strong robustness. Gradually, ACO algorithm was applied in the field of UAV path planning. However, this method also exists a few inherent flaws, including long search time, slow convergence, and easily trapped in a local optimum. In this regard, several researchers have done a lot of researches on the basis of the traditional ACO algorithm and have obtained good results [5]. From the early ACO algorithms, AS and ACS proposed by Dorigo M, the ant colony algorithm has been continuously improved and optimized. Stutzle et al. proposed MMAS [6] to improve the search in the early stages of the algorithm by limiting the concentration range of pheromones. In [7], the ant colony

© Springer Nature Singapore Pte Ltd. 2020
Q. Liang et al. (eds.), *Communications, Signal Processing,
and Systems*, Lecture Notes in Electrical Engineering 516,
https://doi.org/10.1007/978-981-13-6504-1_144

algorithm is combined with the immune system to accelerate the convergence speed. Literature [8] combines the global search capability of genetic algorithm with the positive feedback of ACO algorithm to improve the efficiency of the algorithm.

Through the research on the ACO algorithm, it can be found that the bottleneck for slow search speed is the superfluous time consumed in initial search. At the beginning of the search, ants have few path information and the choices become relatively blind. With the increase of the cycle, the path is prosperous and the speed will be faster after convergence. Therefore, in this paper, aiming at the problem of too long searching time of the ant colony algorithm, the initial distribution of the pheromone is improved. In the initial stage, the ant is guided to reduce the wasted time in the initial search period and speed up the algorithm. In addition, considering that the algorithm is prone to fall into a local optimum, the improved algorithm optimizes the pheromone distribution strategy, by changing the pheromone within the effective range. As a result, the advantages of the excellent path are preserved, and the diversity of the solutions is increased by optimizing the probability transfer rule of global search.

2   System Model

2.1   Grid Map

This paper is based on two-dimensional environmental maps and uses grid method to establish the path planning environment model. In a raster map, each raster represents a region that has been rasterized to abstract the map with a uniform equal grid. Among them, each grid has a color attribute, i.e., white denotes a passable area, black denotes an impassable area and is regarded as an obstacle. The position information of the grid is described by the coordinate pair, in addition to the ID attribute that can be used to uniquely identify it. The map environment model established by the grid method is shown in Fig. 1a.

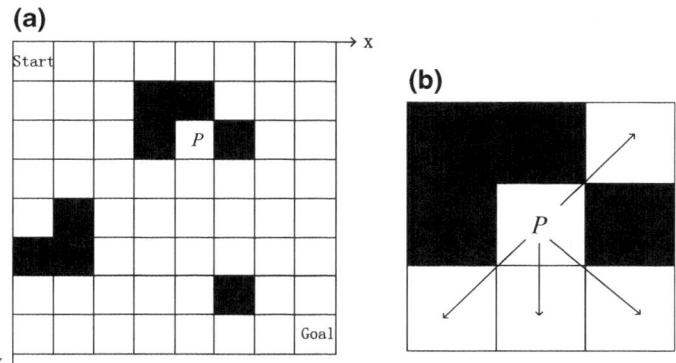

Fig. 1. **a** Grid map **b** Possible routes

As shown in Fig. 1a, there are several obstacles with uneven distribution and different shapes. The goal of path planning is to find an optimal path between the starting point and the end point. It is also required to keep away from obstacles while requiring the shortest distance, to strengthen the safety of the path. In Fig. 1a, the start and end points are: the grid with ID 0 in the upper left corner and the grid with ID (the grid number-1) in the lower right corner. In the path-finding process, each grid can be regarded as a node, and the UAV searched path is a set of line segments connected by the traversed nodes. In this paper, the default UAV flight range is one step at a time, that is, it passes through a grid node in any direction. The direction in which the UAV can fly is eight directions around the current position, allowing the selected node to be a node in the adjacent node that is not an obstacle and has not been traversed. Nodes that are not traversed here refer to nodes marked 0 in the tabu list. For example, in the position of P in Fig. 1a, the next optional flight path is shown in Fig. 1b.

2.2   Path Planning Problem

In the classic ACO algorithm, when the pheromone is initialized, the amount of pheromone in each path is unified to a certain value. Therefore, it takes a lot of time to clear the optimal path in the early stage of searching, which makes the convergence speed of the algorithm slow. In the grid map, there are eight directions for each step. In addition to the obstacles that are not accessible, there are still many types of next routes that can be selected. The same setting of the initial pheromone may cause the ant to choose the direction opposite to the endpoint. The regional nodes either take the route, which will lead to the increase of the search time, the length of the route, and for the physical UAV, the frequent conversion direction during the flight will increase the time consumption.

Besides, in a complex map environment, when an ant's current position in the next selectable node list is an obstacle or a traversed grid, it cannot select a node to transfer. At this time, the ant is caught in a deadlock. Figure 2 shows a schematic of a deadlock. The current path node sequence in the figure is 0–4–5–9–8–12. When the ant is in the 12th grid, there is no next node to reach. A rotor UAV is used in this paper. It is assumed that a certain track has a total of N nodes. As shown in Fig. 3, the distance from section i is l_i, and the total path of the track can be expressed as $L = \sum_{i=1}^{N-1} l_i$.

3   Algorithm Descriptions

3.1   Algorithm Improvements

An ant releases a certain amount of pheromone after passing through a path. Later ants adjust the transition probability according to the number of pheromone on the path and make path selection. Therefore, the distribution of pheromone is very important for the advantages and disadvantages of ant's global path planning. This article makes the following improvements to the pheromone distribution strategy in the original algorithm.

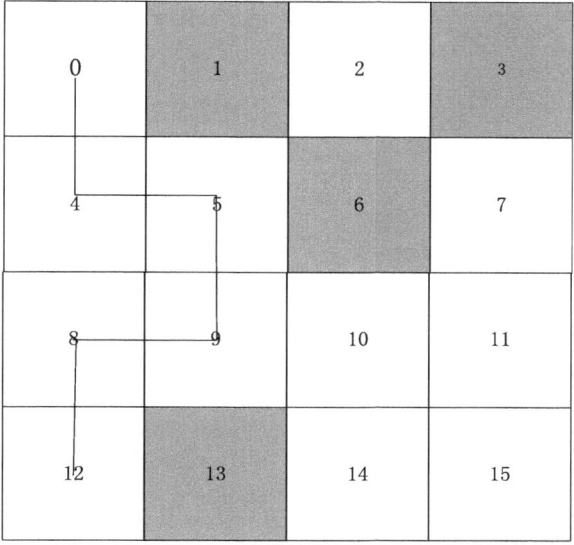

Fig. 2. Stuck in a deadlock

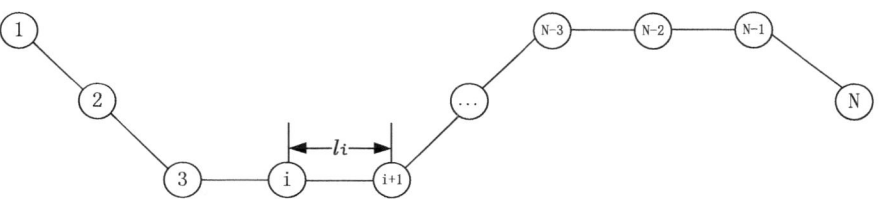

Fig. 3. Path diagram of UAV

Pheromone Distribution Optimization. For maps like the one shown in Fig. 1a, the starting point and the ending point have already been determined, so the global direction of the ant's advance can also be determined. This article uses the direction guidance to distribute the initial pheromone on each path. Take Fig. 1a as an example. The revised rules are as follows:

$$
P_{mn} = \begin{cases} p & n = m - 1 \ or \ n = m - \text{col} \ or \ n = m - \text{col} - 1 \\ 2*p & n = m - \text{col} + 1 \ or \ n = m = \text{col} - 1 \\ 3*p & n = m + 1 \ or \ n = m + \text{col} \ or \ n = m + \text{col} + 1 \\ 0 & \text{else} \end{cases} \tag{1}
$$

where p is the initial amount of pheromone, col is the number of rows in the raster map, and mn is the path from raster m to raster n. According to the location of the grid n in the path relative to the grid m, different initial pheromones are set, and a global direction guidance is given. It can be seen here that $P_{mn} \neq P_{nm}$. The volatilization of pheromone directly relates to the efficiency of the algorithm and the ability of global search. In the volatile of pheromones, we must maintain the advantages of the excellent path and concurrently do not affect the choice of follow-up ants too much. The pheromone volatility rule in the classical algorithm is: $\tau_{ij}(t+n) = \tau_{ij}(t) * (1 - \rho)$, where ρ is the volatility coefficient. This article proposes an adaptive volatility strategy based on path length:

$$\tau_{ij}^* = \begin{cases} 1 - \frac{L_{best}}{L_{better}} * \rho & (a) \\ 1 - \frac{L}{L_{better}} * \rho & (b) \end{cases}. \tag{2}$$

Among them, L represents the length of the path searched by the ant, L_{best} represents the shortest path of the current cycle, and L_{better} represents the shortest path of history. If the path length obtained by the search is the shortest path length of the current period, use the formula (a) to volatilize the path segment on the path. Otherwise, use formula (b). Here, a lower limit of pheromone is given as the pheromone initial value P. The volatilization strategy based on path length gives the corresponding volatilization weight for different path lengths, which not only preserves the advantages of good path, but also guides the subsequent selection.

Pheromone Updating Rule. The pheromone increment calculation formula in the original algorithm is:

$$\Delta\tau_{ij}(t) = \sum_{k=1}^{AntNum} \Delta\tau_{ij}^k(t) \tag{3}$$

in which

$$\Delta\tau_{ij}^k(t) = \begin{cases} \frac{Q}{L} & \text{if ant } k \text{ goes though } (i,j) \\ 0 & \text{else} \end{cases} \tag{4}$$

Among them, Q is the total amount of pheromone, which is a fixed value. AntNum is the ant population. In order to manifest the advantages of the historical best path, preserve the excellent path information and speed up the convergence, the improved algorithm gives different update rules for each path segment in the local optimal solution and the current historical optimal solution. Apply the formula (5) to calculate the pheromone increment for the results of the path search in this cycle.

$$\Delta\tau_{ij}^k = \begin{cases} 2 * \frac{Q}{L} + \frac{Q}{L_{better}} & ij \in \text{optimal path} \\ 0 & \text{else} \end{cases} \tag{5}$$

and

$$\Delta\tau_{ij}(t) = \frac{\sqrt{CityNum}}{4} * \frac{Q}{L_{better}} \quad \text{Historical optimal path} \tag{6}$$

The increment of pheromone is calculated by applying formula (6) to each path segment in the optimal path of history. In the formula, CityNum is the number of grids in the map. Due to different map environments, the number of updates of the optimal path in the iterative process may be very different. Therefore, the weight setting is added to the consideration of scale to make it adapt to the environment.

Improvement of Transfer Probability. The transition probability is an important basis for the choice of the ant to the next arrival point. The transitional probability formula in the original method is as follows:

$$P_{ij}^k(t) = \begin{cases} \dfrac{[\tau_{ij}(t)]^\alpha [\eta_{ij}(t)]^\beta}{\sum\limits_{s \in T_{\text{allowed},k}} [\tau_{is}(t)]^\alpha [\eta_{is}(t)]^\beta}, & j \in T_{\text{allowed},k} \quad \text{(a)} \\ 0, & \text{else} \quad\quad \text{(b)} \end{cases} \tag{7}$$

where $\tau_{ij}(t)$ denotes the amount of pheromone on the road between i-grid and j-grid at time t, $\eta_{ij}(t)$ denotes heuristic information, and the allowed table is the current set of selectable rasters. The value in original method is usually $1/d_{ij}$, which is the inverse of the distance between two grids. However, in this paper, the grids are adjacent to each other and the distance difference is small, so here we use the reciprocal of the Euclidean distance between the grid j and the target grid as the heuristic information.

$$\eta_{ij}(t) = \frac{1}{d_{js}} \tag{8}$$

where S is the target grid node. This allows the heuristic effect of distance length to be reflected.

Aiming at the problem of deadlock in the path planning mentioned in Sect. 2.2, we propose a probability selection strategy based on regional security information. In a grid map, each grid's neighboring grid has two possibilities, either passable or obstructed. If a grid has a majority of obstacles around it, the security of this grid area is relatively poor. Here, we use the variable black to represent the black state of the grid:

$$\text{black}_{\text{city}} = \frac{\text{AroundBlack[city]}}{\text{Around[city]}}. \tag{9}$$

In (9), AroundBlack[city] represents the number of obstacles in the grid adjacent to the current grid, and Around[city] represents the number of grids adjacent to the current grid. Therefore, black is actually the proportion of the barrier grid in the adjacent grid of the current grid. Consider the consideration of regional security factors, and the modified formula is as follows:

$$P_{ij}^k(t) = \begin{cases} \dfrac{[\tau_{ij}(t)]^\alpha [\eta_{ij}(t)]^\beta (1-\text{black}_j)^\sigma}{\sum\limits_{s \in T_{\text{allowed},k}} [\tau_{is}(t)]^\alpha [\eta_{is}(t)]^\beta (1-\text{black}_s)^\sigma}, & j \in T_{\text{allowed},k} \\ 0, & \text{else} \end{cases} \tag{10}$$

where σ is the weight of the security information. Considering that there is such a possibility that although the surrounding obstacles are in the majority, this grid is still a high possibility or even a necessary path. Therefore, the value of σ should not be too high. In this paper, $\partial: \sigma = 2.5 : 1$.

3.2 Algorithm Description

In details, Algorithm 1 provides a pseudocode path planning for a complex environment with obstacles. In the algorithm, the obstacle grid map is generated randomly. According to the formula (1), initialize the path pheromone, update the allowed table of the ant $k(k = 0, 1, 2 \ldots, \text{AntNum})$ at the current location. The algorithm calculates the probability of transfer of every feasible grid in the table through the improved probability transfer formula, and uses roulette to select the next grid. Finally, UAV moves a step of range along the planned path.

Algorithm 1 Path Planning

INPUT: *map*

OUTPUT: *newState, path*

1: *rouletteWheel* $= 0$

2: *states* $=$ ant. getUnvisitedStates()

3: **for** *newState* **in** *states* **do**

4:     *rouletteWheel*$+=$

5:     Math. pow$($getPheromone$(state, newState)$, getParam$('alpha'))$

6:     $*$ Math. pow$($calcHeuristicValue$(state, newState)$, getParam$('beta'))$

7:     $*$ Math. pow$($getBlackValue$(state, newState)$, getParam$('zeta'))$

8: **end for**

9: *randomValue* $=$ random()

10: *wheelPosition* $= 0$

11: **for** *newState* **in** *states* **do**

12:     *wheelPosition* $+=$

13:     Math. pow(getPheromone$(state, newState)$, getParam$('alpha'))$

14:     $*$ Math. pow(calcHeuristicValue$(state, newState)$, getParam$('beta'))$

15:     $*$ Math. pow(getBlackValue$(state, newState)$, getParam$('zeta'))$

16:     *wheelPosition* $= wheelPosition \backslash rouletteWheel$

17:     **if** *wheelPosition* $\geq randomValue$ **do**

18:         **return** *newState*

19: **end for**

---

4 Simulation Results

In this section, we make experimental simulations for evaluations on the proposed algorithm. Two groups of simulation are carried out in MATLAB software. The first group is in a 20 × 20 grid map 1 and the second group is in a 40 × 40 grid map 2. For the convenience of analysis, Table 1 shows the parameters for simulations. The starting point is set to the upper left raster, and the end point is the lower right raster. In Figs. 4 and 5, the black dashed line represents the original ACO algorithm, and the red solid line represents the improved ACO algorithm.

Table 1. Parameters of simulation

Parameter	Parameter specification	Value 1	Value 2
m	Number of ants	50	100
v	Volatile coefficient	0.16	0.16
IterNum	Number of iterations	100	100
β	Weight of heuristic information.	5.0	6.0
∂	Weight of pheromones.	1.5	1.5

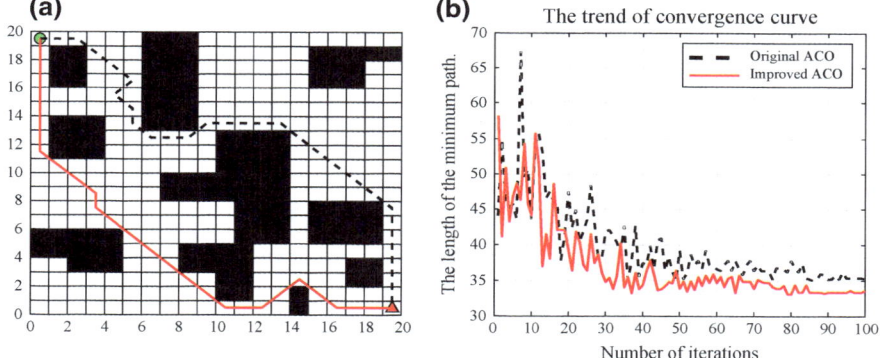

Fig. 4. **a** Comparison results in map 1. **b** Comparison with the original algorithm in map 1

Firstly, we chose value 1 in Table 1 for map 1. As shown in Fig. 4a, it is obvious that the path of the improved algorithm is shorter than that of the original algorithm. Moreover, the improved algorithm has fewer inflection points, so it is more suitable for actual flight scenarios. As shown in Fig. 4b, it can be observed that the improved ACO improves the convergence speed compared with the original algorithm, which shows the effectiveness of improving the convergence speed of the algorithm by optimizing the pheromone distribution strategy in the improved ACO. In Fig. 5, we chose value 2 to simulate randomly generated map with larger and more complex obstacles and obtained the same conclusion. Furthermore, we compare the results of the 50

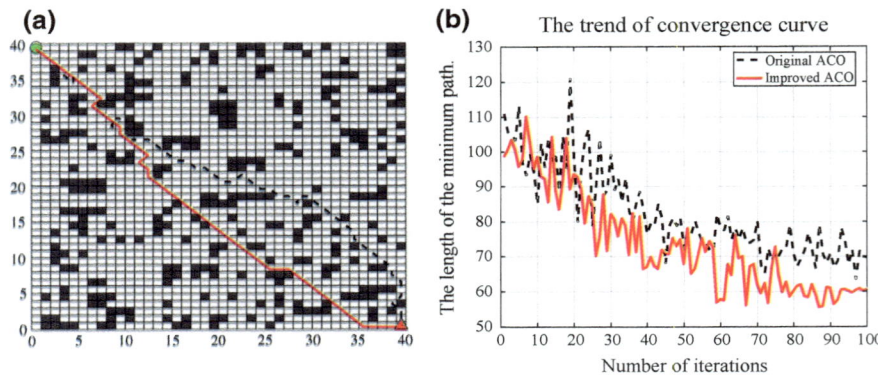

Fig. 5. **a** Comparisons in map 2. **b** Comparison with the original algorithm in map 2

Table 2. Comparison of experimental results

Algorithm	Average time cost	Average path length	Shortest path length	Iteration times when converging	Standard deviation of path length
Original ACO	16.54	35.3847	33.8944	95	3.024
Improved ACO	15.65	32.7990	30.9607	83	1.966

experiments in Map 1, and the specific results are shown in Table 2. From Table 2, it can be concluded that,

- The improvement of ACO greatly reduces the time required for path planning and verifies the effect of improving ACO on search speed.
- The path length obtained by the improved ACO is much shorter than that obtained by the original ACO. So, the effectiveness of directional guidance information optimization is verified.
- The improved ACO is more stable in many experiments.
- The improved ACO converges with fewer iterations.

In Fig. 6, the conditions are randomly distributed obstacles with density from 0.2 to 0.8. We carry out thousand experiments for calculating the success rate of path planning. It is seen that, with a small amount density of obstacles, the improved ACO can reach a probability of 98% by successfully avoiding obstacles. While, the original ACO is obviously lower than the improved method about 16 percent. With density of obstacles increasing, the improved APF method can still reach more than 80% probability by avoiding the obstacles.

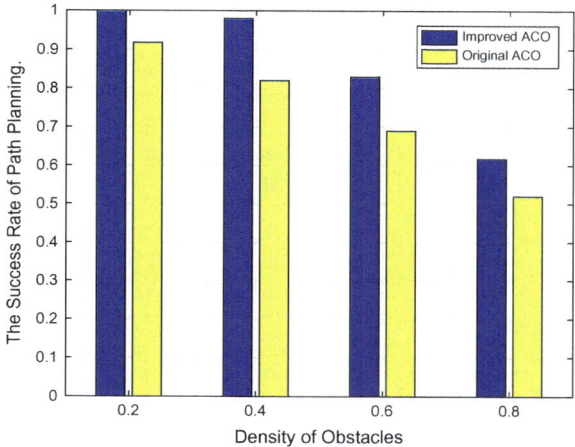

Fig. 6. Success rate of path planning

5 Conclusion

An UAV flight path planning algorithm based on ACO algorithm has been presented in this paper. By comparisons through the simulations, the proposed algorithm achieves a considerable reduction of traveling distance and improves the convergence speed by optimizing the pheromone distribution. Moreover, by adding the regional security factors to improving the probability transfer rules, it also greatly avoiding the possibility of deadlock. Simulation results verified that the method is effective for UAV path planning.

Acknowledgments. The authors would like to express their high appreciations to the supports from the Shenzhen Basic Research Project (JCYJ20150403161923521, JCYJ20170413110004682 and JCYJ20150403161923521).

References

1. Khatib O. Real-time obstacle avoidance for manipulators and mobile robots. Int J Robot Res. 1986;5(1):90–8.
2. Weerakoon T, Ishii K, Nassiraei AAF. An artificial potential field based mobile robot navigation method to prevent From Deadlock. J Artif Intell Soft Comput Res. 2015;5(3):189–203.
3. Lazarowska A. Multi-criteria trajectory base path planning algorithm for a moving object in a dynamic environment. IEEE international conference on innovations in intelligent systems and applications. IEEE; 2017. p. 79–83.
4. Lin S. Computer solutions of the traveling salesman problem. Bell Labs Tech J. 2014;44 (10):2245–69.
5. Zhu QB, Zhang YL. An ant colony algorithm based on grid method for mobile robot path planning. Robot. 2005;27(2):132–6.

6. Dorigo M, Birattari M, Stutzle T. Ant colony optimization. IEEE Comput Intell Mag. 2007;1 (4):28–39.

7. Yuan M, Wang S, Li P. A model of ant colony and immune network and its application in path planning. In: IEEE conference on industrial electronics and applications. IEEE; 2008. p. 102–7.

8. Hu Y, Li D, Ding Y. A path planning algorithm based on genetic and ant colony dynamic integration. In: Intelligent control and automation. IEEE; 2015. p. 4881–6.

Abnormal Event Detection and Localization in Visual Surveillance

Yonglin Mu and Bo Zhang[✉]

College of Information Science and Technology, Dalian Maritime University,
Linghai Road 1, Dalian 116026, China
{yonglinmu,bzhang}@dlmu.edu.cn

Abstract. In this paper, we propose a framework for abnormal event detection and analysis in the field of visual surveillance based on the state-of-the-art deep learning techniques. We train a pair of conditional generative adversarial networks (cGANs) using the normal behavior samples, where one cGAN takes video frames as inputs and generates the corresponding optical flow features. While on the other hand, the other cGANs take optical flow features as inputs and generate the corresponding video frames. By analyzing the differences between the generated frames/optical flow features and the realistic samples, abnormal events can be detected and localized effectively. Moreover, for suspected regions, we adopt the faster RCNN to analyze the abnormal events. Experimental results demonstrate that the proposed framework can detect the abnormal events accurately and efficiently.

Keywords: Conditional GANs · Faster RCNN · Abnormal event detection · Visual surveillance

1 Introduction

With global urbanization and the rapid growth of population, detecting and analyzing abnormal events in public places are crucial in the field of visual surveillance. Despite a lot of research has been done in this area in the past 10 years [1–3], the problem is still open and far from being solved automatically. Two main challenges exist in this field: (1) Very few abnormal samples can be found in public benchmarks. This limitation constrains the applications of current machine learning methods, especially the deep learning-based models, due to the lack of training samples; (2) there is no clear definition on abnormality. For instance, it is very common to see vehicles moving fast on the highway, while cars moving fast on campus or on the squares can be regarded as abnormal. Typical normal/abnormal events in the UCSD and UMN datasets are shown in Fig. 1, where only pedestrians walking in the scenes are labeled as normal, while bicycles, vehicles, and running action appeared in the scenes are considered as abnormal events.

Suspected events always have multimodes and are difficult to model due to the limited number of training samples. The latest trend focuses on modeling merely the normal motion patterns of events using generative methods. Abnormal events can be

© Springer Nature Singapore Pte Ltd. 2020
Q. Liang et al. (eds.), *Communications, Signal Processing, and Systems*, Lecture Notes in Electrical Engineering 516,
https://doi.org/10.1007/978-981-13-6504-1_145

Fig. 1. First row: normal events; second row: abnormal events. These samples are taken from the UCSD and UMN datasets

detected as the outliers of normal event distribution. Since only normal samples are needed in the training phase, it is more feasible in realistic applications. Motivated by the fast development of generative adversarial networks (GANs), we propose a framework based on the conditional generative adversarial networks (cGANs) for abnormal events detection. Furthermore, we adopt the faster RCNN to analyze the suspicious events. The whole framework is shown in Fig. 2.

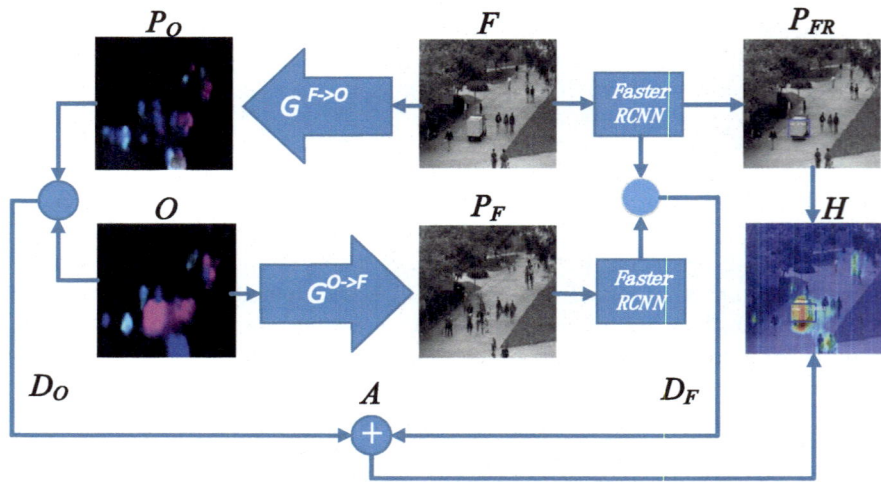

Fig. 2. Whole framework of abnormal events detection

GANs are originally used to generate images [4]. In this work, we propose to exploit cGANs [5] to learn the normal motion and appearance features in the training phase. Abnormal event detection is achieved by measuring the distance between the normal and abnormal distributions in terms of the motion and appearance features in the test period. Once suspicious regions are obtained, we learn the prior knowledge of objects using the faster RCNN [6] based on the VOC2007 dataset, in order to analyze the semantic meaning of events in the scenes.

The main contributions of this paper are listed as follows:

- We propose a framework for abnormal event detection and analysis based on the state-of-the-art deep learning techniques, which does not rely on handcrafted features.
- The cGANs are exploited to learn the motion and appearance features only from normal events; thus, it is not necessary to incorporate negative samples in the training period.
- We adopt the faster RCNN to further analyze the semantic meaning of abnormal events in suspicious regions.

The rest of the paper is organized as follows: Sect. 2 reviews the related work briefly. We present the proposed method in Sect. 3 including the overall framework and the details for anomaly detection, localization, and analysis. The qualitative and quantitative results are provided in Sect. 4. We conclude our work in Sect. 5.

2   Related Works

In recent years, abnormal event detection has developed rapidly in the field of computer vision. Traditional approaches often rely on complex handcrafted features to represent visual information. Ionescu et al. [7] use the one-class SVM to detect abnormal events based on the spatiotemporal motion features extracted in 3-D local cuboids. In [8], a space-time Markov random field model is proposed to localize abnormal events in videos. Based on the concept of proxemics [9], the so-called social force model is exploited to analyze abnormal pedestrian behaviors in crowded scenarios [1]. In [10], the histogram of oriented tracklets is used to detect abnormalities in the scenes.

With the rapid development of deep learning techniques, deep models gradually become effective tools for feature learning and representation. In [11], semantic information (learned by using the existing CNN models) and low-level optical flows are combined to measure the local abnormality in videos. In [12], a FCN-based framework is proposed to detect and localize abnormal events in crowded scenarios. Most recently, generative adversarial networks [4] and its variant [13] have shown great power in learning the latent distribution of data and generating samples. In [14], Ravanbakhsh et al. propose a generative deep learning model for abnormality detection in crowd analysis.

3   Proposed Method

3.1   Feature Learning

We use cGANs to learn the motion and appearance features from training samples. Specifically, let F_t be the tth frame of a video and O_t be the optical flow obtained using

F_t and F_{t+1}, using the method in paper [15].O_t consists of the horizontal, vertical, and the magnitude components.

We train a pair of networks: $(1)N^{O \rightarrow F}$, which generates video frames from the corresponding optical flow; and $(2)N^{F \rightarrow O}$, which generates optical flow from the corresponding frames. We adopt G and D to indicate the generator and discriminator, respectively. In the case of $N^{O \rightarrow F}$, G takes an optical flow O_t and a noise vector z as the input and generates the corresponding frame $P_{Ft} = G(O_t, z)$, while D evaluates the deviation of data generated by G from the real distribution. Similarly, for network $N^{F \rightarrow O}$, G takes a frame F_t and a noise vector z as the input and generates the corresponding optical flow $P_{Ot} = G(F_t, z)$. Taking $N^{F \rightarrow O}$, for example, training set X consists of original video frames and the corresponding optical flow features, denoted as $X = \{F_t, O_t\}$. G and D are trained simultaneously using the loss L_{cGANs} and loss L_1 as in Eqs. (1) and (2), respectively:

$$L_1(F_t, O_t) = ||O_t - G(F_t, z)||_1 \tag{1}$$

$$\begin{aligned} L_{cGANs}(G, D) &= E_{(x,y) \in X}[\log D(x, y)] \\ &+ E_{x \in \{F_t\}, z \in Z}[\log(1 - D(x, (G(x, z))))] \end{aligned} \tag{2}$$

The training procedure of $N^{O \rightarrow F}$ is similar. Since only normal events are used in the training procedure, the model cannot reconstruct abnormal events in the test phase. As shown in Fig. 2, $G^{F \rightarrow O}$ takes an image F as the input, which contains a vehicle moving on campus. However, in the corresponding map P_O, the optical flow in the area that the moving vehicle is passing by cannot reconstructed correctly. Similarly, $G^{O \rightarrow F}$ takes an optical flow O as the input and generates the corresponding frame P_F. Comparing F with P_F, the area where the vehicle is passing by cannot reconstructed correctly

3.2   Abnormality Detection and Analysis

At testing phase, we use the generators $G^{F \rightarrow O}$ and $G^{O \rightarrow F}$ to generate the reconstructed frame P_F and optical flow P_O. Comparing the real optical flow O and the generated optical flow P_O, we obtained $D_O = O - P_O$, where D_O highlights the local differences between the real optical flow and its reconstruction. In abnormal areas, these differences are significantly higher. Similarly, we obtain $D_F = F - P_F$. However, the value of D_F in abnormal regions is not that obvious; thus, we adopt the faster RCNN to obtain the differences. We use the first 13 layers of the faster RCNN to extract feature vector F' and P'_F from image F and P_F and compute $D_F = F' - P'_F$. Then, we upsample D_F in order to obtain D'_F with the same resolution as D_O. Next, for each video V, we compute the maximum value M_O and M_F of D_O and D'_F from all the frames.D'_F and D_O are normalized as in Eqs. (3) and (4):

$$N_O(i, j) = 1/M_O D_O(i, j) \tag{3}$$

$$N_F(i, j) = 1/M_F D'_F(i, j) \tag{4}$$

We create the so-called 'abnormal map' A as in Eq. (5):

$$A = N_O + N_F \tag{5}$$

We adopt the faster RCNN to classify objects in the scene, which takes a frame F as the input and outputs an image P_{FR}, where objects are labeled. P_{FR} need to be upsampled in order to obtain P'_{FR} with the same resolution as A. The labeled abnormal map H is given by Eq. (6):

$$H = \alpha P'_{FR} + (1 - \alpha)A \tag{6}$$

In our experiments, $\alpha = 0.7$. Examples of typical H are shown in Fig. 5, where we can not only observe the abnormal areas intuitively, but also recognize the semantic meanings of objects in the scenes so as to analyze what happened and whether the events are abnormal.

4   Experimental Results

4.1   Benchmarks and Experiment Setup

We evaluate the proposed approach on two standard benchmarks, namely the UCSD dataset and the UMN anomaly detection dataset. The UCSD dataset is divided into two parts: (1) **Ped1**, which contains 34 train and 16 test videos with the resolution of 158 by 238; and (2) **Ped2**, which contains 16 train and 12 test videos with the resolution of 240 by 360. This dataset is very challenging due to the low-resolution images. The crowd density in the walkways ranges from sparse to very dense. The UMN dataset consists of 11 videos that contain escape event in three different indoor and outdoor scenarios, with a total amount of 7700 frames. In our experiments, the video frames are resized to the resolution of 256 by 256. $N^{O \to F}$ and $N^{F \to O}$ are trained by stochastic gradient descent with momentum 0.5 and batch size 1. Each network is trained for ten epochs.

4.2   Results and Discussion

In our experiments, the criterion of detection is described as follows:

$$T(i,j) = \begin{cases} 1 & A(i,j) \geq \text{theta} \\ 0 & A(i,j) < \text{theta} \end{cases} \tag{7}$$

where $T(i,j) = 1$ indicates that the pixel (i,j) is abnormal, while $T(i,j) = 0$ indicates that the pixel is normal, and theta is the threshold for anomaly detection. If the frame contains at least one abnormal pixel, it is labeled as abnormal. We compute the ROC curve according to different thetas in order to evaluate our approach in a quantitative way.

- **UCSD Dataset**

Quantitative results using the equal error rate (EER) and area under curve (AUC) measures are shown in Table 1, and the ROC curves are shown in Fig. 3.

Table 1. Comparison with other approaches on the UCSD dataset.

Method	Ped1		Ped2	
	EER (%)	AUC (%)	EER (%)	AUC (%)
MPPCA [8]	40	59	30	69.3
Social force (SF) [1]	31	67.5	42	55.6
SF + MPPCA [16]	32	68.8	36	61.3
SR [2]	19	–	–	–
MDT [16]	25	81.8	25	82.9
Detection at 150 fps [17]	15	91.8	–	–
AMDN (double fusion) [18]	16	92.1	17	90.8
Proposed method	*13*	*92.5*	*17*	*89.1*

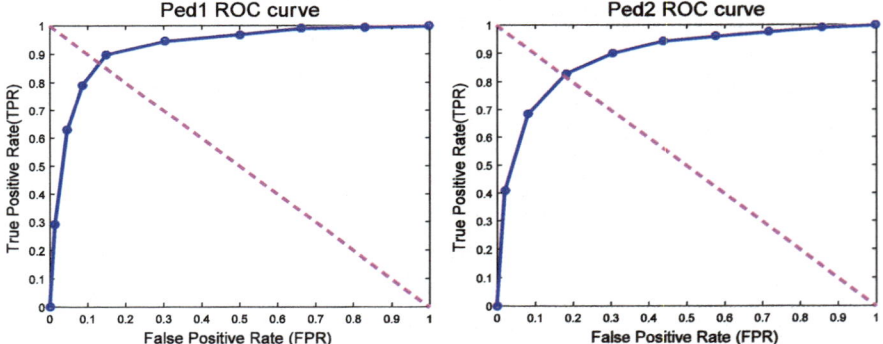

Fig. 3. ROC curves on the UCSD anomaly detection dataset.

- **UMN Dataset**

Quantitative results on the UMN dataset are shown in Table 2, and the ROC curve is shown in Fig. 4.

Typical examples of abnormality localization and analysis on the UCSD anomaly detection dataset are shown in Fig. 5. Our method can not only visualize abnormal regions, but also obtain the semantic labels of objects within the suspicious regions (blue boxes).

Table 2. Comparison with other approaches on the UMN dataset.

Method	AUC
Optical flow [1]	0.84
SFM [1]	0.96
Sparse reconstruction [2]	0.97
Plug-and-Play CNN [11]	0.97
Proposed method	***0.97***

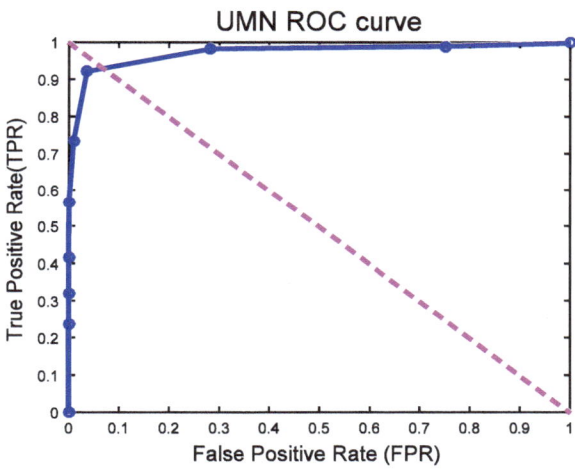

Fig. 4. ROC curve on the UMN anomaly detection dataset.

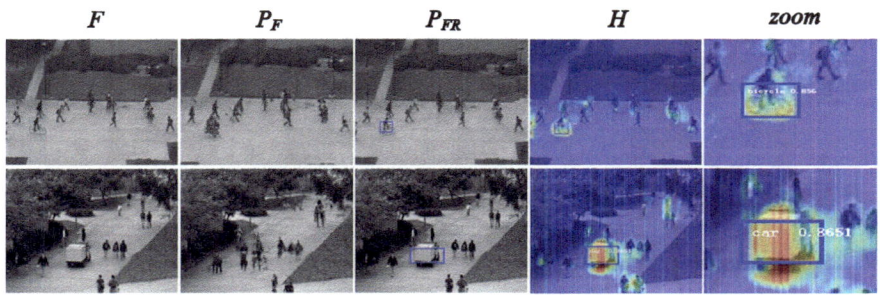

Fig. 5. Some examples of abnormality localization and analysis on the UCSD anomaly detection dataset.

5   Conclusions

In this paper, we propose a framework for abnormal event detection and analysis in crowd scenarios based on the state-of-the-art deep learning techniques. We train the cGANs using ordinary samples, which take video frames and optical flow features as inputs and generate the corresponding optical flow and frames. By analyzing the differences between the generated frames/optical flows and the realistic correspondences, the abnormal events can be detected and localized effectively. Moreover, for suspicious regions, we adopt the faster RCNN to recognize the semantic meanings of objects in the scenes so as to analyze what happened and determine whether the events are abnormal. Experimental results demonstrate the applicability of our proposed approach.

Acknowledgments. This work is partly supported by the National Natural Science Foundation of China (Grant No. 61702073) and the Fundamental Research Funds for the Central Universities (Grant No. 3132018190).

References

1. Mehran R, Oyama A, Shah M. Abnormal crowd behavior detection using social force model. In: IEEE conference on computer vision and pattern recognition; 2009. p. 935–42.
2. Cong Y, Yuan J, Liu J. Sparse reconstruction cost for abnormal event detection. In: IEEE conference on computer vision and pattern recognition; 2011. p. 3449–3456.
3. Hinami R, Mei T, Shin. Joint detection and recounting of abnormal events by learning deep generic knowledge. In: IEEE international conference on computer vision; 2017. p. 3639–47.
4. Goodfellow IJ, Pouget-Abadie J, Mirza M, Xu B, Warde-Farley D, Ozair S, Courville AC, Bengio Y. Generative adversarial nets. In: International conference on neural information processing systems; 2014. p. 2672–80.
5. Isola P, Zhu J, Zhou T, Efros AA. Image-to-image translation with conditional adversarial networks. In: IEEE conference on computer vision and pattern recognition; 2016. p. 5967–76.
6. Ren S, Girshick R, Girshick R, Sun J. Faster RCNN: towards real-time object detection with region proposal networks. IEEE Trans Pattern Anal Mach Intell. 2017;39(6):1137–49.
7. Ionescu RT, Smeureanu S, Popescu M, Alexe B. Detecting abnormal events in video using narrowed motion clusters; 2018. https://arxiv.org/abs/1801.05030.
8. Kim J, Grauman K. Observe locally, infer globally: A space-time MRF for detecting abnormal activities with incremental updates. In: IEEE Conference on Computer Vision and Pattern Recognition; 2015. 2921–7.
9. Choi W, Shahid K, Savarese S. Learning context for collective activity recognition. In: IEEE conference on computer vision and pattern recognition; 2011. p. 3273–80.
10. Mousavi H, Nabi M, Galoogahi HK, Perina A, Murino V. Abnormality detection with improved histogram of oriented tracklets. In: International conference on image analysis and processing; 2015. p. 722–32.
11. Ravanbakhsh M, Nabi M, Mousavi H, Sangineto E, Sebe N. Plug-and-play CNN for crowd motion analysis: an application in abnormal event detection. In: IEEE winter conference on applications of computer vision; 2018.

12. Sabokrou M, Fayyaz M, Fathy M, Klette R. Deep-anomaly: fully convolutional neural network for fast anomaly detection in crowded scenes; 2018. https://arxiv.org/abs/1609.00866.

13. Radford A, Metz L, Chintala S. Unsupervised representation learning with deep convolutional generative adversarial networks. In: International Conference on Learning Representations; 2016.

14. Ravanbakhsh M, Nabi M, Sangineto E, Marcenaro L, Regazzoni C, Sebe N. Abnormal event detection in videos using generative adversarial nets. In: IEEE international conference on image processing; 2017.

15. Brox T, Malik J. Large displacement optical flow: descriptor matching in variational motion estimation. IEEE Trans Pattern Anal Mach Intell. 2011;33(3):500–13.

16. Mahadevan V, Li W, Vasconcelos N. Anomaly detection in crowded scenes. In: IEEE conference on computer vision and pattern recognition; 2010. p. 1975–81.

17. Lu C, Shi J, Jia J. Abnormal event detection at 150 FPS in MATLAB. In: IEEE international conference on computer vision; 2014. p. 2720–7.

18. Xu D, Yan Y, Ricci E, Sebe N. Detecting anomalous events in videos by learning deep representations of appearance and motion. Comput Vis Image Underst. 2016;156:117–27.

Pseudorange Fusion Algorithm for GPS/BDS Software Receiver

Jiang Yi, Fan Yue$^{(\boxtimes)}$, Han Yan, and Shao Han

Information Science and Technology College, Dalian Maritime University,
Dalian 116026, China
598150256@qq.com

Abstract. The multi-mode positioning of Global Navigation Satellite System (GNSS) could improve the positioning accuracy compared with the traditional single-mode positioning, as the number of observed satellites is increased and the geometry distribution of visual satellites is improved. In this paper, pseudorange fusion algorithm is proposed to combine the pseudorange observations of both Global Positioning System (GPS) and BeiDou Navigation System (BDS) to obtain the position equation for dual-mode positioning. Then the weighted least square method is used to solve this position equation. Besides that, the proposed pseudorange fusion algorithm is implemented in a GPS/BDS software receiver. According to positioning result comparison of single-mode and dual-mode positioning, it is concluded that the dilution of precision (DOP) of the dual-mode positioning is smaller and the positioning accuracy is more precision.

Keywords: Pseudorange fusion · Weighted least square method · Multi-mode positioning · Software receiver

1 Introduction

With the development of Global Navigation Satellite System (GNSS) technology and applications, many countries have been developing their own navigation systems. The US Global Positioning System (GPS) has been operated stably for more than 20 years and used widely in the world. The BeiDou Navigation System (BDS) which has been developing independently by China has already provided services for the Asia-Pacific region since 2012 and now is in the globalization phase. Meanwhile, dual-mode or multi-mode positioning technologies have become one of the developing trends and hot topics in GNSS receiver's algorithm research [1]. Besides that, many research groups studied GNSS software receivers based on a generic central processing unit (CPU) or an embedded platform, as software receivers could enhance its performance by replacing the internal software rather than changing the hardware structure [2].

The early research on the multi-mode GNSS positioning mainly focuses on a fusion of GPS and GLONASS or Galileo navigation system. With the development of BDS, more and more researchers began to study the multi-mode positioning with GPS and BDS. The dual-mode positioning of GPS/BDS is more reliable than that of single-mode positioning [3, 4]. In terms of software receivers, nowadays programming languages

© Springer Nature Singapore Pte Ltd. 2020
Q. Liang et al. (eds.), *Communications, Signal Processing,
and Systems*, Lecture Notes in Electrical Engineering 516,
https://doi.org/10.1007/978-981-13-6504-1_146

are mainly MATLAB, C/C++, etc [2, 5, 6]. However, the MATLAB program execution is inefficient and inflexible. In contrast, the software receiver implemented in C++ is faster and more portable. Therefore, the pseudorange fusion algorithm is proposed for GPS/BDS dual-mode positioning in this paper, and a GPS/BDS combined software receiver based on real observation data is realized.

Firstly, the principle of GPS/BDS dual-mode positioning algorithm is investigated in this paper. Pseudorange fusion algorithm is used to combine the pseudorange observations of both GPS and BDS to obtain the position equation. The position of the receiver is estimated by the weighted least square method. Based on the above algorithm, a GPS/BDS software receiver based on C++ is designed and implemented. Position calculation both in single-mode and in GPS/BDS dual-mode positioning has been realized. Finally, the positioning results of GPS, BDS and GPS/BDS output from the software receiver are compared and analyzed. The results of this research support the idea that the dilution of precision (DOP) of GPS/BDS and the position accuracy of GPS/BDS are improved.

The rest of the paper is organized as follows. Section 2 investigates the pseudorange fusion algorithm in GPS/BDS dual-mode positioning technology. In Sect. 3, the proposed algorithm is implemented in the GPS/BDS dual-mode software receiver. In Sect. 4, the positioning results in different positioning mode output from the software are given and analyzed. Finally, conclusions are made in Sect. 5.

2 Pseudorange Fusion Algorithm

The principle of dual-mode GNSS positioning is similar to the single-mode positioning. The key technology of dual-mode positioning is the fusion algorithm. According to different fusion information, the dual-mode positioning can be divided into positioning result fusion algorithm and pseudorange fusion algorithm [7, 8]. In positioning result fusion algorithm, the position estimation from different navigation systems is obtained separately, and then the final positioning result is obtained by weighted averaging of these independent positioning results. In the pseudorange fusion algorithm, the pseudorange observation of different navigation systems is used to set up the position equation, thereby solving the final positioning result. Therefore, the former requires at least four visible satellites for each GNSS, while the latter only requires at least five visible satellites in total without restrictions for each GNSS. In terms of the number of visible satellite, pseudorange fusion algorithm is more applicable. For the GPS/BDS dual-mode receiver, signals from at least five satellites can be received generally under most circumstances.

Generally speaking, the time of arrival (TOA) technology is used in GNSS to determine a receiver's position [3]. Position equation used in single-mode positioning is shown in Eq. (1).

$$\sqrt{(x_i - x_u)^2 + (y_i - y_u)^2 + (z_i - z_u)^2} + c\delta t_u = \rho_{ci} - \varepsilon_{\rho i} \qquad (1)$$

As the position of the satellite (x_i, y_i, z_i) can be computed according to the ephemerides, $(x_u, y_u, z_u, \delta t_u)$ is the coordinates and clock offset of the receiver, ρ_{ci} is corrected pseudorange measurement, and $\varepsilon_{\rho i}$ is the measurement error of the pseudo-range. As time reference of different GNSS systems is different, time unified problem must be considered in dual-mode GNSS positioning. Thus, the system time clock bias δt_{GB} should be added in the GPS/BDS dual-mode position equations. The GPS/BDS dual-mode position equation contains five unknowns $(x_u, y_u, z_u, \delta t_u, \delta t_{GB})$.

Furthermore, in order to ensure that the receiver clock offset δt_u in the position equation of each measurement is the same, all the pseudorange observations should be obtained at the same time in the receiver. Position equation of pseudorange fusion algorithm used in GPS/BDS dual-mode positioning is shown in Eq. (2).

$$\begin{cases} \sqrt{(x_{G_i} - x_u)^2 + (y_{G_i} - y_u)^2 + (z_{G_i} - z_u)^2} + c\delta t_u = \rho_{G_i} - \varepsilon_{\rho_{G_i}} \\ \sqrt{(x_{B_j} - x_u)^2 + (y_{B_j} - y_u)^2 + (z_{B_j} - z_u)^2} + c(\delta t_u + \delta t_{GB}) = \rho_{B_j} - \varepsilon_{\rho_{B_j}} \end{cases} \tag{2}$$

The subscript G and B represent GPS and BDS, respectively. The term i refers to the ith GPS satellite, and the term j refers to the jth BDS satellite. As Eq. (2) is nonlinear, it needs to be linearized firstly before solving it. After expanding with the Taylor formula, Eq. (2) can be expressed as:

$$\delta\boldsymbol{\rho}_G = \mathbf{u}_G \cdot \mathbf{dx_0} + \varepsilon_{\boldsymbol{\rho}_G} \qquad \delta\boldsymbol{\rho}_B = \mathbf{u}_B \cdot \mathbf{dx_0} + \varepsilon_{\boldsymbol{\rho}_B} \tag{3}$$

where defined the vector $\mathbf{u}_G$, $\mathbf{u}_B$ and $\mathbf{dx_0}$ as follows:

$$\begin{cases} \mathbf{u}_G \triangleq \left[\left.\frac{\partial\rho_{G_i}}{\partial x_u}\right|_{x_0}, \left.\frac{\partial\rho_{G_i}}{\partial y_u}\right|_{y_0}, \left.\frac{\partial\rho_{G_i}}{\partial z_u}\right|_{z_0}, \left.\frac{\partial\rho_{G_i}}{\partial\delta t}\right|_{\delta t_0}, \left.\frac{\partial\rho_{G_i}}{\partial t_{GB}}\right|_{t_{GB0}} \right] \\ \mathbf{u}_B \triangleq \left[\left.\frac{\partial\rho_{B_j}}{\partial x_u}\right|_{x_0}, \left.\frac{\partial\rho_{B_j}}{\partial y_u}\right|_{y_0}, \left.\frac{\partial\rho_{B_j}}{\partial z_u}\right|_{z_0}, \left.\frac{\partial\rho_{B_j}}{\partial\delta t}\right|_{\delta t_0}, \left.\frac{\partial\rho_{B_j}}{\partial t_{GB}}\right|_{t_{GB0}} \right] \\ \mathbf{dx_0} \triangleq \left[(x_u - x_0), (y_u - y_0), (z_u - z_0), (\delta t - \delta t_0), (t_{GB} - t_{GB0}) \right]^T \end{cases} \tag{4}$$

where $(x_0, y_0, z_0, \delta t_0, t_{GB0})$ is the initial estimated value, the vector $\delta\boldsymbol{\rho}_G$, $\delta\boldsymbol{\rho}_B$ as follows:

$$\delta\boldsymbol{\rho}_G = [\delta\rho_{G1} \quad \cdots \quad \delta\rho_{Gm}]^T \qquad \delta\boldsymbol{\rho}_B = [\delta\rho_{B1} \quad \cdots \quad \delta\rho_{Bn}]^T \tag{5}$$

and

$$\varepsilon_{\boldsymbol{\rho}_G} = \left[\varepsilon_{\rho_{G1}} \quad \cdots \quad \varepsilon_{\rho_{Gm}}\right]^T \qquad \varepsilon_{\boldsymbol{\rho}_B} = \left[\varepsilon_{\rho_{B1}} \quad \cdots \quad \varepsilon_{\rho_{Bn}}\right]^T \tag{6}$$

where n and m represent the number of BDS and GPS visual satellites, respectively. Equation (5) is generally called pseudorange residual, and the element of it can be expressed by

$$\delta\rho_{Gi} = \rho_{Gi}(x_u) - \rho_{Gi}(x_0) \qquad \delta\rho_{Bj} = \rho_{Bj}(x_u) - \rho_{Bj}(x_0) \tag{7}$$

Position equation in the form of matrix of GPS/BDS dual-mode positioning is:

$$\delta\rho = \mathbf{H}d\mathbf{x_0} + \varepsilon_\rho \tag{8}$$

where

$$\delta_\rho = \begin{bmatrix} \delta_{\rho_G} & \delta_{\rho_B} \end{bmatrix}^T \quad \varepsilon_\rho = \begin{bmatrix} \varepsilon_{\rho_G} & \varepsilon_{\rho_B} \end{bmatrix}^T \quad \mathbf{H} = \begin{bmatrix} \mathbf{U}_{G1}^T & \cdots & \mathbf{U}_{Gm}^T & \mathbf{U}_{B1}^T & \cdots & \mathbf{U}_{Bn}^T \end{bmatrix}^T \tag{9}$$

The dimension of H is $(n + m) \times 5$. We use the weighted least square method to solve the position equation:

$$d\mathbf{x_0} = \left(\mathbf{H}^T\mathbf{W}\mathbf{H}\right)^{-1}\mathbf{H}^T\mathbf{W}\,\delta\rho \tag{10}$$

$d\mathbf{x_0}$ is the initial estimation matrix of iteration. Matrix $\mathbf{W}$ is given by

$$\mathbf{W} = \mathrm{diag}(w_{G1} \cdots w_{Gm}\, w_{B1} \cdots w_{Bn}) \tag{11}$$

w_{Gi} and w_{Bj} are the weighting factor of the ith GPS satellite and the jth BDS satellite, respectively. They can be calculated based on the CN_O of satellite signals and the elevation angles of satellites [9]. The position of the receiver is estimated based on Eq. (10). For the kth iteration, the iteration process can be described as follows:

$$\begin{cases} d\mathbf{x_{k-1}} = (\mathbf{H}_{k-1}^T\mathbf{W}_{k-1}\mathbf{H}_{k-1})\mathbf{H}_{k-1}^T\mathbf{W}_{k-1}\delta\rho_{k-1} \\ \mathbf{x_k} = \mathbf{x_{k-1}} + d\mathbf{x_{k-1}} \end{cases} \tag{12}$$

3 Algorithm Implementation

The structure of the GPS/BDS dual-mode software receiver is shown in Fig. 1. GPS and BDS signals are received from the radio frequency (RF) front end and intermediate frequency (IF) data is processed by the baseband signal processing module. GNSS signals are acquired and tracked by the baseband processing module. After that, the navigation message data of GPS and BDS can both be decoded. The proposed pseudorange fusion algorithm is mainly implemented in the position estimation module. The function of the position mode selection submodule is selecting the positioning mode. Then the result display module could display positioning information, time information and satellite constellation.

The position mode selection submodule selects the positioning mode through the macro definition in C++. The single-mode positioning submodule is shared by GPS and BDS. It will distinguish between GPS and BDS through data interface. When GPS IF data is chosen to process, the single-mode positioning submodule is used to estimate the receiver position using GPS data. When BDS IF data is chosen to process, BDS data is used to estimate the receiver position in the single-mode positioning submodule.

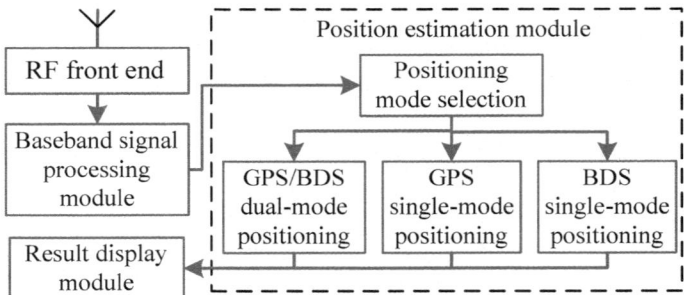

Fig. 1. Structure of GPS/BDS dual-mode software receiver

In the GPS/BDS dual-mode positioning submodule, GPS IF data and BDS IF data are processed simultaneously. The flow chart of the GPS/BDS dual-mode submodule is shown in Fig. 2.

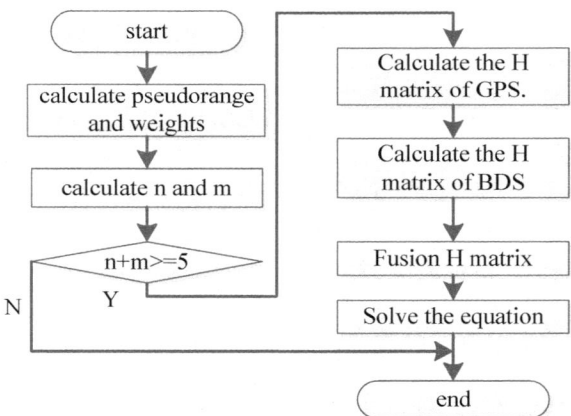

Fig. 2. Flow chart of GPS/BDS dual-mode positioning

Firstly, this submodule computes pseudoranges and weight factors for all tracked GPS and BDS satellites. Then the submodule determines whether the total number of tracked satellites meet the combination positioning requirement. If the condition is met, it will calculate the matrix **H** and matrix **W** given by Eqs. (9) and (11). Finally, the submodule computes the position of the receiver according to iteration process shown in Eq. (12).

4   Results and Analysis

The positioning experiments were carried out on the roof of Science Hall in Dalian Maritime University. With the Use of HG-SOFTGPS04 four-channel GNSS interme-diate frequency signal collector, GPS and BDS signals are received simultaneously.

4.1   GPS Positioning Results

In the GPS single-mode positioning, the result shown by the result display module of the software receiver at a certain moment is illustrated in Fig. 3.

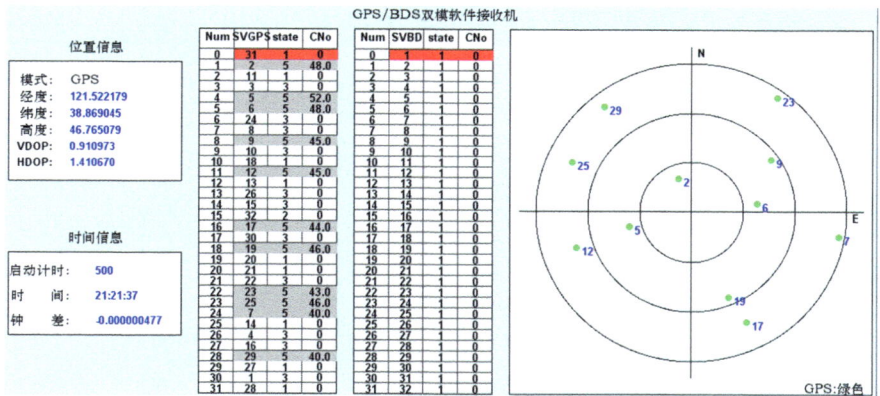

Fig. 3. GPS positioning results

From Fig. 3. It can be observed that a total of 10 satellites are tracked. The dis-tribution of tracked satellites in the Earth-fixed coordinate system (ENU) is also shown. Positioning result is (121.522179°E, 38.869045°N).

4.2   BDS Positioning Results

In the BDS single-mode positioning, the output of the result display module of the software receiver at the same moment is shown in Fig. 4. It can be seen that the software receiver tracked a total of 12 satellites. The longitude of the receiver is 121.522141°E, and the latitude is 38.869015°N.

4.3   Dual-Mode Positioning Results

In GPS/BDS dual-mode positioning, the result given by the software receiver is illustrated in Fig. 5. As can be seen from Fig. 5, the software receiver tracked a total of 22 satellites. The receiver's positioning output from the software receiver is (121.522194°E, 38.869072°N).

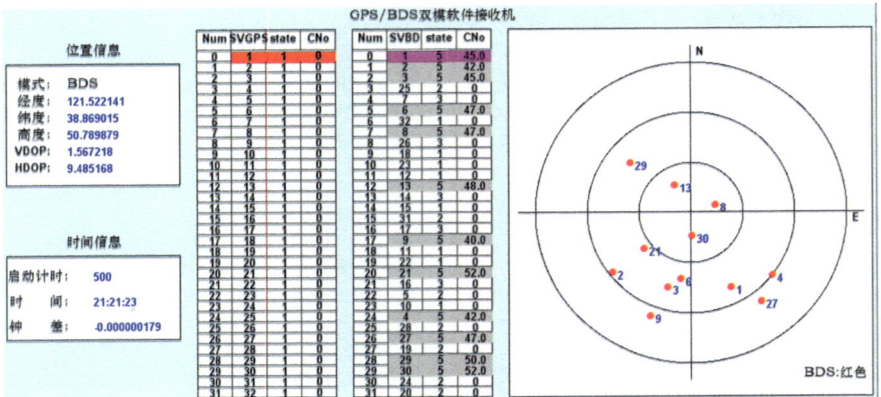

Fig. 4. BDS positioning results

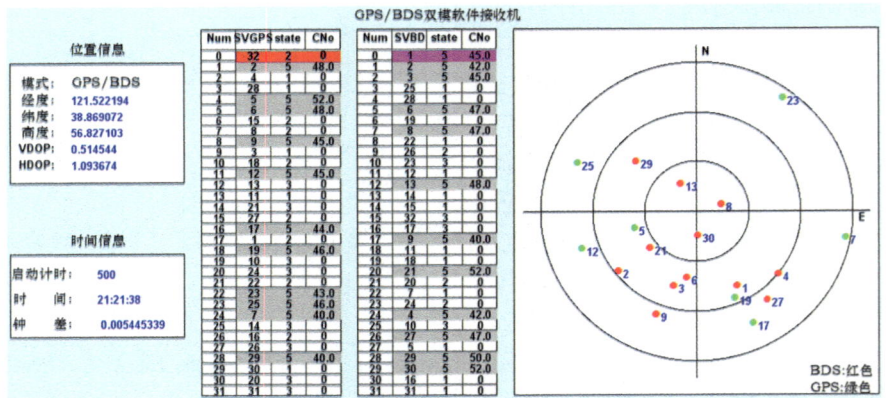

Fig. 5. GPS/BDS dual-mode positioning results

The DOP paramerters are defined as geometry factors that relate parameters of the user position and time bias errors to those of pseudrange errors. It only relates to the geometric distribution of visual satellites. In general, if there are more visual satellites, the corresponding DOP value will be smaller. DOP values are dimensionless. In general, they multiply the range errors corresponding to the position errors.

DOP values of GPS/BDS single-mode positioning and GPS/BDS dual-mode positioning are shown in Fig. 6. Figure 6a shows position dilution of precision (PDOP) of three kinds of positioning mode. Figure 6b shows horizontal dilution of precision (HDOP), and Figure 6c shows vertical dilution of precision (VDOP).

From Fig. 6, although the number of the visible satellites of BDS is larger than those of GPS, as the proportion of the GEO satellites in the tracked satellites is high, the satellite geometry distribution is poor. Therefore, DOP is large and the positioning

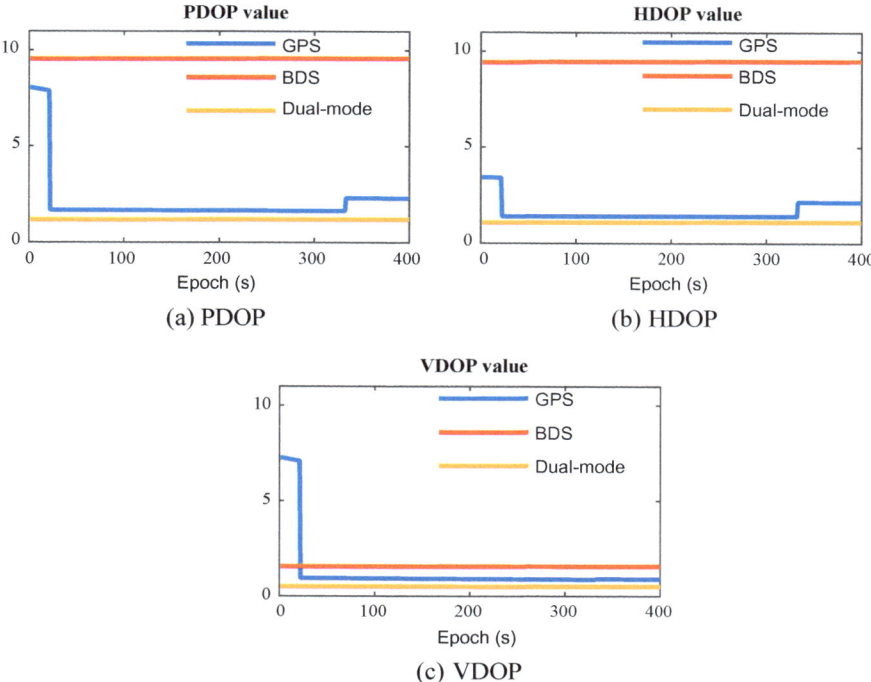

(a) PDOP

(b) HDOP

(c) VDOP

Fig. 6. DOP value of GPS/BDS dual-mode positioning system and single positioning system

accuracy is bad in BDS single-mode positioning. Meanwhile in the GPS single-mode positioning, the geometry distribution of the tracked satellites is better than BDS, and DOP value is relatively small. Additionally, the DOP values of GPS/BDS dual-mode positioning are apparently smaller compared to GPS/BDS single-mode positioning, which reveals that the satellite geometry distribution is further enhanced by the combination of GPS and BDS. As there are nearly 10 GPS satellites in this experiment, GPS positioning is well-geometry distribution. The DOP values of GPS/BDS dual-mode positioning are close to the DOP values of GPS positioning.

5   Conclusion

The pseudorange fusion algorithm is proposed to estimate the receiver's position in the GPS/BDS dual-mode positioning. In order to verify the proposed algorithm, the GPS/BDS software receiver based on C++ has been designed and implemented. This software receiver can achieve GPS and BDS single-mode positioning and BDS/GPS dual-mode positioning. Compared the position results of the dual-mode positioning output from the software receiver in the experiments, it can be concluded that the DOP of the GPS/BDS positioning mode is smaller and the position result of the GPS/BDS dual-mode positioning is more accurate.

Acknowledgments. This research is partially supported by the Chinese National Science Foundation (No. 61501079 and 61231006), Doctoral Scientific Research Starting Foundation of Liaoning Province (No. 2017011243 and No. 2017011149), Remote Sensing Youth Science and Technology Innovative Research, Dalian Technology Star Program, Foundation of Liaoning Educational Committee (No. L2015059) and the Fundamental Research Funds for the Central Universities (No. 3132018182 and 3132016317).

References

1. Zhang P, Zeng Q, Zhu X, Pei L. Research progress of Global Navigation Satellite System software receiver. J Hebei Univ Sci Technol. 2016;3:220–9.
2. Juang J, Tsai C, Chen Y. Development of a PC-based software receiver for the reception of Beidou navigation satellite signals. J Navig. 2013;66:701–18.
3. Lu Y. Beidou/GPS dual-mode software receiver principle and implementation technology. Beijing: Publishing House of Electronics Industry; 2016.
4. Li X, Ge M, Dai X. Accuracy and reliability of multi-GNSS real-time precise positioning: GPS, GLONASS, BeiDou, and Galileo. J Geodesy. 2015;89(6):607–35.
5. Kim G, So H, Jeon S, Kee C, Cho Y, Choi W. The development of modularized post processing GPS software receiving platform. In: 2008 international conference on control, automation and systems, Seoul; 2008. p. 1094–8.
6. Zhang Z. Design and research of GPS software receiver and its key technologies based on VC++. Tianjin: Nankai University; 2013.
7. Wang Q. The design and implementation positioning algorithm for GPS/BD2 dual-mode receiver. Southeast University; 2016.
8. Yu C, Qin H, Jin T. Design of weighted least squares positioning algorithm for BD2/GPS based on CNO. In: China Satellite Navigation Academic annual conference; 2015.
9. Afifi A, El-Rabbany A. Improved dual frequency PPP model using GPS and BeiDou observations. J Geodetic Sci. 2017;7(1):1–8.

An Improved RSA Algorithm for Wireless Localization

Jiafei Fu[1]([✉]), Jingyu Hua[1], Zhijiang Xu[1], Weidang Lu[1],
and Jiamin Li[2]

[1] College of Information Engineering, Zhejiang University of Technology,
Hangzhou 310023, China
hanmaozi@hotmail.com, eehjy@163.com
[2] National Mobile Communications Research Lab, Southeast University,
Nanjing 210096, China

Abstract. Wireless localization has become a hot issue in Internet of things, but the none-line-of-sight (NLOS) propagation will degrade the performance of traditional localization algorithms. Therefore, this paper proposed an improved range scaling algorithm (RSA) in the wireless sensor networks, where we use a two-step improvement to enhance the constrained optimization model. Simulations demonstrate that the proposed algorithm outperforms the compared algorithms, and effectively suppress the localization error caused by the none-line-of-sight propagation.

Keywords: Wireless location · None-line-of-sight propagation · Range scaling algorithm (RSA) · Quadratic programming

1 Introduction

Wireless localization had become a hot topic in recent years. Consequently, the positioning technology had become an important issue in the field of Internet of things (IoT) or communication networks [1].

In previous studies, the positioning schemes included methods based on time of arrival (TOA) [2], angle of arrival (AOA) [3], time difference of arrival (TDOA) [4], and received signal strength (RSS) [5]. However, the traditional location technology did not work effectively in practical wireless networks corrupted by NLOS error and measurement noise, in which the NLOS error acted as the major factor for the bad positioning performance [6].

There had been many studies on the elimination of NLOS effects, mainly including two kinds of methods. The first kind is to eliminate NLOS by appropriate weighting or optimization, such as approximate maximum likelihood algorithm [7], residual weighted algorithm [8], and quadratic programming algorithm [9]. The second kind is to identify NLOS BSs and then use only LOS BSs for localization; viz. the NLOS propagation is recognized [10].

This paper proposes an improved algorithm based on range scaling algorithm (RSA) [11]. This improved algorithm introduces a new constraint condition, and then,

Q. Liang et al. (eds.), *Communications, Signal Processing, and Systems*, Lecture Notes in Electrical Engineering 516,
https://doi.org/10.1007/978-981-13-6504-1_147

the quadratic programming is implemented in two steps. The first step is to use the original RSA algorithm to estimate a rough location and determine which area this rough position belongs to. The second step is to construct a tighter distance constraint based on this rough location region. Finally, we obtain the optimal solution according to this optimal model. Simulation results show that the proposed algorithm yields more accurate positioning performance than other compared algorithms.

2 Measurement Distance Model

We take into account the location algorithm based on TOA position estimation, and then, the distance between BS and MS is expressed as:

$$r_i = \sqrt{(x - x_i)^2 + (y - y_i)^2}, \quad i = 1, \ldots, N \tag{1}$$

where (x_i, y_i) and (x, y) are the coordinate of the ith BS and the MS coordinate of MS position to be solved. If the corresponding measurement distance is R_i, the RSA method uses the following scaling relation:

$$r_i = \alpha_i R_i \tag{2}$$

where α_i represents the scaling factor, which satisfies $0 < \alpha_i < 1$ due to the influence of NLOS error.

Combining Eqs. (1) and (2), we obtain the following expression:

$$(x - x_i)^2 + (y - y_i)^2 = \alpha_i^2 R_i^2, i = 1, \ldots, N \tag{3}$$

We can define the weight vector as $\mathbf{v} = [v_1, \ldots, v_N]^{\mathrm{T}} = [\alpha_1^2, \ldots, \alpha_N^2]^{\mathrm{T}}$ to simplify the next derivations.

If the weight vector is accurately known, then the MS position estimation (x, y) will be solved perfectly according to Eq. (3). Hence, how to obtain the accurate weight vector is the key issue as indicated in [11].

3 The Two-Step Optimization-Based RSA Algorithm

3.1 New Tighter Distance Constraint

From [12], we can derive the lower limit of $\mathbf{v}$, i.e.,

$$\mathbf{v}_{\min} = \{ \, \alpha_{1,\min}^2 \quad \cdots \quad \alpha_{N,\min}^2 \, \}^{\mathrm{T}} \tag{4}$$

where $\alpha_{i,\min}$ obeys,

$$
\left\{
\begin{aligned}
\alpha_{1,\min} &= \max\left\{\frac{L_{1,2} - R_2}{R_1}, \frac{L_{1,3} - R_3}{R_1}, \cdots, \frac{L_{1,N} - R_N}{R_1}\right\} \\
&\qquad\cdots \\
\alpha_{N,\min} &= \max\left\{\frac{L_{1,N} - R_1}{R_N}, \frac{L_{2,N} - R_2}{R_N}, \cdots, \frac{L_{N,N-1} - R_{N-1}}{R_N}\right\}
\end{aligned}
\right.
\tag{5}
$$

where $\max(\bullet)$ means the maximum operation, and $L_{i,j}$, $i \neq j$, refers to the distance between ith BS and jth BS.

In practice, the value of v_i will not beyond one, viz. $\mathbf{v}_{\max} = [1, 1, \ldots, 1]^{\mathrm{T}}$. Hence, the original constraint in [11] is expressed as follows:

$$
\mathbf{v}_{\min} \leq \mathbf{v} \leq \mathbf{v}_{\max}
\tag{6}
$$

In fact, the original constraint above is too loose to eliminate NLOS error enough. In order to suppress the NLOS error completely, we will refine the distance constraint next.

If MS is located in the region of {BS1, BS2, BS3} of Fig. 1a, i.e., MS is in the triangle surrounded by BS1, BS2 and BS3, and denoting that the distance between neighboring BSs is R, we have the inequality (7).

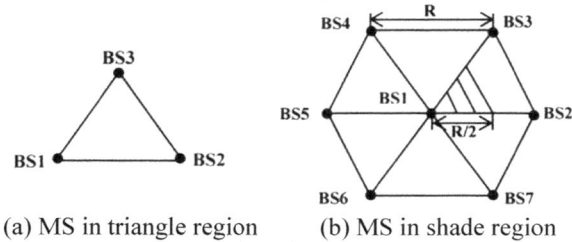

(a) MS in triangle region   (b) MS in shade region

Fig. 1. Seven-BS topology

$$
r_i \leq R, \quad i = 1, 2, 3
\tag{7}
$$

Otherwise, the distance measurement of other BSs must obey the following constraint:

$$
r_i \leq 2R, \quad i = 4, 5, 6, 7
\tag{8}
$$

Moreover, we take into consideration the MS position which is within the regular hexagon as shown in Fig. 1b, and then, we obtain the tighter distance constraint

$$\begin{cases} 0 \le r_1 \le R \\ \dfrac{R}{2} \le r_i \le R, \quad i = 2,3 \\ \dfrac{\sqrt{3}R}{2} \le r_i \le \dfrac{\sqrt{7}R}{2}, \quad i = 4,7 \\ R \le r_i \le \dfrac{3R}{2}, \quad i = 5,6 \end{cases} \tag{9}$$

In fact, Fig. 1b and formula (9) mean that MS belongs to a center BS (BS1 in this example). If we choose BS1 as the center base station, the inequalities above are determined. Finally, we can construct the optimal model with tighter constraints so long as the MS region is recognizable, which will be solved in Sect. 3.2.

3.2   The Two-Step RSA Algorithm

We utilize a two-step processing in our study. First, an original RSA is operated to find the coarse MS region, and then, the second step operated the RSA with refined distance constraint; finally, an improved localization performance can be obtained.

The optimization model of RSA at the first step is as follows:

$$\underset{\mathbf{v}}{\text{Minimize}} \quad F(\mathbf{v}) = \sum_{k=1}^{d} \left| (x - x_k)^2 + (y - y_k)^2 \right| \tag{10}$$

$$\text{s.t.} \quad \mathbf{v}_{\min} \le \mathbf{v} \le \mathbf{v}_{\max}, \ \text{MS} \in \text{FR}$$

where FR denotes the feasible region and (x_k, y_k) is the vertex of FR. We denote the vertex set as $\mathbf{X}_k = \{(x_k, y_k)\}_{k \in \{1,\ldots,d\}}$ with d representing the number of vertexes.

Furthermore, we expand formula (3)

$$\begin{cases} v_1 R_1^2 - x_1^2 - y_1^2 = x^2 + y^2 - 2x_1 x - 2y_1 y \\ \qquad \cdots \\ v_N R_N^2 - x_N^2 - y_N^2 = x^2 + y^2 - 2x_N x - 2y_N y \end{cases} \tag{11}$$

Equation (11) can be written into a matrix form

$$\mathbf{Y} = \mathbf{Ax} \tag{12}$$

where $\mathbf{Y} = \begin{bmatrix} v_1 R_1^2 - x_1^2 - y_1^2 \\ v_2 R_2^2 - x_2^2 - y_2^2 \\ \cdots \\ v_N R_N^2 - x_N^2 - y_N^2 \end{bmatrix}$, $\mathbf{A} = \begin{bmatrix} -2x_1, -2y_1, x^2 + y^2 \\ -2x_2, -2y_2, x^2 + y^2 \\ \cdots \\ -2x_N, -2y_N, x^2 + y^2 \end{bmatrix}$, $\mathbf{X} = \begin{bmatrix} x \\ y \\ 1 \end{bmatrix}$.

Then, we can solve the optimal solution of $\mathbf{x}$ on the basis of the least square method.

Taking $\hat{\mathbf{x}}$ as the position estimation, we rewrite the cost function of RSA as:

$$F(\mathbf{v}) = \sum_{k=1}^{d} \text{norm}(\hat{\mathbf{x}} - \mathbf{X}_k)^2 \tag{13}$$

Finally, from formula (9) and (13), we revise the new optimization model at the second step as:

$$\begin{array}{ll}
\underset{\mathbf{v}}{\text{Minimize}} & F(\mathbf{v}) \\[2mm]
\text{s.t.} & \begin{cases}
\mathbf{v}_{\min} \le \mathbf{v} \le \mathbf{v}_{\max}, \ \text{MS} \in \text{FR} \\[2mm]
\alpha_1 r_1 \le R \\[2mm]
\dfrac{R}{2} \le \alpha_i r_i \le R, \ i = 2, 3 \\[2mm]
\dfrac{\sqrt{3}R}{2} \le \alpha_i r_i \le \dfrac{\sqrt{7}R}{2}, \ i = 4, 7 \\[2mm]
R \le \alpha_i r_i \le \dfrac{3R}{2}, \ i = 5, 6
\end{cases}
\end{array} \tag{14}$$

4 Simulation and Analysis

Assume that MS is uniformly distributed in the shade of Fig. 1b with the classical seven-BS topology, i.e., the BS locates at $(0,0), (R,0), \left(\frac{R}{2}, \frac{\sqrt{3}R}{2}\right), \left(-\frac{R}{2}, \frac{\sqrt{3}R}{2}\right), (-R,0),$ $\left(-\frac{R}{2}, -\frac{\sqrt{3}R}{2}\right), \left(\frac{R}{2}, -\frac{\sqrt{3}R}{2}\right)$. In our study, $R = 1000$ m is employed. In addition, we deal with the measurement distance R_i in terms of two main errors: the measurement error (d_{measure}) and the NLOS error (d_{NLOS}), i.e.,

$$R_i = r_i + d_{\text{NLOS}} + d_{\text{measure}} \tag{15}$$

where d_{NLOS} is uniformly distributed in MIN and MAX, and d_{measure} is a Gaussian variable with zero mean and standard deviation (SD) of 10 m. Moreover, the original RSA algorithm [11], the linear line of position (LLOP) algorithm [13], and the CLS algorithm [14] are used to compare with the proposed algorithm.

4.1 Influence of NLOS Error

We fixed the lower limit of NLOS at MIN = 100 m and study the effect of MAX on positioning performance.

Figure 2 shows that the performance of the proposed algorithm is improved about 12% compared to the original RSA algorithm. Meanwhile, the proposed algorithm outperforms other two comparison algorithms. Moreover, with the increase in MAX, the performance advantage is more obvious. However, the larger MAX leads to the worse accuracy.

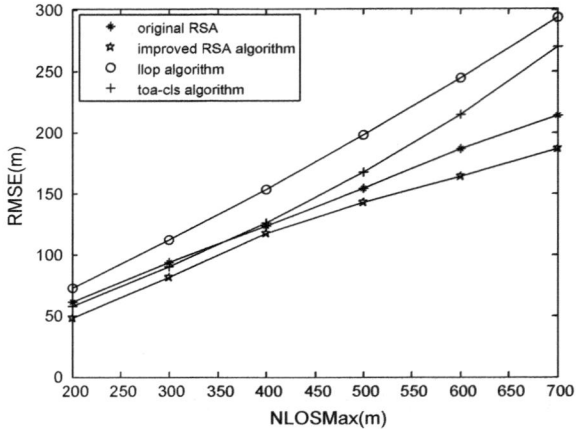

Fig. 2. Localization accuracy versus MAX: MIN = 200 m

4.2   Effect of BS Number

Here, we observe the cumulative probability distribution function (CDF) in Fig. 3.

Explicitly from Fig. 3, we catch the sight of a significant improvement on positioning performance from 3BS topology to 7BS topology. In detail, the proposed algorithm outperforms other three conventional algorithms in all cases, whose CDF of 250 m error approximately equals 0.86 (5BS) and 0.96 (7BS).

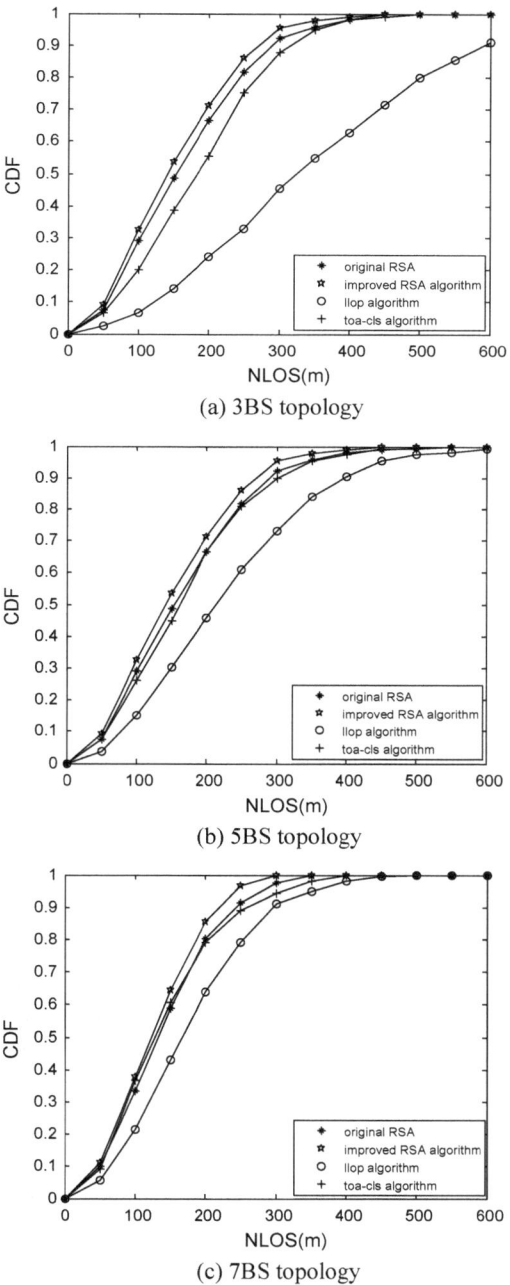

Fig. 3. CDF variations versus different BS numbers

5   Conclusions

It is a key issue to eliminate NLOS error in wireless positioning accuracy. This paper constructs the new distance constraints on the basis of two-step RSA algorithm. Simulation results indicated that the proposed algorithm can effectively eliminate the NLOS error, and the algorithm performance is superior to the conventional algorithm.

Acknowledgments. This paper was sponsored by the National Natural Science Foundation of China under Grant No. 61471322.

References

1. Spirito MA. On the accuracy of cellular mobile station location estimation. IEEE Trans Veh Technol. 2001;50(3):674–85.
2. Ke W, Wu L. Constrained least squares algorithm for TOA-based mobile location under NLOS environments. In: International conference on wireless communications, networking and mobile computing, Beijing; 2009. p. 1–4.
3. Kong F, Wang J, Zheng N, et al. A robust weighted intersection algorithm for target localization using AOA measurements. In: IEEE advanced information management, communicates, automation control conference, Xian; 2016. p. 23–8.
4. Compagnoni M, Pini A, Canclini A, et al. A geometrical–statistical approach to outlier removal for TDOA measurements. IEEE Trans Signal Process. 2017;65(15):3960–75.
5. Popleteev A. Indoor localization using ambient FM radio RSS fingerprinting: a 9-month study. In: 2017 IEEE international conference on computer and information technology (CIT), Helsinki; 2017. p. 128–34.
6. Caffery JJ, Stuber GL. Subscriber location in CDMA cellular networks. IEEE Trans Veh Technol. 1998;47(2):406–16.
7. Gazzah L, Najjar L, Besbes H. Improved hybrid AML algorithm without identification of LOS/NLOS nodes. In: IEEE Vehicular Technology Conference; 2015. p. 1–5.
8. Chen PC. A non-line-of-sight error mitigation algorithm in location estimation. In: IEEE wireless communications and networking conference; 1999. p. 316–20.
9. Zeyuan LI. Constrained weighted least squares location algorithm using received signal strength measurements. China Commun. 2016;13(4):81–8.
10. Muqaibel AH, Landolsi MA, Mahmood MN. Practical evaluation of NLOS/LOS parametric classification in UWB channels. In: IEEE international conference on communications; 2013. p. 1–6.
11. Zheng Z, Hua J, Jiang B, et al. A Novel NLOS mitigation and localization algorithm exploiting the optimization method. Chin J Sens Actuators. 2013;26(5):722–7.
12. Venkatraman S, Caffery JJ, You HR. A novel TOA location algorithm using LOS range estimation for NLoS environments. IEEE Trans Veh Technol. 2004;53(5):1515–24.
13. Zheng X, Hua J, Zheng Z, et al. LLOP localization algorithm with optimal scaling in NLOS wireless propagations. In: IEEE international conference on electronics information and emergency communication, Beijing; 2014. p. 45–8.
14. Wang X, Wang Z, O'Dea B. A TOA-based location algorithm reducing the errors due to non-line-of-sight (NLOS) propagation. J China Inst Commun. 2001;52(1):112–6.

A Quadratic Programming Localization Based on TDOA Measurement

Guangzhe Liu[✉], Jingyu Hua, Feng Li, Weidang Lu,
and Zhijiang Xu

College of Information Engineering, Zhejiang University of Technology,
Hangzhou 310023, China
1599731854@qq.com

Abstract. With the popularity of smart devices, applications based on location services have been widely used, and wireless positioning technology can provide accurate positioning information. However, due to the effect of non-line-of-sight (NLOS) errors, the performance of the system can drop significantly. Accordingly, this paper introduces the theory of quadratic programming optimization based on the research of the time difference of arrival (TDOA) theory and proposes an optimization algorithm that can effectively suppress the influence of NLOS error. Simulation results show that compared with other common wireless location algorithms, the proposed algorithm has more reliable positioning accuracy under different environment models and has better system stability.

Keywords: Wireless localization · The time difference of arrival (TDOA) ·
Quadratic programming · Non-line-of-sight propagation

1 Introduction

In recent years, with the rapid development of wireless sensor networks, wireless positioning technology has become a research hot spot and is widely used in drone navigation, disaster rescue, and environmental monitoring [1]. The traditional localization algorithm has high accuracy when the radio wave is propagating in the ideal environment, but in the actual environment, the signal is affected by obstacles during the propagation process to generate multiple refractions and reflections. Adding non-line-of-sight (NLOS) propagation errors can severely impair the positioning accuracy of the algorithm [1]. Not only that, but the measurement error also produces negative gain.

At present, the use of wireless communication networks for location technology research has achieved a lot of results, and some advanced algorithms have been applied to production practices. The existing positioning algorithm mainly estimates the position of the mobile station through the time of arrival (TOA) [2], angle of arrival (AOA) [3], time difference of arrival (TDOA) [4], and received signal strength (RSS) [5].

© Springer Nature Singapore Pte Ltd. 2020
Q. Liang et al. (eds.), *Communications, Signal Processing,
and Systems*, Lecture Notes in Electrical Engineering 516,
https://doi.org/10.1007/978-981-13-6504-1_148

In order to further reduce the impact of the NLOS error, the literature [6] uses the PDF of the adaptive kernel density estimation error and uses the Newton iteration scheme to improve the positioning accuracy. The strategy proposed in [7] is to first identify the NLOS error and mitigate it, then using the Kalman filter to estimate the location. Fascista [8, 9] proposed a new change-detection algorithm based on the generalized likelihood ratio test (GLRT) to improve the accuracy of position estimation by detecting the arrival angle of a specific signal.

Time difference of arrival (TDOA) positioning may be one of the widely used techniques for signal source positioning, and it has higher accuracy than other conventional algorithms [10, 11]. This article proposes a TDOA optimization strategy that solves the quadratic programming problem by combining the geometric relations between MN and SN and combining optimization theory. Simulations show that the proposed algorithm is obviously superior to the original TDOA algorithm, and it has advantages over other commonly used algorithms.

2 Original TDOA Positioning Algorithm

The TDOA algorithm is an improvement of the TOA algorithm. Assuming that there are N base stations and one base station serves as a reference, the difference between the measurement distances of other base stations and the reference base station can be expressed as:

$$d_i = R_i - R_1, \quad i = 2, 3, \ldots, N \tag{1}$$

where R_i is the measurement distance between the ith base station and the mobile station. Assuming the coordinate of the ith base station is (x_i, y_i), the mobile station is (x_{ms}, y_{ms}), we have

$$R_i = \sqrt{(x_{ms} - x_i)^2 + (y_{ms} - y_i)^2} \tag{2}$$

So, according to Eqs. (1) (2) I can have

$$d_i = \sqrt{(x_{ms} - x_i)^2 + (y_{ms} - y_i)^2} - \sqrt{(x_{ms} - x_1)^2 + (y_{ms} - y_1)^2} \tag{3}$$

Writing (3) in matrix form gives

$$\mathbf{KX} = \mathbf{Y} \tag{4}$$

where $\mathbf{K} = 2 \begin{bmatrix} x_2 - x_1 & y_2 - y_1 & d_2 \\ x_3 - x_1 & y_3 - y_1 & d_3 \\ \vdots & \vdots & \vdots \\ x_N - x_1 & y_N - y_1 & d_N \end{bmatrix}, \quad \mathbf{Y} = \begin{bmatrix} \left[-d_2^2 + (x_2 - x_1)^2 + (y_2 - y_1)^2 \right] \\ \left[-d_3^2 + (x_3 - x_1)^2 + (y_3 - y_1)^2 \right] \\ \vdots \\ \left[-d_N^2 + (x_N - x_1)^2 + (y_N - y_1)^2 \right] \end{bmatrix}$

and $\mathbf{X} = [x_{\mathrm{ms}} - x_1, y_{\mathrm{ms}} - y_1, R_1]^T$.

Finally, according to the least squares theory, we can get the estimated value of $\mathbf{X}$,

$$\hat{\mathbf{X}} = \left(\mathbf{K}^T \mathbf{K} \right)^{-1} \mathbf{K}^T \mathbf{Y} \tag{5}$$

In the case of less environmental interference, the traditional TDOA algorithm has higher accuracy, but when the error is serious, the performance of the algorithm is significantly worse and is no longer applicable.

3   Quadratic Programming to Optimize TDOA Algorithm

3.1   Optimization Model

In order to alleviate the impact of errors in the actual situation, especially the NLOS error which has the most significant side effects [12, 13], we use a quadratic programming model to optimize the algorithm. Assuming there is a linear relationship between the measured distance and the actual distance, it can be expressed as:

$$R_i = \lambda_i \cdot r_i \tag{6}$$

where r_i indicates the measurement distance between the ith base station and the mobile station. Obviously,

$$d_i = \lambda_i r_i - \lambda_1 r_1 \tag{7}$$

Combining Eqs. (4) and (8), we have

$$(x_i - x_1)(x_{\mathrm{ms}} - x_1) + (y_i - y_1)(y_{\mathrm{ms}} - y_1) = \frac{1}{2} \left(\lambda_1^2 r_1^2 - \lambda_i^2 r_i^2 + G \right) \tag{8}$$

where $G = (x_i - x_1)^2 + (y_i - y_1)^2$. And writing (9) in matrix form gives

$$\mathbf{AX} = \mathbf{Y} \tag{9}$$

where $\mathbf{A} = \begin{bmatrix} x_2 - x_1 & y_2 - y_1 \\ x_3 - x_1 & y_3 - y_1 \\ \vdots & \vdots \\ x_N - x_1 & y_N - y_1 \end{bmatrix}, \mathbf{Y} = \frac{1}{2} \begin{bmatrix} \lambda_1^2 r_1^2 - \lambda_2^2 r_2^2 + G \\ \lambda_1^2 r_1^2 - \lambda_3^2 r_3^2 + G \\ \vdots \\ \lambda_1^2 r_1^2 - \lambda_N^2 r_N^2 + G \end{bmatrix}, \mathbf{X} = \begin{bmatrix} x_{\mathrm{ms}} - x_1 \\ y_{\mathrm{ms}} - y_1 \end{bmatrix}.$

Therefore, we can get the LS solution of $\hat{\mathbf{X}}$, and the position estimate of the mobile station is obtained.

3.2   Constraint and Cost Function

In an actual cellular network, the measurement distance between the BS and MS must be greater than the actual distance due to the influence of NLOS error, i.e., $\lambda_i \subset (0, 1)$. For convenience, we define $\mathbf{v}$ as

$$\mathbf{v} = [v_1, v_2, \ldots, v_N] = [\lambda_1^2, \lambda_2^2, \ldots, \lambda_N^2] \tag{10}$$

where $\mathbf{v}_{\text{max}} = [1, 1, \ldots, 1]$.

From [14], in the real cellular network, the lower limit of the λ_i should satisfy the following conditions

$$\lambda_{i,\text{min}} = \max\left\{\frac{l_{i,j} - r_j}{r_j} \middle| i, j \in [1, N], i \neq j\right\} \tag{11}$$

where max ($\bullet$) represents the maximum operation. Therefore,

$$\mathbf{v}_{\text{min}} = [\lambda_{1,\text{min}}, \lambda_{2,\text{min}}, \ldots, \lambda_{N,\text{min}}] \tag{12}$$

Then, we can get the constraint of $\mathbf{v}$

$$\mathbf{v}_{\text{min}} \leq \mathbf{v} \leq \mathbf{v}_{\text{max}} \tag{13}$$

Second, we design the following optimization cost

$$\mathrm{F}(\mathbf{v}) = \sum_{i=2}^{N} \left| \left(\text{norm}\left(\hat{X} - \mathrm{BS}_i\right) - \text{norm}\left(\hat{X} - \mathrm{BS}_1\right)\right) - \left(\mathbf{v}_i r_i^2 - \mathbf{v}_1 r_1^2\right) \right| \tag{14}$$

where norm (x) means the norm of vector x. When the value of $\mathbf{v}$ is closer to the real situation, it is obvious that the cost function is closer to zero.

Now the optimization model can be derived as

$$\begin{aligned} &\underset{\mathbf{v}}{\text{minimize}} \quad \mathrm{F}(\mathbf{v}) \\ &\text{subject to } \mathbf{v}_{\text{min}} \leq \mathbf{v} \leq \mathbf{v}_{\text{max}} \end{aligned} \tag{15}$$

Through the quadratic programming function, the final solution can be obtained by solving the above equations.

4   Simulation and Analysis

Assume that the reference base station coordinates are $(0, 0)$, and other four base stations located at $\left(\frac{R}{2}, \frac{\sqrt{3}}{2}R\right)$, $\left(-\frac{R}{2}, \frac{\sqrt{3}}{2}R\right)$, $\left(-\frac{R}{2}, -\frac{\sqrt{3}}{2}R\right)$, $\left(\frac{R}{2}, -\frac{\sqrt{3}}{2}R\right)$. R represents the side length of the cell, i.e., $R = 1000\,\text{m}$ in our study. In this paper, measurement distance errors are mainly caused by non-line-of-sight propagation ($\varepsilon_{\text{nlos}}$) and measurement error (δ_m). Therefore, the distance r_i can be expressed as

$$r_i = R_i + \varepsilon_{\text{nlos},i} + \delta_{m.i} \tag{16}$$

where $\varepsilon_{\text{nlos}}$ is uniformly distributed between 100 m and MAX, and δ_m is a zero-mean Gaussian with a standard deviation of 20 m.

In order to better reflect the superiority of the proposed algorithm, we compare it with the original TDOA algorithm [15], opt-LLOP [16] algorithm, and TDOA two-step maximum likelihood algorithm (TSML) [17]. We independently operate each simulation for 1000 times.

4.1   Influence of NLOS Error

In order to investigate the impact of NLOS error values on the performance of the algorithm, we set the value range from 100 to MAX. In this simulation, the measurement error variance is set to 20 m, and there are a total of 5 base stations. In order to show the performance of algorithm positioning intuitively, we use the root mean square error (RMSE) to represent.

Figure 1 shows that compared with the initial algorithm, the proposed algorithm has improved the performance greatly, and the optimization effect is obvious. Also, it can be seen that with the increase of NLOS error, the performance of the proposed algorithm has the best performance among these groups of algorithms. Therefore, the proposed algorithm can effectively resist the influence of NLOS errors while ensuring higher positioning accuracy.

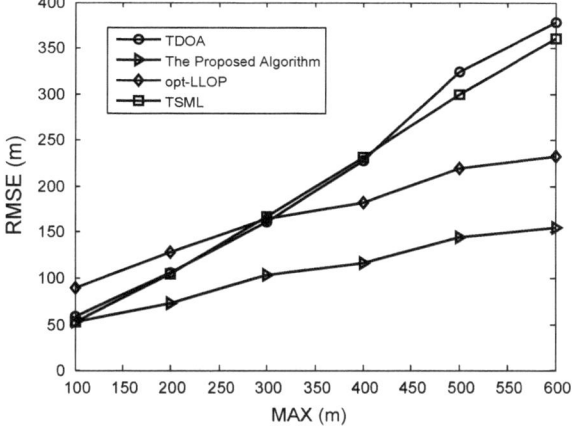

Fig. 1. RMSE variations versus different MAXs

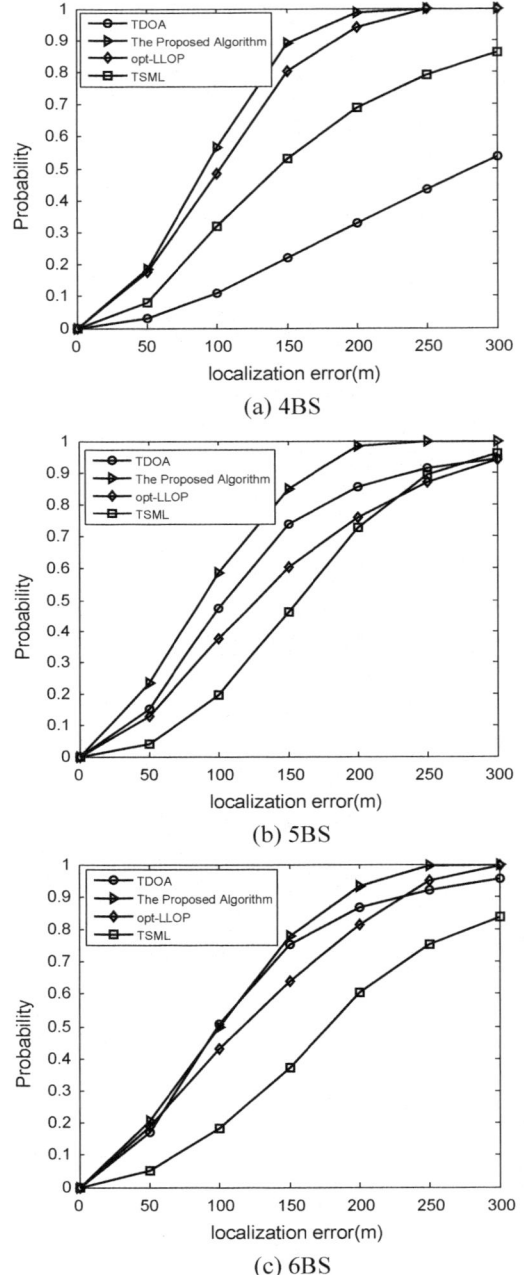

(a) 4BS

(b) 5BS

(c) 6BS

Fig. 2. CDF of each algorithm with different BS number

4.2   Effect of BS Number

In order to explore the effect of the number of base stations on the performance of the algorithm, we control the NLOS error with a value of 300 m and a measurement error of 20 m, and the comparison is shown by cumulative distribution function (CDF)

Figure 2 shows the influence of BS number. It can be seen that the topology formed by the number of different base stations has different degrees of influence on the performance of these algorithms. Obviously, the proposed algorithm has the best positioning accuracy with different number of base stations.

5   Conclusion

In an actual wireless communication network, the most important factor affecting the positioning algorithm performance is the non-line-of-sight propagation error, so how to suppress its influence is the key to research. This paper proposes a new algorithm that combines TDOA and quadratic programming theory. Simulations demonstrate that compared with other common algorithms, the proposed algorithm has a higher ability to resist NLOS errors and better positioning accuracy.

Acknowledgments. This paper was sponsored by the National Natural Science Foundation of China under grant No. 61471322.

References

1. Li H. Study of wireless sensor network applications in network optimization. Sens Transducers. 2013;157(10):180–9.
2. Vaghefi RM, Amuru SD, Buehrer RM. Improving mobile node tracking performance in NLOS environments using cooperation. In: IEEE international conference on communications; 2015. p. 6595–600.
3. Xu W, Quitin F, Leng M, et al. Distributed localization of a RF target in NLOS environments. IEEE J Sel Areas Commun. 2015;33(7):1317–30.
4. Kireev A, Fokin G, Al-odhari AHA. TOA measurement processing analysis for positioning in NLOS conditions. In: Systems of signals generating and processing in the field of on board communications; 2018. p. 1–4.
5. Wang Y, Ho KC. An asymptotically efficient estimator in closed-form for 3-D AOA localization using a sensor network. IEEE Trans Signal Process. 2015;14(12):6524–35.
6. Kim R, Ha T, Lim H, et al. TDOA localization for wireless networks with imperfect clock synchronization. In: International conference on information networking; 2014. p. 417–21.
7. Gholami MR, Vaghefi RM, Ström EG. RSS-based sensor localization in the presence of unknown channel parameters. IEEE Trans Signal Process. 2013;61(15):3752–9.
8. Feng Y, Fritsche C, Gustafsson F, et al. TOA-based robust wireless geolocation and Cramér-Rao lower bound analysis in harsh LOS/NLOS environments. IEEE Trans Signal Process. 2013;61(9):2243–55.
9. Long C, Wang Y, et al. A mobile localization strategy for wireless sensor network in NLOS conditions. China Commun. 2016;13(10):69–78.

10. Fascista A, Ciccarese G, Coluccia A, Ricci G. A change-detection approach to mobile node localization in bounded domains. In: Conference on information sciences and system; 2015. p. 1–6.
11. Martin RK, Yan C, Fan HH, et al. Algorithms and bounds for distributed TDOA-based positioning using OFDM signals. IEEE Trans Signal Process. 2011;59(3):1255–68.
12. Qi Y, Kobayashi H, Suda H. Analysis of wireless geolocation in a non-line-of-sight environment. IEEE Trans Wirel Commun. 2006;5(3):672–81.
13. Caffery JJ, Stuber GL. Subscriber location in CDMA cellular networks. IEEE Trans Veh Technol. 1998;47(2):406–16.
14. Venkatraman S, Caffery JJ, You HR. A novel TOA location algorithm using LOS range estimation for NLOS environments. IEEE Trans Veh Technol. 2004;53(5):1515–24.
15. Cheung KW, So HC, Ma WK, et al. A constrained least squares approach to mobile positioning: algorithms and optimality. EURASIP J Adv Signal Process. 2006. https://doi.org/10.1155/asp/2006/20858.
16. Zheng X, Hua J, Zheng Z, et al. LLOP localization algorithm with optimal scaling in NLOS wireless propagations. In: Proceedings of IEEE international conference on electronics information and emergency communication; 2014. p. 45–8.
17. Chan YT, Ho KC. A simple and efficient estimator for hyperbolic location. IEEE Trans Signal Process. 2002;42(8):1905–15.

Study on Indoor Combined Positioning Method Based on TDOA and IMU

Chaochao Yang, Jianhui Chen[✉], Xiwei Guo, Deliang Liu,
and Yunfei Shi

Missile Engineering Department, Army Engineering University of PLA,
Shijiazhuang Campus, Shijiazhuang 050003, China
ddjqzd@126.com, 1415651643@qq.com

Abstract. This paper studies an indoor positioning method combining wireless sensor network (WSN) and inertial navigation system (INS). Because the positioning error of INS increases with time, the long-term positioning accuracy is poor, so the combination uses the wireless sensor network to measure the distance between unknown node and base station by TDOA method. The dead-reckoning data of inertial measurement unit (IMU) and the distance information of TDOA method are transmitted to the processing terminal, and then the particle filter algorithm is used to smooth the data to obtain the position estimation. The cumulative positioning error of INS is corrected, and the non-line-of-sight (NLOS) error in TDOA positioning method is reduced. The experimental results show that compared with the single TDOA localization method, the accuracy of the combined positioning method is higher.

Keywords: Indoor positioning · TDOA · IMU · Particle filter

1 Introduction

With the deepening of the construction of smart city, the accuracy of location service, especially indoor positioning system, is higher and higher. The importance of indoor positioning in hospitals, shopping malls, and supermarkets is self-evident. The outdoor positioning system is now very mature, such as the US GPS and China's BeiDou Navigation Satellite System (BDS), which can provide efficient and accurate outdoor location services. However, electromagnetic waves can be blocked or reflected by buildings and cannot enter the room, causing the moving objects to receive no satellite signals, so they cannot be located indoors.

At present, the commonly used indoor positioning technologies are ultrasound, Bluetooth, WIFI, radio frequency identification, ZigBEE and so on. Ultra-wideband (UWB) technology is actually a kind of bandwidth that can reach up to GHz, and the transmission rate reaches hundreds of Mbit/s, while the communication distance is about 100 m [1]. The signal bandwidth of UWB system is very wide, so the time resolution is very high, which can reach the nanosecond level, which can reach the

© Springer Nature Singapore Pte Ltd. 2020
Q. Liang et al. (eds.), *Communications, Signal Processing, and Systems*, Lecture Notes in Electrical Engineering 516,
https://doi.org/10.1007/978-981-13-6504-1_149

positioning accuracy of centimeter-level, which is its biggest advantage. However, due to the influence of indoor complex environment, the UWB signal can be refracted and reflected, resulting in the appearance of the non-horizon error (NLOS error). The inertial navigation system (INS) mainly includes accelerometers and gyroscopes, which measure the acceleration a and angular velocity, respectively. It belongs to calculate navigation method, that is, starting from the position ω of known points, through the continuous measurement of sports course angle and acceleration of body, the location of the below point is deduced, and body motion can be continuous measurement of the current position. However, due to the integration of the navigation information, the positioning error increases with time and the long-term accuracy is poor.

This article adopts the method of UWB and INS positioning, which uses the coordinates calculated value of UWB positioning as the main indoor localization method, uses INS correction under the condition of non-line-of-sight its positioning error, and improves the positioning accuracy of integrated positioning system.

2 Combination Positioning System Structure

The mobile node needs to obtain two kinds of information when realizing its positioning, the first is the measurement of inertial parameters, and the second is the estimate of the distance to the base station. For inertial information, using MEMSIC IMU380ZA—409, it was nine degrees of freedom inertial sensor board, can be measured, respectively, three axial linear acceleration and angular rate, there are three optional magnetometers, nine inertial parameters can be based [2].

Mobile node in a moment, on the basis of the optimal estimate position, mainly using UWB system orientation, under the condition of the stadia, precision can reach 5 and 16 cm [3], and under the condition of non-line-of-sight, due to reflection, diffraction and multipath effects, the positioning results will have large errors, then the IMU navigation a calculated as auxiliary positioning method. The self-navigation features of inertial navigation system can effectively compensate for the deficiency of UWB positioning system. Mobile node will be according to the features of UWB positioning signals to judge the moment is in the horizon or non-line-of-sight environments, if is stadia in UWB positioning, if non-line-of-sight, enable the INS system, because the INS positioning accuracy is very high in the short time. The structure diagram of the combined positioning system is shown in Fig. 1.

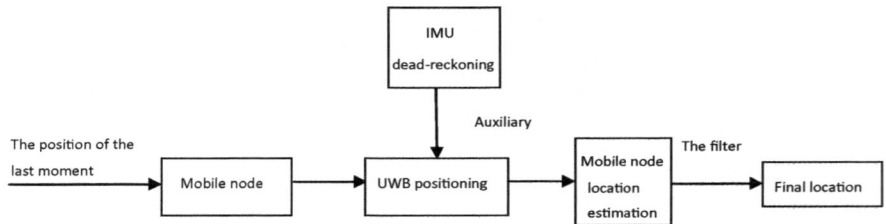

Fig. 1. Structure diagram of the combined positioning system

3 Combination Localization Algorithm

In indoor positioning, the common UWB-based location algorithm has time of arrival (TOA), time difference of arrival (TDOA), angle of arrival (AOA), received signal strength indicator (RSSI), etc. TOA technology through the measured signal transmission time to calculate the distance to the base station, this needs to be strictly accurate time synchronization between two communication nodes, but in the current hardware conditions it is difficult to achieve this goal; AOA ranging technology relies on installing antenna arrays on nodes to obtain angle information, requiring special hardware devices such as antenna arrays or antennas to support them. Signal strength and the space correlation are very strong, according to the intensity of waves in an indoor change model and the measured signal strength can determine the mobile end position, but most of these model parameters are estimated, and the accuracy will or changes in the environment and reduce with the increase of the scope, so the applicable scope is limited.

Different from TOA, TDOA don't need to two of the base station signal absolute time, instead of using time difference positioning, by measuring the time of arrival in the base station signal, can determine the distance of the signal source. Therefore, it only needs the clock synchronization between the base station, and the hardware requirements are low. By comparing the time difference of the signal to multiple base stations, the hyperbola with the focal length of the base station and the distance difference is a long axis, and the intersection point of the hyperbola is the position of the signal.

As shown in Fig. 2, BS1, BS2, BS3 for positioning the base station, the location is known and expressed as $(x1, y1)$, respectively, $(x2, y2)$, $(x3, y3)$, to estimate the position of MS is unknown nodes (x, y), d_i represents the distance between the ith positioning base station and the unknown node [4].

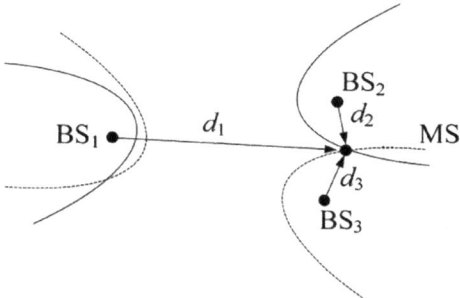

Fig. 2. Schematic diagram of hyperbolic rendezvous positioning

The mathematical relationship is as follows:

$$d_i = \sqrt{(x_i - x)^2 + (y_i - y)^2} \tag{1}$$

$$d_{ij} = d_i - d_j = ct_{i,j} \tag{2}$$

c represents the speed of the light, $t_{i,j}$ represents the measured value of TDOA.

$$d_{21} = d_2 - d_1 = \sqrt{(x_2 - x)^2 + (y_2 - y)^2} - \sqrt{(x_1 - x)^2 + (y_1 - y)^2} \tag{3}$$

$$d_{31} = d_3 - d_1 = \sqrt{(x_3 - x)^2 + (y_3 - y)^2} - \sqrt{(x_1 - x)^2 + (y_1 - y)^2} \tag{4}$$

Equation (3) can be translated into:

$$(d_{21} + d_1)^2 = d_2^2 \tag{5}$$

$$d_{21}^2 + 2d_1 d_{21} = d_2^2 - d_1^2 = x_2^2 + y_2^2 - x_1^2 - y_1^2 - 2(x_2 - x_1)x - 2(y_2 - y_1)y \tag{6}$$

Making $z_i^2 = x_i^2 + y_i^2$, Eq. (6) can be translated into:

$$(x_2 - x_1)x + (y_2 - y_1)y = -d_1 d_{21} + \frac{1}{2}\left(z_2^2 - z_1^2 - d_{21}^2\right) \tag{7}$$

Similarly, the Eq. (4) can be simplified to:

$$(x_3 - x_1)x + (y_3 - y_1)y = -d_1 d_{31} + \frac{1}{2}\left(z_3^2 - z_1^2 - d_{31}^2\right) \tag{8}$$

Simultaneous (7) and (8) are expressed as:

$$Ax_m = Bd_1 + C \tag{9}$$

Among them

$$A = \begin{bmatrix} x_2 - x_1, y_2 - y_1 \\ x_3 - x_1, y_3 - y_1 \end{bmatrix} \quad B = \begin{bmatrix} -d_{21} \\ -d_{31} \end{bmatrix}$$

$$C = \frac{1}{2}\begin{bmatrix} z_2^2 - z_1^2 - d_{21}^2 \\ z_3^2 - z_1^2 - d_{31}^2 \end{bmatrix} \quad x_m = \begin{bmatrix} x \\ y \end{bmatrix}$$

The location of the test node MS is

$$x_m : x_m = A^{-1}Bd_1 + A^{-1}C \tag{10}$$

For the inertial navigation system, assuming that the moving node moves in a two-dimensional plane, the method is based on the dead-reckoning method:

$$\begin{cases} P_{dr}(k+1) = P_{dr}(k) + v_k\Delta t + \dfrac{1}{2}a_{k+1}\Delta t^2 \\ v_{k+1} = v_k + a_{k+1}\Delta t \\ \gamma_{k+1} = \gamma_k + \omega_{k+1}\Delta t \end{cases} \tag{11}$$

Among them $P_{dr}(k+1)$, v_{k+1}, a_{k+1}, γ_{k+1}, respectively, represents the position, velocity, line acceleration, and angular deviation of $k+1$ moment.

Transform the local coordinate system to the reference frame and use the cosine formula:

$$\begin{bmatrix} a_{x_r} \\ a_{y_r} \end{bmatrix} = \begin{bmatrix} \cos\varphi, & -\sin\varphi \\ \sin\varphi, & \cos\varphi \end{bmatrix} \begin{bmatrix} a_{x_l} \\ a_{y_l} \end{bmatrix} \tag{12}$$

$\varphi = \int \omega_{z_r}\, dt$ means the angle between the local coordinate system and the reference frame.

Particle filter is a suboptimal Bayes filter, which provides a non-gaussian environment, with appropriate weights using a set of random samples (often referred to as particles) to represent the state of the posterior distribution [5]. Basic idea of this method is an important choice probability density of random sampling, to get the corresponding weights of random sample, on the basis of state observation to adjust the weight and the size of the particle's position, and then use these sample approximation of the state of the posterior distribution, with these sample weighting and finally as a state estimate.

The state transfer model $p(s_k|s_{k-1})$ and measurement model $p(z_k|s_k)$ are introduced, where s is the state variable of the system and z is the measurement value of the system. For non-linear and non-Gaussian processes, the model can be expressed as:

$$\begin{aligned} s_k &= f(s_{k-1}, v_{k-1}) \\ z_k &= h(s_k, n_k) \end{aligned} \tag{13}$$

where v_{k-1} and n_k are independent and distributed process noise and measurement noise. Our task is to estimate unknown state variable s_k according to the measured value $z_{1:k}$ of noise pollution, where $f()$ and $h()$ are known.

The posterior density of the system at the k moment can be approximately:

$$p(s_{0:k}|z_{1:k}) \approx \sum_{i=1}^{N_s} \omega_k^i \delta(s_{0:k} - s_{0:k}^i) \tag{14}$$

s_k^i is the state value with weight ω_k^i.

At this point, we get a discrete weighted estimation of the true posterior density function.

The option value is based on the importance sampling principle. In general, it's hard to sample from $p(x)$, let's say $p(x) \propto \pi(x)$. We might as well sample the density function $q(x)$ which is easy to sample, namely, $s_i \sim q(x)$, $i = 1, \ldots, N_s$, $q( )$ is called importance density. The system obeys the first order Markov process, $q(s_k|s_{0:k-1}, z_{1:k}) = q(s_k|s_{k-1}, z_k)$ is established, and the weight is fixed as:

$$\omega_k^i \propto \omega_{k-1}^i \frac{p(z_k|s_k^i)p(s_k^i|s_{k-1}^i)}{q(s_k^i|s_{k-1}^i, z_k)} \tag{15}$$

The posterior filtering density can be approximated as:

$$p(s_k|z_{1:k}) \approx \sum_{i=1}^{N_s} \omega_k^i \delta(s_k - s_k^i) \tag{16}$$

Unfortunately, this process leads to degradation. After a series of iterations, the weight of most particles becomes very small, almost trivial, and the weight is concentrated only on a few particles. Effective sample size is introduced.

$$\text{Neff} = \frac{1}{\sum_{i=1}^{N_s} (\omega_k^i)^2} \tag{17}$$

Here ω_k^i is the normalized weight.

In order to solve the degradation problem by resampling, its basic idea is to eliminate the particles with smaller weights and to reproduce the particles with larger weights according to their weight. The specific practice is to resample N_s times from an approximate discrete representation of $p(x_k|z_{1:k})$ and generate $p(x_k|z_{1:k})$ new set of particle $\{s_k^{i*}\}_{i=1}^{N_s}$, all weight of which are set to $\frac{1}{N_s}$.

Algorithm: Particle Filter

$$\left[\{s_k^i, \omega_k^i\}_{i=1}^{N_s}\right] = PF\left[\{s_{k-1}^i, \omega_{k-1}^i\}_{i=1}^{N_s}, z_k\right]$$

FOR $i = 1 : N_s$
- Draw $s_k^i \sim q(s_k|s_{k-1}^i, z_k)$
- Assign the particle a weight ω_k^i
- END FOR
- Calculate total weight: $t = \text{SUM}[\{\omega_k^i\}_{i=1}^{N_s}]$

- FOR $i = 1 : N_s$
- Normalize: $\omega_k^i = t^{-1}\omega_k^i$

END FOR

Calculate Neff

IF Neff $< N_T$

$\quad$ RESAMPLE $\left[\{s_k^i, \omega_k^i\}_{i=1}^{N_s}\right]$

END IF

4   Algorithm Test Results

In order to evaluate and compare the algorithms we proposed, we set the experiment in a rectangular corridor 20 m long and 3 m wide, with corresponding obstacles in the corridor [6]. The six base stations are fixed around the corridor, essentially covering the path of the moving target. We then tested and recorded the estimated results of each algorithm in 48 different locations, using the cumulative distribution function (cdf) to observe the error patterns of all scenarios. Generally, as expected, the TDOA+IMU method performs better than the pure TDOA and inertial navigation methods. After the horizontal coordinate normalization processing, 0–0.5 is captured, as shown in Fig. 3.

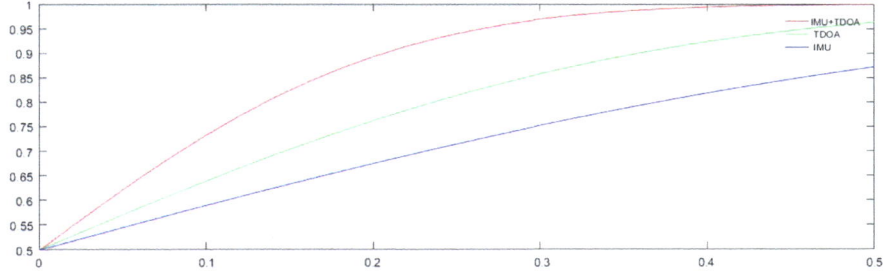

Fig. 3. Experimental result graph

5   Conclusion

In this paper, using a mobile sensor node to the environment of TDOA measurements in combination with the direction and distance from the inertial measurement unit adopts the minimum mean square estimation and maximum a posteriori estimation to get position estimation. TDOA positioning is suitable for passive positioning and does not require a signal time stamp. We make full use of the advantages of UWB positioning and inertial navigation system positioning to complement each other, reduce UWB non-line-of-sight error, and reduce the accelerometer and gyroscope drift and magnetometers, so as to improve the positioning accuracy.

According to the estimate of the unknown nodes with two stations distance through the simulation model of measurement noise, using the minimum mean square estimate and maximum a posteriori estimation algorithm, finally obtains the optimal estimation of unknown node position and suboptimal estimate position. Simulation and experiment show the effectiveness and robustness of this algorithm.

Acknowledgments. This work is supported by the National Natural Science Foundation of China under Grant 61601494.

References

1. Cruz O, Ramos E, Ramírez M. 3D indoor location and navigation system based on Bluetooth. In: 2011 21st international conference on electrical communications and computers (CONIELECOMP). IEEE; 2011.
2. Duan Z, Cai Z. Adaptive particle filter for unknown fault detection of wheeled mobile robots. In: Proceedings of the 2006 IEEE/RSJ international conference on intelligent robots and systems; 2006. p. 1312–15.
3. Hol JD, Dijkstra F, Luinge H. Tightly coupled UWB/IMU pose estimation. In: 2009 IEEE international conference on ultra-wideband; 2009. p. 688–92.
4. Kok M, Hol JD, Schön TB. Indoor positioning using ultrawideband and inertial measurements. IEEE Trans Veh Technol. 2015;64:1293–303.
5. Hellmers H, Norrdine A, Blankenbach J. An IMU/magnetometer-based Indoor positioning system using Kalman filtering. In: Indoor Positioning and Indoor Navigation (IPIN); 2013. p. 1–9.
6. Megalingam RK, Rajendran AP, Dileepkumar D. LARN: implementation of automatic navigation in indoor navigation for physically challenged. In 2012 annual IEEE India conference (INDICON). IEEE; 2012.

Research on the Fast Direction Estimation and Display Method of Far-Field Signal

Rong Liu[1], Jin Chen[1(✉)], Lei Yan[2], Ying Tong[1], Kai-kai Li[1], and Chuan-ya Wang[1]

[1] Tianjin Key Laboratory of Wireless Mobile Communications and Power Transmission, Tianjin Normal University, Tianjin, China
cjwoods@163.com
[2] Beijing Aerospace Measurement and Testing Technology Institute, Beijing, China

Abstract. In the array signal processing, the direction of arrival (DOA) is a very important parameter. It can be used to estimate the spatial parameter or source location of signals, so it has been deeply studied by scholars and has been widely applied in radar detection, underwater operation, and mobile communication. In this paper, an array signal DOA estimation and display system are proposed. It adopts the fast beamforming algorithm, which is an improved algorithm based on the MVDR algorithm. By comparing the output power of different positions, the angle of the arrival of the signal to the array can be obtained. The system can manually set the basic parameters, such as the number of sensors, the frequency of sampling, the parameter of the filter, the transmission speed of the sound wave, and so on. Then, the waveform characteristics of the time and frequency domain of the original acquisition signal and the filtered signal are displayed intuitively, and the real-time waveform of the signal is displayed at the same time. Through theoretical analysis and simulation experiments, we can get that the related algorithm in this system is not only suitable for microphone array, but also can meet the application requirements of radar monitoring, underwater operation, and other fields.

Keywords: Array signal processing · Direction of arrival (DOA) estimation · Beamforming algorithm

1 Introduction

The direction of arrival estimation (DOA) is a very important parameter in the spatial spectrum estimation of the array signal processing. It has been widely used in radar, communication, sonar, earthquake, astronomy, and other scientific and technological fields, which are widely used [1]. In this paper, an array signal DOA estimation and display system are proposed. First, the analog far-field signal is received, and then the DOA is estimated after filtering. In this system, the basic parameters can be set according to the actual conditions, such as the number of sensors, the spacing of the array element, the sampling frequency of the signal, the filter parameters, and so on. In

© Springer Nature Singapore Pte Ltd. 2020
Q. Liang et al. (eds.), *Communications, Signal Processing, and Systems*, Lecture Notes in Electrical Engineering 516, https://doi.org/10.1007/978-981-13-6504-1_150

this way, many applications can be simulated, such as the acoustic signal receiving of the microphone, the radar, and the underwater operation. The system can directly display the time–frequency characteristics of the original acquisition signal, including waveform and phase. After the signal arrives, it will be filtered by the FIR filter to become a narrowband signal. The time and frequency characteristics of the filter can also be displayed in the system. The filtering effect can be easily seen through the comparison of the frequency-domain characteristics of the original signal and the filtered signal. After the basic parameters are set up, we can see the real-time waveform of the signal from a certain direction after running the system. Through theoretical analysis and simulation experiments, the feasibility of the system in the fields of microphone, radar, and underwater operation can be proved.

2 Signal Processing Model

In different application domains, the distance between the source and the receiving array is different. According to the size of the distance, the signal processing model can be divided into two parts: near-field model and far-field model. In the near-field model, the distance between the source and the receiving array is small, and the magnitude of the received signals is quite different. As shown in Fig. 1, the wavefront of the array at this time is the wavefront model of the sphere. The far-field model is the opposite. The distance between the source and the receiving array is far away, so the difference of the amplitude of the signal received by the array element can be ignored. At this time, the array wavefront can be regarded as the front model of the plane, as shown in Fig. 2. The signal received in this paper can be regarded as a far-field signal, and it can be considered that the amplitude of the signals received by each element is the same, which greatly reduces the complexity of the calculation.

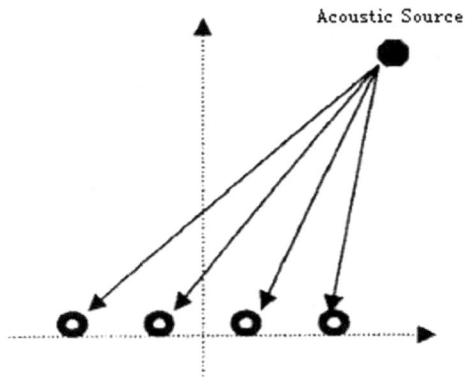

Fig. 1. Near-field model

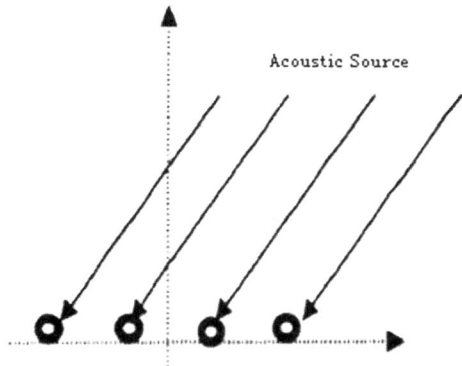

Fig. 2. Far-field model

Assuming that there are arrays of M elements, N far-field signals are received at the same time and then become N far-field narrowband signals after filtering. The signal can be expressed as:

$$S_i(t) = u_i(t)e^{j(w_0 t + \varphi(t))} \tag{1}$$

Then,

$$S_i(t - \tau) = u_i(t - \tau)e^{j(w_0(t-\tau) + \varphi(t-\tau))} \tag{2}$$

Among them, w_0 is the frequency of the signal, $\varphi(t)$ is the phase of the signal, and $u_i(t)$ is the amplitude of the signal. Since the signal is far-field signal, it can be obtained as: $u_i(t - \tau) \approx u_i(t)$, and because the signal is a narrowband signal, it can be obtained as: $\varphi_i(t - \tau) \approx \varphi(t)$. Bring them into (2) and get the following:

$$S_i(t - \tau) \approx S_i(t)e^{-jw_0\tau}, i = 1, 2, \ldots, N; \tag{3}$$

The reception signal of the R element can be expressed as:

$$x_r(t) = \sum g_{ri}S_i(t - \tau_{ri}) + n_r(t), \quad r = 1, 2, \ldots, M; \tag{4}$$

g_{ri} represents the gain of the I signal received by the R element; τ_{ri} represents the delay difference of the I signal to the R element; $n_i(t)$ n represents the interference in the R array element.

Therefore, we can list the signals received by M elements at t time, that is:

$$\begin{bmatrix} x_1(t) \\ x_2(t) \\ \cdot \\ \cdot \\ \cdot \\ x_M(t) \end{bmatrix} = \begin{bmatrix} e^{-jw_0\tau_{11}} & e^{-jw_0\tau_{12}} & \cdots & e^{-jw_0\tau_{1N}} \\ e^{-jw_0\tau_{21}} & e^{-jw_0\tau_{22}} & \cdots & e^{-jw_0\tau_{2N}} \\ \cdot & \cdot & \cdot & \cdot \\ \cdot & \cdot & \cdot & \cdot \\ \cdot & \cdot & \cdot & \cdot \\ e^{-jw_0\tau_{M1}} & e^{-jw_0\tau_{M2}} & \cdots & e^{-jw_0\tau_{MN}} \end{bmatrix} \begin{bmatrix} s_1(t) \\ s_2(t) \\ \cdot \\ \cdot \\ \cdot \\ s_N(t) \end{bmatrix} + \begin{bmatrix} n_1(t) \\ n_2(t) \\ \cdot \\ \cdot \\ \cdot \\ n_M(t) \end{bmatrix}$$

If it is represented as a vector form, it can be obtained as:

$$X(t) = AS(t) + N(t) \tag{5}$$

Among them, $X(t)$ represents the data received by the matrix, A represents the steering vector array, $S(t)$ represents the data emitted by the source, and $N(t)$ represents the noise parameter [2].

3   Far-Field Signal Arrival Estimation and Display System

The system includes the parameter setting module, the far-field signal receiving module, the filtered signal display module, the DOA estimation module, and the real-time waveform display module. The system block diagram is shown in Fig. 3. First, the basic parameters should be set up, including the receiving array parameters, filter parameters, and so on. After the operation, the system will receive the far-field signal; here, the signal can come from the analog signal source and can also come from the actual signal source. These signals will then be filtered to make them narrowband signals. Then, the direction of arrival is estimated and the real-time waveform of the filtered signal will be displayed. The direction of arrival is determined by the difference of color and luminance in different regions of the waveform.

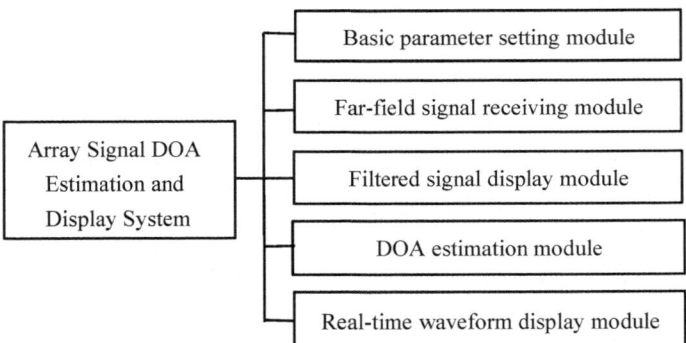

Fig. 3. System block diagram

3.1   Basic Parameter Setting Module

As shown in Fig. 4, the module is set for the basic parameters. Through this module, different parameters can be set according to various actual conditions, including setting parameters for receiving array: the number of sensors in the X-direction and the Y-direction, the spacing between the elements, the sampling frequency of the signal, and the intercepting frequency of the signal. The parameters of the FIR filter used for signal filtering, including its order, upper and lower cutoff frequencies, the gain fluctuation of the passband, and the attenuation multiple of the stopband, can also be set. In addition, it can also change the sound propagation speed V and determine the wavelength of the array with the frequency f. Setting up different parameters can simulate a variety of working environment, so the system is theoretically applicable to the sound reception of the microphone, radar signal detection, underwater operation, and many other fields.

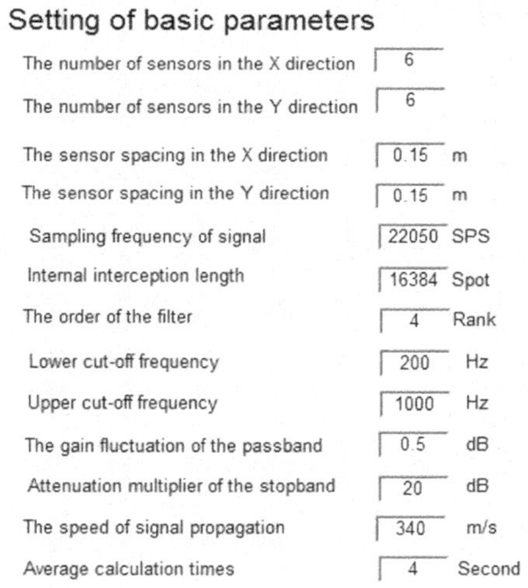

Setting of basic parameters

The number of sensors in the X direction	6
The number of sensors in the Y direction	6
The sensor spacing in the X direction	0.15 m
The sensor spacing in the Y direction	0.15 m
Sampling frequency of signal	22050 SPS
Internal interception length	16384 Spot
The order of the filter	4 Rank
Lower cut-off frequency	200 Hz
Upper cut-off frequency	1000 Hz
The gain fluctuation of the passband	0.5 dB
Attenuation multiplier of the stopband	20 dB
The speed of signal propagation	340 m/s
Average calculation times	4 Second

Fig. 4. Basic parameter setting module

3.2   Far-Field Signal Receiving Module

As shown in Fig. 5, it is the far-field signal receiving module, which can be analog or real signals. The parameters of the X-direction and the Y-direction are set according to the actual conditions. Through the "drawing" button, the time-domain waveform, frequency-domain phase, frequency-domain amplitude, and power density of the original far-field signal can be observed, respectively.

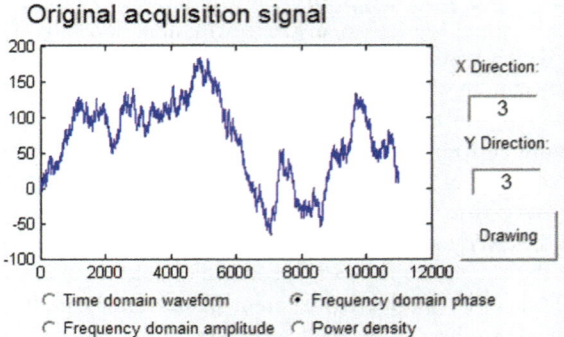

Fig. 5. Far-field signal receiving module

3.3 Filtered Signal Display Module

As shown in Fig. 6, it is the filtered signal display module. It shows the time–frequency characteristics of the narrowband signal formed by the original far-field signal filtered by the FIR filter. Similar to the far-field signal receiving module, this module can intuitively display the time-domain waveform, frequency-domain phase, frequency-domain amplitude, and power density of the filtered signal under different parameters. Compared with the original signal waveform in the frequency domain, it can clearly see its filtering effect. Compared with the original signal waveform in the frequency domain, the filtering effect can be displayed intuitively.

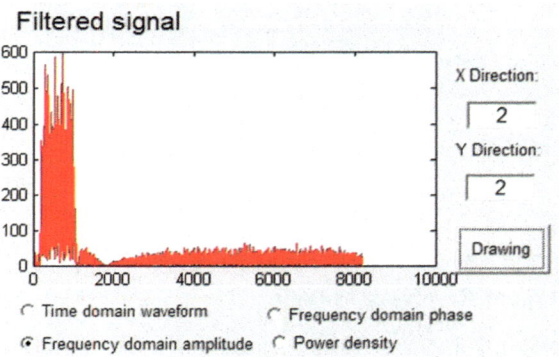

Fig. 6. Filtered signal display module

3.4 DOA Estimation Module

The module is the core module of the system. After processing the filtered data by correlation algorithm, the direction of arrival can be estimated. The beamforming algorithm is used here. The traditional beamforming algorithm is delay-and-sum

(DAS) beamforming. The distance between signal sources to various elements is different, so there will be delays, and each delay corresponds to a direction of arrival. The DAS algorithm is to apply the delay value set in advance to the output signal of each sensor and finally superimpose the output of all sensors. When the corresponding angle of the delay coincides with the direction of the signal source, the output power will reach the maximum value, thus obtaining the direction of the source [3]. The minimum variance beamforming used in this system is based on the DAS algorithm. Because in the DAS algorithm, the predetermined delay value is fixed and does not take into account the characteristics of the input signal. However, the delay in the minimum variance beamforming changes with the input signal, and finally, the output power is minimized without distortion. Finally, the power of the output signal will be represented by different brightness colors, so as to facilitate direct observation.

3.5 Real-Time Waveform Display Module

As shown in Fig. 7, it is the real-time waveform display module. After you have set the basic parameters, click "start," and the module will display the real-time waveform. The waveform is an upward view, where a rectangle window is intercepted. Since the signal is random, the graph is dynamic. The depth of the color and brightness represents the power of the signal in different directions. The deeper the color, the greater the signal power in the direction. By comparing the output power of various positions, the angle of signal arriving at the array can be obtained.

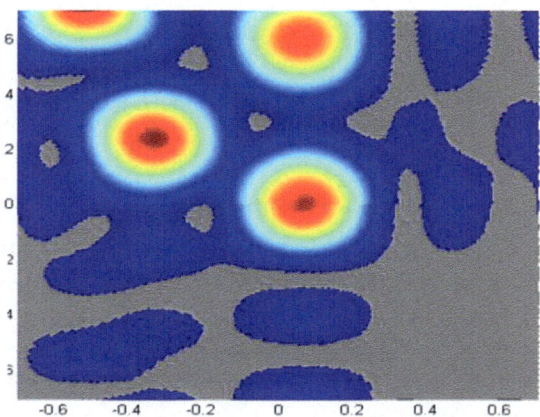

Fig. 7. Real-time waveform display module

3.6   Total Effect Diagram of the System

As shown in Fig. 8, the total effect diagram of the array signal DOA estimation and display system are presented.

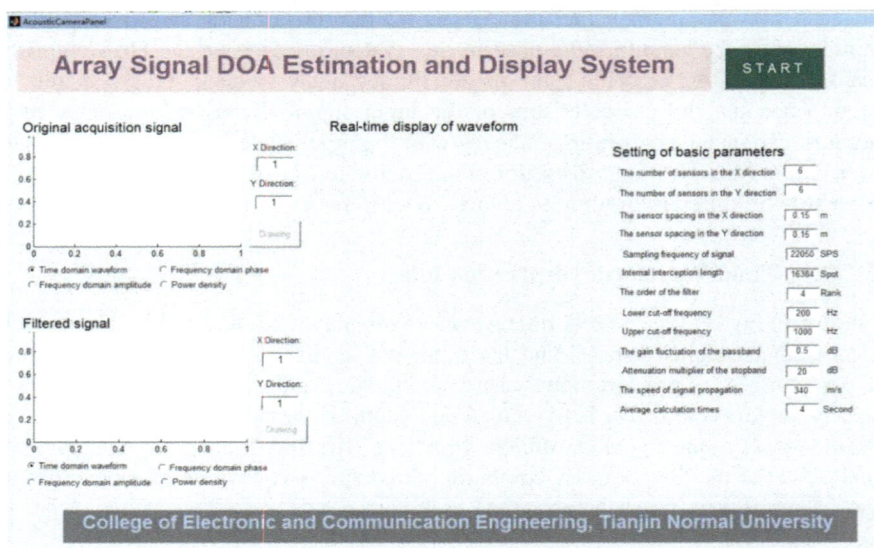

Fig. 8. Array signal DOA estimation and display system

4   Simulation

In order to verify the effectiveness of the system, the following simulation is carried out.

4.1   DOA Estimation Based on Acoustic Sensor

The velocity of the sound wave is $c = 340$ m/s, where the central frequency of the wave signal is set to 1500 Hz, and the wavelength of the obtained wave is 0.23 m according to the $v = \lambda f$. So the spacing of the array element is set to 0.1 m. A two-dimensional (2D) uniform microphone array with a received array of 8 * 8 arrays is set; the sampling frequency is set as 22.05 k SPS; the filter order is set as 4; the upper and lower limits of the cutoff frequency are set as 20 and 500 Hz, respectively; the passband gain is set to 0.5 dB, and the stopband attenuation multiplier is set to 20 dB. The average number of times that the incoming wave is processed by smoothing filtering is set to 4. After the above parameters are set, the system will get the result shown in Fig. 9.

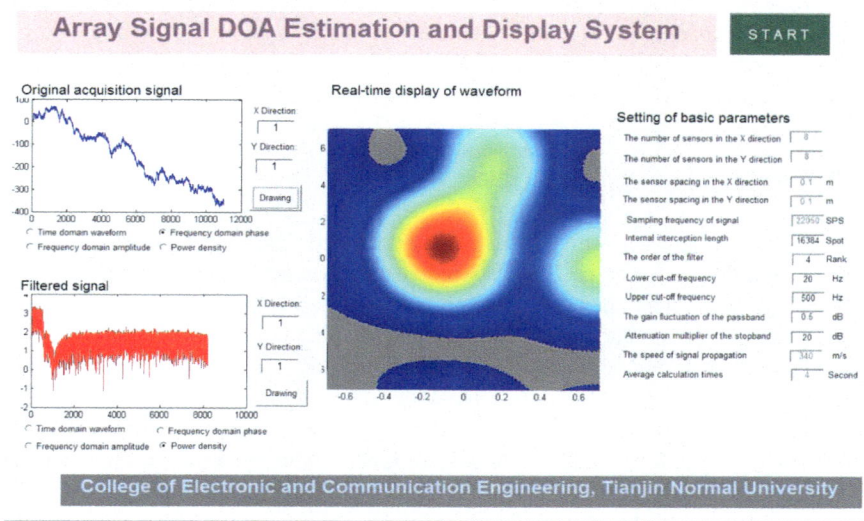

Fig. 9. DOA estimation based on acoustic sensor

4.2 DOA Estimation Based on Sonar Signal

Sonar consists mainly of active sonar and passive sonar. Active sonar mainly uses ultrasonic waves to work, and its working frequency is high, probably in the range of 3 K–97 KHz. The working frequency of passive sonar is relatively low, with a range of 3 Hz–97 KHz. The working environment of passive sonar is simulated. The center frequency of the wave is set as 1 kHz, and the transmission speed of the ultrasonic

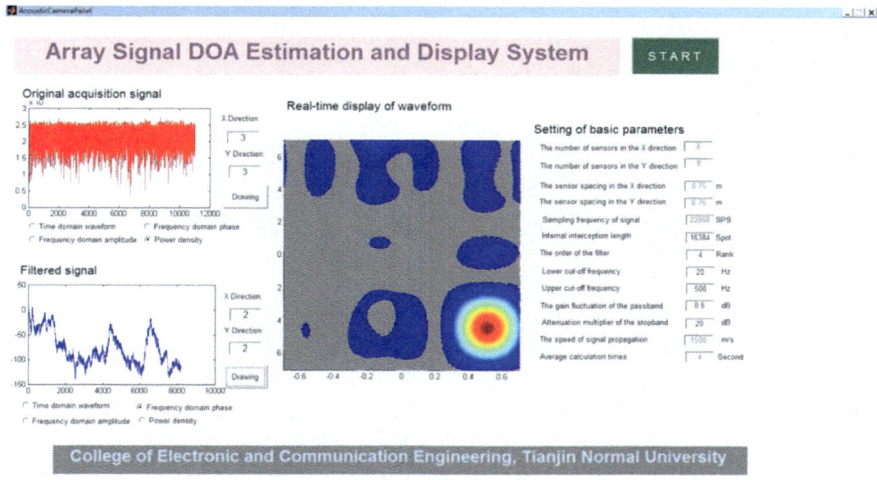

Fig. 10. DOA estimation based on sonar signal

wave is 1500 m/s. Thus, the wavelength of the wave is 1.5 m according to the formula $v = \lambda f$. So the spacing of the array element is 0.75 m, and the settings of other parameters are the same as those in 4.1. After setting the above parameters, the operation result is shown in Fig. 10.

5 Summary

This paper proposes a system which can estimate the DOA: the array signal DOA estimation and display system. The system can simulate various working environments by changing the values of basic parameters, such as acoustic signal reception of microphone array, radar, and sonar. After setting the parameters, the system can receive the analog far-field signal, and the filtered signal is estimated by DOA. Finally, the real-time waveform of the signal will be displayed. The output signal power of different positions can be judged by the difference of the color and luminance of various regions in the waveform graph, so that the direction of the incoming wave can be estimated.

Through theoretical analysis and simulation experiments, it can be proved that the system is feasible in the fields of microphone, radar, and underwater operation. However, there are still some problems that need to be further studied.

References

1. Yan FG, Rong JJ, Liu S, Shen Y, Jin M. Joint cross-covariance matrix based fast direction of arrival estimation. Syst Eng Electron. 2017;39:1–2.
2. Raimondi FED, Farias RC, Michel OJ, Comon P. Wideband multiple diversity tensor array processing. IEEE Trans Signal Process. 2017;65:5334–45.
3. Li A, Masouros C, Sellathurai M. Analog-digital beamforming in the MU-MISO downlink by use of tunable antenna loads. IEEE Trans Veh Technol. 2018;67:3114–27.

Compressive Sensing Approach for DOA Estimation Based on Sparse Arrays in the Presence of Mutual Coupling

Jian Zhang$^{(\boxtimes)}$, Zhenzhen Duan, Yang Zhang, and Jing Liang

School of Information and Communication Engineering, University of Electronic Science and Technology of China, Chengdu, China
zzzjian@outlook.com

Abstract. In the process of direction-of-arrival (DOA) estimation, the difference co-array of sparse arrays can achieve high degrees of freedom, which can be utilized to detect more signal sources than physical sensors based on spatial smoothing (SS) algorithm. In this paper, we present a method for DOA estimation using sparse signal recovery through compressive sensing (CS) approach in the presence of mutual coupling. Compared with SS algorithm, CS approach achieves a lower estimation error. Additionally, simulation results show that the estimation error of CS approach increases with the increase of mutual coupling. Also, it increases with the increase of the grid interval of the entire DOA space.

Keywords: DOA estimation · Sparse arrays · Spatial smoothing · Compressing sensing · Mutual coupling

1 Introduction

Direction-of-arrival (DOA) estimation plays a significant role in radar and wireless communications [1,2]. It is well known that uniform linear arrays (ULAs) can only detect $N - 1$ sources with N sensors using traditional methods like MUSIC [3] and ESPRIT [4]. Recently, many researchers have proposed sparse arrays constructions, which can achieve higher degrees of freedom and detect more signal sources than physical sensors. In [5], minimum redundancy arrays (MRAs) are proposed, but the disadvantage of the MRA is that there is no closed-form expression to calculate the MRA. In addition, in [6,7], the authors proposed nested arrays and coprime arrays. Based on SS algorithm [6,8], coprime arrays can detect $MN + M - 1$ sources with $N + 2M - 1$ sensors (M and N are coprime integers), and nested arrays can detect $N_2(N_1 + 1) - 1$ sources with N_1 elements of the dense ULA part and N_2 elements of the sparse ULA part. However, coprime arrays have holes in the difference co-array and nested arrays have higher mutual coupling than coprime arrays. Considering the mutual coupling,

© Springer Nature Singapore Pte Ltd. 2020
Q. Liang et al. (eds.), *Communications, Signal Processing, and Systems*, Lecture Notes in Electrical Engineering 516,
https://doi.org/10.1007/978-981-13-6504-1_151

Liu in [9] proposed super nested arrays, which can reduce mutual coupling by redistributing the elements of the dense ULA part of the nested array. Therefore, super nested arrays have higher estimation accuracy than nested arrays. As we know, the array aperture affects the estimation performance in the process of estimating DOA. Larger array aperture tends to have smaller estimation error [10,11]. SS algorithm [6,8] requires that the arrays are no-hole. Therefore, SS algorithm does not fully utilize the array aperture achieved in the difference co-array from the sparse arrays, and the utilizable degrees of freedom are approximately halved.

In this paper, we present sparse signal recovery through compressive sensing (CS) approach for DOA estimation in the presence of mutual coupling. The sensor arrays may not be continuous for CS approach, and CS approach fully utilizes the entire virtual array aperture and the degrees of freedom obtained in the difference co-array. Simulation results show that CS approach has a lower estimation error than that of SS algorithm.

The rest parts of this paper are organized as follows. In Sect. 2, we introduce the construction of coprime arrays, nested arrays, and super nested arrays. Then, in Sect. 3, CS approach are proposed. Section 4 shows simulation results of CS approach and SS algorithm. Conclusion and future work are provided in Sect. 5.

2   The Constructions of Three Categories of Sparse Arrays

Through the difference co-array of sparse arrays, we can obtain lots of virtual sensors, which can be utilized in DOA estimation. In this section, three categories of sparse arrays will be introduced.

2.1   Coprime Arrays

Coprime arrays can achieve lots of virtual sensors, which can be utilized in DOA estimation. Assume M and N are a coprime pair of positive integers with $M < N$; the locations of coprime arrays are

$$\mathbb{S} = \{0, M, 2M, \ldots, (N-1)M, N, 2N, \ldots, (2M-1)N\} \tag{1}$$

The number of physical sensors is $N + 2M - 1$, and the difference co-array is

$$\mathbb{D} = \pm \{Mn - Nm\}, 0 \leq n \leq N - 1, 0 \leq m \leq 2M - 1 \tag{2}$$

where $\mathbb{D}$ has $2MN + 2M - 1$ continuous elements between $-(MN + M - 1)$ and $(MN + M - 1)$. In Fig. 1a, it shows an example of coprime arrays with $M = 3$, $N = 5$, and $d = \lambda/2$. We can get the disadvantage of coprime arrays is that the difference co-array has holes.

2.2   Nested Arrays

Compared with coprime arrays, the difference co-array of nested arrays is no-hole. Assume there are N_1 elements in the dense ULA part and N_2 elements in the sparse ULA part. The locations of nested arrays are

$$\mathbb{S} = \{1, 2, \ldots, N_1, (N_1 + 1), 2(N_1 + 1), \ldots, N_2(N_1 + 1)\} \tag{3}$$

The number of physical sensors is $N_1 + N_2$, and the difference co-array is

$$\mathbb{D} = \{m(N_1 + 1) - n\} \tag{4}$$

where $1 \le m \le N_2$ and $1 \le n \le N_1$. We can obtain $2[(N_1 + 1)N_2 - 1] + 1$ continuous elements between $-(N_1 + 1)N_2 + 1$ and $(N_1 + 1)N_2 - 1$. In Fig. 1b, it shows an example of nested arrays with $N_1 = 5$, $N_2 = 5$, and $d = \lambda/2$. As the result of the inter-element spacing of the dense ULA part is small, the mutual coupling between sensors is very large.

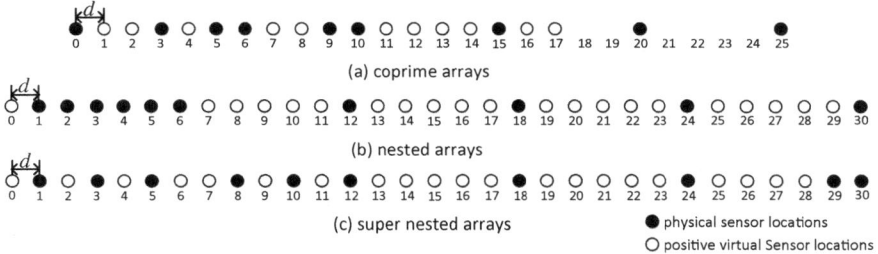

(a) coprime arrays

(b) nested arrays

(c) super nested arrays   ● physical sensor locations
                                  ○ positive virtual Sensor locations

Fig. 1. Three categories of sparse arrays structures for physical sensors is 10

2.3   Super Nested Arrays

The dense ULA part of nested arrays is the main origin of the mutual coupling. In order to alleviate mutual coupling between the sensors. The authors in [9] proposed super nested arrays, which remove some sensors in the dense ULA part and relocate them, and the difference co-array of super nested arrays is no-hole. Assume N_1 and N_2 are integers satisfying $N_1 \ge 4$ and $N_2 \ge 3$. The locations of nested arrays are

$$\mathbb{S} = X_1 \cup X_2 \cup X_3 \cup X_4 \cup X_5 \cup X_6 \tag{5}$$

where

$$X_1 = \{1 + 2\ell | 0 \le \ell \le A_1\}$$
$$X_2 = \{(N_1 + 1) - (1 + 2\ell) | 0 \le \ell \le B_1\}$$
$$X_3 = \{(N_1 + 1) + (2 + 2\ell) | 0 \le \ell \le A_2\}$$
$$X_4 = \{2(N_1 + 1) - (2 + 2\ell) | 0 \le \ell \le B_2\}$$
$$X_5 = \{\ell(N_1 + 1) | 2 \le \ell \le N_2\}$$
$$X_6 = \{N_2(N_1 + 1) - 1\}$$

$$(A_1, \ B_1, \ A_2, \ B_2) = \begin{cases} (r, \ r-1, \ r-1, \ r-2), & if \ N_1 = 4r, \\ (r, \ r-1, \ r-1, \ r-1), & if \ N_1 = 4r + 1, \\ (r+1, \ r-1, \ r, \ r-2), & if \ N_1 = 4r + 2, \\ (r, \ r, \ r, \ r-1), & if \ N_1 = 4r + 3, \end{cases}$$

Super nested arrays increase the spacing between physical sensors, and the difference co-array is equal to nested arrays. In Fig. 1c, it depicts an example of super nested arrays with $N_1 = 5$, $N_2 = 5$, and $d = \lambda/2$.

Based on the three categories of sparse arrays, we can utilize sparse signal recovery through CS approach to detect DOA. In the next section, CS approach is presented.

3   DOA Estimation with CS Approach in the Presence of Mutual Coupling

In this section, we present CS approach used in DOA estimation. Assume D narrowband sources with powers $[\sigma_1^2, \ldots, \sigma_D^2]$ impinge on the three categories of sparse arrays from direction $\theta_i, i = 1, 2, \ldots, D$. The locations of sensors are nd, where n belongs to the set $\mathbb{S}$ and $d = \lambda/2$, where λ is the wavelength of the signal. When mutual coupling is taken into account, the received signal of the arrays is

$$\mathbf{x}_{\mathbb{S}}[k] = \mathbf{CAs}[k] + \mathbf{n}[k] \tag{6}$$

where $\mathbf{C}$ is a mutual coupling matrix and can be approximated by a B-banded symmetric Toeplitz matrix in the ULA configuration [12,13] as follows:

$$\langle \mathbf{C} \rangle_{n_1, n_2} = \begin{cases} c_{|n_1 - n_2|}, & if \ |n_1 - n_2| \le B, \\ 0, & \text{otherwise,} \end{cases} \tag{7}$$

$n_1, n_2 \in \mathbb{S}$, and coupling coefficients $c_0, c_1, \ldots, c_B$ satisfy $1 = c_0 > |c_1| > \cdots > |c_B|$. $\mathbf{A} = [\mathbf{a}_{\mathbb{S}}(\theta_1), \mathbf{a}_{\mathbb{S}}(\theta_2), \ldots, \mathbf{a}_{\mathbb{S}}(\theta_D)]$ denotes the array manifold matrix, and $\mathbf{a}_{\mathbb{S}}(\theta_i) = e^{j(2\pi/\lambda)nd\sin\theta_i}$, $n \in \mathbb{S}$, $\mathbf{s}[k] = [s_1[k], s_2[k], \ldots, s_D[k]]$ is the source signal vector. The $\mathbf{n}[k]$ is assumed to be temporally and spatially white noise, which is also uncorrelated from the sources. The covariance matrix of $\mathbf{x}_{\mathbb{S}}[k]$ is

$$\mathbf{R}_{\mathbb{S}} = \sum_{i=1}^{D} \sigma_i^2 \mathbf{a}_{\mathbb{S}}(\theta_i) \mathbf{a}_{\mathbb{S}}^H(\theta_i) + \sigma^2 \mathbf{I} \tag{8}$$

where $\mathbf{a}_{\mathbb{S}}(\theta_i)\,\mathbf{a}_{\mathbb{S}}^H(\theta_i)$ is $e^{j\pi(n_1-n_2)\sin\theta_i}$, $n_1, n_2 \in \mathbb{S}$. Vectorizing $\mathbf{R}_{\mathbb{S}}$ yields

$$\mathbf{z} = \mathrm{vec}(\mathbf{R}_{\mathbb{S}}) = \mathbf{B}\mathbf{r} + \sigma_n^2 \mathbf{I}_n \tag{9}$$

where $\mathbf{B} = [\mathbf{a}_{\mathbb{D}}(\theta_1),\ldots,\mathbf{a}_{\mathbb{D}}(\theta_D)]$, $\mathbf{a}_{\mathbb{D}}(\theta_i) = e^{j\pi n \sin\theta_i}$, $n \in \mathbb{D}$, $\mathbf{r} = [\sigma_1^2, \sigma_2^2, \ldots, \sigma_D^2]$, and $\mathbf{I}_n = \mathrm{vec}(\mathbf{I})$. $\mathbf{z}$ is received signal of the virtual sensors arrays. The virtual source signal becomes a single snapshot of $\mathbf{r}$, and the rank of $\mathbf{z}$ is 1. Therefore, the authors in [6,7] proposed SS algorithm. However, SS algorithm sacrifices the array aperture and halves the utilizable degrees of freedom. In this paper, we present CS approach to estimate DOA based on the three categories of sparse arrays in the presence of mutual coupling. The objective function of CS approach is defined as

$$\hat{\mathbf{r}} = \arg\min_{\tilde{\mathbf{r}}} \|\tilde{\mathbf{r}}\|_1 \quad \text{s.t.} \left\|\mathbf{z} - \tilde{\mathbf{B}}\tilde{\mathbf{r}}\right\|_2 < \beta \tag{10}$$

We define β as follows:

$$\beta = \sqrt{\frac{\left(\mathbf{x}_L^2 - \|\mathbf{x}_{\mathbb{D}}\|_2\right)^2}{L}} \tag{11}$$

where $\mathbf{x}_L$ is the maximum in $\mathbf{x}_{\mathbb{D}}$ and L is the number of virtual sensors. $\tilde{\mathbf{B}} = \left[\mathbf{a}_{\mathbb{D}}\left(\tilde{\theta}_1\right),\ldots,\mathbf{a}_{\mathbb{D}}\left(\tilde{\theta}_Q\right)\right]$, $\tilde{\theta} = \left[\tilde{\theta}_1,\ldots,\tilde{\theta}_Q\right]$ denotes the entire DOA space region, $Q \gg D$. $\tilde{\mathbf{r}} = [r_1, r_2, \ldots, r_Q]^T$ is a D-sparse vector and $\tilde{\mathbf{r}}$ is expressed as

$$\tilde{\mathbf{r}} = \begin{cases} \sigma_i^2, \tilde{\theta}_i \in [\theta_1, \theta_2, \ldots, \theta_D] \\ 0, \tilde{\theta}_i \notin [\theta_1, \theta_2, \ldots, \theta_D] \end{cases} \tag{12}$$

We can solve (10) by using the convex optimization toolbox CVX [14].

4   Simulations and Analysis

In this section, we will compare the estimation error of CS approach and SS algorithm based on the three categories of sparse arrays in the presence of mutual coupling. In the simulation, we assume there are $D = 7$ uncorrelated sources coming from $[-45°, -35°, -25°, 0°, 10°, 20°, 40°]$. The number of the physical sensors is 10, and the physical sensors' locations of the three types sparse arrays are shown in Fig. 1. The root-mean-squared error (RMSE) is defined as (13). Monte Carlo simulation is 5000, and the grid interval θ^g is 1°. The mutual coupling is based on (7) with $c_1 = 0.3e^{j\pi/3}$, $B = 50$, and $c_\ell = c_1 e^{-j(\ell-1)\pi/4}/\ell$.

$$RMSE = \sqrt{\sum_{i=1}^{D}\left(\hat{\theta}_i - \theta_i\right)^2 \Big/ D} \tag{13}$$

Figure 2 shows RMSE of CS approach and SS algorithm for coprime arrays, nested arrays, and super nested arrays versus SNR. The red lines represent

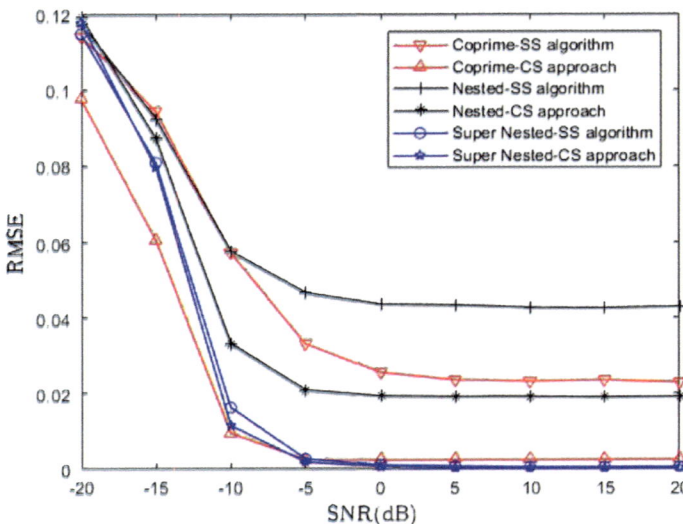

Fig. 2. RMSE of CS approach and SS algorithm in the presence of mutual coupling versus SNR; the number of snapshots is 300

coprime arrays, the black lines represent nested arrays, and the blue lines represent super nested arrays. It shows that the estimation error of CS approach is lower than that of SS algorithm for the three types sparse arrays, because CS approach utilizes all virtual aperture. With the increase of SNR, the estimation error of super nested arrays is the smallest in the presence of mutual coupling, and nested arrays is worst. Figure 3 shows RMSE of CS approach and SS algorithm versus the number of snapshots. We can draw the conclusion that RMSE decreases as the number of snapshots increases and the estimation error of CS approach is lower than that of SS algorithm.

RMSE of CS approach and SS algorithm versus the parameter $c(1)$ of the mutual coupling is plotted in Fig. 4. We choose coprime arrays as an example. It is quite obvious that the estimation performance heavily depends on the mutual coupling. RMSE is smallest if $c1$ is close to 0. The estimation error starts to increase dramatically for CS approach when $c1$ is larger than 0.24, and $c1$ is larger than 0.16 for SS algorithm. This phenomenon indicates that CS approach has a lower estimation error than that of SS algorithm in the presence of mutual coupling. Figure 5 shows RMSE of CS approach versus the grid interval θ^g. RMSE appears a rising trend with the increase of the grid interval θ^g. In order to get a lower estimation error, we trend to choose a smaller θ^g. However, the time needed to do one complete simulation will increase.

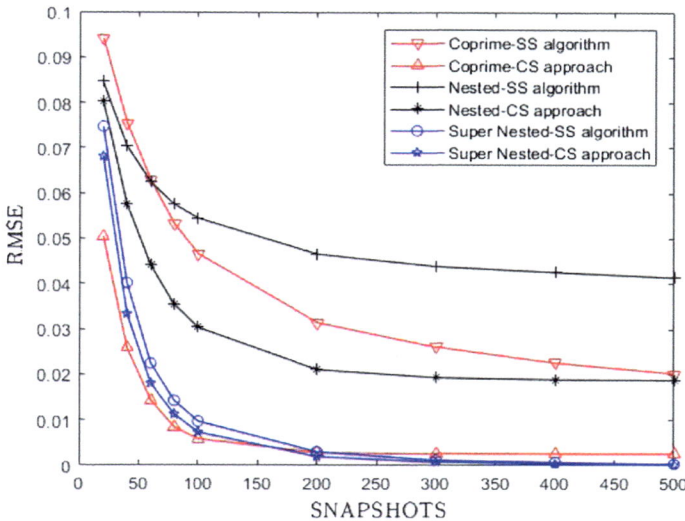

Fig. 3. RMSE of CS approach and SS algorithm in the presence of mutual coupling versus the number of snapshots, $SNR = 0$ dB

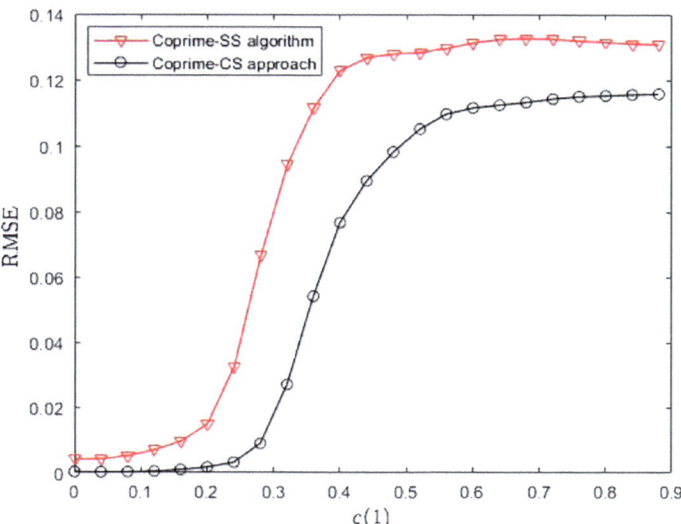

Fig. 4. RMSE of CS approach and SS algorithm for coprime arrays versus the parameter $c(1)$; the number of snapshots is 300 and $SNR = 0$ dB

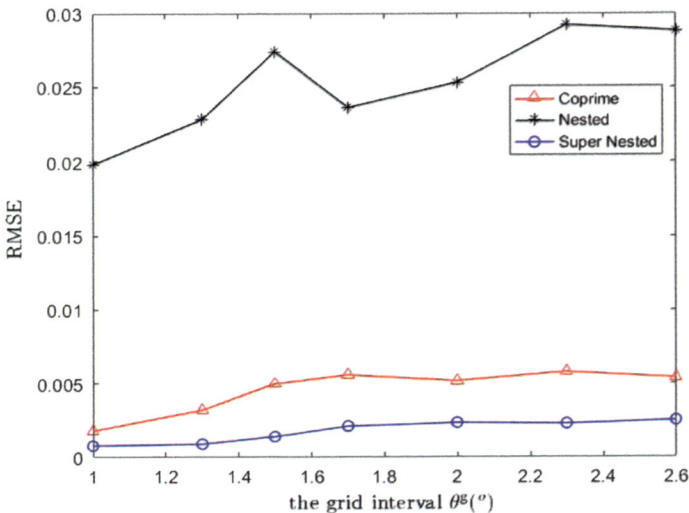

Fig. 5. RMSE of CS approach versus the grid interval θ^g, the number of snapshots is 300 and $SNR = 0$ dB

5   Conclusion and Future Work

This paper utilizes three categories of sparse arrays and proposes a method for DOA estimation using sparse signal recovery through CS approach in the presence of mutual coupling. Compared with SS algorithm, CS approach fully uses all the virtual array aperture, which is extended based on the three categories of sparse arrays structures. The estimation error of CS approach is lower than that of SS algorithm. In addition, we discuss the effect of mutual coupling parameter $c(1)$ on the estimation error. The threshold of $c(1)$ for CS approach is lower than that of SS algorithm. Therefore, CS approach has the advantage over SS algorithm. Besides, the estimation error of CS approach presents a rising trend with the increase of the grid interval θ^g. In the future, we may apply the sensor nodes moving algorithm to the three categories of sparse arrays structures.

Acknowledgments. This work was supported by the National Natural Science Foundation of China (61671138, 61731006) and was partly supported by the 111 Project No. B17008.

References

1. Van Trees HL. Optimum array processing: part IV of detection, estimation and modulation theory. Publishing House of Elec; 2002.
2. Godara LC. Application of antenna arrays to mobile communications. ii. Beamforming and direction-of-arrival considerations. Proc IEEE. 2009;85(8):1195–245.

3. Schmidt R. Multiple emitter location and signal parameter estimation. IEEE Trans Antennas Propag. 1986;34(3):276–80.
4. Roy R, Kailath T. Esprit-estimation of signal parameters via rotational invariance techniques. IEEE Trans Acoust Speech Signal Process. 2002;37(7):984–95.
5. Moffet A. Minimum-redundancy linear arrays. IEEE Trans Antennas Propag. 2003;16(2):172–5.
6. Pal P, Vaidyanathan PP. Nested arrays: a novel approach to array processing with enhanced degrees of freedom. IEEE Trans Signal Process. 2010;58(8):4167–81.
7. Vaidyanathan PP, Pal P. Sparse sensing with co-prime samplers and arrays. IEEE Trans Signal Process. 2011;59(2):573–86.
8. Pal P, Vaidyanathan PP: Coprime sampling and the music algorithm. In: Digital signal processing workshop and IEEE signal processing education workshop; 2011. p. 289–94
9. Liu CL, Vaidyanathan PP. Super nested arrays: linear sparse arrays with reduced mutual coupling part I: fundamentals. IEEE Trans Signal Process. 2016;64(15):3997–4012.
10. Chambers C, Tozer TC, Sharman KC. Temporal and spatial sampling influence on the estimates of superimposed narrowband signals: when less can mean more. IEEE Trans Signal Process. 1996;44(12):3085–98.
11. Pillai SU, Bar-Ness Y, Haber F. A new approach to array geometry for improved spatial spectrum estimation. Proc IEEE. 1985;73(10):1522–4.
12. Friedlander B, Weiss AJ. Direction finding in the presence of mutual coupling. IEEE Trans Antennas Propag. 1991;39(3):273–84.
13. Svantesson T. Mutual coupling compensation using subspace fitting. In: Proceedings of the Sensor array and multichannel signal processing workshop, 2000; 2000. p. 494–8.
14. Grant M. CVX: Matlab software for disciplined convex programming, version 1.21; 2008. p. 155–210.

DBSCAN-Based Mobile AP Detection for Indoor WLAN Localization

Wei Nie, Hui Yuan$^{(\boxtimes)}$, Mu Zhou, Liangbo Xie, and Zengshan Tian

Chongqing Key Lab of Mobile Communications Technology, Chongqing University of Posts and Telecommunications, Chongqing 400065, China
niewei@cqupt.edu.cn, yuanhui0128@foxmail.com, zhoumu@cqupt.edu.cn, xielb@cqupt.edu.cn, tianzs@cqupt.edu.cn

Abstract. The vast market of location-based services (LBSs) has brought opportunities for the rapid development of indoor positioning technology. In current indoor venues, by considering the fact that the wireless local area network (WLAN) infrastructure is widely deployed, the indoor WLAN localization method has become the focus of study. Nowadays, the WLAN module is used widely in a large number of advanced mobile devices, and meanwhile there are a variety of WLAN mobile access points (APs) in indoor environment. In this circumstance, due to the uncertainty of the state of mobile APs, the associated received signal strength (RSS) data are usually lowly dependent on the locations, which will consequently result in the decrease in localization accuracy. To solve this problem, a new method of mobile AP detection based on the density-based spatial clustering of applications with noise (DBSCAN) is proposed. This method aims to identify mobile APs in target area so as to eliminate the adverse impact of mobile APs on localization accuracy.

Keywords: Indoor localization · mobile AP detection · Location dependency · DBSCAN · WLAN

1 Introduction

Nowadays, people have been farther and farther away from the problem of getting lost, which has to be attributed to the rapid development of localization technology. As more and more electronic products such as mobile phones and tablet computer are able to access WLAN, indoor localization based on RSS in WLAN environment has gained a good opportunity for development. The reason why the traditional RSS-based indoor localization technology obtains high-accuracy lies in the high location dependency of RSS data. However, more and more electronic devices have mobile hotspot function, which leads to the existence of many mobile APs in the target area. The widespread existence of mobile APs significantly reduces the location dependency of the collected RSS data, causing large fluctuations in the collected RSS and serious localization errors. In this way, removing the impact of mobile APs on indoor

© Springer Nature Singapore Pte Ltd. 2020
Q. Liang et al. (eds.), *Communications, Signal Processing, and Systems*, Lecture Notes in Electrical Engineering 516,
https://doi.org/10.1007/978-981-13-6504-1_152

localization is particularly important. To this end, we propose a new approach of mobile AP detection based on density-based spatial clustering of applications with noise (DBSCAN) to identify mobile APs in WLAN. In this way, in the RSS-based localization process, after the construction of position fingerprint database in the offline phase, the RSS data from the mobile APs are removed, and the RSS data from the stationary APs are used as fingerprint database for matching location in the online phase. The structure of this paper is as follows. Localization-related technologies are introduced in Sect. 2. Mobile AP detection and localization system are described in Sect. 3. Then, Sect. 3 shows the experimental results. Finally, we conclude this paper in Sect. 4.

2   Related Work

In the outdoor environment, global positioning system (GPS) [1] can provide us with efficient localization and navigation services. However, GPS signals are not strong enough to provide high-accuracy localization service indoors due to the building obstruction. In this case, there are still a lot of achievements in indoor localization without GPS signals. Among the numerous approaches, the WLAN-based indoor localization has been widely welcomed by scholars because of its low infrastructure input costs and convenient RSS signal acquisition [2]. Indoor localization based on WLAN RSS is divided into offline phase and online phase [3]. In offline phase, a large number of reference points (RP) will be set in the target area and the RSS data collected from all APs at each RP constitute a piece of position fingerprint information. After that, the position fingerprint information at all RPs constitutes the position fingerprint database. In online phase, we will select several known location test points and newly collect RSS data at test points. Then, by comparing the online newly collected RSS data with the preconstructed fingerprint database, we will estimate the location of test points.

3   Approach Overview

3.1   RSS Sequence Generation

Previous research shows that indoor propagation model can well reflect the relationship between physical location and RSS data [4]. RSS sequences corresponding to the motion path are generated based on the indoor propagation model, as shown in Fig. 1.

For the convenience of the following discussion, the RPs is uniformly selected on every known location path of motion. Next, the RSS at each RP is calculated by the indoor propagation model. We assume that there are APs and L motion path in the target area. On each motion path, N_l ($l = 1, \ldots, L$) RPs are been selected. Therefore, the set of RSS sequences, **RSS**, can be represented as

$$\begin{cases} \mathbf{RSS} = \{RSS_l, \; l = 1, \ldots, L\} \\ RSS_l = \begin{bmatrix} rss_{l1}^1 & \cdots & rss_{lM}^1 \\ \vdots & \ddots & \vdots \\ rss_{l1}^{N_l} & \cdots & rss_{lM}^{N_l} \end{bmatrix} \end{cases} \quad (1)$$

where rss_{lm}^n $(n = 1, \ldots, N_l; m = 1, \ldots, M)$ is the RSS at the nth RP of the lth motion path from the mth AP, which is calculated by

$$rss_{lm}^n = p_0 - 10\eta \lg(d_{lm}^n d_0) - p_{\text{wall}} \tag{2}$$

where p_0 is the RSS value whose location is d_0 (usually set to 1m) away from the AP, η is the path loss factor, p_{wall} is the path loss caused by walls, and d_{lm}^n is the distance between nth RP on the lth motion path and the mth AP. It can be seen that if the AP is mobile, the RSS generated by the indoor propagation model changes continuously with the movement of the AP. Similarly, it is not hard to imagine that the area where RPs can receive signals from mobile APs will be larger than a stationary one.

3.2   Mobile AP Detection

For each AP, we select the RPs whose RSS r_m^{sel} is in the following range

$$\max\{r_{lm}^n\} - p_{\text{diff}}^m \leq r_m^{sel} \leq \max\{r_{lm}^n\} \tag{3}$$

where $\max\{r_{lm}^n\}$ is the maximum of RSS from the mth AP corresponding different RPs and p_{diff}^m can be calculated by

$$p_{\text{diff}}^m = -10\eta \lg\left(1 + d_{\text{diff}} d_{\text{min}}^m\right) \tag{4}$$

where d_{diff} is the signal reference distance (SRD) and d_{min}^m is the minimum distances between the mth AP and different RPs. According to the electromagnetic field coherent diffraction theory [5], we usually set d_{diff} to 10 m.

Then, we cluster the position coordinates of the RP selected for each AP through DBSCAN [6]. At this time, if the maximum distance, d_{max}, between two RPs in a cluster is larger than $2d_{\text{diff}}$, the corresponding AP is identified as a mobile one, and otherwise it is identified as a stationary one. Here, the reason why the maximum distance between two RPs in the mobile AP's cluster is larger than the stationary one can be interpreted by the fact that if the AP is mobile, as the AP moves, more RPs can receive signals from the mobile AP and the RSS is much stronger. Therefore, for mobile APs, we will select more RPs for clustering, and the cluster will be larger. By comparing the sizes of d_{max} corresponding different APs and $2d_{\text{diff}}$, we can accurately determine whether the AP is mobile or stationary. To illustrate this process clearer, the pseudo-code of the proposed approach is shown in Algorithm 1.

4   Experimental Result

4.1   Environmental Layout

For this target area, we deploy five stationary APs and two mobile APs in five rooms and two corridors, as shown in Fig. 1.

Algorithm 1 Pseudo-code of mobile AP detection

Input: m-th AP's RSS sequence
Output: m-th AP's mobility
1: $d_{\text{diff}} \leftarrow$ 10m; // Initialize SRD
2: Calculate p_{diff}^n;
3: Initialize L; //Number of motion paths
4: **for** $l = 1{:}L$ **do**
5:     **for** $n = 1{:}N_l$ **do**
6:         **if** $\max\{r_{lm}^n\} - p_{\text{diff}}^m \leq r_m^{sel} \leq \max\{r_{lm}^n\}$ **then**
7:             r_m^{sel} is selected;
8:         **end if**
9:     **end for**
10: **end for**
11: $C \leftarrow 0$; //Initialize clusters number
12: $\Gamma \leftarrow r_m^{sel}$; //Initialize unvisited RP set
13: **while** $\Gamma \neq \Phi$ **do**
14:     **for** each unvisited RP, $p \in \Gamma$ **do**
15:         $\Psi = $ getNeighbors(p, Eps); //Find p's neighbors with the radius Eps
16:         **if** sizeof(Ψ) ¡ MinPts **then**
17:             Label p as noise;
18:         **else**
19:             $C \leftarrow C + 1$
20:             Add p into the cluster with the cluster ID C;
21:             **for** each point p' in Ψ **do**
22:                 Label p' as visited RP;
23:                 $\Psi' = $ getNeighbor(p, Eps);
24:                 **if** sizeof(Ψ')$\geq$ MinPts **then**
25:                     $\Psi = \Psi \cup \Psi'$;
26:                 **end if**
27:                 **if** p' is not included in any cluster **then**
28:                     Add p' into the cluster with the cluster ID C;
29:                 **end if**
30:             **end for**
31:         **end if**
32:     **end for**
33: **end while**
34: Calculate $l_{\max}$;
35: **if** $l_{\max} > 2l_{\text{diff}}$ **then**
36:     The n-th AP is in mobile state;
37: **else**
38:     The n-th AP is in stationary state;
39: **end if**

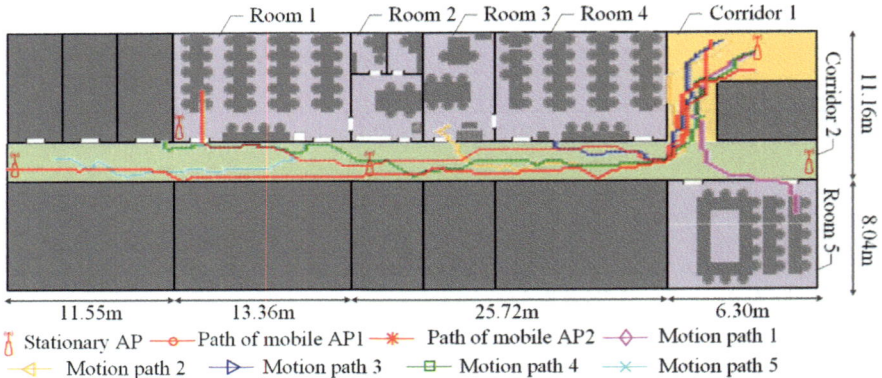

Fig. 1. Layout of experiment

4.2 Mobile AP Detection Results

For ease of testing, we randomly select five known motion paths in the target area as shown in Fig. 1. Then, we select one RP every second on this five paths and generate corresponding RSS sequences based on the indoor propagation model [4]. Next, we cluster the filtered RPs' position based on DBSCAN and obtain clustering results, as shown in Figs. 2 and 3.

After getting the clustering results, we calculate each AP's maximum distance between two RPs and the result is shown in Table 1. We preset the threshold 20 m to detect mobile AP. From Table 1, we can see that our approach can accurately detect mobile APs.

Table 1. Maximum of distance between different RPs in the cluster

AP IDs	Maximum of distance (m)
AP1	8.49
AP2	9.30
AP3	13.15
AP4	8.58
AP5	11.68
Mobile AP1	21.65
Mobile AP2	24.36

4.3 Contrast of Localization Accuracy with and Without Mobile AP Detection

In order to verify the improvement of mobile AP detection for indoor localization accuracy, we use the traditional KNN algorithm to locate

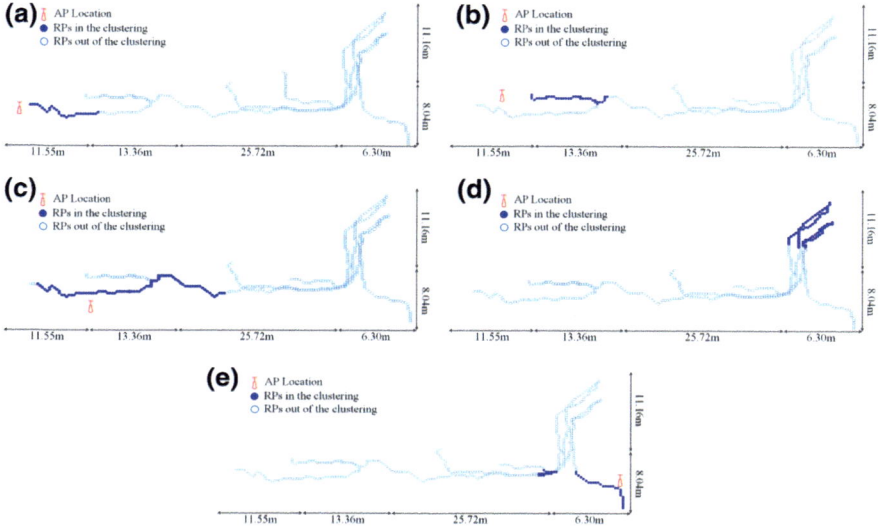

Fig. 2. Clustering result by the DBSCAN for stationary AP.

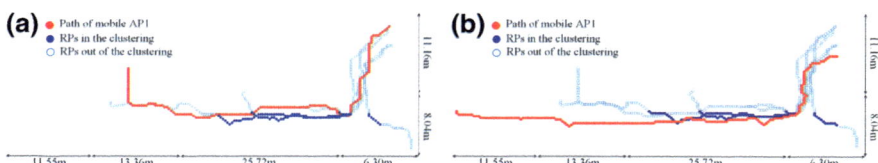

Fig. 3. Clustering result by the DBSCAN for mobile AP.

[7]. From Fig. 4, we can find that with the use of mobile AP detection, localization accuracy has been increased by about one time. Therefore, mobile AP detection can effectively improve the indoor localization accuracy.

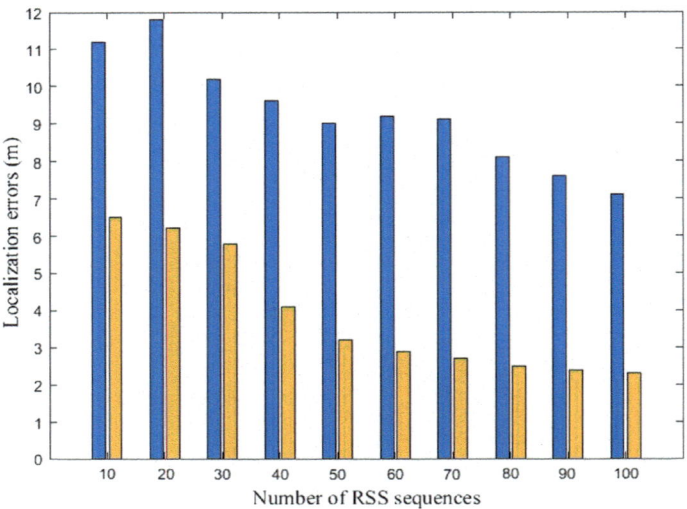

Fig. 4. Mean of localization errors with and without mobile AP detection.

5   Conclusion

We have proposed a new approach to detect mobile APs with the purpose of removing the RSS data from them and meanwhile claimed that the RSS data from stationary APs are featured with strong location dependency. At the same time, the KNN algorithm is applied to verify that the localization accuracy after detecting mobile APs can be significantly improved compared with the one without mobile AP detection.

Acknowledgments. This work is supported in part by the National Natural Science Foundation of China (61771083, 61704015), Program for Changjiang Scholars and Innovative Research Team in University (IRT1299), Special Fund of Chongqing Key Laboratory (CSTC), Fundamental Science and Frontier Technology Research Project of Chongqing (cstc2017jcyjAX0380, cstc2015jcyjBX0065), Scientific and Technological Research Foundation of Chongqing Municipal Education Commission (KJ1704083), and University Outstanding Achievement Transformation Project of Chongqing (KJZH17117).

References

1. Saha HN, Basu S, Auddy S, et al. A low cost fully autonomous GPS (Global Positioning System) based quad copter for disaster management. In: IEEE annual computing and communication workshop and conference; 2014. p. 654–60.
2. Chan F, Chan YT, Inkol R. Path loss exponent estimation and RSS localization using the linearizing variable constraint. In: Military communications conference; 2016. p. 225–9.
3. Zhou M, Tang Y, Tian Z, et al. Semi-supervised learning for indoor hybrid fingerprint database calibration with low effort. IEEE Access. 2017;5(99):4388–400.
4. Chai P, Zhang L. Indoor radio propagation models and wireless network planning. In: IEEE international conference on computer science and automation engineering; 2012. p. 738–41.
5. Cheung KW, Sau JHM, Murch RD. A new empirical model for indoor propagation prediction. IEEE Trans Veh Technol. 1998;47(3):996–1001.
6. Wang J, Tan N, Luo J, et al. WOLoc: WiFi-only outdoor localization using crowdsensed hotspot labels. In: INFOCOM 2017—IEEE conference on computer communications; 2017. p. 1–9.
7. Markom MA, Adom AH, Shukor SAA, et al. Scan matching and KNN classification for mobile robot localisation algorithm. In: IEEE international symposium in robotics and manufacturing automation; 2017. p. 1–6.

Error Bound Estimation for Wi-Fi Localization: A Comprehensive Survey

Mu Zhou, Yanmeng Wang$^{(\boxtimes)}$, Shasha Wang, Hui Yuan, and Liangbo Xie

Chongqing Key Lab of Mobile Communications Technology, Chongqing University
of Posts and Telecommunications, Chongqing 400065, China
{zhoumu,xielb}@cqupt.edu.cn, hiwangym@gmail.com,
{w20ss08,yuanhui0128}@foxmail.com

Abstract. Applications on location-based services (LBSs) have driven
the increasingly demand for indoor localization technology. Motivated by
the widely deployed wireless local area network (WLAN) infrastructure
and the corresponding easily accessible WLAN received signal strength
(RSS) data, the Wi-Fi signal-based localization has become one of the
superior positioning techniques in GPS-denied scenes. Meanwhile, the
error bound estimation for the Wi-Fi localization has been attracting
much attention due to its significant guidance meaning in practice. In
this survey, the error bound estimation approaches for different categories
of Wi-Fi localization approaches are overviewed and compared, including
the error bound estimation with temporal and spatial signal features, and
that with the RSS characteristics. Regarding the temporal and spatial
signal feature-based Wi-Fi localization, we present how to utilize the time
of arrival (TOA), the time difference of arrival (TDOA) as well as the
arrival of angle (AOA) to analyze the error bound of localization systems.
Regarding the received signal strength (RSS) characteristic-based Wi-Fi
localization, we clarify the error bound estimation approaches for both
the wireless signal propagation-based and location fingerprinting-based
localization schemes. In addition, some future directions with respect to
the error bound estimation for Wi-Fi localization are also discussed.

Keywords: Error bound estimation · Wi-Fi localization · Signal
features

1 Introduction

For well over a decade, the rapid development of wireless communication technology has driven the increasing demand for the location-based services (LBSs).
Because of the complicated indoor building structure and multipath effect,
the performance of the outdoor positioning systems such as global positioning will dramatically deteriorate in indoor environment. Meanwhile, with the

© Springer Nature Singapore Pte Ltd. 2020
Q. Liang et al. (eds.), *Communications, Signal Processing,
and Systems*, Lecture Notes in Electrical Engineering 516,
https://doi.org/10.1007/978-981-13-6504-1_153

wide deployment of wireless Wi-Fi infrastructure, Wi-Fi has become one of the priorities for indoor positioning.

Wireless technology used for indoor positioning has been reviewed [1], while few of these works focus on the Wi-Fi localization error estimation, which has not yet been properly reviewed but has a significant guidance meaning in real practice. In this paper, we first introduce some typical localization techniques in Wi-Fi environment from following aspects.

(1) **Temporal and spatial feature-based Wi-Fi localization**. The temporal feature-based Wi-Fi localization includes the time of arrival (TOA) and time difference of arrival (TDOA)-based approach, while the spatial feature based localization contains the localization approach with the arrival of angle (AOA) information.
(2) **Received signal strength (RSS) characteristic-based Wi-Fi localization**. The RSS characteristic-based Wi-Fi localization utilizes the RSS from access points (APs) to estimate the collecting locations and can be divided into two categories including the wireless signal propagation based as well as the location fingerprinting.

Then, the error bound estimation approaches for different categories of Wi-Fi localization approaches are reviewed and compared from the following categories.

(1) **Error bound estimation with temporal and spatial signal features**. By utilizing the TOA, TDOA as well as the AOA information, the Wi-Fi localization based on temporal and spatial features is clearly clarified.
(2) **Error bound estimation with RSS features**. With the easily hearable RSS from APs, the error bound estimation for both the wireless signal propagation-based as well as the location fingerprinting-based Wi-Fi localization approaches is analyzed.

This survey is organized as follows. In Sect. 2, we give a comprehensive introduction about the existing Wi-Fi localization algorithms. In Sect. 3, the error bound estimation of the temporal and spatial signal feature-based Wi-Fi localization approaches is clearly discussed. In Sect. 4, the error bound estimation with respect to the error bound of the RSS-based Wi-Fi localization approaches is reviewed and compared. Finally, Sect. 5 concludes the paper and gives some future directions.

2   Review of Wi-Fi Localization Approaches

2.1   Temporal and Spatial Feature-Based Wi-Fi Localization

The temporal and spatial feature-based Wi-Fi localization utilizes the TOA, TDOA as well as AOA of the signals from APs to estimate the collecting locations and can be divided into two categories including the temporal feature-based as well as the spatial feature-based Wi-Fi localization approaches.

Wi-Fi localization with temporal features. Both TOA and TDOA are the spatial features in Wi-Fi localization. For the localization with TOA, according to the TOA from AP_i to the target under noiseless environment, t_i, the distance between AP_i and the target, r_i, can be calculated through multiplying t_i by the signal propagation velocity C. Then, the location of target is at the intersection of circles as shown in Fig. 1a. However, the TOA-based Wi-Fi localization has extremely strict requirement for the time synchronization between APs and mobile terminal. In addition, the target is often in the non-line-of-sight (NLOS) range round APs, and this will cause large measurement deviation and make the circles in Fig. 1a cannot intersect at a point as shown in Fig. 1b.

To overcome the strict requirement for time synchronization, some researches utilize the TDOA between different APs to localize target. As shown in Fig. 2, the target is at the interaction of hyperbolas, where focal points are the locations of APs, and focal distance is the distance difference from different APs to target. This method just needs the time synchronization between APs, without demand for the time synchronization between APs and mobile terminals. However, the localization approach with TOA is also influenced by the NLOS factors.

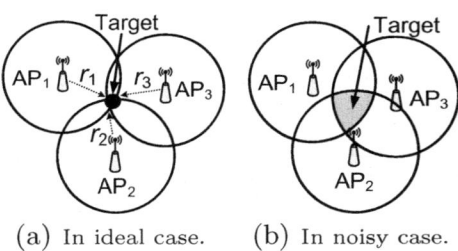

(a) In ideal case.   (b) In noisy case.

Fig. 1. Localization with TOA.

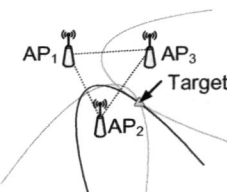

Fig. 2. Localization with TDOA.

Wi-Fi localization with spatial features. The AOA-based localization approaches utilize the directional antenna or array antennas to measure the direction angles of the signal arrived at APs from the target, which are used to estimate the targets location in Fig. 3. This kind of approach can achieve high even sub-meter level localization accuracy with good superiority in positioning principle and operability. However, the estimation of AOA requires special antenna array, which will increase the cost and the complexity of the localization system. Besides, the localization with AOA is also influenced by NLOS factors.

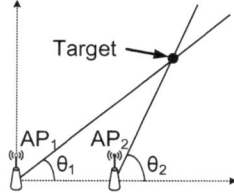

Fig. 3. Localization with AOA.

2.2 RSS Characteristic-Based Wi-Fi Localization

Signal propagation model-based Wi-Fi localization. In this approach, the wireless signal propagation model is established to describe the RSS value at each collecting location from APs, based on which the distance between collecting location and each hearable AP can be calculated, and the corresponding location of mobile terminal is estimated with triangulation method. However, due to the complicated building structure and multipath effect in indoor environment, the constructed signal propagation model is often unstable, which may dramatically decrease the localization accuracy.

Location fingerprinting-based Wi-Fi localization. Due to the easily accessible WLAN RSS, the location fingerprinting-based WLAN indoor localization systems have been widely researched. Generally, the location fingerprinting-based Wi-Fi localization consists of two phases, including the offline and online phases. Specifically, in the offline phase, the RSS value from APs collected at each reference point (RP) is regarded as the RPs' fingerprint, which is stored with the RPs' position together in the location fingerprint database. In the online phase, the newly collected RSS signal by mobile terminal is sent to the server and matched with the stored fingerprints to obtain the estimated location. However, due to the influential factors such as the distance interval of RPs, the signal fluctuation indoor environment as well as the reliability of location fingerprint database construction, the distance error between the real and estimated positions is inevitable in practical applications.

3 Error Bound Estimation with Temporal and Spatial Signal Features

3.1 Error Bound Estimation with Temporal Signal Features

As shown in Table 1, there are a batch of studies focusing on the error bounds of temporal signal feature-based Wi-Fi localization. With the TOA-based distance estimation, the authors in [2] develop an analytical framework for the relationships between multiple-access design parameters and localization error in Wi-Fi environment with a spread-spectrum physical layer, which is further utilized to

Table 1. Different methods utilizing temporal and spatial signal features

Category	Scheme	Signal feature	Key algorithm	Limitations
Temporal signal features	[2]	APs' locations, TOA	CRLB	Targets should be located in the LOS range from APs
	[3]	APs' locations, TOA	CRLB	FIM may be singular and incalculable
	[4]	TOA	CRB	Cannot be applied in the NLOS case
	[5]	APs' locations, TOA	Generalized CRLB	FIM may be singular and incalculable
	[6]	APs' locations, TOA	CRB	Targets should be located in the LOS range from APs
Spatial signal features	[5]	APs' locations, AOA	CRB	Require the prior information of distance range distribution from AP to target
	[7]	APs' locations, AOA	CRB	Cannot be applied in the time-varying environment
	[8]	APs' locations, AOA	CRB	Cannot be applied in the NLOS environment

extract the optimal network parameters for maximizing the localization accuracy. The authors in [3] compute the Cramer–Rao lower bound (CRLB) on positioning accuracy under different assumptions on the network synchronization to assess the ultimately achievable accuracy in practice, and the corresponding numerical examples show that the collaboration of multi-APs can eliminate the influence caused by AP deployment. The authors in [4] derive two kinds of error bound with respect to the TOA-based Wi-Fi localization, where the first kind utilizes the geographic information to calculate the upper and lower Cramer–Rao bound (CRB) of localization error with the prior information about at least three APs locations, and the second kind derives the approximate CRB for Wi-Fi localization without any prior knowledge of the APs.

However, these approaches all assume that the targets are localized in the line-of-sight (LOS) range round APs, which may be inconsistent with the real environment especially the complicated indoor environment. To overcome this, the authors in [5] propose a generalized lower CRB-based localization error estimation in both the LOS and NLOS environments, and the experimental results prove that the multipath information of Wi-Fi signal can effectively improve the localization performance. In addition, in the environment with dense AP deployment, some APs' locations cannot be accurately obtained, thus will cause with derived Fisher Information Matrix (FIM) in CRB singular and incalculable. To solve this, the authors in [6] construct an equivalent FIM to derive the localization error bound of the TOA-based system.

3.2 Error Bound Estimation with Spatial Signal Features

As one of the most representative studies, the authors in [4] construct the Fisher Information Matrix (FIM) as a function of the angles between the target and APs to derive the upper and lower CRB of localization error, respectively. Specifically, according to the AOA from the ith mobile terminal to the kth AP, α_{ik}, the FIM for AOA-based localization can be constructed as $J = \frac{1}{\sigma^2} \left[\frac{W + \sum_{k=1}^{W} \cos(\alpha_{ik})}{2} \quad \frac{\sum_{k=1}^{W} \sin(\alpha_{ik})}{2}; \frac{W + \sum_{k=1}^{W} \cos(\alpha_{ik})}{2} \quad \frac{W - \sum_{k=1}^{W} \cos(\alpha_{ik})}{2} \right]$, where W is the AP number, and σ^2 is the variance of the distance range from each AP to the target. Then, with the unbiased estimator, the lower CRB with respect to the localization error of the target, $\tilde{d}$, can be obtained by $\tilde{d}^2 \geq \frac{4W\sigma^2}{W^2 - (\sum_{k=1}^{W} \cos(\alpha_{ik}))^2 - (\sum_{k=1}^{W} \sin(\alpha_{ik}))^2} \geq \frac{4\sigma^2}{W}$.

However, this method requires the prior assumption that the distance ranges from APs to the target are i.i.d. Gaussian with zero mean and common variance σ^2, which may be inconsistent with the real case and make the derived error bound cannot effectively describe the localization error in target environment. Different from this, the authors in [7] utilize AOA measurements to analyze the localization accuracy of moving target and extract the lower bound of distance error between real and estimated positions via CRLB. However, this approach requires to assume that the AOA of received signals at a location is stationary, which may be unsuitable for the real environment. Besides, the authors in [8] derive the lower error bound for AOA-based passive source localization.

4   Error Bound Estimation with RSS Features

The existing studies on the error bound estimation of RSS-based Wi-Fi localization are usually based on the wireless signal propagation model, kernel density estimation, and other approaches. Some typical works are shown in Table 2 and are described as follows.

Up to now, the overwhelming majority of studies working on the error bound of the RSS-based Wi-Fi localization are based on the CRB. For example, the

Table 2. Different methods utilizing RSS characteristics.

Scheme	Signal feature	Key algorithm	Limitations
[9]	APs' location, SSD, AOA	Joint PDF of SSD, CRLB	Need to assume that the RSS follows Gaussian distribution
[10]	APs' location, RSS, AOA	Joint PDF of RSS, CRLB	Cannot be applied in the time-varying environment
[11]	APs' location, RSS, AOA	Joint PDF of RSS sum from APs, CRLB	Require accurate APs' locations
[12]	APs' location, RSS, AOA	Log-normal model of RSS, CRLB	Cannot be applied in the time-varying environment
[13]	APs' location, RSS, AOA	CRLB	RSS should follow Gaussian distribution
[14]	APs' location, RSS, AOA	Joint PDF of different distributed RSS, CRLB	Cannot be applied in the time-varying environment
[15]	RSS	Nonparametric kernel density estimation	Require accurate PDF of RSS
[16]	RSS	Hyper-parameters estimation	RSS should follow Gaussian distribution

authors in [9] rely on the CRLB to analyze the error bound of Wi-Fi localization using the signal strength difference (SSD) data as the location fingerprint. In concrete terms, based on wireless signal propagation model, the SSDs expression can be obtained as $[p_{ik_1}/p_{ik_2}]_{dB} = -10\beta \log(d_{ik_1}/d_{ik_2}) + n_{ik_1} - n_{ik_2}$, where p_{ik_1} and p_{ik_2} denote the RSS value at RP_i from AP_{k_1} and AP_{k_2}, respectively, in mW scale, d_{ik_1} and d_{ik_2} are the distances from AP_{k_1} and AP_{k_2}, and n_{ik_1} and n_{ik_2} are the corresponding signal noise with the same variance σ^2. Utilizing SSD expression, the joint probability density function (PDF) of the independent SSD measurements with W APs can be written as $f(p_i) = \prod_{k=1}^{W-1} \frac{1}{\sqrt{2\pi W \sigma^2}} \frac{10}{In10} \frac{p_{ik_1}}{p_{ik_2}} \times \exp\{-[10\log(p_{ik_1}/p_{ik_2}) + 10\beta\log(d_{ik_1}/d_{ik_2})]^2/2\sigma^2\}$. Based on this, the FIM of the localization is constructed, and the CRLB of target's location can be obtained.

Similarly to this, the authors in [10] adopt the joint PDF with respect to the RSS value from each AP to calculate the CRLB of Wi-Fi localization errors. The authors in [11] rely on the sum of RSS from APs to establish a CRLB-based error bound estimation criterion for Wi-Fi localization. With the log-normal model for RSS measurement, the authors in [12] derive a closed-form solution to the CRLB of the variance with respect to the distance between real and estimated locations. The authors in [13] obtain the CRLB of the distance between real and estimated locations in dynamic environment. In addition, the authors in [14] construct the joint PDF of RSS by considering three basic RSS distributions (i.e., Gaussian, Rayleigh, and Rice distributions) in Wi-Fi environment, and derive the error bound of Wi-Fi localization under both the LOS and NLOS scenarios.

However, due to the difficulty of acquiring the accurate AP locations as well as complexity and dynamics of indoor signal propagation property, the CRLB-based approaches are most often ineffective and unstable for indoor WLAN localization error bound estimation [10]. Different from above estimation methods, the authors in [15] utilize the nonparametric kernel density function to estimate the PDF of errors and as well as the corresponding confidence regions for the generalized location fingerprint-based positioning, based on which the localization error bounds are derived. The authors in [16] derive the CRLB of location fingerprint-based Wi-Fi localization via analyzing the hyper-parameters of kernel function for RSS distribution. However, the kernel function-based error bound estimation schemes often require exact PDF expression of RSS as well as some prior information about the Wi-Fi signal distribution in the target environment.

5   Conclusion

In this paper, the error bound estimation approaches for different categories of Wi-Fi localization approaches are comprehensively reviewed and compared, including the error bound estimation with temporal and spatial signal features, and that with the RSS characteristics. For the temporal and spatial signal feature-based Wi-Fi localization, we present how to utilize the TOA, TDOA as well as the AOA to calculate the error bound of localization systems. For the RSS characteristic-based Wi-Fi localization, we clarify the error bound estimation approaches for both the wireless signal propagation-based and location fingerprinting-based localization schemes. However, the assumptions of the target environment in most existing error bound estimation approaches are ideal and simple and cannot be widely spread. Therefore, developing the localization error estimation approaches for more complicated and time-varying Wi-Fi environment forms an interesting work in future.

Acknowledgments. This work was supported in part by the National Natural Science Foundation of China (61771083, 61704015), Program for Changjiang Scholars and Innovative Research Team in University (IRT1299), Special Fund of Chongqing Key Laboratory (CSTC), Fundamental and Frontier Research Project of Chongqing (cstc2017jcyjAX0380, cstc2015jcyjBX0065), University Outstanding Achievement Transformation Project of Chongqing (KJZH17117), and Postgraduate Scientific Research and Innovation Project of Chongqing (CYS17221).

References

1. He S, Chan SHG. Wi-fi fingerprint-based indoor positioning: recent advances and comparisons. IEEE Commun Surv Tutorials. 2017;18(1):466–90.
2. Venkatesh S, Buehrer RM. Multiple-access insights from bounds on sensor localization. Pervasive Mob Comput. 2008;4(1):33–61.
3. Larssom E. Cramer-Rao bound analysis of distributed positioning in sensor netwroks. IEEE Signal Process Lett. 2004;11(3):334–7.
4. Chang C, Sahai A. Estimation bounds for localization. In: IEEE SECON; 2004, p. 415–24.
5. Qi Y, Kobayashi H, Suda H. On time-of-arrival positioning in a multipath environment. IEEE Trans Veh Technol. 2006;55:1516–26.
6. Shen Y, Wymeersch H, Win MZ. Fundamental limits of wideband cooperative localization via fisher information. In: IEEE WCNC; 2007. p. 3951–5.
7. Hejazi F, Norouzi Y, Nayebi MM. Lower bound of error in AOA based passive source localization using single moving platform. In: IEEE East-west design and test international symposium; 2013. p. 1–4.
8. Luo J, Zhang XP, Wang Z. A new passive source localization method using AOA-GROA-TDOA in wireless sensor array networks and its Cramer-Rao bound analysis. In: IEEE ICASSP; 2013. p. 4031–5.
9. Hossain AKMM, Soh WS. Cramer-Rao bound analysis of localization using signal strength difference as location fingerprint. In: IEEE INFOCOM; 2010. p. 1–9.
10. Stella M, Russo M, Begusic D. RF localization in indoor environment. Radioengineering. 2012;21(2):557–67.
11. Laitinen E, Lohan ES. Access Point topology evaluation and optimization based on Cramer-Rao lower bound for WLAN indoor positioning. In: International conference on localization and GNSS; 2016. p. 1–5.
12. Mazuelas S, Bahillo A, Lorenzo RM, et al. Robust indoor positioning provided by real-time RSSI values in unmodified WLAN networks. IEEE J Sel Top Signal Process. 2009;3(5):821–31.
13. Zhou M, Xu K, Tian Z, et al. Error bound analysis of indoor Wi-Fi location fingerprint based positioning for intelligent access point optimization via Fisher information. Comput Commun. 2016;86:57–74.
14. Kaemarungsi K. Indoor localization improvement via adaptive RSS fingerprinting database. In: International conference on information networking, vol. 19; 2013. p. 412–6.
15. Jin Y, Soh WS, Wong WC. Error analysis for fingerprint-based localization. IEEE Commun Lett. 2010;14(5):393–5.
16. Kumar S, Hegde RM, Trigoni N. Gaussian process regression for fingerprinting based localization. Ad Hoc Netw. 2016;51:1–10.

Indoor WLAN Localization Based on Augmented Manifold Alignment

Liangbo Xie, Yaoping Li[(⊠)], Mu Zhou, Wei Nie, and Zengshan Tian

Chongqing Key Lab of Mobile Communications Technology,
Chongqing University of Posts and Telecommunications,
400065 Chongqing, People's Republic of China
{xielb,zhoumu,niewei,tianzs}@cqupt.edu.cn
liyaopingna@foxmail.com

Abstract. With the dramatic development of location-based service (LBS), indoor localization techniques have been widely used in recent years. Among them, the indoor wireless local area network (WLAN) localization technique is recognized as one of the most favored solutions due to its low maintenance overhead and high localization accuracy. In this paper, we propose a new received signal strength (RSS)-based indoor localization approach using augmented manifold alignment. First of all, we construct the objective function in manifold space for indoor localization. Second, the optimal transform matrix is used to transform the coordinates of reference points (RPs) and the corresponding RSS vectors into manifold space. Finally, we locate the target at the RP with the transformed coordinates nearest to the transformation of the newly collected RSS vector in manifold space. The experimental results demonstrate that the proposed approach is able to achieve satisfactory localization accuracy with low overhead.

Keywords: Indoor localization · Augmented manifold alignment ·
Transform matrix · Lagrange multiplier · WLAN

1 Introduction

With the rapid development of wireless communication technology and the widespread popularity of mobile devices, the demand for location-based service (LBS) is continuously increasing, which has brought unprecedented development space to LBS. Therefore, the indoor localization systems have attracted great attention for domestic and foreign scholars. Among them, wireless local area network (WLAN) indoor localization system [1] is widely used due to the widespread deployment and low overhead of WLAN.

Currently, the measurements used in WLAN indoor localization systems are mainly composed of time of arrival (TOA) [2], time difference of arrival (TDOA) [3], angle of arrival (AOA) [4], andreceived signal strength (RSS) [5] according

© Springer Nature Singapore Pte Ltd. 2020
Q. Liang et al. (eds.), *Communications, Signal Processing,*
and Systems, Lecture Notes in Electrical Engineering 516,
https://doi.org/10.1007/978-981-13-6504-1_154

to its basic principles and methods. For RSS-based localization technology, all it needs is to install the corresponding software on the mobile device to store and read the RSS values at different locations without adding any additional equipment, which results in high portability and feasibility. Therefore, RSS-based localization technology is preferred in most WLAN indoor localization systems.

In this paper, to further improve the indoor localization accuracy, a new WLAN RSS-based indoor localization approach using augmented manifold alignment is proposed. Through the objective function construction, we can get the optimized transform matrix that can transform RP coordinates and corresponding RSS vectors into manifold space and still maintains their neighborhood relations. Afterward, the target location can be estimated by finding the nearest RSS to the newly collected RSS vectors in manifold space.

The rest of this paper is organized as follows. In Sect. 2, we give some related work on manifold alignment. Section 3 describes the proposed approach in detail. The experimental results are presented in Sect. 4. Finally, Sect. 5 concludes the paper and provides an outlook of future work.

2   Related Work

Generally, semi-supervised learning approach is always used to improve the problem of time-consuming and costly site survey, which can achieve satisfactory localization accuracy through a small number of calibrated points. The authors in [6] propagate the labels to the unlabeled data by embedding labeled data and unlabeled data in common low-dimensional manifold to reduce calibration effort for localization. Without indoor space measurement, the authors in [7] automatically find the best correspondence between floor plan and RSS through graph matching-based manifold alignment process, which can achieve high localization and tracking accuracy.

Different from the works mentioned before, firstly, we augment the physical coordinates of RPs into the same dimension as the corresponding RSS vectors. Then the objective function of manifold alignment is constructed to obtain the optimal transform matrix. Finally, we convert the online RSS vector collected at the target location into manifold space and choose the nearest coordinate as the estimated target location.

3   System Overview

3.1   Algorithm Description

As shown in Fig. 1, the proposed algorithm is composed of two phases, namely offline phase and online phase, respectively. In offline phase, we first collet RSS vectors $\mathbf{r} = (\mathbf{r}_1, \ldots, \mathbf{r}_m)^{\mathrm{T}}$ from m RPs, where $\mathbf{r}_i = (r_{i1}, \ldots, r_{in})$ is the RSS measure collected at the ith RP and r_{ij} $(j = 1, \ldots, n)$ denotes the RSS measure from the jth AP and n is the number of APs. Then, we construct the objective function and map RP coordinates and corresponding RSS vectors into manifold space, such that we can get the optimized transform matrix through solving the objective function.

In online phase, firstly the online RSS vector is mapped into manifold space and then we choose the nearest physical coordinate to it as the estimated location in manifold space.

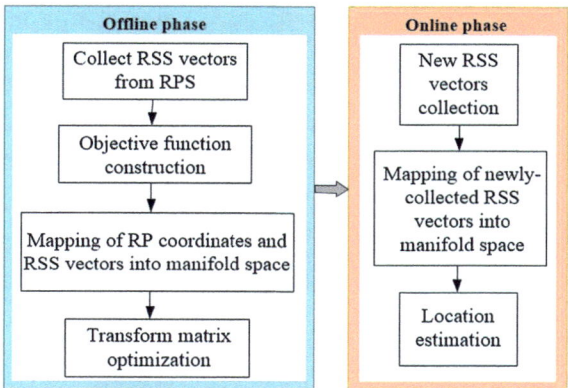

Fig. 1. Algorithm process

3.2   Objective Function

Let $\mathbf{c}_i' = (x_i, y_i)^{\mathrm{T}}$ be the physical coordinate of the ith RP. According to the concept of manifold alignment that data should be connected in the same dimension [8], we firstly augment the two-dimensional coordinate into the same dimension as the corresponding RSS vector $\mathbf{r}_i = (r_{i1}, \ldots, r_{in})$. Thus, we can get

$$
\mathbf{c}_i = \begin{cases} \left(\underbrace{x_i, y_i, \cdots, x_i, y_i}_{n2 \text{ paris of } (x_i, y_i)} \right)^{\mathrm{T}}, & \text{when } n \text{ is even} \\ \left(\underbrace{x_i, y_i, \cdots, x_i, y_i}_{(n-1)2 \text{ paris of } (x_i, y_i)}, x_i \right)^{\mathrm{T}}, & \text{when } n \text{ is odd} \end{cases} \tag{1}
$$

It can be seen that the above formula still preserves the relative distance between different coordinates of RPs. Based on this, we construct the objective function of the proposed algorithm for indoor localization as

$$
\arg\min_{\mathbf{P_c},\mathbf{P_r}} \left\{ \sum_{i\in\{1,\cdots,m\}} \left\| \mathbf{P_c}^\mathrm{T}\mathbf{c}_i - \mathbf{P_r}^\mathrm{T}\mathbf{r}_i \right\|_2^2 + \sum_{i,i'\in\{1,\cdots,m\};i\neq i'} \left(\left\| \mathbf{P_c}^\mathrm{T}\mathbf{c}_i - \mathbf{P_c}^\mathrm{T}\mathbf{c}_{i'} \right\|_2^2 S_{ii'}^c \right) \right.
$$
$$
\left. + \sum_{j,j'\in\{1,\cdots,m\};j\neq j'} \left(\left\| \mathbf{P_r}^\mathrm{T}\mathbf{r}_j - \mathbf{P_r}^\mathrm{T}\mathbf{r}_{j'} \right\|_2^2 S_{jj'}^r \right) \right\}
$$

$$(2)$$

where the notation "$\|\cdot\|_2$" represents the two-norm operation, the transform matrices $\mathbf{P_c}$ and $\mathbf{P_r}$ are optimized to transform $\mathbf{c}_i$ and $\mathbf{r}_i$ into $\mathbf{a}_i = \mathbf{P_c}^\mathrm{T}\mathbf{c}_i$ and $\mathbf{b}_i = \mathbf{P_r}^\mathrm{T}\mathbf{r}_i$, respectively, in the n-dimensional manifold space, $S_{ii'}^c = \exp\left(-\|\mathbf{c}_i - \mathbf{c}_{i'}\|_2^2\right)$, $S_{jj'}^r = \exp\left(-\|\mathbf{r}_j - \mathbf{r}_{j'}\|_2^2\right)$. The first term stands for the proximity between the physical coordinates of RPs and corresponding RSS vectors in manifold space. The second and third term are to preserve proximity between coordinates and RSS vectors, respectively. Since

$$
\sum_{i\in\{1,\cdots,m\}} \left\| \mathbf{P_c}^\mathrm{T}\mathbf{c}_i - \mathbf{P_r}^\mathrm{T}\mathbf{r}_i \right\|_2^2 = \mathrm{Tr}(\mathbf{P_c}^\mathrm{T}\mathbf{CC}^\mathrm{T}\mathbf{P_c} + \mathbf{P_r}^\mathrm{T}\mathbf{RR}^\mathrm{T}\mathbf{P_r} - \mathbf{P_c}^\mathrm{T}\mathbf{CR}^\mathrm{T}\mathbf{P_r} - \mathbf{P_r}^\mathrm{T}\mathbf{RC}^\mathrm{T}\mathbf{P_c})
$$

$$(3)$$

$$
\sum_{i,i'\in\{1,\cdots,m\};i\neq i'} \left(\left\| \mathbf{P_c}^\mathrm{T}\mathbf{c}_i - \mathbf{P_c}^\mathrm{T}\mathbf{c}_{i'} \right\|_2^2 S_{ii'}^c \right) = 2\mathrm{Tr}\left(\mathbf{P_c}^\mathrm{T}\mathbf{CH_c}\mathbf{C}^\mathrm{T}\mathbf{P_c} \right) \qquad (4)
$$

$$
\sum_{j,j'\in\{1,\cdots,m\};j\neq j'} \left(\left\| \mathbf{P_r}^\mathrm{T}\mathbf{r}_j - \mathbf{P_r}^\mathrm{T}\mathbf{r}_{j'} \right\|_2^2 S_{jj'}^r \right) = 2\mathrm{Tr}\left(\mathbf{P_r}^\mathrm{T}\mathbf{CH_r}\mathbf{C}^\mathrm{T}\mathbf{P_r} \right) \qquad (5)
$$

where the notation "$\mathrm{Tr}\,(\cdot)$" is the trace of matrix, $\mathbf{C} = (\mathbf{c}_1, \cdots, \mathbf{c}_m)$, $\mathbf{R} = (\mathbf{r}_1, \cdots, \mathbf{r}_m)$, $\mathbf{H_c} = \mathbf{M_c} - \mathbf{N_c}$, $\mathbf{H_r} = \mathbf{M_r} - \mathbf{N_r}$, $\mathbf{N_c} = \left\{ N_c^{ii'} \right\}$, $N_c^{ii'} = \exp\left(-\|\mathbf{c}_i - \mathbf{c}_{i'}\|_2^2\right)$, $\mathbf{N_r} = \left\{ N_r^{jj'} \right\}$, $N_r^{jj'} = \exp\left(-\|\mathbf{r}_j - \mathbf{r}_{j'}\|_2^2\right)$, $\mathbf{M_c} = diag\left(\sum_{i'=1}^m N_c^{1i'}, \cdots, \sum_{i'=1}^m N_c^{mi'}\right)$, $\mathbf{M_r} = diag\left(\sum_{j'=1}^m N_r^{1j'}, \cdots, \sum_{j'=1}^m N_r^{mj'}\right)$, we have

$$
\arg\min_{\mathbf{P_c},\mathbf{P_r}} \left\{ \mathrm{Tr}\left(\mathbf{P_c}^\mathrm{T}\mathbf{CC}^\mathrm{T}\mathbf{P_c} + \mathbf{P_r}^\mathrm{T}\mathbf{RR}^\mathrm{T}\mathbf{P_r} - \mathbf{P_c}^\mathrm{T}\mathbf{CR}^\mathrm{T}\mathbf{P_r} - \mathbf{P_r}^\mathrm{T}\mathbf{RC}^\mathrm{T}\mathbf{P_c} \right) \right.
$$
$$
\left. + 2\mathrm{Tr}\left(\mathbf{P_c}^\mathrm{T}\mathbf{CH_c}\mathbf{C}^\mathrm{T}\mathbf{P_c} \right) + 2\mathrm{Tr}\left(\mathbf{P_r}^\mathrm{T}\mathbf{RH_r}\mathbf{R}^\mathrm{T}\mathbf{P_r} \right) \right\}
$$

$$(6)$$

Considering the noise interference $\boldsymbol{\delta}$ on RSS, $\mathbf{R}$ is modified into $\mathbf{R} + \boldsymbol{\delta}$, thus the objective function is rewritten as

$$
\arg\min_{\mathbf{P_c},\mathbf{P_r}} \left\{ \mathrm{Tr}\left(\mathbf{P_c}^\mathrm{T}\mathbf{CC}^\mathrm{T}\mathbf{P_c} + \mathbf{P_r}^\mathrm{T}\mathbf{RR}^\mathrm{T}\mathbf{P_r} - \mathbf{P_c}^\mathrm{T}\mathbf{CR}^\mathrm{T}\mathbf{P_r} - \mathbf{P_r}^\mathrm{T}\mathbf{RC}^\mathrm{T}\mathbf{P_c} \right) \right.
$$
$$
+ \mathrm{Tr}\left(\mathbf{P_r}^\mathrm{T}\mathbf{R}\boldsymbol{\delta}^\mathrm{T}\mathbf{P_r} + \mathbf{P_r}^\mathrm{T}\boldsymbol{\delta}^\mathrm{T}\mathbf{RP_r} + \mathbf{P_r}^\mathrm{T}\boldsymbol{\delta}\boldsymbol{\delta}^\mathrm{T}\mathbf{P_r} - \mathbf{P_c}^\mathrm{T}\mathbf{C}\boldsymbol{\delta}^\mathrm{T}\mathbf{P_r} - \mathbf{P_r}^\mathrm{T}\boldsymbol{\delta}\mathbf{C}^\mathrm{T}\mathbf{P_c} \right)
$$
$$
+ 2\mathrm{Tr}\left(\mathbf{P_c}^\mathrm{T}\mathbf{CH_c}\mathbf{C}^\mathrm{T}\mathbf{P_c} \right) + 2\mathrm{Tr}\left(\mathbf{P_r}^\mathrm{T}\mathbf{RH_r}\mathbf{R}^\mathrm{T}\mathbf{P_r} \right) + 2\mathrm{Tr}\left(\mathbf{P_r}^\mathrm{T}\mathbf{RH_r}\boldsymbol{\delta}^\mathrm{T}\mathbf{P_r} \right)
$$
$$
\left. + 2\mathrm{Tr}\left(\mathbf{P_r}^\mathrm{T}\boldsymbol{\delta}\mathbf{H_r}\mathbf{R}^\mathrm{T}\mathbf{P_r} \right) + 2\mathrm{Tr}\left(\mathbf{P_r}^\mathrm{T}\boldsymbol{\delta}\mathbf{H_r}\boldsymbol{\delta}^\mathrm{T}\mathbf{P_r} \right) \right\}
$$

$$(7)$$

Let $\mathbf{P} = \begin{pmatrix} \mathbf{P_c} \\ \mathbf{P_r} \end{pmatrix}$, $\mathbf{V} = \begin{pmatrix} \mathbf{C} & \mathbf{0} \\ \mathbf{0} & \mathbf{R} \end{pmatrix}$ and $\mathbf{G} = \begin{pmatrix} \mathbf{I} + 2\mathbf{H_c} & -\mathbf{I} \\ -\mathbf{I} & \mathbf{I} + 2\mathbf{H_r} \end{pmatrix}$, where $\mathbf{I}$ is the $m \times m$ unit matrix, (7) can be converted into

$$
\arg\min_{\mathbf{P}} \left\{ \mathrm{Tr} \left(\mathbf{P}^T \mathbf{V} \mathbf{G} \mathbf{V}^T \mathbf{P} + \mathbf{P}^T \mathbf{V} \mathbf{G} \begin{pmatrix} \mathbf{0} & \mathbf{0} \\ \mathbf{0} & \boldsymbol{\delta} \end{pmatrix}^T \mathbf{P} + \mathbf{P}^T \begin{pmatrix} \mathbf{0} & \mathbf{0} \\ \mathbf{0} & \boldsymbol{\delta} \end{pmatrix} \mathbf{G} \mathbf{V}^T \mathbf{P} + \right.\right.
$$
$$
\left.\left. \mathbf{P}^T \begin{pmatrix} \mathbf{0} & \mathbf{0} \\ \mathbf{0} & \boldsymbol{\delta} \end{pmatrix} \mathbf{G} \times \begin{pmatrix} \mathbf{0} & \mathbf{0} \\ \mathbf{0} & \boldsymbol{\delta} \end{pmatrix}^T \mathbf{P} \right) \right\} \quad \text{s.t. } \mathbf{P}^T \mathbf{V} \mathbf{V}^T \mathbf{P} = \mathbf{I}, \ \mathbf{P}^T \mathbf{V} \mathbf{e} = \mathbf{0}
\tag{8}
$$

where $\mathbf{e}$ is the $2m \times n$ all-one matrix. By using the Lagrange multiplier approach and letting the partial derivatives of the above formula with respect to $\mathbf{P}$ equal to 0, we can obtain

$$
2\mathbf{V}\mathbf{G}\mathbf{V}^T\mathbf{P} - 2\lambda\mathbf{V}\mathbf{V}^T\mathbf{P} - \mu\mathbf{V}\mathbf{e}
$$
$$
+2 \begin{pmatrix} -\mathbf{C}\boldsymbol{\delta}^T \\ -\mathbf{C}\boldsymbol{\delta}^T\mathbf{P_c} + \mathbf{R}\left(\mathbf{I} + 2\mathbf{H_r}\right)\boldsymbol{\delta}^T + \boldsymbol{\delta}\left(\mathbf{I} + 2\mathbf{H_r}\right)\mathbf{R}^T + \boldsymbol{\delta}\left(\mathbf{I} + 2\mathbf{H_r}\right)\boldsymbol{\delta}^T \end{pmatrix} \mathbf{P} = 0
\tag{9}
$$

where λ and μ are the Lagrange coefficients. By multiplying both sides of (9) with $\mathbf{P}'\mathbf{P}^T$, where $\mathbf{P}'\mathbf{P}^T = \mathbf{E}$ is the $2n \times 2n$ unit matrix and $\mathbf{P}'$ is the invertible matrix of $\mathbf{P}^T$, we have

$$
\left(\mathbf{V}\mathbf{G}\mathbf{V}^T + \begin{pmatrix} \mathbf{0} & -\mathbf{C}\boldsymbol{\delta}^T \\ -\mathbf{C}\boldsymbol{\delta}^T & \mathbf{R}\left(\mathbf{I} + 2\mathbf{H_r}\right)\boldsymbol{\delta}^T + \boldsymbol{\delta}\left(\mathbf{I} + 2\mathbf{H_r}\right)\mathbf{R}^T + \boldsymbol{\delta}\left(\mathbf{I} + 2\mathbf{H_r}\right)\boldsymbol{\delta}^T \end{pmatrix} \right) \mathbf{P}
$$
$$
= \lambda\mathbf{V}\mathbf{V}^T\mathbf{P}
\tag{10}
$$

By multiplying both sides of (10) with , we can obtain

$$
\mathbf{P}^T \left(\mathbf{V}\mathbf{G}\mathbf{V}^T + \begin{pmatrix} \mathbf{0} & -\mathbf{C}\boldsymbol{\delta}^T \\ -\mathbf{C}\boldsymbol{\delta}^T & \mathbf{R}\left(\mathbf{I} + 2\mathbf{H_r}\right)\boldsymbol{\delta}^T + \boldsymbol{\delta}\left(\mathbf{I} + 2\mathbf{H_r}\right)\mathbf{R}^T + \boldsymbol{\delta}\left(\mathbf{I} + 2\mathbf{H_r}\right)\boldsymbol{\delta}^T \end{pmatrix} \right) \mathbf{P}
$$
$$
= \lambda\mathbf{P}^T\mathbf{V}\mathbf{V}^T\mathbf{P} = \lambda\mathbf{I}
\tag{11}
$$

Let $\mathbf{J} = \mathbf{V}\mathbf{G}\mathbf{V}^T + \begin{pmatrix} \mathbf{0} & -\mathbf{C}\boldsymbol{\delta}^T \\ -\mathbf{C}\boldsymbol{\delta}^T & \mathbf{R}\left(\mathbf{I} + 2\mathbf{H_r}\right)\boldsymbol{\delta}^T + \boldsymbol{\delta}\left(\mathbf{I} + 2\mathbf{H_r}\right)\mathbf{R}^T + \boldsymbol{\delta}\left(\mathbf{I} + 2\mathbf{H_r}\right)\boldsymbol{\delta}^T \end{pmatrix}$ and $\mathbf{K} = \mathbf{V}\mathbf{V}^T$, we convert (11) into

$$
\mathbf{J}\mathbf{P} = \lambda\mathbf{K}\mathbf{P}
\tag{12}
$$

Therefore, the optimal solution $\mathbf{P_{opt}} = \begin{pmatrix} \mathbf{P_{optc}} \\ \mathbf{P_{optr}} \end{pmatrix}$ is composed of n non-zero minimum generalized eigenvalues corresponding to n generalized eigenvectors, where $\mathbf{P_{optc}}$ and $\mathbf{P_{optr}}$ are the optimal solutions to $\mathbf{P_c}$ and $\mathbf{P_r}$ respectively.

During online phase, firstly, map the RSS vector $\mathbf{r_{tar}} = (r_1, \ldots, r_n)^T$ collected at the target location into manifold space as $\mathbf{P_{optr}}^T \mathbf{r_{tar}}$. And then, we select the coordinate of RP closet to $\mathbf{P_{optr}}^T \mathbf{r_{tar}}$ in manifold space as the estimated location of the target, i.e.,

$$
\arg\min_{\mathbf{c_i}} \left\{ \left\| \mathbf{P_{optc}}^T \mathbf{c_i} - \mathbf{P_{optr}}^T \mathbf{r_{tar}} \right\|_2^2 \right\}
\tag{13}
$$

4   Experimental Results

4.1   Environmental Layout

As shown in Fig. 2, the experimental environment is selected on the fifth floor in a building with the dimensions of 57 m × 25 m. There are five APs fixed in target environment. The 73 RPs are uniformly distributed and there are several RSS vectors collected at each of them.

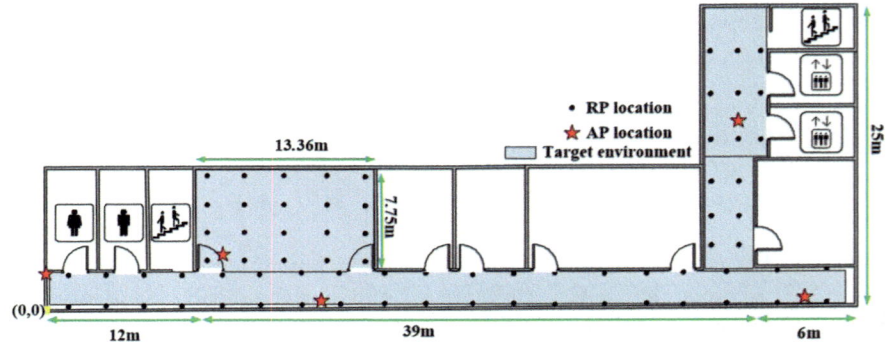

Fig. 2. Environmental layout

4.2   Localization Performance

We learn the noise interference on localization performance through experiment with measured RSS vectors. The experimental result in Fig. 3 shows that the increase in noise causes the deterioration of localization errors.

In Fig. 4, we compare the mean of localization errors of the proposed localization approach with CIMLoc [9] and WILL [10], respectively. As can be seen, our approach performs better than the other two existing approaches, although the mean of localization errors of CIMLoc is close to the proposed one under large number of RSS sequences. And there is more location information in large size of RSS vectors, which finally improves the performance of localization.

5   Conclusion

We propose an indoor WLAN localization approach based on augmented manifold alignment. By preserving the neighborhood relations of the transformed RP coordinates and the corresponding RSS vectors, we achieve satisfactory localization accuracy for the target. In addition, the impact of environmental noise and size of RSS vectors is also discussed through the experiments conducted in the actual indoor WLAN environment. In future, the application of the proposed approach in more complicated indoor environment forms an interesting direction.

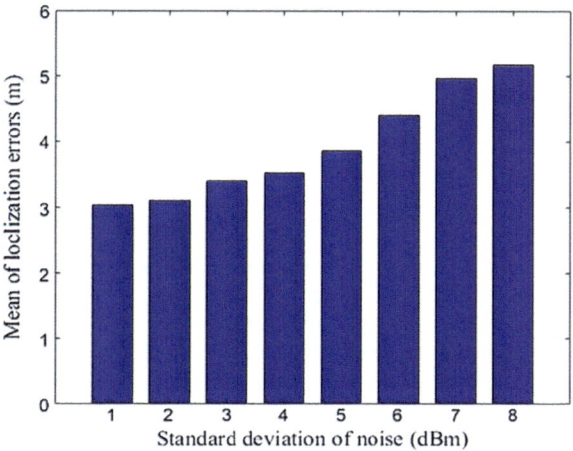

Fig. 3. Localization errors under different standard deviation of noise

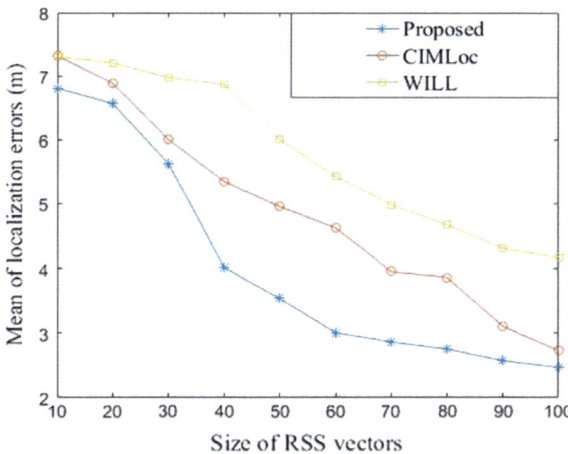

Fig. 4. Localization errors under different standard deviation of noise

Acknowledgments. This work is supported in part by the National Natural Science Foundation of China (61771083, 61704015), Program for Changjiang Scholars and Innovative Research Team in University (IRT1299), Special Fund of Chongqing Key Laboratory (CSTC), Fundamental Science and Frontier Technology Research Project of Chongqing (cstc2017jcyjAX0380, cstc2015jcyjBX0065), Scientific and Technological Research Foundation of Chongqing Municipal Education Commission (KJ1704083), and University Outstanding Achievement Transformation Project of Chongqing (KJZH17117).

References

1. Nguyen GK, Nguyen TV, Shin H. Learning dictionary and compressive sensing for WLAN localization. In: IEEE WCNC; 2014. p. 2910–5.
2. Zheng Y, Wang H, Wan L, et al. A placement strategy for accurate TOA localization algorithm. In: Annual communication networks and services research conference; 2009. p. 166–70.
3. Zhang L, Yu X. A Kernel-based TDOA localization algorithm. In: International conference on computer application and system modeling; 2010. p. 412–5.
4. Dogancay K, Hmam H. Optimal angular sensor separation for AOA localization. Signal Process. 2008;88(5):1248–60.
5. Bahl P, Padmanabhan VN. RADAR: an in-building RF-based user location and tracking system. In: IEEE INFOCOM; 2000. p. 775–84.
6. Wang H, Zhang V, Zhao J, et al. Indoor localization in multi-floor environments with reduced effort. In: IEEE international conference on pervasive computing and communications; 2010. p. 244–52.
7. Jiang Z, Zhao J, Han J, et al. Wi-Fi fingerprint based indoor localization without indoor space measurement. In: IEEE international conference on mobile ad-hoc and sensor systems; 2013. p. 384–92.
8. Zhou M, Zhang Q, Tian Z. et al. Indoor WLAN localization using high-dimensional manifold alignment with limited calibration load. In: IEEE ICC; 2017. p. 1–6.
9. Zhang X, Jin Y, Tan HX, et al. CIMLoc: a crowdsourcing indoor digital map construction system for localization. In: IEEE International conference on intelligent sensors, sensor networks and information processing; 2014. p. 1–6.
10. Wu C, Yang Z, Liu Y, et al. WILL: wireless indoor localization without site survey. IEEE Trans Parallel Distrib Syst. 2013;24(4):839–48.

Trajectory Reckoning Method Based on BDS Attitude Measuring and Point Positioning

Liangbo Xie, Shuai Lu$^{(\boxtimes)}$, Mu Zhou, Yi Chen, and Xiaoxiao Jin

Chongqing Key Lab of Mobile Communications Technology, Chongqing University of Posts and Telecommunications, Chongqing 400065, People's Republic of China
`xielb@cqupt.edu.cn,lushuai.139@163.com,zhoumu@cqupt.edu.cn,`
`751796746@qq.com,jinxiaoxiaosx@163.com`

Abstract. The traditional outdoor integrated positioning and navigation system is normally suffered by the disadvantages of accumulative error and high power consumption. To solve this problem, we propose a new trajectory reckoning method which use the BeiDou system (BDS) to conduct the attitude measuring and point positioning with respect to the target. In concrete terms, the target location is estimated by solving the pseudo-range observation equation, while the attitude angle is obtained from the dual-difference pseudo-range and carrier phase observation equations. Then, the trajectory of the target is constructed based on the estimated location and associated attitude angle. Finally, the extensive experimental results demonstrate the effectiveness of the proposed trajectory reckoning method with the BDS attitude measuring and point positioning.

Keywords: BeiDou System · Trajectory reckoning · Attitude measuring · Point positioning · Carrier phase

1 Introduction

The BDS has been used in all aspects of daily life, such as automatic driving, unmanned navigation, and air transportation. In recent years, with the rapid development of BDS, navigation satellites are used to obtain reliable and accurate attitude information, which has become a high-income, low-cost technology. Fan et al. [1] use a centralized extended Kalman filter to achieve attitude measuring based on GPS/gyroscope combination, conduct static and vehicle-borne dynamic experiments. Aboelmagd Noureldin achieves GPS/INS combined attitude measuring and positioning navigation by constructing a neural network model [2]. The traditional integrated navigation systems rely on satellites and

© Springer Nature Singapore Pte Ltd. 2020
Q. Liang et al. (eds.), *Communications, Signal Processing,*
and Systems, Lecture Notes in Electrical Engineering 516,
https://doi.org/10.1007/978-981-13-6504-1_155

inertial sensors for integrated navigation to obtain track information, but inertial sensors have disadvantages such as high cost and accumulated error.

In order to solve the problems of the traditional integrated navigation system, we propose a trajectory reckoning method using BDS for positioning and attitude measuring. In attitude measuring phase, by solving the ephemeris data and observation data, the dual-difference pseudo-range and carrier phase observation equations are solved to obtain the attitude information. In point positioning phase, we establish the pseudo-range observation equations based on the satellite ephemeris data to solve the location of receiver.

The rest of the paper is organized as follows. In Sect. 2, we describe the proposed method in detail. In Sect. 3, we introduce the framework of the proposed trajectory reckoning system and also present the algorithm principle of the BDS attitude measuring and point positioning. Section 4 shows the experimental results and finally the conclusion of this paper is given in Sect. 5.

2   Related Work

In [3], the authors propose a GPS/BDS-combined baseline solution method and focus on the issue of cycle slips in relative positioning, and conduct a detailed study of how to detect cycle slips. Wu et al. [4] propose a multi-information fusion directional attitude measuring method, and use a micro-electro-mechanical System (MEMS) sensor and a GPS dual antenna to perform fusion attitude measuring. In [5], the authors use GPS and dead reckoning (DR) combined positioning method based on federated Kalman filter structure to carry out research on vehicle-mounted technology based on geographic information system, they use the unscented Kalman filter to achieve GPS/DR combined positioning to further improve the accuracy of the car navigation system.

Different from the works mentioned before, we firstly use the BDS pseudo-range observation equations to solve the location of moving target. Second, the carrier phase observation equations are solved by using the least-squares ambiguity decorrelation adjustment (LAMBDA) algorithm, and the algorithm is also used to solve heading angle. Finally, we use the path deduction method to output estimated trajectory of moving target.

3   System Description

As shown in Fig. 1, the proposed system contains heading angle and location calculation modules. In our system, we use BDS antenna to receive the original data. According to the BDS original data, we select BDS dual-antenna attitude measuring algorithm and point positioning algorithm to calculate the location and heading angle of carrier at each moment. Then, the location and heading angle information are used to calculate the trajectory of moving target.

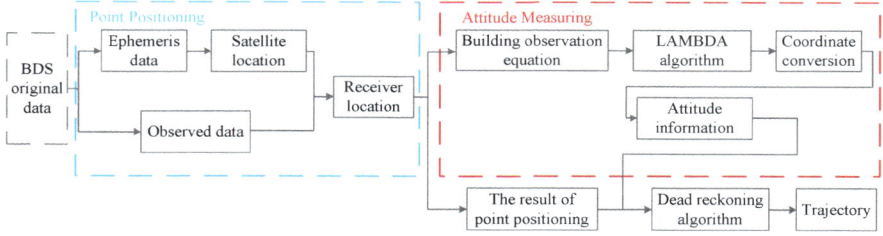

Fig. 1. System structure of the proposed BDS trajectory reckoning.

3.1 Attitude Measuring

In this paper, we mainly study single-baseline attitude measuring system, which is composed of dual antennas and the two BDS receiving antennas are fixed in the longitudinal direction of the carrier. As shown in Fig. 2, Antenna 1 is the main antenna and Antenna 2 is the secondary antenna.

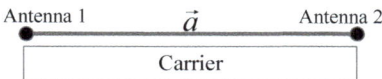

Fig. 2. Sketch map of single baseline attitude measuring.

Antenna 1 and Antenna 2 simultaneously perform pseudo-range and carrier phase measurements. As shown in Fig. 3, according to the attitude measuring algorithm, the location of baseline in the Earth-fixed Coordinate System (ENU) can be obtained, so as to calculate the attitude angles yaw and pitch.

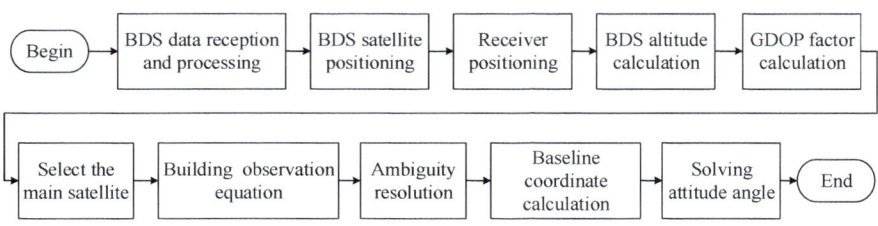

Fig. 3. Scheme of single baseline attitude measuring.

3.1.1 Double-Difference Observation Equation

BDS broadcast satellite ephemeris. By analyzing the ephemeris, the location of satellite orbital can be calculated as

$$\begin{cases} X_{\mathrm{k}} = x_k \cdot \cos \Omega_k - y_k \cdot \cos i_k \cdot \sin \Omega_k \\ Y_{\mathrm{k}} = x_k \cdot \sin \Omega_k - y_k \cdot \cos i_k \cdot \cos \Omega_k \\ Z_{\mathrm{k}} = y_k \cdot \sin i_k \end{cases} \tag{1}$$

where $[X_k, Y_k, Z_k]$ is the location of satellite and $[x_k, y_k, z_k]$ is the coordinates of satellite in orbital plane.

According to the geometric dilution Of precision (GDOP) factor selection results, the four minimum satellite combinations of GDOP factors are selected. Combined with the principle of double-difference observation model, six double-difference observation models are constructed to establish the following observation equation

$$Y = AX \tag{2}$$

where A is the receiver-to-satellite unit vector matrix and X is the required baseline vector and ambiguity solution. According to the least squares theory, X [6,7] can be obtained as

$$X = (A^T \cdot P \cdot A)^{-1} \cdot A^T \cdot P \cdot Y \tag{3}$$

where P is weight matrix.

Then, we set the double-difference integer ambiguity vector is a and baseline correction vector is $b = [\delta X_j, \delta Y_j, \delta Z_j]^T$, the least squares result can be calculated by

$$\hat{X} = \begin{bmatrix} \hat{b} \\ \hat{a} \end{bmatrix} \tag{4}$$

3.1.2 Gesture Solution

By searching for the integer ambiguity by the Lambda algorithm, a double-difference integer ambiguity vector can be obtained. The integer accuracy of the integer ambiguity can be used to further improve the baseline vector estimation accuracy [8]

$$\breve{b} = \hat{b} - Q_{\hat{b}\hat{a}} \cdot Q_{\hat{a}}^{-1} \cdot (\hat{a} - \breve{a}) \tag{5}$$

where the baseline $\breve{b}$ is based on the CGCS2000 (China Geodetic Coordinate System 2000), $\breve{b}_n$ is obtained by converting the coordinate transformation matrix to the ENU coordinate system.

$$\breve{b}_n = \begin{bmatrix} x_n \\ y_n \\ z_n \end{bmatrix} = R_L^n \cdot \breve{b}_l = \begin{pmatrix} -\sin\lambda & \cos\lambda & 0 \\ -\cos\lambda\sin\phi & -\sin\lambda\sin\phi & \cos\phi \\ \cos\cos\varphi & \sin\lambda\cos\varphi & \sin\phi \end{pmatrix} \begin{bmatrix} \delta X \\ \delta Y \\ \delta Z \end{bmatrix} \tag{6}$$

where λ is the longitude and ϕ is the latitude. The attitude rotation matrix C_n^b is expressed as

$$C_n^b = R_x \cdot R_y \cdot R_z = \begin{pmatrix} 1 & 0 & 0 \\ 0 & \cos\theta & \sin\theta \\ 0 & -\sin\theta & \cos\theta \end{pmatrix} \begin{pmatrix} \cos\gamma & 0 & -\sin\gamma \\ 0 & 1 & 0 \\ \sin\gamma & 0 & \cos\gamma \end{pmatrix} \begin{pmatrix} \cos\psi & \sin\psi & 0 \\ -\sin\psi & \cos\psi & 0 \\ 0 & 0 & 1 \end{pmatrix} \tag{7}$$

$$\begin{bmatrix} x_n \\ y_n \\ z_n \end{bmatrix} = C_n^b \cdot \begin{bmatrix} x_b \\ y_b \\ z_b \end{bmatrix} = \begin{pmatrix} \cos\varphi & \sin\varphi & 0 \\ -\cos\theta\sin\varphi & \cos\theta\cos\varphi & \sin\theta \\ \sin\theta\sin\varphi & -\sin\theta\cos\varphi & \cos\theta \end{pmatrix} \begin{bmatrix} 0 \\ 1 \\ 0 \end{bmatrix} \tag{8}$$

$$\begin{cases} x_n = -\cos\theta\sin\varphi \\ y_n = \cos\theta\cos\varphi \\ z_n = \sin\theta \end{cases} \tag{9}$$

The heading angle can be calculated by

$$\varphi = \arctan\frac{x_n}{y_n} \tag{10}$$

3.2   Point Positioning

According to the pseudo-range observation equation [9]

$$\rho^{(n)} = r^{(n)} + \delta\,t_u - \delta\,t^{(n)} + I^{(n)} + T^{(n)} + \varepsilon\,\rho^{(n)} \tag{11}$$

where $n = 1, 2, \ldots, N$ is temporary number of satellite measurements, $r^{(n)}$ is the geometry distance from satellite to receiver, $\delta\,t_u$ is receiver clock error, $\delta\,t^{(n)}$ is satellite clock error, $I^{(n)}$ is ionosphere delay, $T^{(n)}$ is tropospheric delay, and pseudo-range measurement noise is $\varepsilon\,\rho^{(n)}$. Error-corrected pseudo-range measurements $\rho_c^{(n)}$ is expressed as

$$\rho_c^{(n)} = \rho^{(n)} + \delta\,t^{(n)} - I^{(n)} - T^{(n)} \tag{12}$$

The corrected pseudo-range observation equation is described as

$$r^{(n)} + \delta\,t_u = \rho_c^{(n)} - \varepsilon\,\rho^{(n)} \tag{13}$$

where $r^{(n)}$ is geometric distance from the receiver to the satellite n that can be expressed as

$$r^{(n)} = \left\| x^{(n)} - x \right\| = \sqrt{\left(x^{(n)} - x\right)^2 + \left(y^{(n)} - y\right)^2 + \left(z^{(n)} - z\right)^2} \tag{14}$$

where $x = [x, y, z]^T$ is an unknown receiver position coordinate vector, $\left[x^{(n)}, y^{(n)}, z^{(n)}\right]^T$ is the location coordinate vector of satellite n [10]. Then, we subtract the pseudo-range error $\varepsilon\,\rho^{(n)}$ in Eq. (13) to obtain a quaternary nonlinear system of equations

$$\begin{cases} \sqrt{\left(x^{(1)} - x\right)^2 + \left(y^{(1)} - y\right)^2 + \left(z^{(1)} - z\right)^2} + \delta\,t_u = \rho_c^{(1)} \\ \sqrt{\left(x^{(2)} - x\right)^2 + \left(y^{(2)} - y\right)^2 + \left(z^{(2)} - z\right)^2} + \delta\,t_u = \rho_c^{(2)} \\ \cdots \\ \sqrt{\left(x^{(N)} - x\right)^2 + \left(y^{(N)} - y\right)^2 + \left(z^{(N)} - z\right)^2} + \delta\,t_u = \rho_c^{(N)} \end{cases} \tag{15}$$

By solving the four unknowns of the above equations, the location of receiver can be solved. Then, the iterative process is repeated through Newton iteration to obtain accurate positioning results. The iteration process is shown as below.

Step (1) Preparing data and setting initial solution.
Step (2) Linearizing nonlinear equations.
Step (3) Solving linear equations.
Step (4) Updating the root of the nonlinear equations.
Step (5) Determining the convergence of Newton's iteration.

4 Experimental Results

In this part, we verify the feasibility of the proposed method in a real-world environment. The platform consists of a BDS receiver, a personal computer (PC), and two BDS antennas, which is as shown in Fig. 4. During the test, the moving target carries the dual-antenna masts move in the direction of the predetermined trajectory, and saves the ephemeris data and observation data received by the BDS receiver on PC.

In the open environment (N 29.5389° and E 106.6039°), we set the heading angle in the north 0°, east 90°, west −90°, and south −180°. The real trajectories of moving target are L-shape and rectangle, respectively.

The L-shaped trajectory starts from direction of the east, and goes straight ahead by 10 m turn to the north. The positioning coordinates of moving target are shown in Fig. 5 and the heading angle is shown in Fig. 6. From Figs. 5 and 6, we can find that the actual positioning trajectory of L-shape test trajectory roughly consistent with the actual trajectory, the heading angle output clearly jumps from 90° to 0° at the corner.

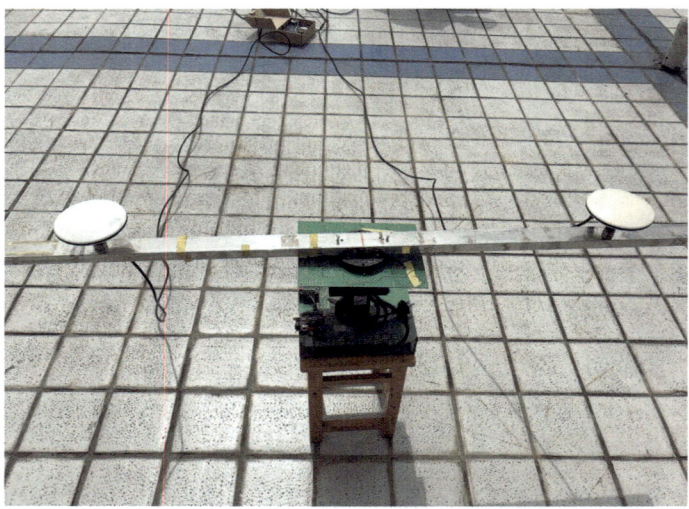

Fig. 4. Platform of BDS dual-antenna test.

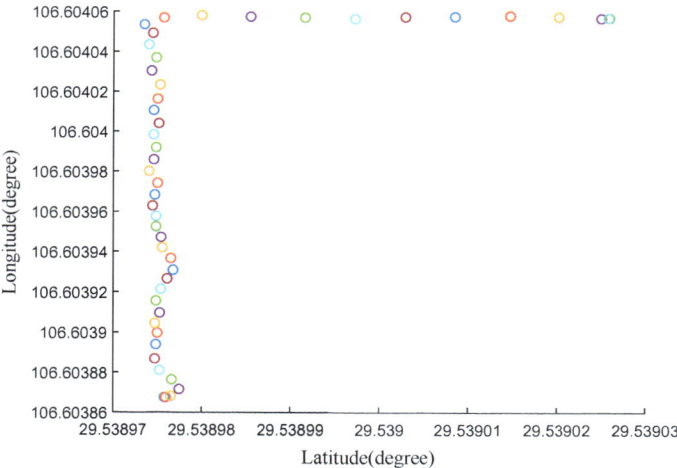

Fig. 5. Positioning result of L-shape trajectory.

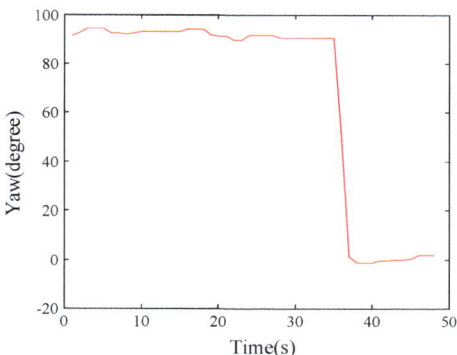

Fig. 6. Estimated heading angle of L-shape trajectory.

The rectangle test trajectory starts from direction of the east, and goes straight ahead 10 m turn to the north, then turns straight to the west and goes straight for 10 m, at last, turn left 90°, heading angle is −180°, which is a closed rectangle. The positioning coordinates are shown in Fig. 7 and the heading angle output is shown in Fig. 8. From the two figures, we can find that the estimated trajectory almost fits the true trajectory.

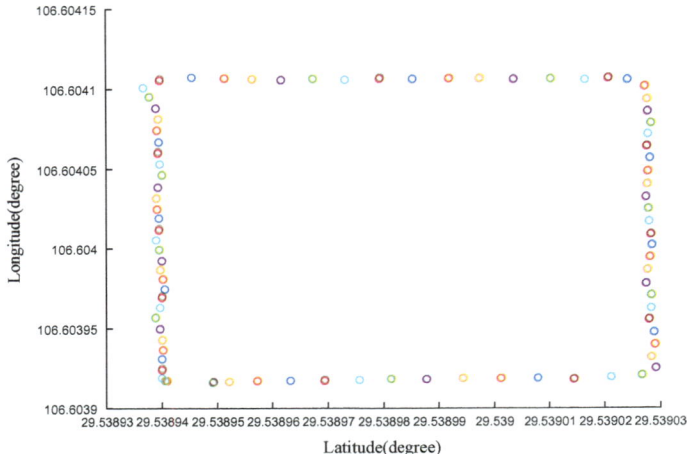

Fig. 7. Positioning result of rectangle trajectory.

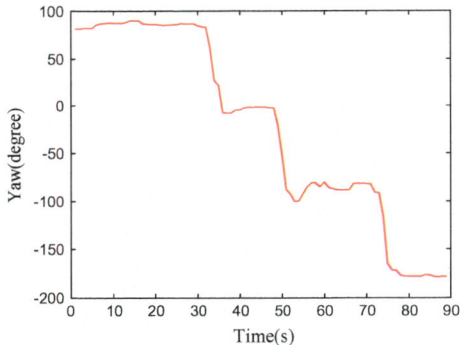

Fig. 8. Estimated heading angle of rectangle trajectory.

5   Conclusion

In this paper, we rely on the BDS single-baseline attitude measuring and point positioning to effectively improve the accuracy of target trajectory reckoning. By constructing the carrier phase and pseudo-range observation equations, we are capable of estimating the target location and the associated attitude angle. Compared with the traditional integrated positioning and navigation system, the proposed method is featured with high flexibility and low cost.

Acknowledgments. This work is supported in part by the National Natural Science Foundation of China (61771083, 61704015), Program for Changjiang Scholars and Innovative Research Team in University (IRT1299), Special Fund of Chongqing Key Laboratory (CSTC), Fundamental Science and Frontier Technology Research Project of Chongqing (cstc2017jcyjAX0380, cstc2015jcyjBX0065), Scientific and Technological Research Foundation of Chongqing Municipal Education Commission (KJ1704083), and University Outstanding Achievement Transformation Project of Chongqing (KJZH17117).

References

1. Fan S, Zhang K, Wu F. Ambiguity resolution in GPS-based, low-cost attitude determination. Positioning. 2005;4(1):207–14.
2. Noureldin A, El-Shafie A, Bayoumi M. GPS/INS integration utilizing dynamic neural networks for vehicular navigation. Inf Fusion. 2001;12(1):48–57.
3. Chen J, Yue DJ, Zhao XW, Wang J. BDS/GPS combined single epoch baseline solution method. J Surv Mapp Sci Technol. 2017;34(3):232–5.
4. Wu JJ, Qian F. Research on directional posture method of multiple information fusion. Electron Meas Technol. 2012;35(2):41–5.
5. Yang C, Li RZ, Li Y. The application of GIS-based positioning correction in vehicle-borne inertial navigation system. Telemetry Remote Control. 2015;36(3):70–4.
6. Mosavi MR, Azarshahi S, Emamgholipour I, et al. Least squares techniques for BDS receivers positioning filter using pseudo-range and carrier phase measurements. Iran J Electr Electron Eng. 2014;10(1):18–26.
7. Park C, Teunissen PJG. Integer least squares with quadratic equality constraints and its application to GNSS attitude determination systems. Int J Control Autom Syst. 2009;7(4):566–76.
8. Wang JM, Ma TM, Zhu HZ. Improved LAMBDA algorithm to quickly resolve BDS dual-frequency integer ambiguities. Syst Eng Theory Pract. 2010;37(3):768–72.
9. Xie G. GPS principle and receiver design. Electronic Industry Press; 2011. p. 101–2.
10. He JL, Liu ZM. Beidou navigation satellite position calculation method. Glob Positioning Syst. 2013;38(5):5–10.

An Adaptive Passive Radio Map Construction for Indoor WLAN Intrusion Detection

Yixin Lin[✉], Wei Nie, Mu Zhou, Yong Wang, and Zengshan Tian

Chongqing Key Lab of Mobile Communications Technology, Chongqing University of
Posts and Telecommunications, Chongqing 400065, China
260150244@qq.com, linyixin_cqupt@foxmail.com,
{niewei,zhoumu,tianzs}@cqupt.edu.cn, dr_ywang@hotmail.com

Abstract. Indoor WLAN intrusion detection technique for the
anonymous target has been widely applied in many fields such as the
smart home management, security monitoring, counterterrorism, and
disaster relief. However, the existing indoor WLAN intrusion detection
systems usually require constructing a passive radio map involving a lot
of manpower and time cost, which is a significant barrier of the deployment of WLAN intrusion detection systems. In this paper, we propose
to use the adaptive-depth ray tree model to automatically construct an
adaptive passive radio map for indoor WLAN intrusion detection. In
concrete terms, the quasi-3D ray-tracing model is enhanced by using the
genetic algorithm to predict the received signal strength (RSS) propagation feature under the indoor silence and intrusion scenarios, which
improves the computational efficiency while preserving the accuracy of
passive radio map. Then, the RSS mean, variance, maximum, minimum,
range, and median are allied to increase the robustness of passive radio
map. Finally, we conduct empirical evaluations on the real-world data to
validate the high intrusion detection rate and low database construction
cost of the proposed method.

Keywords: Indoor intrusion detection · Adaptive ray-tracing ·
Passive radio map · Genetic algorithm · WLAN

1 Introduction

With the wide deployment of wireless local area network (WLAN) and general
support of WLAN protocol by various intelligent terminals, the intrusion detection with respect to the indoor target can be realized by using the existing WLAN
infrastructure. Among the existing anonymous target intrusion detection techniques, the wireless local area network (WLAN) indoor target intrusion detection
system [1–4] proposed by the University of Maryland performs outstandingly

© Springer Nature Singapore Pte Ltd. 2020
Q. Liang et al. (eds.), *Communications, Signal Processing,
and Systems*, Lecture Notes in Electrical Engineering 516,
https://doi.org/10.1007/978-981-13-6504-1_156

because it can effectively protect the user's location privacy and work stably under non-line-of-sight and without special hardware at the same time. However, the main problem with this kind of algorithms is that the construction of the prior passive radio map takes a lot of manpower and time, which is a major barrier of WLAN intrusion detection systems deployment. On this basis, the WLAN indoor target intrusion detection algorithm proposed in this paper uses the adaptive-depth ray tree-based quasi-3D ray-tracing model to construct the passive radio map automatically, which requires less labor overhead compared with the traditional RSS feature database construction method. In addition, six signal characteristics of the passive radio map are constructed, which results in the better pattern recognition ability and learning convergence. The rest of this paper is structured as follows. In Sect. 2, we describe the proposed indoor WLAN intrusion detection method in detail, and the related experimental results are shown in Sect. 3. Finally, we conclude this paper in Sect. 4.

2 System Description

The overall flow of the system is shown in Fig. 1. First, a number of WLAN access points (APs) and monitor points (MPs) are arranged in the target area. Second, the GA algorithm is used to optimize the limited number of depth of the ray tree adaptively and the RSS characteristics under the indoor silence, and intrusion scenarios are constructed according to the optimized ray-tracing model. Then, the obtained RSS characteristics are used for probabilistic neural network (PNN) training. Finally, the trained PNN is used to classify the new observation RSS data by multiple classifications, so as to realize the intrusion detection and area localization.

2.1 Signal Prediction

Considering the limitations of the existing 2D and 3D ray-tracing models [5, 6] on the accuracy of the signal prediction and the complexity of the algorithm respectively, the quasi-3D ray-tracing model used in this paper first carries out the ray-tracing in the 2D projection plane, and then transforms it into the propagation path in the 3D space, and this process significantly improves the computational efficiency while guaranteeing the accuracy of prediction. In this case, as shown in Fig. 2, a quasi-3D ray-tracing model based on the adaptive-depth ray tree is proposed in this paper, considering two factors: the model accuracy and calculation efficiency.

The import of environmental information. Figure 3 gives a 3D modeling of a simple environment and the corresponding 2D projection results. The gray and black parts of the diagram represent the boundary face of environment and the indoor facilities, respectively. In addition, in order to ensure the integrity of the imported environmental information, the 3D vertex coordinates, height information and relative permittivity, conductivity and permeability of the corresponding material will be recorded.

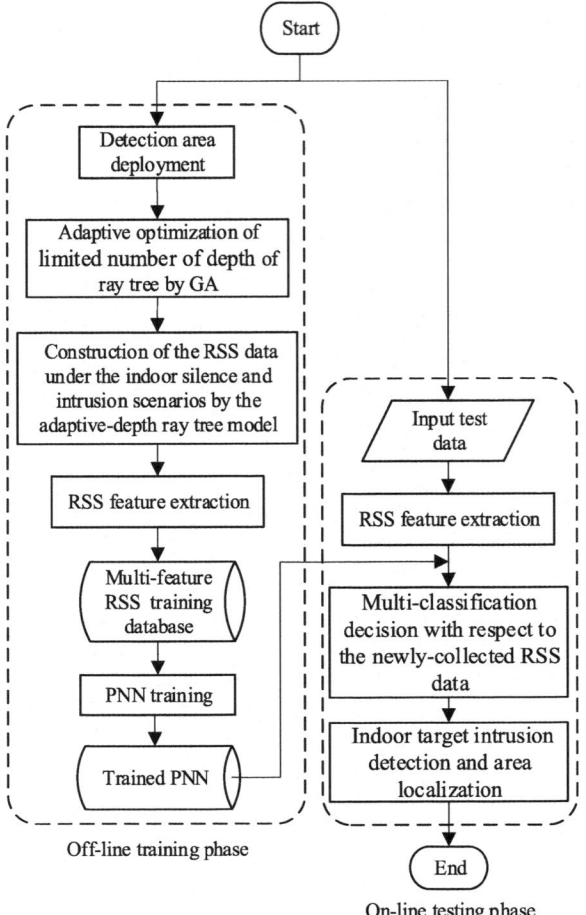

Fig. 1. Overall system flowchart

Optimization of the limited number of depth. In order to significantly improve the computing efficiency of the ray tree, the GA algorithm is used to optimize the limited number of depth of the ray tree in different environments. Specifically, first, the limited number of depth is initialized to 1; secondly, all the vertical planes and vertical lines of the 3D modeling of the environment are numbered; besides, the number of the functional parts of each ray is spliced into a chromosome in chronological order, and the field strength of each ray that reaches the MP is used as the fitness of its corresponding chromosome; then, the contribution rate of the ray to the field strength at MP under the condition of the current limited number of depth is calculated by Algorithm 1;

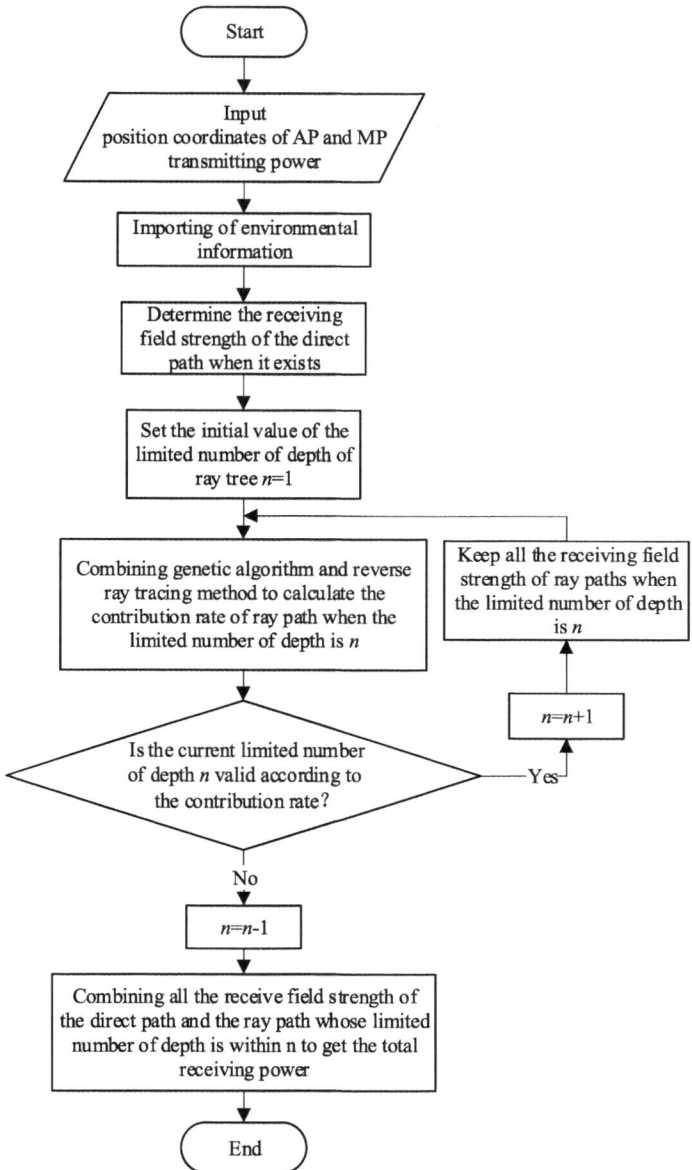

Fig. 2. Signal prediction flowchart

finally, determine whether the contribution rate of the ray under the current limited number of depth is greater than the preset threshold, and if so, add 1 to the limited number of depth and repeat the above steps, otherwise the current limited number of depth is the optimal limited number of depth (or the

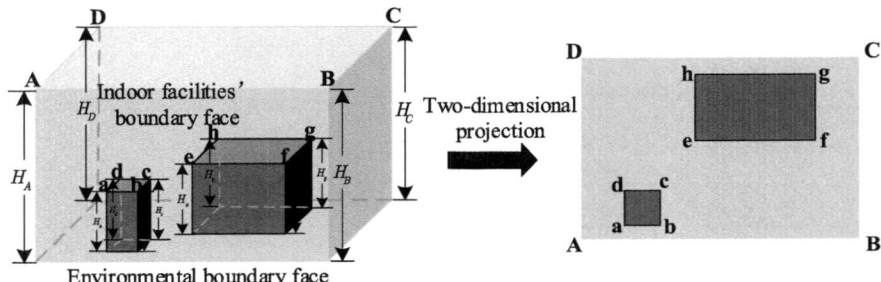

Fig. 3. 2D projection from 3D modeling of environment

Algorithm 1 Calculation of contribution rate of ray to the field strength

---

Input: n (the limited number of depth); $D_k (k=1, \ldots, N)$(the unique number of the kth functional part); N (the total number of the functional parts); P_c (crossover probability); P_m (mutation probability); M (population size); ρ_{th}(calculation rate threshold); e_{n-1}(maximum field strength of the n-1 order ray)

Output: C_n(contribution rate of the ray to the n order field strength)

1: The first generation population T_1 is randomly generated according to the limited number of depth n (that is, M random number sequences with a length of n), and set the current population $T = T_1$

2: **while** $\rho < \rho_{th}$ **do**

3:     Use inverse ray-tracing method to calculate the fitness of each chromosome in current population T, namely $e_1,...,e_M$ (as described in Algorithm 2)

4:     **for** $i= 1{:}M$ **do**

5:        2 chromosomes were selected from T by fitness ratio selection algorithm[7]

6:        **if** random(0,1)$< P_c$ **then**

7:           implement crossover operation on the selected 2 chromosomes

8:        **end if**

9:        **if** random(0,1)$< P_m$ **then**

10:          implement mutation operation on the selected 2 chromosomes

11:       **end if**

12:       Add this 2 new chromosomes to the updated population T_{new}

13:    **end for**

14:    $T \Leftarrow T_{new}$

15:    Compute $\rho = N_C/N^n$, and N_C is the number of chromosomes types that have appeared from the initial population to the current population

16: **end while**

17: Compute $T_f = e_{n-1}/2$

18: Compute m namely the number of chromosomes whose fitness$> T_f$ in the current population T

19: Compute $C_n = m/M$

---

optimal ray order). In Algorithm 1, the fitness of each chromosome in the current population is calculated by the reverse ray-tracing method, and its calculation process is described in Algorithm 2.

Algorithm 2 Calculation of the n order chromosome fitness

Input: $D_k(k=1, \ldots, N$ (the unique number of the kth functional part); N (the total number of the functional parts); P_k and H_k $(k = 1, \ldots, N_1)$ (vertex coordinate and height information of the kth vertical lines); N_1 (the total number of the vertical lines); c_k ε_k and μ_k $(k = 1, \ldots, N_2)$ (relative permittivity, conductivity and permeability of the kth vertical planes); N_2 (the total number of the vertical planes); λ (working wavelength); n (the limited number of depth); P_{AP} and P_{MP} (position coordinates of AP and MP); P_t (transmitting power of AP)

Output: e_i(chromosome fitness of the ith n order ray)

1: Assign B_k $(k = 1, \ldots, n+2)$ according to the number sequence of functional part of the ith $(i = 1, \ldots, M)$ n order ray, in which $B_1 = 1$ and $B_{n+2} = 1$ is the initial position of the ray AP and the termination position MP respectively; $B_k=0$ and 1 $(k = 2, \ldots, n+1)$represents reflection and diffraction occurs in the kth functional part respectively.

2: The 2D projection coordinates of AP and MP are L_1 and L_{n+2}, respectively

3: **for** $k = 2: n+1$ **do**

4:     **if** $B_k = 0$ **then**

5:         $L_k \Leftarrow$ 2D projection coordinates of the mirror points of L_{k-1} with respect to the kth vertical planes

6:     **else if** $B_k = 1$ **then**

7:         $L_k \Leftarrow$ 2D projection coordinates of the kth vertical planes

8:     **end if**

9: **end for**

10: **while** there exists $k \in (1, ..., n)$ which makes $B_k = 1$ **do**

11:     **if** $B_k = 0$ and $B_{k+1} = 1$ **then**

12:         $L_k \Leftarrow$ 2D projection coordinates of the intersection point of the line connecting L_k and L_{k+1} and the kth vertical planes

13:         $B_k \Leftarrow 1$;

14:     **end if**

15: **end while**

16: **for** $k = 1 : n+1$ **do**

17:     Set $T_{k,k+1}$ as the line connecting L_k and L_{k+1}

18:     **if** there exists the intersection of $T_{k,k+1}$ and any functional part **then**

19:         $e_i \Leftarrow 0$

20:         break;

21:     **end if**

22: **end for**

23: **if** $e_i \neq 0$ **then**

24:     The 2D projection of the ith n order ray extends to 3D space according to the Fermat principle, and then the fitness of the corresponding chromosome e_i is calculated

25: **end if**

Calculation of received signal power. In order to calculate the received signal power of MP, direct and non-direct rays are considered respectively. All the direct and non-direct rays within the n order are superimposed on the signal field strength, and the received signal power at MP can be obtained by the ray power summation method [8] as

$$P_{total} = \sum_{i=1}^{l} \left(\frac{\lambda |E_i|}{4\pi |E_0|} \right)^2, \qquad (1)$$

in which E_0 is the arrival signal field strength at 1 m from AP, and E_i is the arrival signal field strength of the ith ray, l is the total number of rays.

2.2 Intrusion Detection

In this paper, the kernel density estimation method based on Bayesian decision theory is applied to train the PNN feature data under the indoor silence and intrusion scenarios.[1] In particular, the kernel density function is used to estimate the conditional probability of different states, and then the state of the maximum posterior probability is used as the PNN output [9] according to the Bias decision theory. In order to ensure the stability of RSS characteristic data between each pair of AP and MP, this paper uses a sliding window function to segment the original RSS data[2] and calculates the mean, variance, maximum, minimum, maximum, and middle value of each segment data. On the basis of these six signal characteristics, six PNN structures are trained respectively. Finally, according to the voting criterion, the indoor target detection and location are realized by the multiclassification decision of the newly acquired RSS data.

3 Experimental Result

3.1 Environmental Layout

Figure 4 shows an experimental environment, in which two APs (AP1 and AP2 with model D-Link DAP 2310) and three MPs (MP1, MP2, and MP3 with model SAMSUNG GT-S7568) are placed at 2 and 0.5 m high, respectively. At each MP, 5 min of RSS data from each AP are collected separately under the indoor silence and intrusion scenarios.

3.2 GA Optimization Result

Figure 5 shows the change of the overall fitness of each generation of population under the conditions of different values of ρ_{th} when the GA was used to calculate the ray contribution rate. The overall fitness is defined as the ratio of m to the population size M, and m is the number of chromosomes whose fitness is greater than the threshold value T_f in the population. It can be seen that with the increasing of population algebra, the overall fitness is on the rise and tends to be the same when the population algebra reaches 30. In addition, Table 1

---

[1] Considering the content of water in the human body more than 70%, the human body is modeled as a 3D water column [10] with a certain height.

[2] The length of each segment of the RSS data is the width of the sliding window.

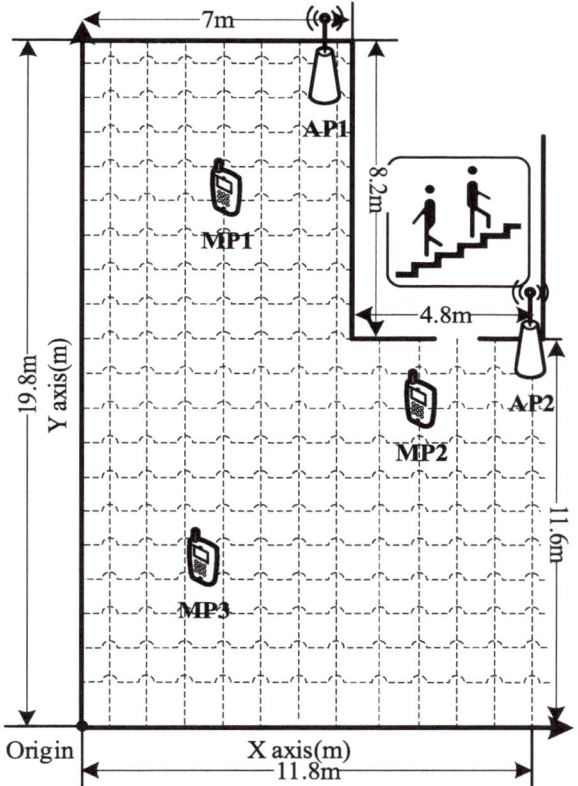

Fig. 4. Structure of experimental environment

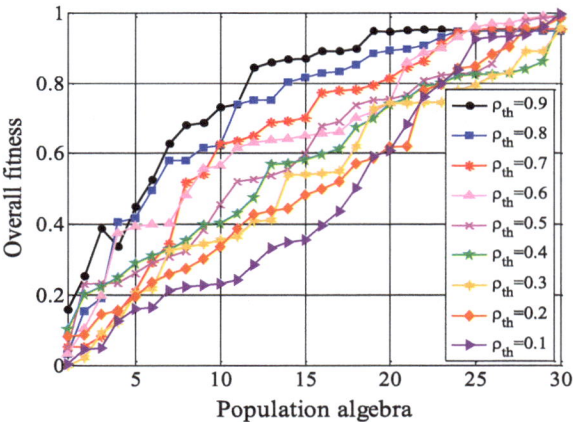

Fig. 5. Change of overall fitness

compares the average time overhead required by the 3D ray-tracing model [5], the traditional 2D ray-tracing model [6], and the proposed method for the ray modeling between each pair of AP and MP under the condition of the limited number of depth of 3. It can be seen from the table that this method performs obviously better than the methods used in the literature [5,6] in terms of time overhead.

Table 1. Average time cost for ray modeling between each pair of AP and MP

Performance index	Paper [6]	Paper [5]	The proposed
Time overhead (s)	6.03	7.25	3.41

3.3 Signal Prediction Result

Figures 6 and 7 compare the cumulative density function (CDF) of RSS prediction errors by the proposed method and the ones in [5,6] under the limited number of depth of 3, from which we can find that the proposed method performs better than the others.

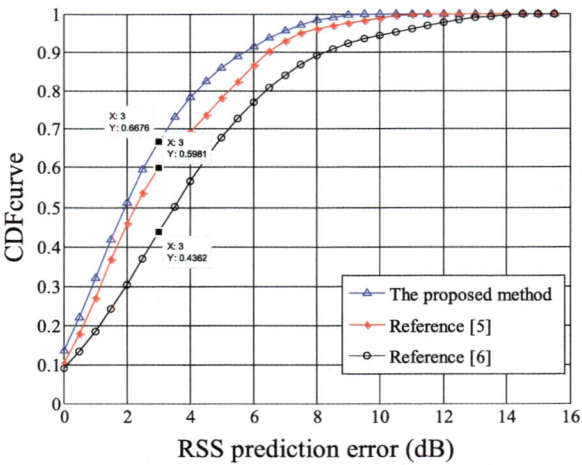

Fig. 6. CDF of errors for AP1

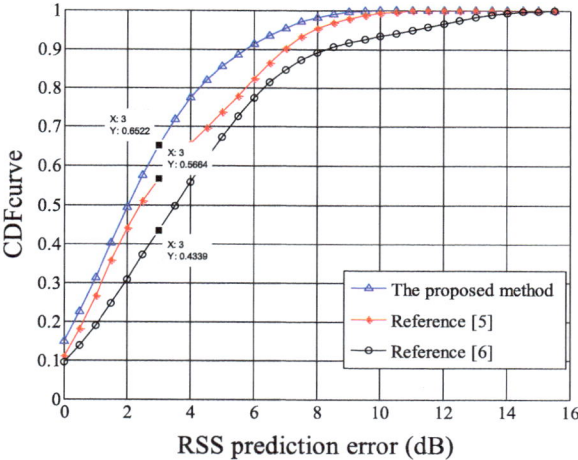

Fig. 7. CDF of errors for AP2

4 Conclusion

In this paper, we propose the adaptive-depth ray tree model, which can be used to adaptively construct a passive radio map for indoor WLAN intrusion detection. For one thing, we use the genetic algorithm to enhance the traditional quasi-3D ray-tracing model to depict the RSS variation under the indoor silence and intrusion scenarios with low labor and time cost. For another, six common signal features are allied to ensure the stability of RSS data and robustness of passive radio map. In future, we will continue to investigate a more effective passive radio map construction method to accurately locate multiple targets in the anonymous indoor WLAN environment.

Acknowledgments. This work is supported in part by the Fundamental Science and Frontier Technology Research Project of Chongqing (cstc2017jcyjAX0380).

References

1. Youssef M, Mah M, Agrawala A. Challenges: device-free passive localization for wireless environments. In: ACM international conference on mobile computing and networking; 2007. p. 222–9.
2. Jin S, Choi S. A seamless handoff with multiple radios in IEEE 802.11 WLAN. IEEE Trans Veh Technol. 2014;63(3):1408–18.
3. Wang Q, Yigitler H, Jantti R, et al. Localizing multiple objects using radio tomographic imaging technology. IEEE Trans Veh Technol. 2016;65(5):3641–56.
4. Deak G, Curran K, Condell J, et al. Detection of multi-occupancy using device-free passive localization. IET Wirel Sens Syst. 2014;4(3):130–7.
5. Liu Z, Guo L, Tao W. Full automatic preprocessing of digital map for 2.5D ray tracing propagation model in urban microcellular environment. Waves Random Complex Media. 2013;23(3):267–78.

6. Jong YLCD, Herben MAHJ. Prediction of local mean power using 2-D ray-tracing-based propagation models. IEEE Trans Veh Technol. 2001;50(1):325–31.

7. Sabar NR, Ayob M, Kendall G, et al. A dynamic multiarmed bandit-gene expression programming hyper-heuristic for combinatorial optimization problems. IEEE Trans Cybern. 2015;45(2):217–28.

8. Erceg V, Rustako AJ, Roman R. Diffraction around corners and its effects on the microcell coverage area in urban and suburban environments at 900 MHz, 2 GHz, and 6 GHz. IEEE Trans Veh Technol. 1994;43(3):762–6.

9. Dutt V, Chaudhry V, Khan I. Different approaches in pattern recognition. Comput Sci Eng. 2011;1(2):32–5.

10. Queiroz A, Trintinalia LC. An analysis of human body shadowing models for ray-tracing radio channel characterization; 2016. p. 1–5.

An Iris Location Algorithm Based on Gray Projection and Hough Transform

Baoju Zhang$^{(\boxtimes)}$ and Jingqi Fei

Tianjin Key Laboratory of Wireless Mobile Communications and Power Transmission, Tianjin Normal University, Tianjin 300387, China
wdxyzbj@163.com

Abstract. In order to improve the performance of the existing iris location algorithm, a transform algorithm based on gray projection and Hough is proposed. The algorithm uses the grayscale transformation of the binary image to obtain a graph of the gray projection. At the same time, according to the value of the peak or trough in the graph, the maximum radius of the circle is obtained. The result of experiment shows that: The algorithm can get the parameters needed in Hough transform, which greatly improves the speed and accuracy of iris positioning.

Keywords: Iris recognition · Canny edge detection · Binarization · Grayscale projection · Hough transform

1 Introduction

With the development of society, the importance of identity recognition is increasingly evident. In recent years, in the fields of maintaining national security, aviation safety, financial security, social security, and network security, there is a need for more accurate, reliable, and more practical authentication methods for identifying and authenticating identity. However, relying on identity documents, user names, and identity authentication in the form of passwords is far from meeting the requirements of the information age for the validity of authentication and the accuracy of identification. As an important identification feature, iris has the advantages of uniqueness, stability, collectability, and non-invasion. One of the keys to the iris recognition algorithm is to accurately locate the iris region from the acquired iris image. It mainly includes the edge position of the pupil and the iris, the iris and the sclera, hereinafter referred to as the inner edge and the outer edge.

There are many articles on iris recognition at domestic and foreign, including Daugman [1], who used the characteristics of the inner and outer edge of the iris as an approximate circle, a circular difference operator is proposed to extract the edge of the iris. Wildes [2] uses a two-step method combining edge detection and Hough transform to locate the iris region. Both of these iris location methods have high accuracy, but both searching in three-dimensional space, positioning speed is slow and is unable to meet the requirements of real-time systems. Tisse proposes to extract iris features using a time-phase technique. Weiqi Yuan [3] uses the traditional gray projection method to roughly locate the pupil center and position and then uses Hough transform to

© Springer Nature Singapore Pte Ltd. 2020
Q. Liang et al. (eds.), *Communications, Signal Processing, and Systems*, Lecture Notes in Electrical Engineering 516, https://doi.org/10.1007/978-981-13-6504-1_157

accurately position. When the iris image contains thick eyelashes, the gray projection method can easily misplace the eyelash position as the pupil position, so that the accuracy of this method is lower.

In order to overcome the above defects and further improve the iris recognition system, according to the characteristics of the iris itself, an iris location algorithm based on gray projection and Hough transform is proposed. This method can not only suppress noise interference, but also improve the accuracy of iris location.

2   Iris Localization Algorithm

2.1   Iris Image Smoothing and Edge Extraction

The collected iris images have different levels of interference. Filtering before the iris boundary location can help to eliminate the effect of interference on the boundary location. The image should be smoothed. On the one hand, it can highlight a large area, a low-frequency component and a trunk part, on the other hand, smoothed can suppress image noise and interfere with high-frequency components. At the same time, it can reduce a sudden gradient, and improve image quality [4]. In addition, taking into account the edge extraction of the grayscale image of the iris, the inner and outer boundaries of the iris belong to the detailed information. In order to not only retain the iris edge information but also effectively eliminate high-frequency interference, Gaussian filter templates are used for smoothing. The formula is as follows:

$$H_{ij} = \frac{1}{2\pi\delta^2} \exp\left(- \frac{[i - (k+1)]^2 + [j - (k+1)]^2}{2\delta^2} \right) \tag{1}$$

When Gaussian template filtering is used, the point farther from the center point will have less effect on the smoothing effect. According to the distance, selecting a weighting coefficient can construct Gaussian templates of different sizes whose filter parameters can be determined by the variance of the Gaussian function. The Gaussian template can well protect the contour information of the iris region while filtering the noise, so as not to cause the edges to be too fuzzy. Figure 1a is the original. The result is shown in Fig. 1b.

For effective edge extraction, Canny edge detection is used in this paper. In the first step, the gradient intensity and direction of each pixel in the image are calculated for the smoothed image. Edges in the image can point in all directions, so the Canny algorithm uses four operators to detect horizontal, vertical, and diagonal edges in the image. The operator of the edge detection (such as Roberts, Prewitt, and Sobel) returns the first derivative value of the horizontal G_x and vertical G_y directions, thereby determining the gradient G and the direction theta of the pixel [5]. The formula is as follows:

$$G = \sqrt{G_x^2 + G_y^2} \tag{2}$$

(a)   **(b)**

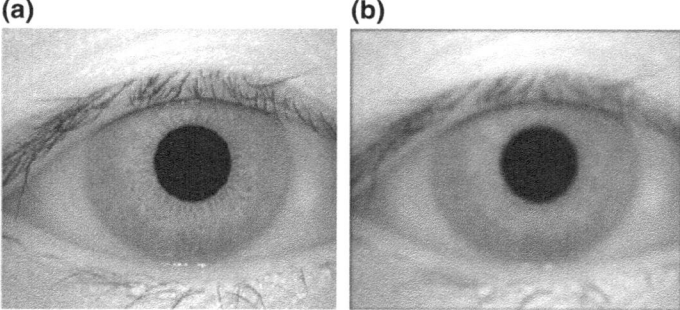

Fig. 1. (a) Original image (b) Gaussian image

$$\theta = \arctan\left(G_y/G_x\right) \tag{3}$$

In the second step, non-maximum suppression is performed to compare the gradient intensity of the current pixel with two pixels in the positive and negative gradient directions. If the gradient intensity of the current pixel is the largest compared to the other two pixels, the pixel remains as an edge point, otherwise the pixel will be suppressed. The third step, it is dual threshold detection. After non-maximal suppression, the remaining pixels can more accurately represent the actual edges in the image. However, there are still some edge pixels caused by noise and color changes. In order to solve this type of edge pixels, weak gradient values are used to filter edge pixels, and edge pixels with high gradient values are retained, which is achieved by selecting high and low thresholds [6]. If the edge pixel's gradient value is higher than the high threshold, it is marked as a strong edge pixel; if the edge pixel's gradient value is less than the high threshold and greater than the low threshold, it is marked as a weak edge pixel; if the edge pixel's gradient value is less than low thresholds, it will be suppressed. The choice of threshold depends on the content of a given input image. The result is shown in Fig. 2.

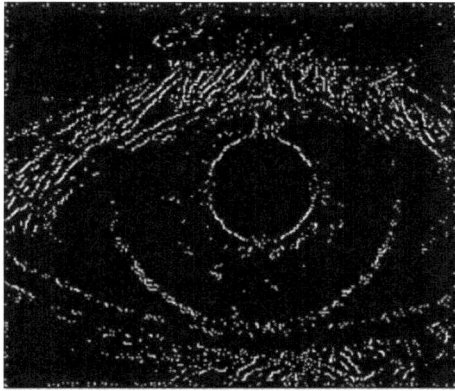

Fig. 2. Edge contour extraction map of the entire eye

2.2   Iris Boundary Localization

To locate the inner circle, we need to extract the inner circle. By setting the threshold, the image is binarized and the pupil is extracted. The formula is as follows:

$$\begin{cases} B(m,n) = 1, & \text{if} \quad I(m,n) > \alpha \\ B(m,n) = 0, & \text{if} \quad I(m,n) \le \alpha \end{cases} \tag{4}$$

According to the formula, in this paper, α has a value of 47, the binarized picture obtained, at the same time, we should extract the outline, as shown in Fig. 3a, b.

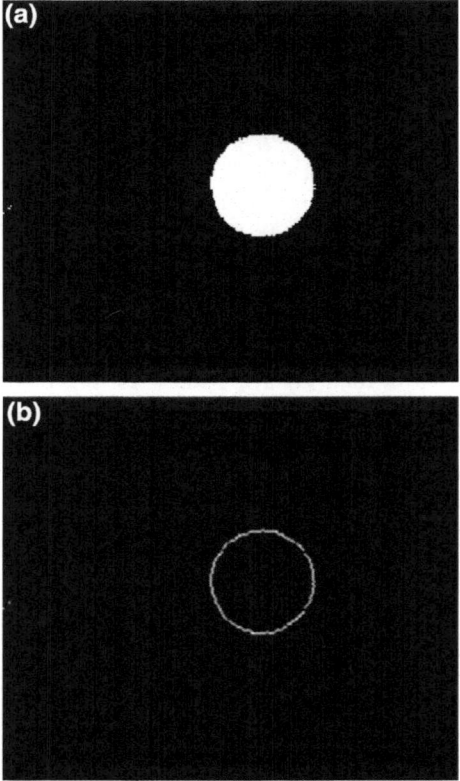

Fig. 3. (a) and (b) show further extraction of contours by binarization

Thus leaving only the outline of the inner circle. Then we can use the information of the gray change to determine the parameters of the pupil. The algorithm is as follows:

Step 1. Find the horizontal grayscale projection curve. The image is represented by a matrix of $M * N$, then M can be regarded as a matrix, i and j are the matrix rows and columns, and the horizontal grayscale projection curve of the image, then the abscissa is the row number of each row of the matrix, and the ordinate is a matrix [7]. The sum of the row elements, that is, the sum of the pixel values for each row in the picture. Actually, the curve contains the change in the sum of the pixels in each row of the image.

Step 2. Based on the calculation of the gray curve, the peaks and troughs of the curve are obtained, as shown in Fig. 4a. Since the black gray value is 0 and the white gray value is 255, the region coordinates information with a smaller gray value is included in the trough.

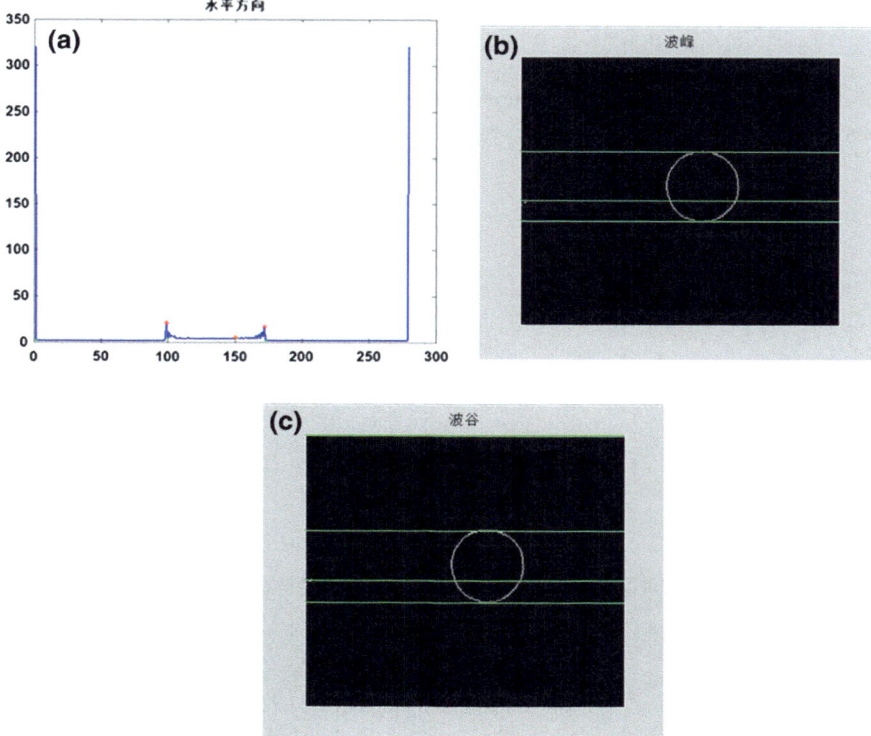

Fig. 4. (a) Shows a chart of pixels in the horizontal direction (b) and (c) show the peaks and valleys of the outline according to the chart

Step 3. Drawing the line information of the image represented by the abscissa of the wave trough, as shown in Fig. 4b, c, and the maximum value of the radius can take the maximum difference r_max in the crest or trough. According to the prior knowledge, the minimum radius can be obtained by subtracting 13 from the maximum difference r_max.

The Hough transform is used to locate the circle parameter. A circle is a typical and regular geometric shape with fewer parameters, namely the center coordinates and radius [8]. Therefore, detection of a circle becomes a process of voting on a parameter group. Define a three-dimensional array as

$$H(x_m, y_n, r) = \sum_{j=1}^{n} h(x_m, y_m, x_n, y_n, r) \qquad (5)$$

Here, $H(x_m, y_n, r)$ is an accumulator corresponding to the parameter group (x_m, y_j, r) formed by the center point coordinate and the radius size, and the obtained value is used to accumulate the votes of the group of parameters. The number of votes is represented by the number of boundary points passed by the circle drawn by the parameter. If the edge point falls on the circle corresponding to the parameter group, it is equivalent to the edge point casting a vote for the parameter group, and the corresponding array element value is increased by 1, otherwise the corresponding array element value is not changed [9]. The formula is as follow:

$$h(x_m, y_m, x_n, y_n, r) = \begin{cases} 1 & g(x_m, y_m, x_n, y_n, r) = 0 \\ 0 & g(x_m, y_m, x_n, y_n, r) = 1 \end{cases} \qquad (6)$$

In the formula, $g(x_m, y_m, x_n, y_n, r) = (x_m - x_n)^2 + (y_m - y_n)^2 - r^2$ is a decision function that satisfies the circular equation of the parameter (x_n, y_n, r). When $g(x_m, y_m, x_n, y_n, r) = 0$, the center of the circle is (x_n, y_n), and the boundary circle with the radius r passes through the edge point (x_m, y_m), indicating that the point will vote for the parameter (x_n, y_n, r). After passing through all votes of columnar-(5) statistic, the circle with the most votes will be used to determine the circle of the boundary equation [10]. In the end, the results obtained are shown in Fig. 5.

Fig. 5. Inner edge contour extraction image

3   Experiments Results

The computer CPU used in the experiment was clocked at 2.01 GHz and the memory was 1 G. The programming tool was MATLAB 2014a. The iris images used were all from the CASIA 1.0 database of the Chinese Academy of Sciences. According to the method mentioned above, the iris is positioned inside and outside the circle. The final result is shown in Fig. 6.

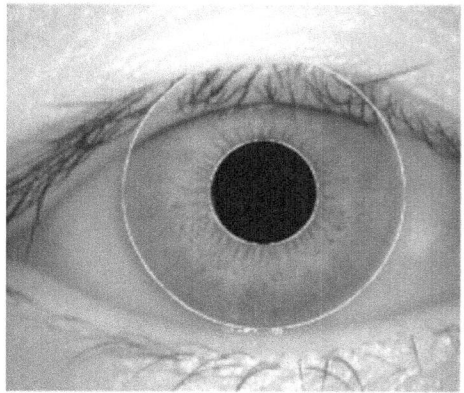

Fig. 6. The final result

4   Conclusions

Based on previous study of iris recognition, this paper proposes an algorithm which can locate iris by grayscale changes. Firstly, image which should be detected is binarized and the outline is extracted. Then, grayscale graph is drawn by taking advantage of gray change of the image. Finally, according to the peaks and valleys in the graph, the parameters required by the Hough transform are obtained. Experiments show that the method used in this paper reduces Hough transform traversal, which can make iris localization improve efficiently.

Acknowledgments. This paper is supported by Natural Youth Science Foundation of China (61501326, 61401310), the National Natural Science Foundation of China (61731006) and Natural Science Foundation of China (61271411). It also supported by Tianjin Research Program of Application Foundation and Advanced Technology (15JCZDJC31500), and Tianjin Science Foundation (16JCYBJC16500). This work was also supported by the Tianjin Higher Education Creative Team Funds Program.

References

1. Daugman J. How iris recognition work. IEEE Trans Circ Syst Video Technol. 2004;14 (1):21–30.
2. Wildes RP. Iris recognition an emerging biometric technology. Proc IEEE. 1997;85 (9):1348–63; 2000;15(10):939–57.
3. Yuan W, Lin Z, Xu L. A rapid iris location method based on the structure of human eyes. In: 27th annual international conference; 2006. p. 17–8.
4. Daugman J. New methods in Iris recognition. IEEE Trans Syst Man Cybern Part B Cybern. 2007;37(5):1167–75.
5. Kawaguchi T, Rizon M. Iris detection using intensity and edge information. Pattern Recogn. 2003;36:549–62.
6. Wang J-G, Sung E. Study on eye gaze estimation. IEEE Trans Syst Man Cybern Part B. 2002;32(3):332–50.
7. Daugman J. High confidence visual recognition of person by a test of statistical independence. IEEE Trans Pattern Anal Mach Intell. 1993;15(11):1148–61.
8. Canny J. A computational approach to edge detection. IEEE Trans Pattern Anal Mach Intell. 1986;8(6):679–714.
9. Williams GO. Iris recognition technology. IEEE Aerosp Electron Syst Mag. 1997;12(4): 23–9.
10. Park KR, Kim J. A real-time focusing algorithm for Iris recognition camera. IEEE Trans Syst Man Cybern Part C. 2005;35(3):441–4.

Robust Tracking via Dual Constrained Filters

Bo Yuan$^{(\boxtimes)}$, Tingfa Xu, Bo Liu, Yu Bai, Ruoling Yang, Xueyuan Sun, and Yiwen Chen

School of Optics and Photonics, Beijing Institute of Technology,
Beijing 100081, China
{bityuanbo,ciom_xtf1}@bit.edu.cn

Abstract. In this paper, we propose a novel correlation filter framework constrained by dual filters. The Minimum Output Sum of Squared Error (MOSSE) filter is the unbiased estimate of the filter which easily to cause overfitting. The trained filter by linear ridge regression is the biased estimate of the filter which can deal with the overfitting. We combine the advantages of the two filters to constrain the trained filter which optimizes our model. To deal with background clutter, clipping background patches around the target position up, down, left, and right, we add the cropped background patches to the learning filter. To overcome the challenge of occlusion, we introduce a novel criterion, Average Peak-to-Correlation Energy (APCE). Extensive experiments on the CVPR 2013 Benchmark well demonstrate that our tracker can effectively solve the background clutter and occlusion. Both quantitative analysis and qualitative analysis show that our tracker outperforms some state-of-the-art trackers.

Keywords: Dual filters · Background clutter · Occlusion ·
CVPR 2013 Benchmark

1 Introduction

In recent years, visual tracking has attracted the attention of researchers as an important part of machine vision field. A large number of tracking algorithms have been proposed [1–6]. In general, visual tracking is divided into single-target tracking and multi-target tracking. The paper focuses on the theoretical exploration of single-target tracking. The problem of object tracking is to estimate the trajectory of an object in a video sequence when given object's initial state (position and size) in the first frame. There are still many challenges that will affect the target tracking effect although some trackers have demonstrated superior performance in both precision and success rate, such as illumination variation,

© Springer Nature Singapore Pte Ltd. 2020
Q. Liang et al. (eds.), *Communications, Signal Processing,
and Systems*, Lecture Notes in Electrical Engineering 516,
https://doi.org/10.1007/978-981-13-6504-1_158

scale variation, occlusion, deformation, background clutter, and fast motion. So how to design a robust target tracking algorithm is still a difficult problem in the current target tracking field. In this paper, we are mainly focused on designing a novel visual tracking approach to correlation filtering and solving the problems of background clutter and occlusion.

Correlation Filter (CF) tracking algorithms have attracted the attention of researchers in recent years, due to their excellent performance. Bolme et al. [1], proposed the MOSSE tracker, it introduces correlation filtering into the target tracking field for the first time. The Fourier transform is used to directly convert the operation to the frequency domain, which greatly reduces the amount of computation. Henriques et al. [2], proposed Circulant Structure with Kernel (CSK) has superior performance. Furthermore, many CF-based trackers have been proposed to solve a kind of tracking challenges such as DSST proposed in [3] to deal with scale variation and KCF [4] proposed a CF-based tracker combining multi-channel features which improves tracking performance. Although these methods above have greatly improved in both precision and success rate, they still can not solve the tracking drift, due to scale variation, occlusion, and deformation.

Current correlation filter trackers still have some shortcomings; the first one is that tracking box is fixed, and the tracker cannot adapt to the scale change of the target. In other words, the model will be polluted due to the introduction of a lot of background information. And model training is incorrect due to the introduction of few of background information. The second shortcoming is that model update learning rate is fixed. That is, the filter parameters would be learned from occluded objects when occlusion occurs, which may lead to wrongly update of the model. As time passes, it will be tracking drift.

To solve the challenging problem of occlusion, Bolme et al. [1] proposed to adaptively update the model based on Peak-to-Sidelobe Ratio (PSR), which measures the strength of a correlation peak and can be used to detect occlusions or tracking failure. The confidence degree (APCE) is proposed in [6]; in the method, both the response peaks and the APCE are a certain proportion and greater than their respective historical averages. Then update the tracking model.

In this paper, we mainly focus on designing a robust CF-based tracker to solve the above-mentioned problems which are the occlusion and background clutter. We aim to build a comprehensive correlation filter that can handle the background clutter and occlusion. We propose a novel correlation filter framework constrained by dual filters. The Minimum Output Sum of Squared Error filter is the unbiased estimate of the filter which easily to cause overfitting. The trained filter by linear ridge regression is the biased estimate of the filter which can deal with the overfitting. We combine the advantages of the two filters to constrain the trained filter which optimize our model. To deal with background clutter, clipping background patches around the target position up, down, left, and right, we add the cropped background patches to the learning filter. To overcome the challenge of occlusion, we introduce a novel criterion, APCE. Our method shows superior tracking performance than the existing CF, especially when dealing with the tracking problems of occlusion and background clutter.

2   Proposed Method

2.1   Dual Constrained Filters

Bolme et al. [1], proposed the MOSSE tracker, which trained filter by the Minimum Output Sum of Squared Error. Henriques et al. [2], proposed the CSK tracker, which trained filter by ridge regression. We propose a novel correlation filter framework constrained by dual filters, which combined the advantages of the above filters. Then we will introduce our model in detail. The goal of the CF trackers is to learn a discriminative correlation filter which can be applied to the region of interest in consecutive frames to infer the location of the target. It can be seen to find a linear regression function $f(x) = \mathbf{w}^T \mathbf{x}$ to minimize the training sample error. Where x is a training sample, and the learned correlation filter is represented by the vector $\mathbf{w}$. In order to achieve this with dual constrained filters, the model can be optimized as follows:

$$\min_{\mathbf{w_1}} \sum_i \left(f(x_i) - y_i \right)^2 + \lambda_1 \parallel \mathbf{w}_1 \parallel^2 + \lambda_2 ||\mathbf{w}_1 - \mathbf{w}_2||^2 \tag{1}$$

where $L(y_i, f(x_i)) = (y_i - f(x_i))^2$ express as a Loss function, and y_i is the corresponding regression label to the sample x_i. And λ_1 is a regularization coefficient to prevent the model from overfitting. λ_2 represents penalty coefficients for double-constrained filters. $\mathbf{w}_2$ is the trained filter by the Minimum Output Sum of Squared Error. The optimization model can be seen as a ridge regression problem, which has a closed-loop solution. Equation 1 can be solved as follows:

$$\mathbf{w_1} = (\mathbf{X}^T\mathbf{X} + \lambda_1\mathbf{I} + \lambda_2\mathbf{I})^{-1}(\mathbf{X}^T\mathbf{y} + \lambda_2\mathbf{w_2}) \tag{2}$$

where $\mathbf{X}$, $\mathbf{y}$ represent the sample matrix and the label matrix, respectively. $\mathbf{I}$ represents a unit matrix. However, the inversion process in Eq. 2 is a problem to solve. We can solve Eq. 2 easily by introducing the property (Eq. 3) that the circulant matrix is diagonalizable in the Fourier domain.

$$\mathbf{x} = \mathbf{F}diag\left(\widehat{x}\right)\mathbf{F}^H \tag{3}$$

The solution to Eq. 2 is as follows:

$$\widehat{\mathbf{w}}_1 = \frac{\widehat{\mathbf{x}}^* \odot \widehat{\mathbf{y}} + \lambda_2\widehat{\mathbf{w}}_2}{\widehat{\mathbf{x}}^* \odot \widehat{\mathbf{x}} + \lambda_1 + \lambda_2} \tag{4}$$

In the case of no-linear regression, kernel trick $f(\mathbf{z}) = \mathbf{w}^T\mathbf{z} = \sum_{i=1}^n \alpha_i k(\mathbf{z}, \mathbf{x}_i)$ is applied to allow more powerful classifier. For the most commonly used kernel functions, the circulant matrix trick can also be used. Therefore, the dual domain resolves as follows:

$$\widehat{\alpha} = \frac{\widehat{\mathbf{y}} + \lambda_2\widehat{\beta}}{\widehat{\mathbf{k}} + \lambda_1 + \lambda_2} \tag{5}$$

where $\widehat{\beta}$ is the solution of the dual domain of $\widehat{\mathbf{w}}_2$. And $\widehat{\beta} = \frac{\widehat{\mathbf{y}}}{\widehat{\mathbf{k}}}$. Therefore, Eq. 5 can be rewritten as follows:

$$\widehat{\alpha} = \frac{\widehat{\mathbf{k}}^* \odot \widehat{\mathbf{y}} + \lambda_2 \widehat{\mathbf{y}}}{\widehat{\mathbf{k}}^* \odot \widehat{\mathbf{k}} + \lambda_1 \widehat{\mathbf{k}} + \lambda_2 \widehat{\mathbf{k}}} \tag{6}$$

In order to predict the position of target in the next frame and the location corresponding to the peak response is the position of the target, the target detection formula can be expressed as follows:

$$\widehat{f} = \widehat{\mathbf{k}} \odot \widehat{\alpha} \tag{7}$$

The above is a correlation filter general framework proposed by us, suitable for other related filter trackers. To solve the background clutter, we introduce a context-aware correlation filter tracker [8] in our model. Clipping background patches around the target position up, down, left, and right, we add the cropped background patches to the learning filter. Equation 1 can be rewritten as:

$$\min_{\mathbf{w}_1} = ||\mathbf{A}_0 \mathbf{w}_1 - \mathbf{y}||^2 + \lambda_1 ||\mathbf{w}_1||^2 + \lambda_2 ||\mathbf{w}_1 - \mathbf{w}_2||^2 + \lambda_3 \sum_i^k ||\mathbf{A}_i \mathbf{w}_1||^2 \tag{8}$$

where $\mathbf{A}_0$ represents the cyclic matrix generated by the cyclic shift of the base sample. $\mathbf{A}_i$ is the cyclic matrix generated by the cyclic shift of the background patch. The number of background patches is represented as k. Similar as Eqs. 1, 8 can be solved as:

$$\mathbf{w_1} = (\mathbf{A}_0^T \mathbf{A}_0 + \lambda_1 \mathbf{I} + \lambda_2 \mathbf{I} + \lambda_3 \sum_i^k \mathbf{A}_i^T \mathbf{A}_i)^{-1} (\mathbf{A}_0^T \mathbf{y} + \lambda_2 \mathbf{w}_2) \tag{9}$$

which can also be rewritten as:

$$\widehat{\mathbf{w}}_1 = \frac{\mathbf{a}_0^* \odot \mathbf{y} + \lambda_2 \widehat{\mathbf{w}}_2}{\mathbf{a}_0^* \odot \mathbf{a}_0 + \lambda_1 + \lambda_2 + \lambda_3 \sum_i^k \mathbf{a}_i^* \odot \mathbf{a}_i} \tag{10}$$

2.2   Model Updating

The traditional CF-based trackers update the target model with a fixed learning rate. However, the model will be degenerated if a fixed high learning rate to update the model when the target is occluded. To overcome this problem, we introduce a novel criterion, APCE, which indicates the degree of oscillation of the response map. And the confidence degree (APCE) is defined as follows:

$$APCE = \frac{|F_{\max} - F_{\min}|^2}{mean\left(\sum_{w,h} (F_{w,h} - F_{\min})^2\right)} \tag{11}$$

where $F_{\max}$ represents the peak of the response map. $F_{\min}$ is the minimum value of the response map. The response value on the coordinate (w, h) is expressed as $F_{\omega,h}$. We update the model only when the response peaks and APCE of the current frame are greater than their respective historical averages by a certain percentage β_1, β_2.

$$\begin{cases} response_{\max} \geq \beta_1 \cdot mean\_response \\ APCE \geq \beta_2 \cdot mean\_APCE \end{cases} \tag{12}$$

where $mean\_response$ and $mean\_APCE$ represent the historical average of the response peak and the historical average of APCE, respectively. Then we update the model with a learning rate parameter η as

$$\begin{cases} \widehat{\alpha}_t = (1 - \eta) \cdot \widehat{\alpha}_{t-1} + \eta \cdot \widehat{\alpha}_t \\ \widehat{\mathbf{x}}_t = (1 - \eta) \cdot \widehat{\mathbf{x}}_{t-1} + \eta \cdot \widehat{\mathbf{x}}_{tt} \end{cases} \tag{13}$$

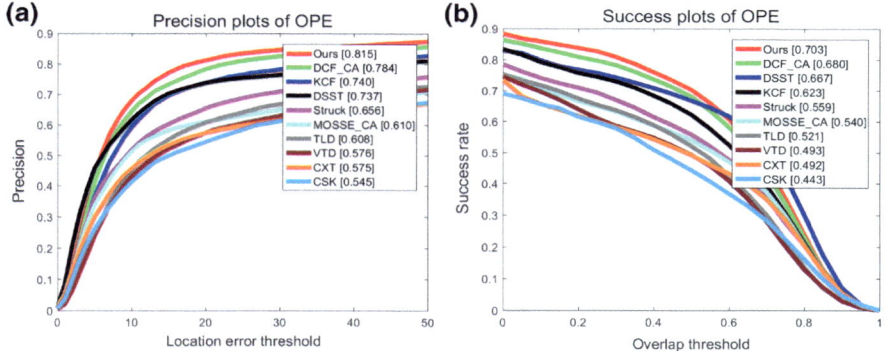

Fig. 1. Overall quantitative evaluation on CVPR 2013 Benchmark: precision plots (**a**) and success plots (**b**).

3   Experimental

3.1   Experimental Setup

We consider doing some work on feature extraction; the powerful features we use is HOG features. The proposed method in this paper is implemented in MATLAB 2016a. We perform the experiments on a PC with Intel i5-6500 CPU (3.2 GHz) and 8 GB RAM memory. We test the performance of the proposed tracker with the total 51 video sequences using in the CVPR 2013 benchmark [7] and compare with the top nine state-of-the-art trackers which include DSST [3], DCF_CA [8], KCF [4], CSK [2], MOSSE_CA [8], Struck [9], TLD [10], VTD

Fig. 2. Qualitative results of ten trackers on the video sequence with Coke, Jogging-1, and Lemming, which in the challenges of occlusion.

[11], and CXT [12], where Struck, VTD, and CXT are the three best-performed ones demonstrated in the CVPR 2013 Benchmark.

The parameters are set as follows. The regularization parameters λ_1, λ_2, λ_3 are set to be 10^{-4}, 0.0045 and 25. The learning rate parameter η and search box (padding) are set to 0.015 and 2, respectively. β_1, β_2 are set to be 0.5 and 0.4.

3.2   Experimental Results

Figure 1 contains the precision plots and success plots that overall quantitative evaluation with One-Pass Evaluation (OPE). Both precision plots and success plots show that our tracker is more robust than some state-of-the-art trackers in terms of the total 51 video sequences in CVPR 2013 Benchmark. Our trackers have achieved 0.703 and 0.815 in success rate and precision, respectively, which both rank the first. The baseline tacker KCF has achieved 0.623 and 0.740 in success rate and precision, respectively. Our tracker is better than the baseline tracker.

Seen in Fig. 2 the qualitative results of 10 trackers on the video sequence with Coke, Jogging-1, and Lemming in the attribute of occlusion.

As seen in Fig. 3, we give the results of our tracker in CVPR 2013 Benchmark on the attributes. Our tracker have achieved 0.684 and 0.804 in success rate and precision in the attribute of occlusion, respectively, which both rank the first.

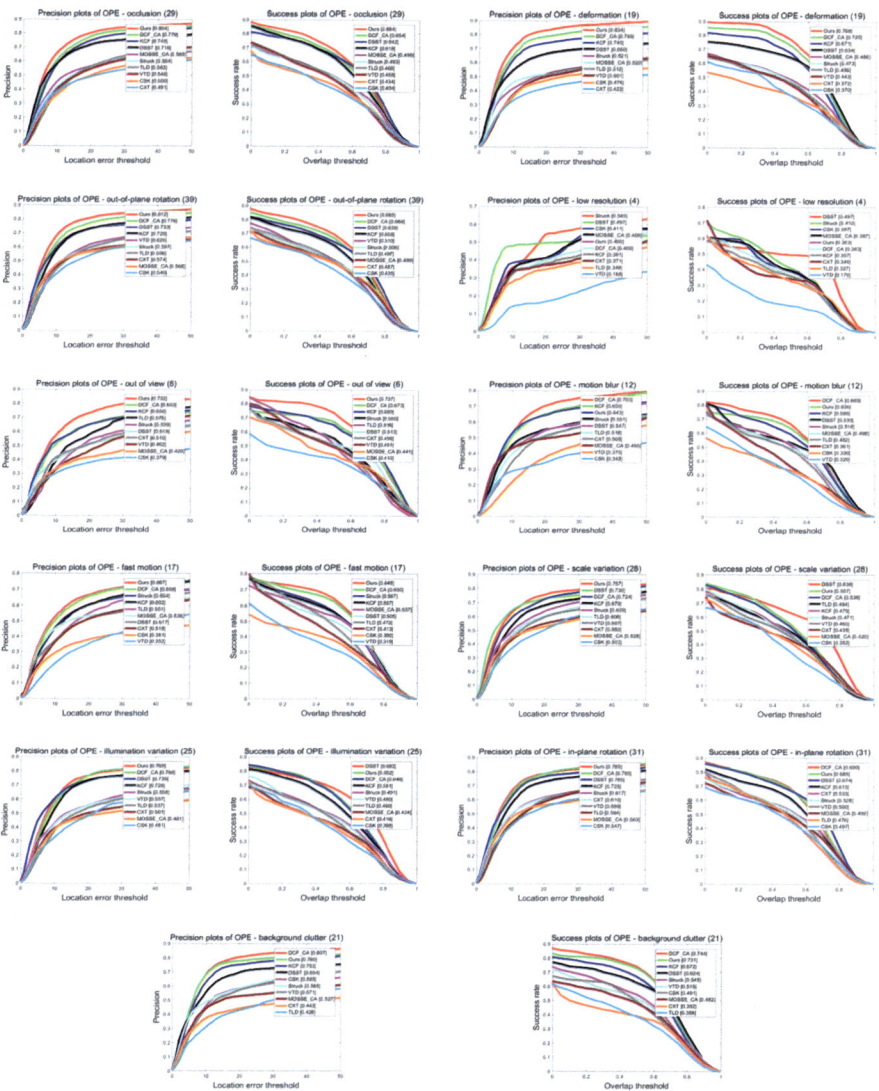

Fig. 3. Attribute-based comparison of our tracker with some state-of-the-art trackers in precision rate and success rate.

4   Conclusion

This paper has shown that the correlation filter framework constrained by dual filters, which combine the advantages of the MOSSE filter and the linear ridge regression filter, has better robustness with excellent performance. With adding the cropped background patches to the learning filter and introducing a novel criterion, APCE, our designed tracker can effectively solve the background clutter and occlusion. Extensive experiments on the CVPR 2013 Benchmark well demonstrate that our tracker outperforms some state-of-the-art trackers in both quantitative and qualitative analysis. The correlation filter framework constrained by dual filters can also be extended to other similar CF-based trackers to improve the robustness and performance.

Acknowledgments. This work was supported by the Major Science Instrument Program of the National Natural Science Foundation of China under grant 61527802, and the General Program of National Nature Science Foundation of China under grants 61371132 and 61471043.

References

1. Bolme DS, Beveridge JR. Visual object tracking using adaptive correlation filters. In: IEEE computer vision and pattern recognition; 2010. p. 2544–50.
2. Henriques JF, Rui C, Martins P. Exploiting the circulant structure of tracking-by-detection with kernels. In: Computer vision—ECCV. Berlin: Springer; 2012. p. 702–15.
3. Danelljan M, Gustav H. Discriminative scale space tracking. IEEE Trans Pattern Anal Mach Intell. 2017;39(8):1561–75.
4. Henriques JF, Rui C, Martins P. High-speed tracking with kernelized correlation filters. IEEE Trans Pattern Anal Mach Intell. 2015;37(3):583–96.
5. Li Y, Zhu J. Adaptive kernel correlation filter tracker with feature integration, vol. 8926. Springer; 2014. p. 254–65.
6. Wang MM, Liu Y, Huang ZY. Large margin object tracking with circulant feature maps. In: IEEE computer vision and pattern recognition; 2017. p. 4800–8.
7. Wu Y, Lmi J, Yang MH. Online object tracking: a benchmark. In: IEEE computer vision and pattern recognition; 2013. p. 2411–8.
8. Mueller M, Smith N, Ghanem B. Context-aware correlation filter tracking. In: IEEE computer vision and pattern recognition; 2017. p. 1387–95.
9. Sunando S. Struck: structured output tracking with kernels. IEEE Trans Pattern Anal Mach Intell. 2015;38(10):2096–109.
10. Kalal Z, Mikolajczyj K, Matas J. Tracking-learning-detection. IEEE Trans Pattern Anal Mach Intell. 2012;34(7):1409–22.
11. Kwon J, Lee KM. Visual tracking decomposition. In: IEEE computer vision and pattern recognition; 2010. p. 1269–76.
12. Medioni G. Exploring supporters and distracters in unconstrained environments. In: IEEE computer vision and pattern recognition; 2011. p. 1177–84.

Grid-Based Monte Carlo Localization for Mobile Wireless Sensor Networks

Qin Tang$^{(\boxtimes)}$ and Jing Liang

School of Information and Communication Engineering, University of Electronic
Science and Technology of China, Chengdu, China
tangqin0228@163.com

Abstract. Localization is an important requirement for wireless sensor networks (WSNs), but the inclusion of GPS receivers in sensor network nodes is often too expensive. Therefore, many solutions focus on static networks and do not consider mobility. In this paper, we analyze the Monte Carlo location (MCL) algorithm and propose an improved method—grid-based MCL. It applies the mobility of nodes to reduce the sampling area and to build an internal grid to predict the behavior of nodes. We investigate the properties of our technology and analyze its performance. The simulation and analysis show that the proposed grid-based MCL not only reduces localization error, but also improves sampling efficiency.

Keywords: WSNs · Grid-based MCL · Mobility

1 Introduction

Location awareness is important for wireless sensor networks since many applications such as environment monitoring, vehicle tracking, and mapping depend on knowing the locations of sensor nodes [1,2]. Localization denotes the process of identifying the own position in space, but placing GPS receivers in every node or manually configuring locations is not cost-effective. Therefore, localization schemes for sensor networks typically let a small number of seed nodes whose location and protocols are already known(e.g., they are all equipped with GPS receivers) [3] to broadcast their location messages around and then estimate the location of other nodes according to the messages they received.

Solutions for localization in WSNs can be classified into two groups of algorithms: range-based and range-free [4–6]. Range-based localization means that distances between sensor nodes are estimated by using some physical properties of communication signals [7,8], while range-free algorithms are often based only on connectivity. A prominent example of range-free is the Monte Carlo localization (MCL). MCL is designed for applications in which all nodes are able to move freely in the deployment area. In certain applications, one can assume that nodes

© Springer Nature Singapore Pte Ltd. 2020
Q. Liang et al. (eds.), *Communications, Signal Processing,
and Systems*, Lecture Notes in Electrical Engineering 516,
https://doi.org/10.1007/978-981-13-6504-1_159

are mainly moving on a group of paths [9]. The authors in [10] proposed an path-oriented approach (PO_MCL). It directly applies the original MCL algorithm. PO_MCL builds grids for static paths but requires additional hardware (magnetometer) on the nodes, and it may cost many times of sampling to find a valid sample. MCB [11,12] was proposed to reduce the sampling area. In this paper, we propose grid-based mobility behavior and present an adapted solution based on the MCL algorithm. We not only use the idea of MCB to improve sampling efficiency, but also use a forecast grid to divide the application area into cells to prognosticate the node's direction of movement. The grid is updated dynamically based on new observations from seed nodes. We show that by scheming anchor boxes and the grid, a better sample prediction of node localization can be achieved.

The rest of the paper is organized as follows. In Sect. 2, we present the main idea of MCL and proposes grid-based MCL. In Sect. 3, we present our experimental results. Finally, we conclude our work in Sect. 4.

2   Grid-Based MCL

2.1   MCL Algorithm

There are two main steps of MCL localization algorithms.

• Prediction step: In this step, a new sample is drawn from a circular sampling area with radius $r_{sarea} = v_{\max} \times t_{check}$ around its current position given by a transition equation $p(l_t|l_{t-1})$. The probability of the current location based on the previous location estimation is given by a uniform distribution (1).

$$p\left(l_t|l_{t-1}\right) = \begin{cases} \frac{1}{\pi \times v_{\max}{}^2}, & d\left(l_t, l_{t-1}\right) \in [0, v_{\max}) \\ 0, & d\left(l_t, l_{t-1}\right) \notin [0, v_{\max}) \end{cases} \tag{1}$$

• Filtering step: In this step, the set generated by the prediction step is put into the filtering step which uses the observations to filter out impossible node locations from the sample set. Each node keeps track of its rst-hop neighbor seeds S and of its second-hop neighbor seeds T. The filtering condition for a sample l is given in Eq. (2).

$$filter\left(l\right) = \forall\, s \in S,\; d\left(l, s\right) \leq r \wedge \forall s \in T,\; r < d\left(l, s\right) \leq 2r \tag{2}$$

However, the MCL algorithm has a large sampling area and the sampling efficiency is not high. In order to obtain a sufficient number of valid samples, continuous sampling is required, which reduces positioning efficiency. For making better use of new observations from seed nodes, we propose a grid-based Monte Carlo positioning algorithm in the next subsection. In Sect. 3, we will show the results of the MCL algorithm simulation and compare it with the algorithm (centroid and APIT) in the static network.

2.2 The Key Idea of the Grid-Based MCL

The main process of this algorithm is similar to the MCL algorithm, with the main difference being the selection of the sampling area and the establishment of the prediction grid. The choice of sampling area lies in the construction of anchor boxes and sample boxes and then the use of node mobility to reduce the sampling area. The sampling area will be divided into cells by using the prediction grid. Based on the observation from the seed node, the grid is updated. As long as seed node information is available, the original MCL algorithm is performed except that the sample is assigned the weight of its corresponding grid cell. The grid-based MCL algorithm is shown in Algorithm 1.

Algorithm 1 Grid-based MCL algorithm

Require: Initial $L_t = \{\ \}$
Require: Initial $|O_t| > 0$
  if $|O_t| > 0$ **then**
    MCL();
    UpdateGrid();
  end if
  EstOnGrid();

2.3 Theoretical Analysis of the Grid-Based MCL

• Selection of the sampling area: We use the idea of the MCB algorithm [6] to select the sampling area and then rely on the mobility of the node to narrow down the sampling area (Fig. 1).

(a)   **(b)**

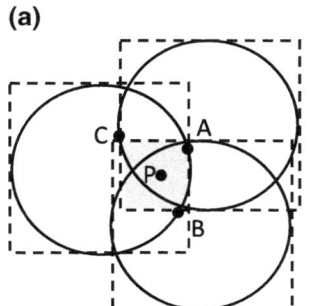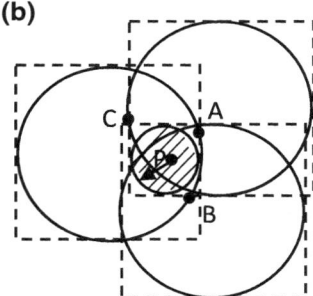

Fig. 1. Selection of the sampling area: **a** build sample boxes based on seed nodes; **b** reducing the sampling box based on node mobility

As shown in Fig. 1a, the superposition region formed by the seed nodes A, B, and C contains the node P to be measured, and then, the four endpoint coordinate expressions of the superposition region are calculated according to Eq. 3. The maximum moving speed of the node P to be measured is $v_{\max}$, and the other nodes to be tested are the same. Therefore, each node can construct a circular sampling area with its own position as the center and radius $v_{\max}$, as shown in Fig. 1b. These nodes have Markov characteristics, and the current position of each node is only related to the previous moment.

$$
\begin{cases} X_{\max} = \max_i^n (x_i - r) \\ Y_{\max} = \max_i^n (y_i - r) \end{cases} \quad \begin{cases} X_{\min} = \min_i^n (x_i - r) \\ Y_{\min} = \min_i^n (y_i - r) \end{cases} \tag{3}
$$

• The construction of prediction grid: A node can only be located in one cell, and the entire grid cell can be divided into eight adjacent cells exactly. The neighboring cells are marked with the corresponding basic directions. We assign values for all grid cells, which indicate the probability of moving to this cell next, as shown in Fig. 2.

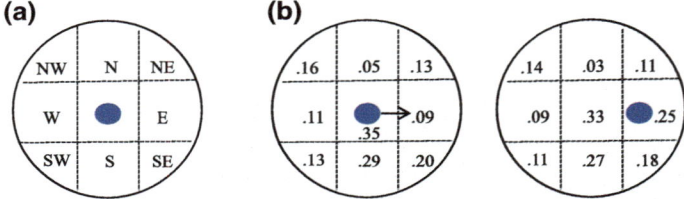

Fig. 2. The construction of prediction grid: **a** grid directions; **b** grid update process.

The size of the grid cells is an important parameter because they mainly determine the memory overhead of the grid-based MCL. We are adjusting the size of the grid cell based on $v_{\max}$ and t_{check}. Since the maximum distance that a node can travel between two location estimates is $d = v_{\max} \times t_{check}$, we also define the size of the grid unit as $2d/3$, as shown in Fig. 2a. Obviously, as the value of d is smaller, the resolution of the grid is higher, and the positioning of the nodes can be more accurately mapped to the grid.

At the beginning, each of the grid cell is given an initial value of 0.1, because the information about the route has not yet been collected and any block of the nine cells of the grid totals 1. Based on the observation from the seed node, the grid is updated so that the probability of the cell to which the node has moved increases and the values of all other neighbor cells decrease. The value of the probability increase Δ_{inc} and the decrease Δ_{dec} is determined according to the values of cell c_t in Eq. 4, where α is a tunable parameter for adjusting the increase level. A smaller value of α leads to a slower rate of increase in the probability that the grid convergence rate will slow down.

$$
\Delta_{inc} = \frac{\alpha}{value}, \quad \Delta_{dec} = \frac{\Delta_{inc}}{8} \tag{4}
$$

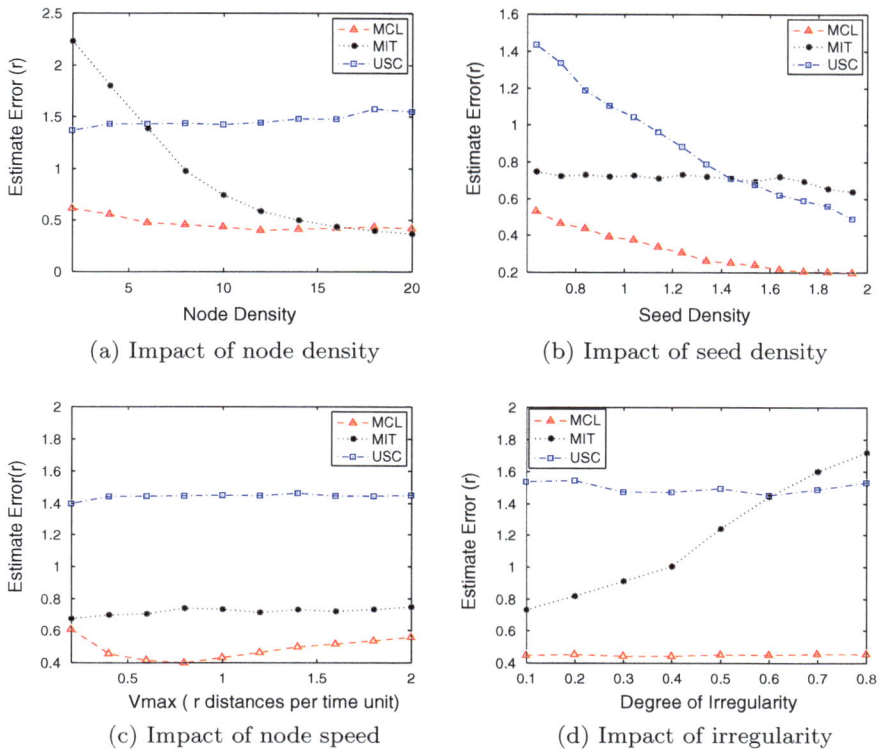

Fig. 3. Comparison of the estimated error of different localization techniques

3   Simulation Results

3.1   Analysis of MCL

We compare the MCL algorithm with the positioning algorithms (centroid algorithm and APIT algorithm) in the static network under the main performance indicators, such as node speed, seed density, node density, and degree of irregularity, as shown in Fig. 3.

Figure 3 shows that mobility can improve the accuracy in WSNs. Our simulation experiments reveal that the MCL technique can provide accurate localization even when the seed density is low, and network transmissions are highly irregular.

3.2   Simulation Results of Grid-Based MCL

We compare the grid-based MCL with the MCL in terms of node speed, seed density, node density, and degree of irregularity. As shown in Fig. 4, the trend of the simulation curve of the grid-based MCL is similar to the MCL, because the

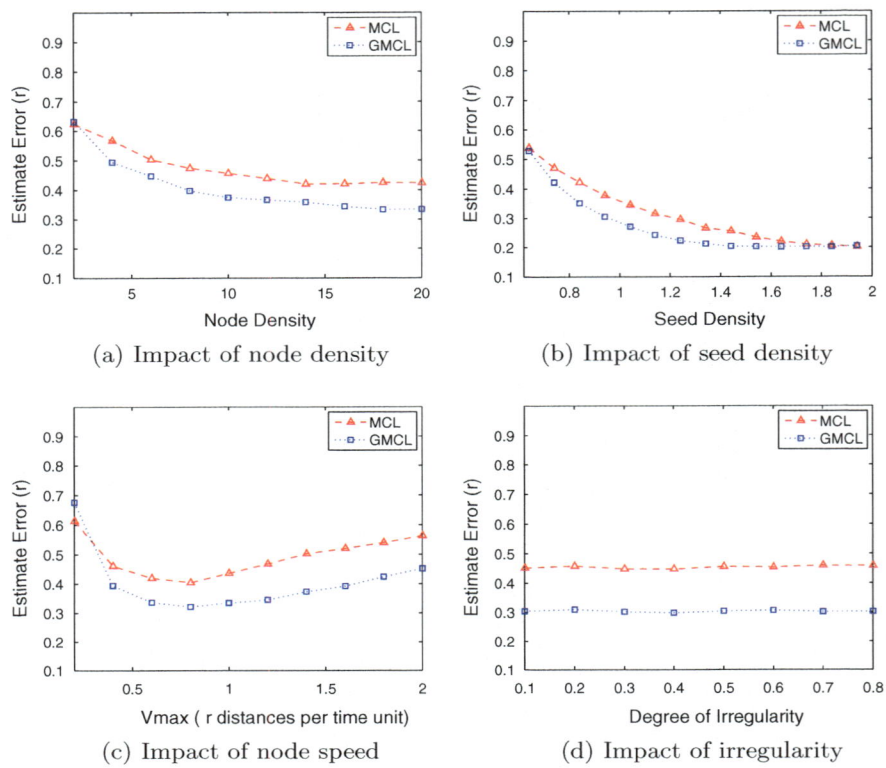

Fig. 4. Comparison of the estimated error of grid-based MCL and MCL algorithm

grid-based MCL still follows the main idea of the MCL, although the grid-based MCL has higher positioning accuracy and faster convergence speed.

Figure 4a illustrates the impact of node density. It performs poorly when network density is below 6, but performs best when network density is larger than 10. This is due to the fact that network density has a great impact on the accuracy of hop count. Figure 4b shows that the accuracy of both grid-based MCL and MCL improves as seed density increases since nodes will receive more location announcements. The localization error of MCL and grid-based MCL will almost converge to a single curve, although grid-based MCL still benefits little from its improved particle weighting. In Fig. 4c, we investigate different node velocities. Both algorithms benefit from an increasing node velocity in the beginning, as periods without seed information are getting shorter as nodes are moving faster. However, since the node velocities are larger than .8r, the localization error is growing as nodes lose contact to seed nodes more often. We use the degree of irregularity (*doi*) to denote the maximum radio range variation in the direction of radio propagation. For example, if *doi* = 0.1, then the actual radio range in each direction is randomly chosen from [0.9r, 1.1r]. Figure 5d shows the grid-based MCL and MCL techniques are not substantially affected by *doi*.

4   Conclusion and Future Work

In this paper, we analyzed the MCL algorithm and proposed grid-based MCL, a localization solution for mobility applications in WSNs. The grid-based MCL uses node's mobility to reduce the sampling area and builds grids to predict the node's movement behavior. In our approach, either the estimated error or the sampling area can be reduced. The proposed grid-based MCL approach can be applied to, but not limited to applications where nodes are moving on predefined paths as in wildlife monitoring, car traffic, and so on. Future work may include how different types of movement will affect localization and how our technologies will be extended to provide security.

Acknowledgments. This work was supported by the National Natural Science Foundation of China (61671138, 61731006) and was partly supported by the 111 Project No. B17008.

References

1. Chien CH, Hsu CC, Wang WY, Kao WC, Chien CJ: Global localization of Monte Carlo localization based on multi-objective particle swarm optimization. In: 2016 IEEE 6th international conference on consumer electronics-Berlin (ICCE-Berlin). IEEE; 2016. p. 96–7.
2. Saarinen J, Andreasson H, Stoyanov T, Lilienthal AJ. Normal distributions transform Monte-Carlo localization (NDT-MCL). In: 2013 IEEE/RSJ international conference on intelligent robots and systems; Nov 2013. p. 382–9.
3. Nuss D, Yuan T, Krehl G, Stuebler M, Reuter S, Dietmayer K. Fusion of laser and radar sensor data with a sequential Monte Carlo Bayesian occupancy filter. In: 2015 IEEE intelligent vehicles symposium (IV); June 2015. p. 1074–81.
4. Doherty L, Pister KSJ, Ghaoui LE. Convex position estimation in wireless sensor networks. In: Proceedings IEEE INFOCOM 2001. Conference on computer communications. Twentieth annual joint conference of the IEEE computer and communications society (Cat. No. 01CH37213), vol. 3; 2001. p. 1655–63.
5. Li CY, Li IH, Chien YH, Wang WY, Hsu CC. Improved Monte Carlo localization with robust orientation estimation based on cloud computing. In: 2016 IEEE congress on evolutionary computation (CEC); July 2016. p. 4522–7.
6. Adewumi OG, Djouani K, Kurien AM. RSSI based indoor and outdoor distance estimation for localization in WSN. In: 2013 IEEE international conference on industrial technology (ICIT); Feb 2013. p. 1534–9.
7. Bellili F, Amor SB, Affes S, Samet A. A new importance-sampling ml estimator of time delays and angles of arrival in multipath environments. In: 2014 IEEE international conference on acoustics, speech and signal processing (ICASSP); May 2014. p. 4219–23.

8. Al-Jazzar SO, Strangeways HJ, McLernon DC. 2-d angle of arrival estimation using a one-dimensional antenna array. In: 2014 22nd European signal processing conference (EUSIPCO); Sept 2014. p. 1905–9.

9. Militzer B, Driver KP. Development of path integral Monte Carlo simulations with localized nodal surfaces for second-row elements. Phys Rev Lett. 2015;115:176403. https://doi.org/10.1103/PhysRevLett.115.176403.

10. Hartung S, Kellner A, Rieck K, Hogrefe D. Monte Carlo localization for path-based mobility in mobile wireless sensor networks. In: 2016 IEEE wireless communications and networking conference; Apr 2016. p. 1–7.

11. Baggio A, Langendoen K. Monte Carlo localization for mobile wireless sensor networks. Ad Hoc Netw. 2008;6(5):718–33. http://www.sciencedirect.com/science/article/pii/S1570870507001242

12. Zhu H, Mao J, Wang L, Fu L, Guo N. The study on point average energy consumption by Monte Carlo in large-scale wireless sensor networks. In: 2015 IEEE international conference on information and automation; Aug 2015. p. 1700–3.

WalkSLAM: A Walking Pattern-Based Mobile SLAM Solution

Lin Ma[1](✉), Tianyang Fang[1], and Danyang Qin[2]

[1] School of Electronics and Information Engineering,
Harbin Institute of Technology, Harbin, China
malin@hit.edu.cn
[2] Electrical Engineering College, Heilongjiang University, Harbin, China

Abstract. In indoor localization scenarios, a sheer coordinate with respect to a basis is insufficient to indicate the users' situation due to a lack of information about landmarks distributed in the environments. To extract landmarks' information manually, however, is inefficient and thus vulnerable to changes of the environments. Simultaneous localization and mapping can solve the localization and landmarks' information extracting problems. This paper presents Walk-SLAM, a SLAM solution that estimates both the path taken by the user and the locations of Wi-fi devices in the indoor space, using a smartphone. This solution extends the previous work by introducing human walking patterns into the specific SLAM problem. Experiments demonstrate that the improvement consists of increased efficiency of the particle filter, and hence, of the overall algorithm, and a better estimation of the user's location and path.

Keywords: Indoor localization · Simultaneous localization and mapping · Walking pattern · Walk ratio

1 Introduction

In the last decades, both military and civil have seen increasing need for efficient and accurate location-based service. Location-based service, as its name indicates, controls features and provides functions regarding location data of the users. Thus, the basis of location-based service is to provide a location estimate of the user according to the measurements of devices distributed in the environment taken by the user's device, which requires mapping measurements into exact locations.

In order to provide massive location-based service in large-scale indoor scenario both efficiently and economically, the idea of using a smartphone to run SLAM algorithm while the user carries it and walk around the indoor [1–3] space comes out and draws researchers' immediate attention. But the inaccurate nature of smartphone sensors will lead to an undesirable uncertainty of both the location estimates and landmark estimates.

© Springer Nature Singapore Pte Ltd. 2020
Q. Liang et al. (eds.), *Communications, Signal Processing, and Systems*, Lecture Notes in Electrical Engineering 516,
https://doi.org/10.1007/978-981-13-6504-1_160

1.1   Related Work

Previous work has addressed this accuracy issue of smartphone SLAM in different levels. SmartSLAM [3] used path-smoothing model and building orientation model to supervise the heading orientation of the user in order to smooth the noisy path estimate; simultaneously localization and configuring [1] used fixed step length model and EKF to overcome the unreliability of motion measurements. However, fixed step length is still not optimal. Some researches also focused on elimination of Gaussian noise in a RSSI-based localization scenario [4].

This paper presents WalkSLAM, a SLAM solution runs on smartphones. Walk-SLAM contributes previous work by adding human walking pattern elements into the SLAM algorithm, both on prior stage and posterior stage. The evaluation on live data shows a noticeable improvement on the path estimation and landmark estimation.

The rest of the paper is organized as follows: In Sect. 2, the problem statement and the model are discussed. In Sect. 3, we introduce a new SLAM algorithm named WalkSLAM. The implantation of WalkSLAM and some refining maneuvers are introduced shortly after. In evaluation section, live data is analyzed and the average errors of both path estimation and landmark estimation are discussed separately with a comparison between WalkSLAM solution and common FastSLAM solution.

2   Problem Definition and Modeling

The original aim of the research is to map an indoor environment using a smartphone without any prior information from the environment. The smartphone is capable of motion sensing and RSSI observing, and in a common SLAM system, the motion sensing ability guarantees the control info of a robot, where location estimation can be extracted, and the RSSI observing achieved by Wi-fi module inside the smartphone can obtain observation of the accessing points working as the landmarks. After this thought, a hidden Markov model [5–7] can be applied to our scenario.

Figure 1 represents a hidden Markov model. The poses make the core Markov chain, and in our situation, they are the locations on the path during the navigation; the state transition means a step of the user is taken. Because we do not know the actual coordinate of the locations, they become the hidden states.

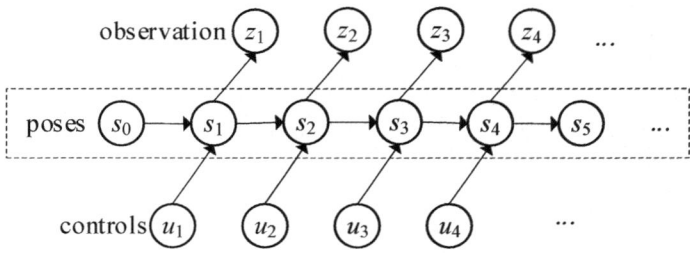

Fig. 1. Hidden Markov model in SLAM problems

Now, we can effectively describe our problem as a SLAM problem. The poses are the user's location, and the current pose will be denoted s_t. Poses evolve according to the motion model [7]:

$$s_t \sim p(s_t|u_t, s_{t-1}) \qquad (1)$$

Among which s_t is the current pose, s_{t-1} the previous, and u_t the current control. The control consists of motion measurements from the device, which in our case is the smartphone. After initializing each pose, to map the surroundings, the device senses landmarks using sensors. Sensor observations follow the measurement model:

$$z_t \sim p(z_t|s_t, \theta) \qquad (2)$$

where θ is a set consisting of all the landmarks. Now, the SLAM problem can be formulated. SLAM problem is to make a refined estimation of all landmarks θ and all poses $s^t = \{s_0, s_1, s_2, \ldots\}$ along the user's path from the controls and observations, which can be described as $p(s^t, \theta|z^t, u^t)$. The controls, as stated above, are the motion measurements from motion sensors on the phone, which indicates a step's heading and orientation. After an initialization of the starting pose, given previous and current control, an estimate of the current state can be calculated. However, the motion sensors are alarmingly inaccurate; thus, in further discussion, we decide to add some prior information into the controls for better performance.

3   WalkSLAM

3.1   Particle Path Sampling

With the control and the observation of our SLAM system defined, we can present our SLAM problem solution in the following steps with implantation details. As shown in Fig. 2, the algorithm begins with the path particle sampling after the input of motion measurements u_t from motion sensors. Due to the nature of human walking pattern, we do not sample the pose s_t directly but rather estimate the transition from s_{t-1} to s_t which is denoted $\overrightarrow{(s_t - s_{t-1})}$.

The transition can be described as a combination of the fixed step length and the heading. For the filtering of the heading direction, instead of the path-smoothing method, we present a new filtering model:

$$\omega_t = \begin{cases} \omega_{t-1}, & |\omega_{t-1} - \omega_t| > \varepsilon_\omega \\ \omega_t, & |\omega_{t-1} - \omega_t| \leq \varepsilon_\omega \end{cases} \qquad (3)$$

The rationale behind this model is that first the motion sensor is pretty noisy and unreliable according to previous reports [1, 8], and because of that, in some degree we can safely ignore small direction changes without worrying too much about losing the details of heading since Monte Carlo method involved in the sampling phase can guarantee the coverage of most minor direction changes.

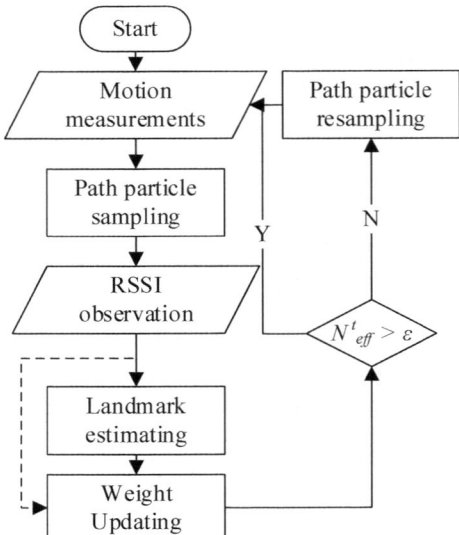

Fig. 2. SLAM algorithm loop. Notice that the dotted line connecting RSSI observation with weight updating step means the observation takes part in the weight's calculation

After the path sampling, we get a set of path particles, and according to our design, the next step should be landmark estimation, where we initialize and update the landmarks' locations using the observations. However, since our observation is the 1D RSSI measurements, a single observation is not enough to initialize a landmark's location. Thankfully this problem can be solved using the range-only localization model presented in [9], and the major requirement of this method is several consecutive poses and relative observations.

The updating of landmarks is guided by the observations, so we need a filter to refine the landmark location, in this situation a 2D coordinate, with 1D RSSI as the observations. An extended Kalman filter can be applied to this situation, using the method mentioned in [1].

3.2 Importance Weight Updating

In robotic SLAM systems, the only indicator of the reliability of a particle is how the observation matches up the landmark estimations, and it is called importance weight factor:

$$w_t^{[m]} = \frac{\text{target distribution}}{\text{proposal distribution}} = \frac{p\left(s^{t,[m]}\big|z^t,u^t\right)}{p\left(s^{t,[m]}\big|z^{t-1},u^t\right)} \tag{4}$$

However, in our WalkSLAM scenario, another factor should be taken into consideration to evaluate the degree of the particle path fitting human walking pattern [10], specifically the walk ratio indicator [11]. For convenience, we call the original importance weight factor observation factor (OF), and the walk ratio indicator walk factor (WF). Then, the new weight becomes:

$$w_t^{[m]} = \text{OF} \cdot \text{WF} \tag{5}$$

As the matter of implantation, for the calculation of OF, we can use the conclusion from [7]:

$$\text{OF} = \frac{p\left(s^{t,[m]}|z^t, u^t\right)}{p\left(s^{t,[m]}|z^{t-1}, u^t\right)} \overset{\text{EKF}}{\approx} \sum_k p\left(z_t|\theta_k^{[m]}, s_t^{[m]}\right) p\left(\theta_k^{[m]}\right) \tag{6}$$

Since we are always sure about the identity of each RSSI scan result, the equation can be simplified to:

$$\text{OF}^{[m]} = \sum_k p\left(z_t|\theta_k^{[m]}, s_t^{[m]}\right) \tag{7}$$

Then, we can calculate OF using the Gaussian function:

$$\text{OF}^{[m]} = \sum_k f\left(z_t \left|\left\|\theta_k^{[m]} - s_t^{[m]}\right\|\right., \delta_\omega\right) \tag{8}$$

For the calculation of WF, since the proposal distribution is made to be a Gaussian distribution, according to the Bayesian theory we can simplify the equation as:

$$\text{WF}^{[m]} \propto p\left(d^{t,[m]}|\Delta^{t,[m]}, \delta_{\text{WR}}\right) \tag{9}$$

For implementation purpose, we first calculate the average WR of the path particle,

$$\overline{\text{WR}} = \frac{1}{n} \sum_{k=1}^n d_k \cdot \Delta_k \tag{10}$$

Then, through a Gaussian function, we can get the WF of the particle as:

$$\text{WF}^{[m]} = f\left(\overline{\text{WR}}|R, \delta_{\text{WR}}\right) \tag{11}$$

where d_k is the step length and Δ_k the time that step costs. R is the average walk ratio of a healthy man, and δ_{WR} the variance of walk ratio.

To sum up, the new weight updating model becomes:

$$w_t^{[m]} = f\left(\frac{1}{n} \sum_{k=1}^n d_k \cdot \Delta_k|R, \delta_{\text{WR}}\right) \sum_k p\left(z_t|\theta_k^{[m]}, s_t^{[m]}\right) \tag{12}$$

Then, for further use, we normalize the weights:

$$w_t^{[m]} = w_t^{[m]} \left(\sum_{k=1}^{n} w_t^k \right)^{-1} \tag{13}$$

3.3   Resampling Particles

After getting the weights of the particle set, we now resample the particle set due to the weight of each particle. To evaluate the effectiveness of a particle set using the weight factor, the effectiveness index is introduced:

$$N_{\text{eff}}^t = \left(\sum_{k=1}^{n} w_t^k \right)^{-1} \tag{14}$$

As shown in Fig. 2, every time the value of N_{eff}^t gets low enough, the resampling step is triggered; this way we can resample only when we need to, to effectively use all the particles and lower the time consumption.

4   Implementation and Evaluation

In the test, we choose a typical indoor scenario which is a floor of office with a long corridor. The longest part of the corridor is 47.3 m at length, and the overall corridor is 3 m at width. Thirteen access points were accessible along the corridor. A live test in comparison with the ground truth is given in Figs. 3 and 4. Since our goal is to estimate both the path and the landmarks, the localization biases of them are evaluated.

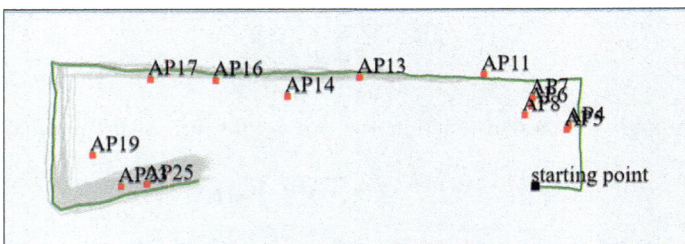

Fig. 3. Replay of a live test

Considering the additive nature of the pedometer bias, tests were run on different lengths to indicate the difference made by navigating different lengths and the results are given in Table 1, although due to the variance of step length the tests can only be grouped by a range of distance they cover.

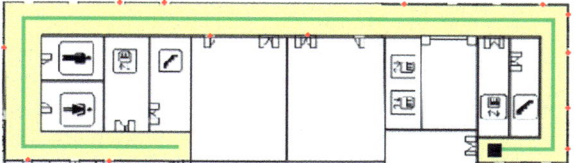

Fig. 4. Ground truth of the same test with a floorplan

Table 1. Result of three long tests and three short tests

Path length	Test	Localization error (m)	Standard deviation
90–100 m	1	2.97	0.37
	2	2.34	0.29
	3	2.78	0.30
35–40 m	4	1.17	0.11
	5	1.41	0.17
	6	0.94	0.10

We can still conclude from the results that for a short run of the WalkSLAM, the average localization error falls below 3.0 m, which is good enough for a room-level localization scenario. Another essential factor of the evaluation is the localization error of the landmarks. In the three long runs, 12 access points are all observable; thus, we use them to calculate the error.

To better demonstrate the overall performance when localizing the landmarks, we take each landmark estimation in each run as an individual landmark estimation and use the cumulative distribution function to bring out the result. When compared to a common FastSLAM run with an S1 model applied and one without, as shown in Fig. 5, the improvement in performance of WalkSLAM is quite noticeable.

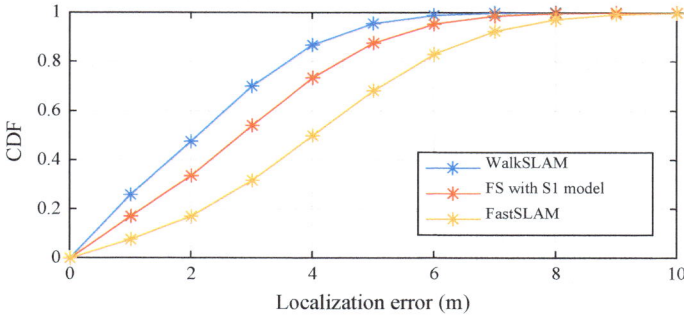

Fig. 5. Comparison of landmark estimating performance between WalkSLAM and traditional FastSLAM

5   Conclusion

In this research, a simultaneous localization and mapping solution that focuses on running on smartphones was developed. Compared to traditional SLAM solutions, in this new solution, the human walking pattern was considered into the algorithm, with a new sampling model and an advanced weight updating model presented, and the implantation methods given in detail. Six live tests were run in a typical indoor environment. The result indicated that the accuracy could fit a room-level localization scenario and the landmark estimation performance of WalkSLAM is noticeably better than that of a traditional solution.

Acknowledgments. This paper is supported by National Natural Science Foundation of China (61571162), Ministry of Education—China Mobile Research Foundation (MCM20170106).

References

1. Bulten W, Van Rossum AC, Haselager WFG. Human SLAM, indoor localisation of devices and users. In: 2016 IEEE first international conference on internet-of-things design and implementation (IoTDI). IEEE; 2016. p. 211–22.
2. Ferris BD, Fox D, Lawrence N. WiFi-SLAM using Gaussian process latent variable models. Science. 2007;7:2480–5.
3. Shin H, Chon Y, Cha H. Unsupervised construction of an indoor floor plan using a smartphone. IEEE Trans Syst Man Cybern Part C Appl Rev. 2012;42(6):889–98.
4. Patri A, Rath SP. Elimination of Gaussian noise using entropy function for a RSSI based localization. In 2013 IEEE second international conference on image information processing (ICIIP-2013). IEEE; 2013. p. 690–4.
5. Arulampalam MS, Maskell S, Gordon N, Clapp T. A tutorial on particle filters for online nonlinear/non-Gaussian Bayesian tracking. IEEE Trans Signal Process. 2002;50(2):174–88.
6. Dellaert F, Fox D, Burgard W, Thrun S. Monte Carlo localization for mobile robots. In: Proceedings 1999 IEEE international conference on robotics and automation (Cat. No.99CH36288C), vol. 2; May, 1999. p. 1322–8.
7. Montemerlo M, Thrun S, Koller D, Wegbreit B. FastSLAM: a factored solution to the simultaneous localization and mapping problem. In: Proceedings of 8th national conference on artificial intelligence/14th conference on innovative applications of artificial intelligence, vol. 68, issue 2; 2002. p. 593–8.
8. Zhao N. Full-featured pedometer design realized with 3-axis digital accelerometer. Analog Dialogue. 2010;44:1–5.
9. Olson E, Leonard JJ, Teller S. Robust range-only beacon localization. IEEE J Oceanic Eng. 2006;31(4):949–58.
10. Sekiya N, Nagasaki H. Reproducibility of the walking patterns of normal young adults: test-retest reliability of the walk ratio(step-length/step-rate). Gait Posture. 1998;7(3):225–7.
11. Rota V, Perucca L, Simone A, Tesio L. Walk ratio (step length/cadence) as a summary index of neuromotor control of gait: application to multiple sclerosis. Int J Rehabil Res. 2011;34(3): 265–9.

Time-Frequency Spatial Smoothing MUSIC Algorithm for DOA Estimation Based on Co-prime Array

Aijun Liu[✉], Zhichao Guo, and Mingfeng Wang

Harbin Institute of Technology, Weihai, China
mylaj@hitwh.edu.cn

Abstract. In this paper, the time-frequency spatial smoothing MUSIC algorithm (TF-SSMUSIC) for DOA estimation based on co-prime array is proposed. The spatial smoothing MUSIC (SSMUSIC) is a typical DOA estimation algorithm based on co-prime array. TF-SSMUSIC replaces SSMUSIC's data covariance matrix with a time-frequency distribution matrix, which leads to a better DOA estimation performance. By selecting points in the time-frequency domain, not only the signal-to-noise ratio (SNR) can be improved effectively, but the signal interference in different time-frequency domains can be isolated. The improvement of SNR makes TF-SSMUSIC have a more accurate DOA estimation than SSMUSIC in the case of low SNR. Especially, if source signals are separable in the time-frequency domain, TF-SSMUSIC can process them solely. In this way, the angle resolution and the number of predictable source signals can be improved greatly.

Keywords: DOA estimation · Time-frequency · Co-prime array · Spatial smoothing MUSIC

1 Introduction

In recent years, co-prime array has attracted extensive attentions because of its own characteristics: simple construction, large degree of freedom, etc. However, the DOA estimation of co-prime array is faced with difficulties. Because the distance between array elements includes multiple scales, the traditional algorithm is no longer applicable. The spatial smoothing MUSIC (SSMUSIC) algorithm has solved this problem, which is introduced in detail in [4–7]. Because of the increased array aperture, the angle resolution is improved. By making effective use of the degree of freedom, the number of predictable source signals can be improved obviously.

In order to further improve the performance of the SSMUSIC algorithm, the time-frequency analysis method was introduced. The spatial time-frequency distribution matrix is applied to the subspace-based algorithm and used to replace the covariance matrix in the SSMUSIC algorithm for DOA estimation, which is called the time-frequency spatial smoothing MUSIC algorithm. PWVD enhances the signal-to-noise ratio and sort source signals in different time-frequency domains, which greatly

© Springer Nature Singapore Pte Ltd. 2020
Q. Liang et al. (eds.), *Communications, Signal Processing, and Systems*, Lecture Notes in Electrical Engineering 516,
https://doi.org/10.1007/978-981-13-6504-1_161

improve the performance of DOA estimation [1–3, 8]. In the case of low SNR, the SNR of the useful signal can be effectively improved by selecting points on the time-frequency ridge of the signal, so the DOA estimation is more accurate. For source signals which are separable in the time-frequency domain, TF-SSMUSIC can process them solely. In this way, the angle resolution and the number of predictable source signals in the time-frequency domain can be improved greatly.

The manuscript is organized as follows. The system model and the SSMUSIC algorithm are briefly introduced in Sect. 2. In Sect. 3, we shall introduce the TF-SSMUSIC algorithm in detail. In order to verify the performance of the TF-SSMUSIC algorithm, the simulation results are given in Sect. 4. Lastly, the paper is concluded in Sect. 5.

2   System Model

Figure 1 shows a co-prime array, consisting of a co-prime pair of uniform linear sub-arrays. The first uniform linear array includes N array elements with an array element spacing of Md. The second uniform array includes $2M - 1$ arrays, and the element spacing is Nd, where M and N are co-primes. The basic array element spacing is $d = \lambda/2$, where λ is the wavelength of signals.

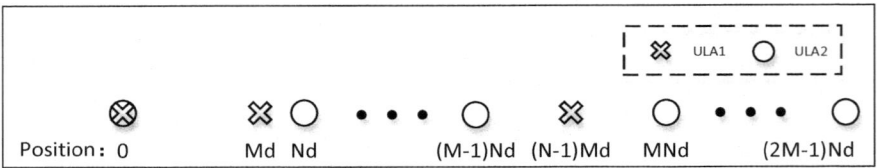

Fig. 1. System model

Since M and N are prime numbers, other array elements will not overlap except the first array element. So, there are $N + 2M - 1$ array elements in total, which will result in a continuation position from $-MNd$ to MNd sets at least. In this way, we can get $2MN + 1$ degrees of freedom from $N + 2M - 1$ physical elements. These elements are located at:

$$S = \{Mnd, 0 \leq n \leq N - 1\} \cup \{Nmd, 0 \leq m \leq 2M - 1\} \tag{2.1}$$

Assuming there are P narrowband signals in the far field, the data received by the co-prime array can be expressed as:

$$\mathbf{X}[k] = \mathbf{A}\mathbf{S}[k] + \mathbf{N}[k] \tag{2.2}$$

Among them, $\mathbf{A} = [a(\theta_1)\ a(\theta_2)\ \cdots\ a(\theta_p)]$ denotes the array manifold matrix, $a(\theta)$ denotes the $N + 2M - 1$ dimension steering vector and $\mathbf{S}[k] = [s_1(k)\ s_2(k)\ \cdots\ s_p(k)]^{\mathrm{T}}$

denotes the kth snapshot of the source vector. The noise $\mathbf{N}[k]$ is assumed to be white Gaussian noise.

Therefore, the array's covariance matrix of the received data $\mathbf{X}(k)$ can be expressed as:

$$\mathbf{R}_{xx} = E\left[\mathbf{X}(k)\mathbf{X}^H(k)\right] \tag{2.3}$$

Vectorize $\mathbf{R}_{xx}$:

$$\mathbf{z} = \mathrm{vec}(\mathbf{R}_{xx}) = \mathrm{vec}\left[\sum_{i=1}^{P} \sigma_i^2 a(\theta_i) a^H(\theta_i)\right] + \rho^2 l_n = \mathbf{B}(\theta_1, \ldots, \theta_P)\mathbf{p} + \rho^2 \mathbf{L} \tag{2.4}$$

where

$$\mathbf{B} = [b(\theta_1)\, b(\theta_2) \cdots b(\theta_P)] \tag{2.5}$$

$$b(\theta_i) = a^*(\theta_i) \otimes a(\theta_i) \tag{2.6}$$

$$\mathbf{p} = \left[\sigma_1^2, \ldots, \sigma_P^2\right]^T \tag{2.7}$$

$$\mathbf{L} = \left[\mathbf{e}_1^T, \ldots, \mathbf{e}_L^T\right] \tag{2.8}$$

Among them, $\mathbf{e}_i$ ($i = 1, \ldots, L$) is a column vector where the ith position is 1 and the other positions are 0. z can be seen as a received signal model whose array manifold is B, where p denotes the source vector and $\rho\, 2L$ can be seen as a noise term.

At this time, each column of B is a longer array manifold, and its specific position is determined by the position differences, including the position differences between the sub-arrays:

$$\{\pm(Mn - Nm)d, 0 \leq n \leq N - 1, 0 \leq m \leq 2M - 1\} \tag{2.9}$$

and the position differences of the two sub-arrays:

$$\{(Mn_1 - Nn_2)d, 0 \leq n_1, n_2 \leq 2M - 1\} \tag{2.10}$$

$$\{(Nm_1 - Nm_2)d, 0 \leq m_1, m_2 \leq 2M - 1\} \tag{2.11}$$

Sort these differences and reject the excesses. From these differences, continuous position differences can be obtained from $-MN$ to MN. This results in a new array manifold $B1$, which is a subset of B, and the dimension is $(2MN + 1)K$. $B1$ can be seen as an array manifold corresponding to an array containing $2MN + 1$ array elements whose position difference is from $-MNd$ to MNd. Then, it is decomposed into $MN + 1$ overlapping sub-arrays, where the position of the ith sub-array is: $\{(-i + 1 + n)d \mid n = 0, 1, \ldots, MN\}$ and the degree of freedom of each sub-array is $MN + 1$. Assuming that Ai represents the array manifold of the ith sub-array, the corresponding sub-array receiving model is: $z_{li} = \mathbf{A}_i\mathbf{p} + \sigma_n^2\mathbf{e}_i$. The ith position of $\mathbf{e}_i$ is 1, and the rest is 0. The covariance matrix of each sub-array is: $\mathbf{R}_i = z_{li}z_{li}^H$. Take the average of them to get the final expression:

$$\mathbf{R} = \frac{1}{MN+1} \sum_{i=1}^{MN+1} \mathbf{R}_i \tag{2.12}$$

This matrix is called a spatial smoothing matrix, which is full rank. Therefore, the subsequent steps can be performed with the MUSIC algorithm. In this way, MN source signals can be estimated.

3   Time-Frequency Spatial Smoothing MUSIC

Before introducing the time-frequency spatial smoothing MUSIC algorithm, we first introduce the spatial time-frequency distribution matrix. In discrete Cohen class bilinear time-frequency transforms, express $\mathbf{x}(t)$ as $\mathbf{x}_p(t)$ and $\mathbf{x}_q(t)$ ($p, q = 1, 2, \ldots, M$), and then we can obtain the mutual time-frequency distribution among elements. We can construct the spatial time-frequency distribution matrix by using these elements.

$$\mathbf{C_{xx}}(t,f) = \sum_{l=-\infty}^{+\infty} \sum_{k=-\infty}^{+\infty} \phi(k,l)\ \mathbf{x}(t+k+l)\mathbf{x}^H(t+k-l)\mathrm{e}^{-j4\pi fl} \tag{3.1}$$

where t denotes time, f denotes frequency and $\phi(k,\ l)$ is a kernel function, which is a function of time and delay. Substituting the received array signal model (2.2) into (3.1) and taking the statistical average:

$$E\{\mathbf{C_{xx}}(t,f)\} = \mathbf{A}\mathbf{C_{ss}}(t,f)\mathbf{A}^H + \sigma^2\mathbf{I_M} \tag{3.2}$$

The formula (3.2) has a similar structure to the covariance matrix of the data received from arrays, so it is called an array reception model in the time-frequency domain.

When the kernel function $\phi(k,\ l) = \delta(k-m)$, the time-frequency distribution is Pseudo-Wigner–Ville distribution (PWVD), which has a good time-frequency flocculability. Spatial PWVD is defined as:

$$\mathbf{C_{xx}}(t,f) = \sum_{\tau=-(L-1)/2}^{(L-1)/2} \mathbf{x}(t+\tau)\ \mathbf{x}^H(t-\tau)\mathrm{e}^{-j4\pi f\tau} \tag{3.3}$$

where L denotes the length of the window function.

The PWVD of the signal is expressed as:

$$\mathbf{C_{ss}}(t,f) = \sum_{\tau=-(L-1)/2}^{(L-1)/2} \mathbf{s}(t+\tau)\,\mathbf{s}^H(t-\tau)e^{-j4\pi f \tau} \tag{3.4}$$

The spatial time-frequency distribution matrix can still be expressed by formula (3.2).

The model shown in Eq. (3.2) is suitable for all the time-frequency points, and we can use the time-frequency average method to choose points. For chirp signals which contain noise, the energy of the signal in the time-frequency plane concentrates on the vicinity of instantaneous frequency. The noise is evenly distributed in the entire time-frequency plane. Therefore, selecting the time-frequency points along the signal time-frequency ridge can increase signal-to-noise ratio effectively.

The ith diagonal element of the signals' spatial time-frequency distribution matrix $\mathbf{C}_{ss}$ (t, f) is:

$$\mathbf{C_{d_i d_i}}(t,f) = \sum_{\tau=-(L-1)/2}^{(L-1)/2} A_i^2 e^{j[\varphi_i(t+\tau)-\varphi_i(t-\tau)]} e^{-j4\pi f \tau} \tag{3.5}$$

where $\varphi_i(t)$ denotes the phase of the ith signal and A_i denotes the amplitude of the ith signal.

For each signal, take $N - L + 1$ time-frequency points along the time-frequency ridge, and then average them to obtain the average spatial time-frequency matrix, as shown in the formula:

$$\hat{\mathbf{C}} = \frac{1}{n_0(N-L+1)} \sum_{q=1}^{n_0} \sum_{i=1}^{N-L+1} \mathbf{C_{xx}}\left(t, f_{q,i}(t)\right) \tag{3.6}$$

$f_{q,i}(t)$ represents the instantaneous frequency of the qth signal at the ith sampling point, taking a mathematical expectation on $\hat{\mathbf{C}}$:

$$\mathbf{C} = E\left(\hat{\mathbf{C}}\right) = \frac{L}{n_0} \mathbf{A^0 R_{ss}^0 (A^0)}^H + \sigma^2 \mathbf{I} \tag{3.7}$$

where

$$\mathbf{R_{ss}^0} = \mathbf{diag}\left[A_i^2,\ i=1,2,\ldots,n_0\right] \tag{3.8}$$

$$\mathbf{A^0} = [\mathbf{a_1, a_2, \ldots, a_{n_0}}] \tag{3.9}$$

only, n_0 signals are considered here.

It has been proved that the spatial time-frequency distribution matrix has a similar structure as a traditional array covariance matrix. Therefore, the STFD matrix can be applied to the subspace-based algorithm to replace the covariance matrix in the spatial smoothing MUSIC algorithm. The spatial spectrum estimation of the signal can be obtained by processing the STFD matrix. We call this the time-frequency smoothing MUSIC algorithm.

4 Simulation Results

In this section, we present several numerical examples to evaluate the proposed method. We will compare the performance of SSMUSIC and TF-SSMUSIC under different conditions by simulation. Construct a co-prime array. In this array, $M = 2$, $N = 3$, basic array element spacing $d = \lambda/2$.

First, we consider a single source coming from $21°$. Using the Monte Carlo method, 500 simulations, take 512 snapshots, and the SNR is varied from -20 to 10 dB in increments of 2 dB. The comparison of SSMUSIC and TF-SSMUSIC is shown in Fig. 2. In DOA estimation success rate, the estimated angle is rounded to an integer.

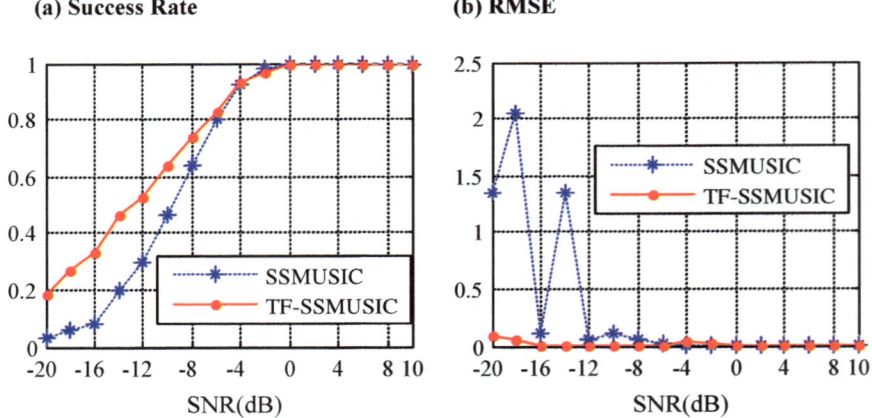

Fig. 2. DOA estimation in the case of single source signal

As shown in the simulation results, the TF-SSMUSIC algorithm is more accurate and stable, especially in the case of low SNR. The energy of the source signal has a good aggregation on the time-frequency ridge, but the noise is evenly distributed on the time-frequency domain. By selecting points on the time-frequency ridge, the actual SNR can be effectively improved. In this way, TF-SSMUSIC can have a better DOA estimation performance.

In order to analyze the angle resolution, set two chirp sources and make the direction of them approach gradually. Keep the SNR constant (SNR = 0 dB), and the start and end frequencies of the two sources are: 0.2–0.3 and 0.3–0.4.

From Fig. 3, we can see that when the direction of arrivals is 0° and 4°, respectively, the SSMUSIC algorithm cannot distinguish them, but the TF-SSMUSIC algorithm can get the DOA clearly. The reason is that the source signals are separable in the time-frequency domain, and the interference of others can be avoided by selecting points in the time-frequency domain. So, as long as the two source signals are separable in the time-frequency domain, they can be distinguished using the TF-SSMUSIC algorithm no matter how close they are.

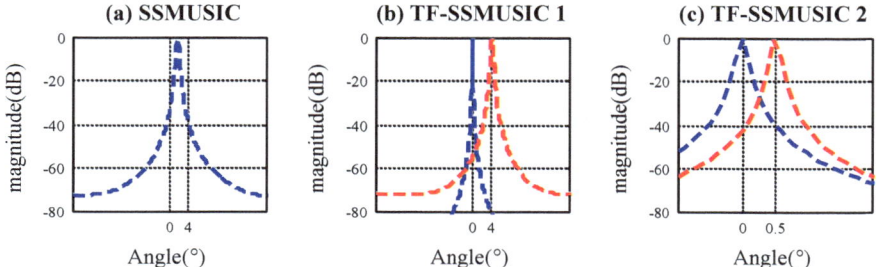

Fig. 3. Angle resolution of SSMUSIC and TF-SSMUSIC

For co-prime array, using the SSMUSIC algorithm, continuous position differences can be obtained from $-MN$ to MN at least. In this part, we consider $MN = 6$ uncorrelated chirp sources with same SNR. The direction and the begin–end frequency of them are shown in Table 1.

Table 1. Source signal settings

Signals	1	2	3	4	5	6
Frequency	0.1–0.15	0.15–0.2	0.2–0.25	0.25–0.3	0.3–0.35	0.35–0.4
DOA	−55	−35	−15	0	20	40

The result is shown in Fig. 4, and it can be seen from the simulation that both algorithms can estimate the DOA of six source signals well. But, for the TF-SSMUSIC algorithm, as long as the signal is separable in the time-frequency domain, it can estimate more source signals regardless of M and N. There are eight sources that estimate the DOA of them using TF-SSMUSIC. It can be seen from Fig. 4c TF-SSMUSIC 2 that the DOA of them can be well estimated.

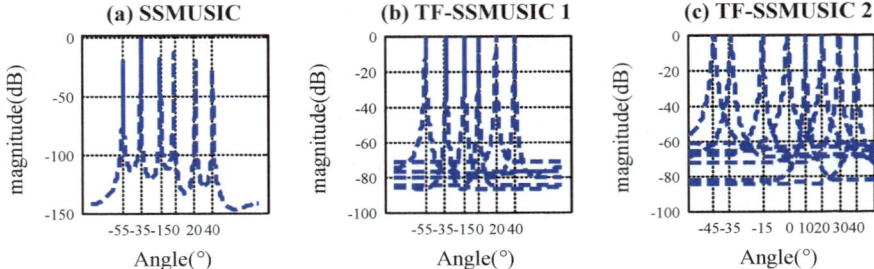

Fig. 4. Maximum number of predictable source signals

Consider six sources as shown in Table 1, **and** change the spacing between the elements. The influence of the reduction in element spacing on TF-SSMUSIC is shown in Fig. 5. For the SSMUSIC algorithm, the interference between source signals will be more serious when the spacing between array elements is reduced, which makes it difficult to reduce the element spacing. By using the frequency division feature of the TF-SSMUSIC algorithm, high angle resolution can be maintained on the premise of narrower array element spacing. However, if different source signals are inseparable in the time-frequency domain, the DOA estimation performance of the TF-SSMUSIC algorithm will be drastically reduced.

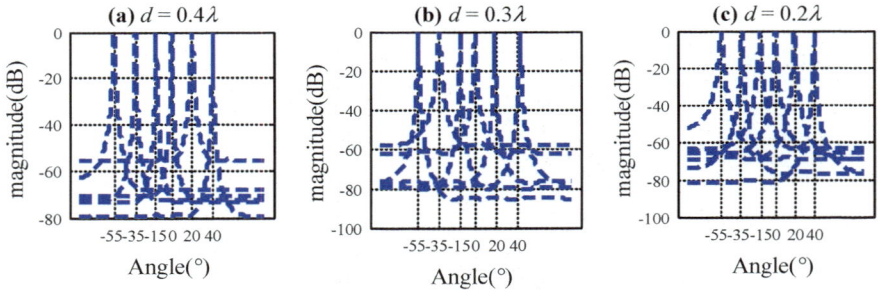

Fig. 5. TF-SSMUSIC

It can be seen from the simulation that the reduction in element spacing leads to a decrease in the DOA estimation performance of SSMUSIC, and it has been unable to estimate the direction of arrival correctly. However, TF-SSMUSIC can still estimate the DOA correctly. When the spacing between the elements is further reduced, the results of the TF-SSMUSIC algorithm are shown in Fig. 5.

5   Conclusion

In this paper, the TF-SSMUSIC algorithm is proposed by replacing SSMUSIC's data covariance matrix with a time-frequency distribution matrix. TF-SSMUSIC algorithm has the function of SNR enhancement and signal screening, which has been verified through theoretical analysis and simulation. Therefore, TF-SSMUSIC algorithm has a better estimation performance than SSMUSIC algorithm. However, TF-SSMUSIC algorithm has its own limitation too. The signals processed by TF-SSMUSIC should be separable in the time-frequency domain.

Acknowledgments. This work was supported in part by the National Key R&D Program of China under Grant 2017YFC1405202, in part by the National Natural Science Foundation of China under Grant 61571159 and Grant 61571157, and in part by the Public Science and Technology Research Funds Projects of Ocean under Grant 201505002.

References

1. Belouchrani A, Amin MG. Time-frequency MUSIC. IEEE Signal Process Lett. 2002;6(5): 109–10.
2. Chen S, Li X, Shao Z. Study on the performance of DOA estimation algorithms. In: IEEE international conference on communication problem-solving. IEEE; 2016. p. 475–7.
3. Li F. Time-frequency DOA estimation and application comprehensive polarization information. Harbin: Harbin Institute of Technology; 2016.
4. Liu CL, Vaidyanathan PP. Remarks on the spatial smoothing step in coarray MUSIC. IEEE Signal Process Lett. 2015;22(9):1438–42.
5. Pal P, Vaidyanathan PP. Co-prime sampling and the music algorithm. In: Digital signal processing workshop and IEEE signal processing education workshop. IEEE; 2011. p. 289–94.
6. Tan Z, Eldar YC, Nehorai A. Direction of arrival estimation using co-prime arrays: a super resolution viewpoint. IEEE Trans Signal Process. 2014;62(21):5565–76.
7. Xiong J, Chen S. Direction of arrived estimation based on the co-prime arrays. In: Fourth international conference on information science and industrial applications. IEEE; 2016. p. 21–4.
8. Zhang Y, Mu W, Amin M. Subspace analysis of spatial time-frequency distribution matrices. IEEE Trans Signal Process. 2001;49(4):747–59.

Non-uniform Sampling Scheme Based on Low-Rank Matrix Approximate for Sparse Photoacoustic Microscope System

Ting Liu$^{(\boxtimes)}$ and Yongsheng Zhao

Department of Control Science and Engineering, Dalian Maritime University,
Dalian 116026, China
{liuting0910,yszhao}@dlmu.edu.cn

Abstract. Optical-resolution photoacoustic microscopy (OR-PAM) has rapidly emerging as tool for label-free morphology and function imaging of the microvasculature in vivo with a high resolution. However, it is difficult to achieve real-time imaging due to the limitation of data acquisition time. Therefore, a sparse PAM (SPAM) has been proposed to obtain a high-resolution PAM image with relatively low sampling density. In order to successfully set up a SPAM system, the two key problems that we need to keep focus on are designation of the compressive sampling scheme and the corresponding image recovery algorithm. Typically, a random uniform sampling scheme is adopted. In this paper, a non-uniform sampling scheme based on low-rank matrix approximate is proposed to replace the conventional point-by-point scanning scheme to implement fast data acquisition. The effectiveness of the proposed non-uniform scanning scheme is validated using both numerical analysis and PAM experiments. As compliments for SPAM system, the total sampling points are dramatically decreased for a relatively high-resolution PAM vascular image and to implement accelerated data acquisition. Thus, OR-PAM is of great potential to find board biomedical applications in the pathophysiology studies of tumor and treatments for anti-angiogenesis.

Keywords: Tumor angiogenesis · Optical-resolution photoacoustic microscopy · Low-rank matrix approximate

1 Introduction

Angiogenesis is a hallmark of tumor growth, invasion, metastasis, etc. all [1–3]. Take tumor growth as an example, regulated by a number grown factors secreted by tumor cells, the remodeling of existing vasculature and forming of new microvascular results in the morphology alteration of microvascular network [3, 4].

Medical imaging modalities such as magnetic resonance imaging and positron emission tomography have been employed to study angiogenesis, but the micrometer-level resolution limits their capacity to visualize the fine feature changes [3, 5, 6]. To clearly visualize the finest micro-vessels in tumor angiogenesis, a spatial resolution of ~ 5 μm is required, which can be provided by optical-resolution photoacoustic

© Springer Nature Singapore Pte Ltd. 2020
Q. Liang et al. (eds.), *Communications, Signal Processing,*
and Systems, Lecture Notes in Electrical Engineering 516,
https://doi.org/10.1007/978-981-13-6504-1_162

microscopy (OR-PAM)—a technique that uses a tightly focused laser beam for photoacoustic excitation [7, 8].

OR-PAM images are typically acquired with a point-by-point sampling scheme in the field of view [9–11]. For OR-PAM, more researches are focused on the resolution enhancement. Unfortunately, most of the methods sacrifice the imaging speed, such as employing more measurements to get high-resolution images [12]. For example, the spatially Fourier-encoded photoacoustic microscopy [13] proposed by Liang et al. In their system, the measurements required are twice as many as the conventional OR-PAM. Besides, the more the number of measurements the great challenge for the system memory and the more is the radiation to tissues. Thus, a fast OR-PAM with a relatively high resolution is preferred.

Like many other medical images [14–16], the sparsity of the PAM has been proven [16–18] and thoughtfully explored by incorporated the technique of compressed sensing. The final image is reconstructed from far less measurements [19–21]. Based on that analysis, we have proposed a SPAM system to achieve fast imaging [11]. The effectiveness of a SPAM system depends on many critical techniques, the most important one being the compressive sampling scheme.

Typically, a random uniform scanning has been adopted. In our research [11], the random uniform sampling mask is generated by a random uniform expander graph, for it can be perfectly recovered by low-rank matrix completion (LRMC) [22, 23]. However, the significant part of the PAM image is ignored for uniform scanning. Thus, a non-uniform sampling scheme is preferred, which could be acquired based on the edge expander graphs. Therefore, different sampling density could be utilized for the region of interest and other background areas to put more measurements on the significant region of the image. This operation can further reduce the measurements.

However, the region of interest and other background areas could not be separated. In this situation, more measurements may be focused on the principal components [24]. Based on that analysis, low-rank matrix approximate (LRMA) based non-uniform sampling scheme is proposed for vascular imaging by OR-PAM.

In this paper, a LRMA-based non-uniform sampling scheme is investigated for vascular PAM imaging. The paper is organized as follows. The proposed method is presented in Sect. 2, including the theory of LRMA and the LRMA based non-uniform sampling scheme. In Sect. 3, the results of both numerical and real PAM system have been analyzed and discussed to validate the non-uniform sampling scheme. Conclusions are drawn in the final section.

2 Methods

2.1 Low-Rank Analysis for PAM Vascular Images

The sparsity of the photoacoustic image has been verified in many researches. More recently, the low-rank property has been incorporated into the field of photoacoustic imaging.

To our knowledge, photoacoustic imaging is typically used in tumor and vasculature imaging. As for the low-rank and non-uniform properties for PAM vascular images, compressive random sampling scheme has been proposed to cope with the redundancy (sparsity) of the observed photoacoustic data to realize rapid data acquisition.

Therefore, the image reconstruction process is formulated into a LRMC problem. Based on that analysis, the compressive sampling model for our system is defined as follow:

$$D = Kx, \tag{1}$$

where D is the incomplete compressive measurement data, x is the final desired PAM image, and K is the random sampling mask. The goal is to recover x from D, which is a typically LRMC problem. The problem is defined as:

$$\min_{x} \quad \text{rank}(x) \\ \text{s.t.} \quad D = Kx, \tag{2}$$

Unfortunately, the rank minimization problem in (2) is NP-hard, and so we solve the following convex relaxation problem instead:

$$\min_{x} \quad \|x\|_* \\ \text{s.t.} \quad D = Kx, \tag{3}$$

where $\|x\|_*$ is the nuclear norm of x.

According to the compressive sampling model, our interest is focused on two key problems: the sampling scheme and the corresponding image recovery method. The major contribution of our paper is the former problem. For the designation of K, the edge expander has been incorporated to fully explore the low-rank and non-uniform properties. However, the non-uniform property is not obvious for vascular image, which means the region of interest cannot be point out from the background. But the edge expander can only be used as uniform random sampling scheme in this situation. Therefore, a LRMA-based sampling scheme has been proposed for PAM vascular images in this paper.

2.2 Low-Rank Matrix Approximation

LRMA method is a fast way to solve the LRMC problem. The method is utilized to obtain the non-uniform sampling mask to focus more measurements on the principal component of vascular.

Given a matrix $A \in R^{m \times n}$ and an integer r, $r \leq \min(m, n)$, finding its LRMA matrix B is a critical task in many technical fields. The problem has been described as:

$$\min_{x} \quad \|A - B\|_F^2 \\ \text{s.t.} \quad \text{rank}(B) < r, \tag{4}$$

Thus, the constraint here is the same for low-rank matrix completion problem. Typically, the problem of LRMA can be solved by the truncated singular value decomposition (TSVD). In order to obtain B with rank $(B) = r$ of A, the SVD of A should be calculated first:

$$A = \sum_{i=1}^{n} u_i \sigma_i v_i^T \tag{5}$$

For TSVD, the first-r singular values approximation is chosen as these of LRMA results for A, which is

$$B = A_r = \sum_{i=1}^{r} u_i \sigma_i v_i^T \tag{6}$$

Take a numerical vascular image, for example, the dimension for the full rank image, which is shown in Fig. 1a, is 256 × 256. The LRMA image obtained by TSVD with $r = 50$ is shown in Fig. 1b. From Fig. 1, all the principal components of the vascular were exacted by TSVD-based LRMA.

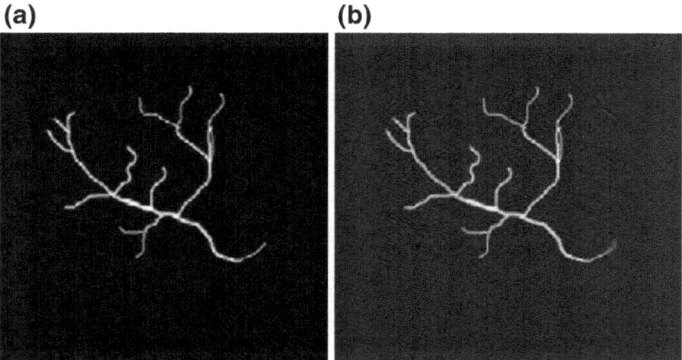

Fig. 1. Low-rank approximation image of photoacoustic image. **a** Original image. **b** Low-rank approximation image, rank is 50.

With the help of low-rank matrix approximate B, the non-uniform sampling mask K could be obtained by the comparison between A and B,

$$K_{ij} = \begin{cases} 1 & \text{if } |A_{ij} - B_{ij}| \geq T \\ 0 & \text{otherwise} \end{cases} \tag{7}$$

where T is the threshold, which is determined by experience.

2.3   LRMA-Based Fast Non-uniform Sampling Scheme

In general, TSVD method is used in LRMA. However, it is time-consuming that all the singular values should be calculated for TSVD. In this section, the bilateral random projections (BRP) of matrix $A \in R^{m \times n}$ are used to replace the SVD computation to significantly reduce the time cost.

Assume that $\tilde{A}$ is a fast rank r BRP approximation of A, the bilateral random projection is

$$
\begin{aligned}
Y_1 &= AA_1 \\
Y_2 &= A^T A_2
\end{aligned}
\tag{8}
$$

where both $A_1 \in R^{n \times r}$ and $A_2 \in R^{m \times r}$ are random matrixes. Therefore, the fast rank r BRP approximation of A is

$$
\tilde{A} = Y_1 \left(A_2^T Y_1 \right)^{-1} Y_2^T
\tag{9}
$$

The computation for $\tilde{A}$ included a matrix inversion operation and three matrix multiplication. Therefore, the computation cost for $\tilde{A}$ is $r^2(2n + r) + mnr$ which is dramatically less than TSVD.

However, the decreasing rate of the singular values for vascular images is not fast enough. In this case, the effect of Eq. (9) will be worsened. Therefore, a modified version of BRP is utilized.

For the modified BRP, $A' = (AA^T)^q A$ is used to instead of A. Because these two matrixes have the same singular values, while they have equal eigenvalues $\lambda_i(A') = \lambda_i(A)^{2q+1}$. However, the decreasing rate of the singular values for A' is much faster than A. Based on that analysis, the BRP for A' is defined as

$$
\begin{aligned}
Y_1 &= A' A_1 \\
Y_2 &= A' A_2^T
\end{aligned}
\tag{10}
$$

And the fast rank r matrix approximate for A' is

$$
\tilde{L} = Y_1 \left(A_2^T Y_1 \right)^{-1} Y_2^T
\tag{11}
$$

Here, the QR decomposition of Y_1 and Y_2 are computed to obtain $\tilde{L}$. Thus, the LRMA is

$$
\begin{aligned}
L &= (\tilde{L})^{2q+1} = Q_1 \left[R_1 \left(A_2^T Y_1 \right)^{-1} R_2^T \right]^{\frac{1}{2q+1}} Q_2^T \\
Y_1 &= Q_1 R_1 \\
Y_2 &= Q_2 R_2
\end{aligned}
\tag{12}
$$

In order to show a comparison study, Fig. 1a is still used to test the performance of BRP-based LRMA method. The LRMA image obtained by BRP with $r = 50$ is shown in Fig. 2b. From the result, all the principal components of the vascular were exacted by BRP-based LRMA. Even though there are many artifacts in BRP-based LRMA image, there is no influence on the non-uniform sampling scheme with the help of threshold T.

(a)   **(b)**

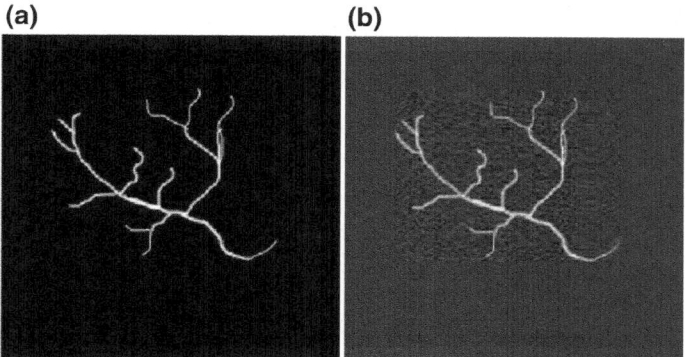

Fig. 2. Low-rank approximation image of photoacoustic image. **a** Original image. **b** BRP-based Low-rank approximation image, rank is 50.

3 Experimental Results

The performance of the proposed non-uniform sampling mask based on LRMA was investigated both by simulated and real PAM images involving quantitative and qualitative analysis.

3.1 Results from Simulated Study

The simulated study is conducted to provide both qualitative and quantitative analysis. Besides, the reference image has been carefully selected to simulate the property of vascular, as shown in Fig. 3a. The dimension of the image is 500×500. The same with our previous results, the compressive sampling image is acquired by downsampling on the reference image according to sampling mask K [11].

In our paper, a non-uniform sampling mask based on LRMA has been proposed to effectively improve the imaging speed by dramatically reduce the sampling points with relative high resolution. As for vascular, the region of interest is not obvious, the edge expander can only be used as uniform sampling mask. Our method focused on the principal components is more appropriate.

Figure 3 shows the non-uniform sampling mask and the comparison results between LRMA-based non-uniform scheme and edge expander-based uniform schemes. Figure 3b shows the non-uniform sampling mask base on LRMA with the sampling rate fixed at 0.5. From the sampling mask, the outline of the vascular could be

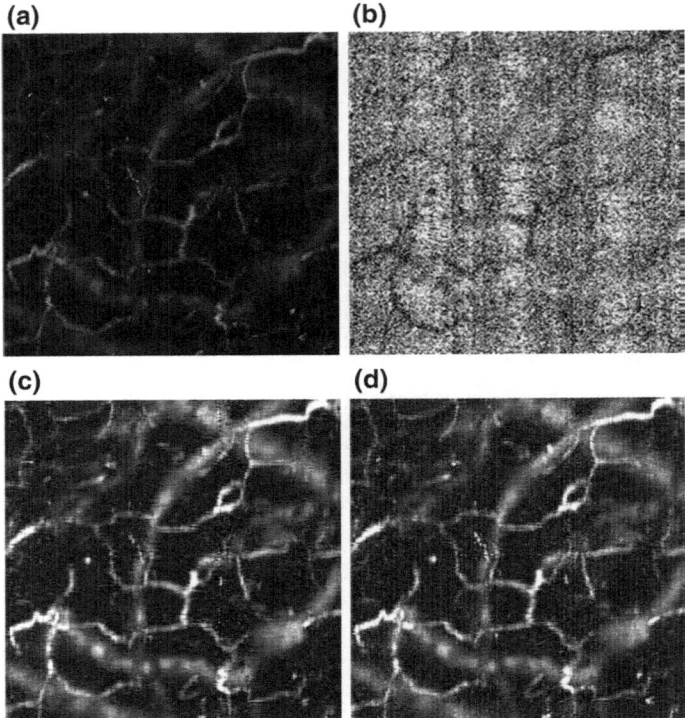

Fig. 3. Result of non-uniform sampling based on low-rank matrix approximation with BRP. **a** Original image. **b** Non-uniform sampling mask. **c** Recovered image with non-uniform sampling scheme. **d** Recovered image with uniform sampling scheme.

found for more sampling points are focused on the principal components. The recovered image under two sampling schemes is shown in Fig. 3c, d. The results show that the two sampling schemes could acquire considerable results.

In order to better compare the two schemes, three quantitative parameters are evaluated to illustrate the performance, mean square error (MSE), peak signal to noise ratio (PSNR), and structural similarity (SSIM), which are defined as

$$\text{MSE} = \sum_{i=1}^{m} \sum_{j=1}^{n} \left(\|X - Y\|^2 / mn \right) \tag{13}$$

$$\text{PSNR} = 10 \log_{10} \left(\frac{255^2}{\text{MSE}} \right) \tag{14}$$

$$\mathrm{SSIM} = l(X, Y) \cdot C(X, Y) \cdot S(X, Y)$$

$$l(X, Y) = \frac{2\mu_x\mu_y + C_1}{\mu_x^2 + \mu_y^2 + C_1},$$

$$C(X, Y) = \frac{2\sigma_x\sigma_y + C_2}{\sigma_x^2 + \sigma_y^2 + C_2} \tag{14}$$

$$S(X, Y) = \frac{\sigma_{xy} + C_3}{\sigma_x\sigma_y + C_3}$$

where X is the reference image and Y is the recovered image. μ_x and μ_y are the mean value of X and Y, while σ_x and σ_y are the variance. Besides, σ_{xy} is covariance between X and Y. The results are shown in Table 1. From all the three parameters, the proposed sampling scheme shows a better result.

Table 1. Performance indices for comparison between uniform and non-uniform sampling mask

	Uniform	Non-uniform
SSIM	0.9949	0.9957
MSE	3.5351e−06	2.8853e−06
PSNR	54.5160 dB	55.3980 dB

3.2 Results from Real PAM System

In this section, the feasibility of our non-uniform sampling scheme in practice has been verified with our system. Figure 4 shows a schematic of the experimental OR-PAM system, adapted from a conventional OR-PAM system. In order to achieve non-uniform sampling, a compressive sampling scheme is utilized to realize rapid data acquisition, and the corresponding image reconstruction problem is formulated as a low-rank matrix completion problem.

As shown in Fig. 4, our system is modified from an OR-PAM system and can be transferred to many other existing OR-PAM systems. The main modification is the compressive sampling mode achieved by the optical scanner. The compressive sampling employs a sampling mask K, which is a 0×1 matrix generated by the PC with a user-defined sampling density k. Based on the sampling mask K, the X–Y linear stage is controlled by a controller to achieve sparse optical scanning.

The mouse ears have complete vascular structure. Moreover, imaging tumor angiogenesis on the mouse ear could avoid the respiratory and cardiac motion artifacts from living animals [3]. So, the mouse ears are chosen as our imaging sample to image the vascular as other previous studies.

The full sampling image (Fig. 5a) has been acquired to be a reference image to better compare the non-uniform and uniform scheme in real system. The resolution of conventional OR-PAM system could reach microvascular level. For comparison between non-uniform and uniform sampling, both schemes are utilized to acquire the incomplete observed data, as shown in Fig. 5b, c. The sampling rate is both fixed at 0.48. From the sampling mask, more sampling pints are focused on vascular for non-

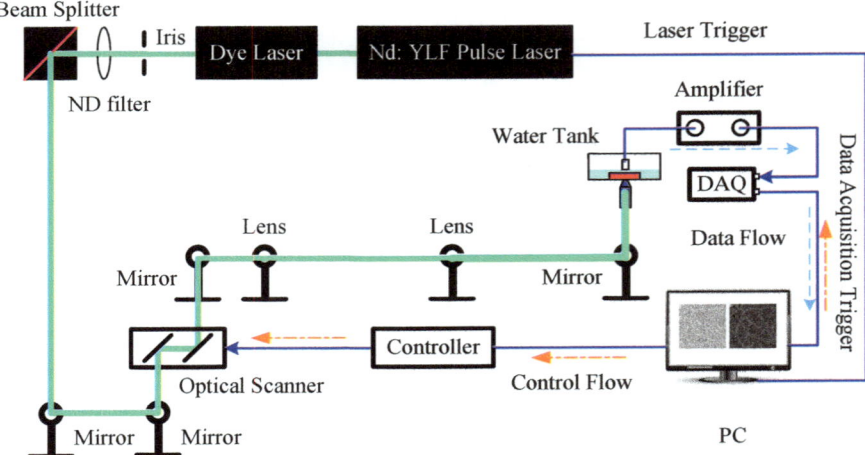

Fig. 4. Schematic of the experimental sparse PAM system.

uniform sampling mask. The corresponding recovery image is shown in Fig. 5e, f. The recovery method we used is OPTSPACE. All the microvascular could be recovered by both schemes. But there are more artifacts shown in Fig. 5e, which is non-uniform sampling scheme.

The same quantitative analysis with numerical study, SSIM and PSNR are also studied. The SSIM and PSNR for uniform sampling are 0.9172 and 39.5907 dB, respectively. Further, the SSIM and PSNR for non-uniform sampling are 0.9898 and 47.0978 dB, respectively. Both parameters have shown significantly improvement.

In summary, non-uniform sampling scheme proposed for photoacoustic microscopy imaging on vascular has obtained more complete sparse sampling data than the uniform random sampling in real PAM experiments. Besides, in terms of the final image recovery performance, non-uniform scheme still has achieved a better result.

4   Conclusion

In this paper, a LRMA-based non-uniform sampling scheme has been proposed. The main idea of the paper is to obtain higher resolution image than uniform sampling scheme by giving more observation to the principal component and recovery the images from limited number of acquisitions. For the designation of LRMA-based non-uniform sampling, BRP has been incorporated to accelerate the speed of sampling mask generation by avoid the computation of SVD. Finally, the numerical and PAM experiment results have validated that the proposed method can achieve fast data better resolution imaging than uniform sampling in the compressive OR-PAM system.

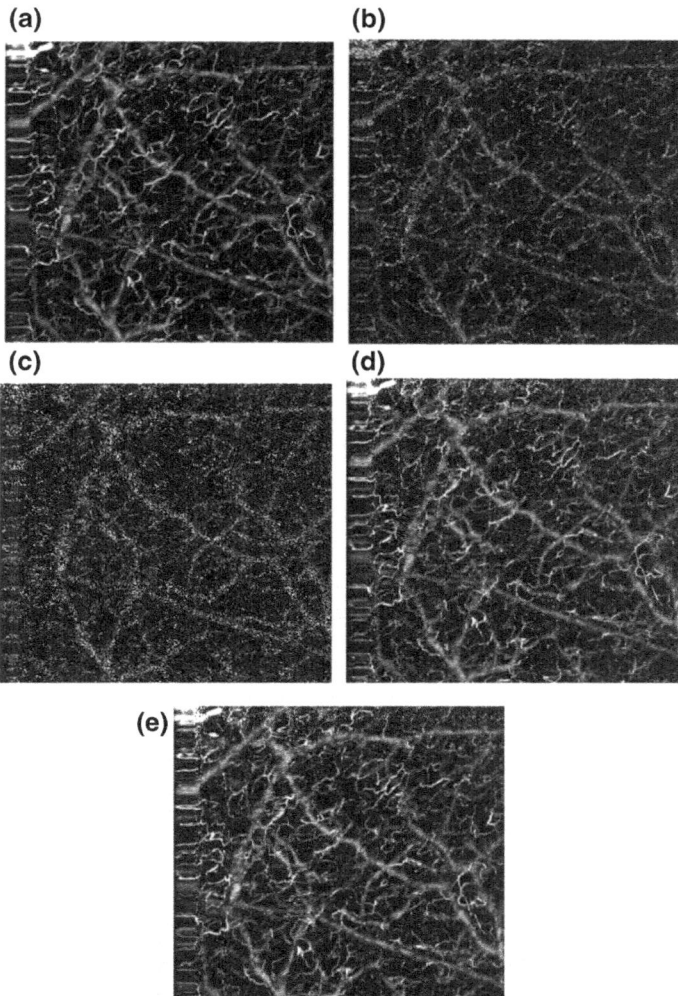

Fig. 5. Results of non-uniform sampling based on low-rank matrix approximation. **a** The full sampling PAM image. **b** The uniform sampling PAM data. **c** Non-uniform sampling PAM data with low-rank matrix approximation method. **d** Recovered image by uniform sparse sampling. **e** Recovered image by non-uniform sparse sampling.

Acknowledgments. This work is supported by Fundamental Research Funds for the Central Universities (Grant No. 3132017127).

References

1. Weis SM, Cheresh DA. Tumor angiogenesis: molecular pathways and therapeutic targets. Nat Med. 2011;17:1359–70.
2. Carmeliet P, Jain RK. Molecular mechanisms and clinical applications of angiogenesis. Nature. 2011;473:298–307.
3. Lin R, Chen J, Wang H, et al. Longitudinal label-free optical-resolution photoacoustic microscopy of tumor angiogenesis in vivo. Quant Imaging Med Surg. 2015;5(1):23–9.
4. Albini A, Tosetti F, Li VW, Noonan DM, Li WW. Cancer prevention by targeting angiogenesis. Nat Rev Clin Oncol. 2012;9:498–509.
5. Emblem KE, Mouridsen K, Bjornerud A, Farrar CT, Jennings D, Borra RJ, Wen PY, Ivy P, Batchelor TT, Rosen BR, Jain RK, Sorensen AG. Vessel architectural imaging identifies cancer patient responders to anti-angiogenic therapy. Nat Med. 2013;19:1178–83.
6. Haubner R, Beer AJ, Wang H, Chen X. Positron emission tomography tracers for imaging angiogenesis. Eur J Nucl Med Mol Imaging. 2010;37:S86–103.
7. Maslov K, Zhang HF, Hu S, Wang LV. Optical-resolution photoacoustic microscopy for in vivo imaging of single capillaries. Opt Lett. 2008;33:929–31.
8. Rao B, Li L, Maslov K, Wang L. Hybrid-scanning optical-resolution photoacoustic microscopy for in vivo vasculature imaging. Opt Lett. 2010;35:1521–3.
9. Lihong VW, Liang G. Photoacoustic microscopy and computed tomography: from bench to bedside. Annu Rev Biomed Eng. 2014;16:155–85.
10. Zhenghua W, Mingjian S, Qiang W, Ting L, Naizhang F, Jie L, Yi S. Photoacoustic microscopy image resolution enhancement via directional total variation regularization. Chin Opt Lett. 2014;12(12):121701.
11. Liu T, Sun M, Meng J, et al. Compressive sampling photoacoustic microscope system based on low rank matrix completion. Biomed Signal Process Control. 2016;26:58–63.
12. Meng J, Wang LV, Ying L, Liang D, Song L. Compressed-sensing photoacoustic computed tomography in vivo with partially known support. Opt Express. 2012;20:16510.
13. Liang J, Gao L, Li C, Wang LV. Spatially Fourier-encoded photoacoustic microscopy using a digital micro mirror device. Opt Lett. 2014;39:430–4.
14. Provost J, Lesage F. The application of compressed sensing for photoacoustic tomography. IEEE Trans Med Imaging. 2009;28(4):585–94.
15. Liang D, Zhang HF, Ying L. Compressed sensing photoacoustic imaging based on random optical illumination. Int J Funct Inform Pers Med. 2009;2:394.
16. Liu X, Dong P, Wei G, Xibo M, Xin Y, Jie T. Compressed sensing photoacoustic imaging based on fast alternating direction algorithm. Int J Biomed Imaging. 2012;2012:1–8.
17. Guo Z, Li C, Song L, et al. Compressed sensing in photoacoustic tomography in vivo. J Biomed Opt. 2010;15(2):021311–021311-6.
18. Sun M, Feng N, Shen Y, et al. Photoacoustic imaging method based on arc-direction compressed sensing and multi-angle observation. Opt Express. 2011;19(16):14801–6.
19. Meng J, Wang LV, Liang D, Song L. In vivo optical-resolution photoacoustic computed tomography with compressed sensing. Opt Lett. 2012;37:4573.
20. Meng J, Liu CB, Zheng JX, Lin RQ, Song L. Compressed sensing based virtual-detector photoacoustic microscopy in vivo. J Biomed Opt. 2014;19:036003.
21. Zhenghua W, Mingjian S, Qiang W, Naizhang F, Yi S. Compressive sampling photoacoustic tomography based on edge expander codes and TV regularization. Chin Opt Lett. 2014;12(10):101102.
22. Wu Z, Wang Q, Liu J, Sun M, Shen Y. Compressive sensing theory based on edge expander graphs. Acta Autom Sin. 2014;12:2824–35.

23. Li W, Lei Z, Zhijie L, Duanqing X, Dongming L. Non-local image inpainting using low-rank matrix completion: non-local image inpainting using low-rank matrix completion. Comput Graph Forum. 2014;00:1–12.
24. Liu T, Sun M, Feng N, et al. Non-uniform sampling photoacoustic microscope system based on low rank matrix completion. In: Instrumentation & measurement technology conference. IEEE; 2016.

An Improved Monte Carlo Localization Algorithm in WSN Based on Newton Interpolation

Lanjun Li$^{(\boxtimes)}$ and Jing Liang

School of Information and Communication Engineering,
University of Electronic Science and Technology of China, Chengdu, China
frankl29@163.com

Abstract. In recent years, with the development of sensor technology and wireless communication technology, wireless sensor network (WSN) as the technology for information acquisition and processing is widely applied in many fields. It is important for nodes to know their localizations for further applications. In this article, a range-free localization algorithm in WSN that builds upon the Monte Carlo Localization (MCL) algorithm is proposed. It concentrates on improving the sampling efficiency by changing the weights of samples. More specifically, mobility is used to improve the sampling efficiency to make sure MCL can perform well even when the sample number is low.

Keywords: Wireless sensor network · Monte Carlo Localization algorithm · Newton interpolation method · Sample weight

1 Introduction

WSN is a new technology filed based upon wireless communication technology, sensor technology, and microelectromechanical system. WSN system includes sensor node, sink node, and management node. Plenty of nodes can be set randomly inside or near the sensor filed by throwing them down from a plane or a rocket. Nodes in the sensor filed can form wireless networks through self-organizing [1].

Nowadays, there are many applications of WSN. For instance, since WSN can be set easily and the cost is not high, people are able to set sensors in those places that are hard for humans to reach, sensors in WSN can collect the information human needs to explore the area. However, receiving monitoring information of nodes without knowing the position is meaningless, so it is crucial for nodes in WSN to know their locations.

© Springer Nature Singapore Pte Ltd. 2020
Q. Liang et al. (eds.), *Communications, Signal Processing, and Systems*, Lecture Notes in Electrical Engineering 516,
https://doi.org/10.1007/978-981-13-6504-1_163

In order to determine the locations of sensor nodes, the global positioning system (GPS) is used, it provides with 3D localization based on direct line-of-sight with at least four satellites. Theoretically, people can put GPS on every single sensor nodes to get the location information. Those nodes with GPS which knows their locations are called beacon nodes, and those nodes which have no idea of their locations are called unknown nodes. However, if every node is equipped with GPS, the price is often too high. As a result, we need to use as few beacon nodes as possible to determine more unknown nodes locations.

The rest of the paper is organized as follows. Section 2 presents several localiztion methods adapted in WSN. Section 3 presents the main idea of MCL and the improvement for MCL using Newton interpolation. Section 4 presents our simulation results. Finally, we conclude our work in Sect. 5.

2 Localization in WSNs

In this section, we mainly introduce several localization protocols designed for wireless sensor networks. Those protocols are traditionally classified in two types, range-free and range-based.

2.1 Range-Based Localization Methods

Range-based localization methods mainly depend on special hardware to determine the distance or angles between different nodes, then use mathematical methods to determine the location of the unknown node.

In common, there are several mathematical methods to determine the position of unknown nodes using angle or distance information between nodes, for example, trilateration method and triangulation method. As a result, it is crucial to obtain the distance and angle information in WSN using other techniques.

Mostly, distance and angles can be determined using four techniques (AOA [2], TOA [3], TDOA [4], and RSSI [5]). Although range-based localization methods always have high accuracy, most range-based methods depend on special hardware making the cost pretty high. Also, in real environment, nodes in WSN can be affected easily, making those methods unsuitable when nodes are moving. In order to be independent of hardware and counter range inaccuracies, researchers developed range-free methods that depend uniquely on the information a node receive from its neighbors.

2.2 Range-Free Localization Methods

Compared to range-based localization methods, range-free localization methods do not rely on distance information or angle information. They can finish localization with data as network connectivity and hop numbers. Range-free methods can achieve high localization accuracy and do not require special hardware or expensive facilities, making them one of the most potential localization techniques in the future. For example, centroid method [6], DV-HOP method [7], and APIT method [8], are commonly used in localization.

3   Improved Monte Carlo Localization Algorithm Based on Newton Interpolation

3.1   Monte Carlo Localization Algorithm

In 2004, Hu and Evans firstly come up with the idea that using Monte Carlo method in WSN localization [9]. Normally, Monte Carlo method is used in determining location of robots. It is a range-free method so that it is low cost and does not have high requirement for hardware. More importantly, Monte Carlo localization (MCL) performs much better than other localization methods when beacon nodes and unknown nodes are moving, which suits the real world environment.

In MCL, the time is divided into discrete intervals. A sensor node realizes in each time interval (the location can be described as L_t). The whole algorithm has three steps, which are prediction step, filtering step, and sampling step.

In prediction step, a node starts its moving from its location in $t - 1$, which is L_{t-1}. Assume the node has no idea of its moving speed and direction, but is aware of its maximum velocity, v_{max}. If a node position in $t - 1$ is L_{t-1}, then L_t should be contained in the circle with a radius v_{max}. Once the location of a node in previous step is known, we are able to obtain the possible area the node may be in the next time interval.

However, only knowing the possible area of nodes is not enough, so in filtering step, we use the relationship between unknown nodes and beacon nodes to obtain filtering conditions, hoping to restrain the possible area. In MCL, we assume that all messages nodes send are received instantly. Hence, at time t, every node within the radio range R of beacon nodes, is able to hear announcement from the beacon nodes. If a beacon node can communicate directly with unknown nodes, which means it is within the radio range R, it is called one-hop node. If a beacon node cannot communicate with unknown nodes directly by itself, but it is able to communicate with nodes via one-hop neighbors, then it is called two-hop node. Formula (1) is used to present the filtering condition. In a time interval, we can obtain one-hop list, S, and two-hop list, T from a node. According to S and T, we are able to get filtered area from (1).

$$filter\,(l) = \forall\, s \in S,\; d\,(l,s) \le R \wedge \forall s \in T,\; R < d\,(l,s) \le 2R \qquad (1)$$

$d(l, s)$ stands for the Euclidean distance between unknown node and beacon node. With the filtering condition, the circle in prediction step can be reduced.

In sampling step, we use importance sampling, if a sample meets the filtering condition, the weight is set as 1, otherwise, the weight is 0. After collecting enough samples MCL needs, the average coordinate of samples are regarded as the coordinate of the estimating position of unknown node in a time interval. However, the degeneracy of the importance sampling is unavoidable since the unconditional variance if the importance weight will increase. It is necessary to do re-sampling if the number of samples is not enough. In MCL, a sample threshold N_{eff} is set as a standard for sampling.

In the simulation, they adopt the random waypoint mobility model and vary several parameters for both sensor nodes and WSN. In MCL, when localization error reaches a balance phase, error in MCL is less than centroid algorithm and amorphous algorithm. Maintaining more samples for MCL algorithm can improve accuracy, but needs additional memory. From this simulation, we find that MCL performs well in WSN when nodes are moving, and mobility can even improve accuracy and reduce the costs of localization.

However, since MCL has high requirement for sample numbers, when the area after filtering is small, it is difficult to obtain enough effective samples to meet the threshold N_{eff}, which means the algorithm will do re-sampling over and over again. It increases the cost of the algorithm and cause time waste. As a result, we suggest an improved MCL algorithm based on Newton interpolation, fixing the problem by using the mobility of nodes.

3.2 Improved Monte Carlo Localization Algorithm Based on Newton Interpolation

In MCL, the weight of samples is either 1 or 0. In our algorithm, we change sample weight into numbers between 0 and 1 to increase localization accuracy. Prediction step and filtering step stay the same. Because of inertia, the moving track of nodes is continuous. Theoretically, a present position of node L_t can be predicted by its previous location L_{t-1}, L_{t-2}, etc. In numerical analysis, interpolation is often used to predict data from the information of its previous value. So we adopt Newton interpolation in our prediction. Assume the position of a node in L_{t-1}, L_{t-2}, L_{t-3} is known, we use second phase Newton interpolation (2) to determine its previous position L_t.

$$f(x) = f(x_0) + (x - x_0)f[x_1, x_0] + (x - x_0)(x - x_1)f[x_2, x_1, x_0] \qquad (2)$$

$f[x_0, x_1]$ stands for difference quotient of x_0 and x_1, the same as $f[x_0, x_1, x_2]$. After computation, the predicted ndoe's coordinates are shown as (3) and (4).

$$x_{pre} = 3x_{t-1} - 3x_{t-2} + x_{t-3} \qquad (3)$$

$$y_{pre} = 3y_{t-1} - 3y_{t-2} + y_{t-3} \qquad (4)$$

Since the predicted position follows the mobility track, a node has higher probability of moving to the position near the predicted position. By using this feature, we can change samples' weights. The nearer a node is to the predicted position, the more contribution it has for localization, which means the importance weight is also higher. In order to make the computation easy, we consider the reciprocal of the distance of samples and predicted location as the sample weight as shown in (5). After normalization shown in (6), the sample weights are changed to numbers between 0 and 1.

$$d_i = \sqrt{(x_i - x_{pre})^2 + (y_i - y_{pre})^2} \qquad (5)$$

$$w_i = \frac{\frac{1}{d_i}}{\sum_{i=1}^{N} \frac{1}{d_i}} \tag{6}$$

The final coordinate the algorithm provides is shown as (7) and (8)

$$x_t = \sum_{i=1}^{N} x_i w_i \tag{7}$$

$$y_t = \sum_{i=1}^{N} y_i w_i \tag{8}$$

If the predicted position does not follow the filtering condition, we still use traditional MCL algorithm to finish localization in this time interval. And we also have to set a sample number threshold in case of degeneration. However, the threshold can be smaller than traditional MCL, for after fixing the importance weights, a sample has more contribution to localization. This method improves sampling efficiency by shortening the time MCL wastes on doing re-sampling. When initializing the algorithm, the location of nodes, L_1, L_2, and L_3 are determined by traditional MCL.

4   Evaluation

In WSN, doing simulation for nodes localization is an essential way to see the advantages and disadvantages of an algorithm. In this paper, we use MATLAB to run the test to show the localization result for improved MCL based on Newton interpolation.

4.1   Simulation Parameters

In our experiment, the simulation parameters are shown as follows.

- Sensor nodes are randomly distributed in a 500 m × 500 m rectangular region.
- The number of unknown nodes is 320, while number of beacon nodes is 96.
- Both beacon nodes and unknown nodes can move in a velocity between 0 and vmax. In our experiment, a node cannot move faster than 50 m in a time interval, which means vmax equals to 50 m.
- The radio range of nodes is 100 m, and all nodes are able to communicate with other nodes if are within the radio range.
- The mobility model of nodes is random waypoint mobility model.

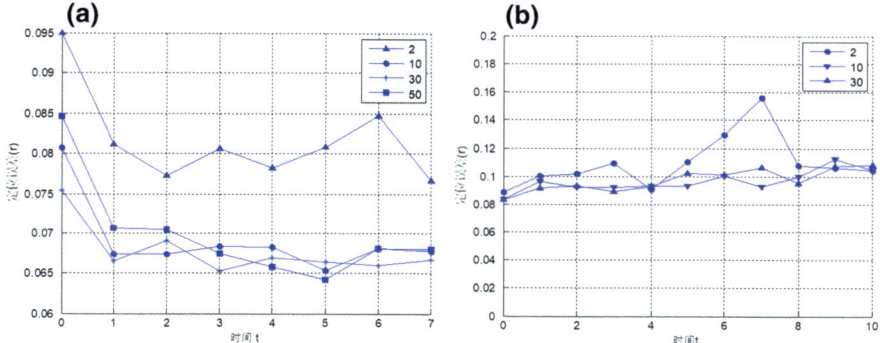

Fig. 1. Impact of sample number: **a** for traditional MCL; **b** for improved MCL

4.2 Number of Samples

As we mentioned in the last section, improved MCL increases sampling efficiency. We are going to prove that by comparing the influence the sample number has for localization accuracy. For traditional MCL, samples numbers threshold are set as four groups, 2 samples, 10 samples, 30 samples, and 50 samples. When the sample number is lower than threshold, do re-sampling until sample number equals to (or is higher than) threshold. The simulation result is shown as Fig. 1a.

The figure shows that maintaining more samples for MCL can definitely increase localization accuracy. However, when sample number is around 30, the effect of sample number is not obvious, especially when localization reaches its balance phase. And if there is too many samples in sampling phase, the accuracy is certainly high, but the computation cost can be high. Also, we cannot sacrifice localization accuracy only to make the computation simple. For improved MCL, sample numbers are set as three groups, 2 samples, 10 samples, and 30 samples. In order to make the effect of Newton interpolation more obvious, we assume that nodes move in certain track. The simulation result is shown as Fig. 1b.

The figures show that localization with high accuracy can also be achieved even when sample number is low. However, when sample number is too small, localization result may not be stable. This problem can be fixed when sample number is high, for example, over 10 samples are enough for a stable localization. This method fixes the problem that too much re-sampling process decreases the sample efficiency.

4.3 Localization Accuracy Comparison

In last section, we can see that improved MCL algorithm increases sampling efficiency. However, they are simulated in two different environments. In this section, we will simulate two algorithms in the same environment, which means nodes in both two algorithms follow the same moving track, to see whether the accuracy has been sacrificed to decrease the sampling efficiency. The simulation result is shown as Fig. 2.

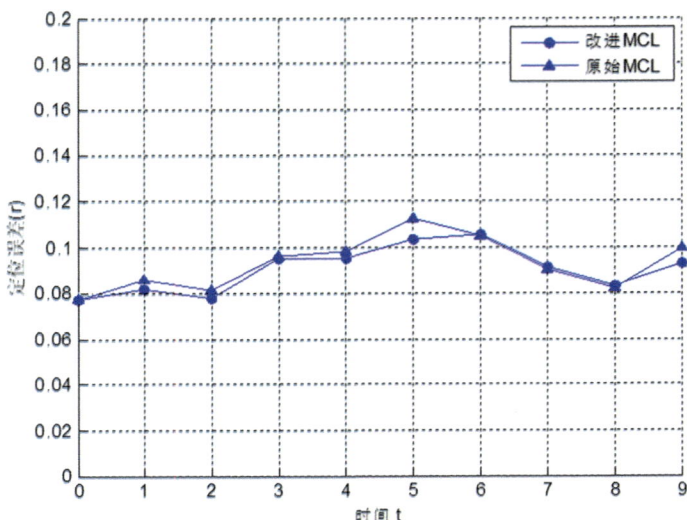

Fig. 2. Comparison of both algorithms

As is shown in Fig. 2, we can see that improved MCL does not affect the localization accuracy, even increases it, although the improvement in accuracy is not obvious.

5   Conclusion

Monte Carlo Localization algorithm is the first work to study range-free localization in the presence of mobility, and it shows that mobility improves localization accuracy, which is essential for localization in moving WSN.

However, MCL suffers from low sampling efficiency, mostly because of resampling. Our work, Monte Carlo Localization based on Newton interpolation increases sampling efficiency by adjusting sample weights due to their contribution to localization. While increasing sampling efficiency, the improved MCL algorithm still has high localization accuracy, even higher than traditional MCL.

However, there are still some disadvantages in the improved MCL algorithm. For example, computing the distance between a sample and the predicted position can be complex, which will increase the cost of the network. And when predicted position does not follow the filtering condition, the algorithm is useless. As a result, many issues remain to be explored in future work to improve those disadvantages mentioned above.

Acknowledgments. This work was supported by the National Natural Science Foundation of China (61671138, 61731006), and was partly supported by the 111 Project No. B17008.

References

1. Akyildiz IF, Su W, Sankarasubramaniam Y. Wireless sensor networks: a survey. Comput Netw. 2002;38:393–441.
2. Niculescu D, Nath B. Ad hoc positioning system (APS) using AoA. In: IEEE INFO-COM, 2003, San Francisco, vol. 3; 2003. p. 1734–43.
3. Girod L, Estrin D. Robust range estimation using acoustic and multimodal sensing. In: IEEE/RSJ international conference on intelligent robots and systems (IROS 01), Maui, vol. 3; 2001. p. 1312–20.
4. Priyantha NB, Miu AKL, Balakrishnan H, et al. The ricket compass for context-aware mobile applications. In: 7th annual international conference on mobile computing and networking, Rome; 2001. p. 1–14.
5. Girod L, Bychovskiy V, Elson J, et al. Locating tiny sensors in time and space: a case study. In: 2002 IEEE international conference on computer design: VLSI in computers and processors, Freiburg; 2002. p. 214–9.
6. Bulusu N, Heidemann J, Estrin D. GPS-less low cost outdoor localization for very small devices. IEEE Pers Commun Mag. 2000;7(5):28–34.
7. Nicolescu D, Nath B. Ad-Hoc positioning systems (APS). In: Proceedings of the 2001 IEEE global telecommunications conference, vol. 5. San Antonio: IEEE Communications Society; 2001, p. 2926–31.
8. He T, Huang C, Blum BM et al. Range-free localization schemes for large scale sensor networks. In: Proceedings of the 9th annual international conference on mobile computing and networking. New York, NY, USA: ACM Press; 2003, p. 81–95.
9. Hu L, Evans D. Localization for mobile sensor networks. In: Proceedings of the 10th annual international conference on mobile computing and networking. ACM Press; 2004. p. 45–7.

UAV Autonomous Path Optimization Simulation Based on Multiple Moving Target Tracking Prediction

Bo Wang$^{(\boxtimes)}$, Jianwei Bao, and Li Zhang

Nanjing University of Aeronautics and Astronautics,
Nanjing 21001, Jiangsu, China
wangbo_nuaa@nuaa.edu.cn

Abstract. In the UAV path planning study, due to the relative movement of multiple targets and the drone, long-term and large-scale UAV autonomous tracking has not been achieved. Therefore, aiming at this problem, this paper uses multiple moving target tracking algorithm to provide a real-time feedback on target position, estimates the later motion state of the target according to its position, and then performs the dynamic path planning by combining the feedback data and the state estimation result. Finally, The UAV path is optimized in real time. Experiments show that the proposed scheme can better plan the UAV path when multiple targets are in motion, thus improving the intelligence of the drone and the capability of long-time tracking.

Keywords: UAV · Target tracking · Motion estimation · Path planning

1 Introduction

Relying on the flexibility of UAV with the fast and low-cost remote sensing of the high-definition cameras on board, UAVs have achieved a wide application in the field of military and civil aviation remote sensing, communication relay, and so on. In the aspect of moving target tracking by UAV, the existing autonomous tracking of the single moving target has been extensively studied and applied [1]. However, as to the problem of multiple moving target tracking, the UAV operation still relies on operator's observing to control the UAV flight, and the autonomous flight path design has not been realized. Therefore, this paper combines the tracking, prediction, and path planning of single moving target by UAV, so as to realize UAV autonomous path planning for multiple targets. In order to achieve long-term and large-scale UAV autonomous tracking of multiple moving targets, this paper presents a solution. There are N moving targets to be tracked, and a drone starts from a point to reach the positions of these targets. After each target is visited once, the drone will return to the starting point. The UAV path problem [2] means how to find the shortest path. To realize the UAV autonomous path planning, this paper divides the scheme into three steps.

The first step is to obtain the specific motion equation of the selected moving targets. Because of the continuous relative movement of the drone and the targets, the

© Springer Nature Singapore Pte Ltd. 2020
Q. Liang et al. (eds.), *Communications, Signal Processing, and Systems*, Lecture Notes in Electrical Engineering 516,
https://doi.org/10.1007/978-981-13-6504-1_164

target's position will constantly change. To obtain the motion equation of the targets, this paper adopts the method of target tracking. Correlation filter method is widely used in the field of target tracking, due to its speed and accuracy, such as the KCF algorithm proposed by Henriques [3]. KCF constructs a training sample of a circulant matrix structure and transforms the solution of the problem into a discrete Fourier domain, avoiding the process of matrix inversion, reducing the complexity of the algorithm, and improving the tracking real-time performance. Therefore, the target position can be effectively updated by the KCF algorithm.

The second step is to predict the positions of moving targets. Because the position of a target is constantly changing, tracking its present position will lead to longer tracking time. Therefore, this paper uses the method of estimating the target position to optimize the UAV path. Due to the state change and uncertainty of moving targets, Kalman filtering method [4] is used in this paper. Kalman filtering method is a time-domain method which solves the problem based on state space. It has high applicability to the estimation of moving targets that frequently change motion states, thus becoming a universal machine learning method.

The last step is to use predicted positions to plan the UAV path, which means to quickly and accurately design a non-repeating shortest path after obtaining the positions of the targets. This problem is called the traveling salesman problem (TSP) in the unified research [5]. Due to the complexity of TSP, the time complexity of the exact solution method is long. Therefore, ant colony algorithm is adopted instead of the exact algorithm. The basic principle of ant colony algorithm uses pheromone to control the direction of ant's movement and autonomously and effectively approach the optimal path ultimately [6]. Then, Stutzle [7] proposed the maximum and minimum ant colony algorithm. Only the pheromone of the ant with the optimal algorithm is updated; thus, the convergence speed of the algorithm is improved. Li [8] uses prior knowledge to limit the increment of pheromone and increase the road weight factor to effectively avoid the stagnation of the algorithm. Therefore, this paper combines the improved methods of scholars to optimize the ant colony algorithm and thus solves the problem of long solution time and falling into the local optimum in the route plan using ant colony algorithm.

In general, to solve the problem of UAV autonomous flight path planning, this paper presents a solution. First, the KCF algorithm is used to track multiple moving targets to obtain the position and motion information of multiple moving targets. Then, the Kalman filter method is used to predict the positions of the moving targets. Finally, the improved ant colony algorithm is used to process the acquired information to design a shortest UAV flight path to track multiple moving targets.

The scheme can realize the near-optimal drone flight path by means of computer vision under the scene of UAV multiple moving targets tracking, thereby reducing manual operations and improving the UAV autonomous flying capability.

2   Algorithm Implementation

2.1   Overall Algorithm Flow

This paper presents an algorithmic process for UAV autonomous planning. Firstly, the targets to be tracked are selected on the video sequence returned by the drone. Then, the tracking algorithm is used to obtain the positions and state changes of the moving targets, and the positions of the targets after a certain period of time are estimated according to the position changes. After obtaining the estimated motion states, the present path is optimized based on the present UAV path setting and the motion estimation of the moving targets using the ant colony algorithm. Finally, when the drone performs the calculated path flight, it updates the positions of the targets and the path in real time to achieve the UAV path planning requirement for multiple moving targets. The specific flowchart is shown in Fig. 1.

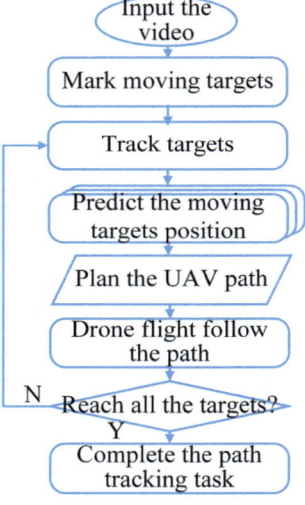

Fig. 1. Flowchart of the algorithmic process

2.2   Target Tracking Algorithm

The KCF tracking algorithm completes target tracking by designing a ridge regression classifier. The purpose of training the ridge regression classifier is to find a function $f(x) = w^T x$ to obtain the minimum loss function:

$$\min_{w} \sum_{i} \left(f(x_i - y_i)^2 + \lambda \|w\|^2 \right) \tag{1}$$

where $x = (x_1, x_2, \ldots, x_n)$ represents a sample of the tracking model, λ represents a regular term which prevents over-fitting, and w represents a solution parameter. Solve the above equation and obtain:

$$w = (X^T X + \lambda I)^{-1} X^T y \tag{2}$$

where X is the cyclic matrix, and y is a set of tag values for each sample. For samples that cannot be classified in the original space, the kernel function is needed to map the linearly inseparable patterns in the low-dimensional space to the high-dimensional space. The format of the kernel function is as follows:

$$k(x, z) = \varphi(x)\varphi(z) \tag{3}$$

where $k(x, z)$ is the kernel function, and $\varphi(x)$ and $\varphi(z)$ are mapping functions from low-dimensional space to high-dimensional space. When the kernel function is used to map the sample x to $\varphi(x)$, in $f(x) = w^T x$, the coefficient w is converted into a in the dual space. Combine with the above formula and obtain:

$$a = (K + \lambda I)^{-1} y \tag{4}$$

where K represents the mapped kernel matrix, $K_{ij} = k(x_i, x_j)$. In the tracking phase, dense sampling is performed near the targets to obtain different tracking samples, and then the ridge regression classifier is used to determine the final position information of the targets. After that, the position information is further trained by the ridge regression classifier again. The process is performed repeatedly to complete the tracking of targets. The specific process is shown in Fig. 2.

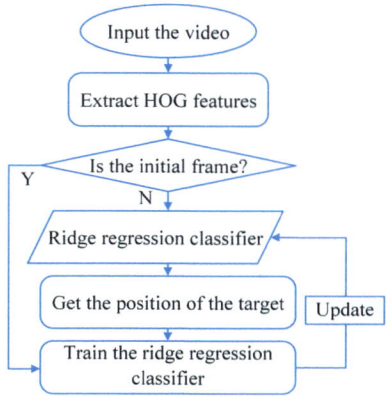

Fig. 2. Flowchart of the KCF tracking algorithm

2.3   Target Position Prediction

Kalman filter is a recursive estimator that implements the estimation of motion states through the principle of feedback control. From the dynamic target tracking in the previous step, the motion vector of the targets can be obtained. According to the historical motion states of the targets, the current states of the targets can be predicted. The basic state model of the targets includes the observation equation and the state equation as follows:

$$\begin{cases} Z_k = H_k X_k + V_k \\ X_{k,k-1} = A_{k,k-1} X_{k-1} + B_{k-1} U_{k-1} \end{cases} \tag{5}$$

where Z_k represents the observed value of the target state, H_k represents the observation matrix, X_k represents the real state at time k, $X_{k,k-1}$ represents the target state estimated from the previous state, $A_{k,k-1}$ represents the state transition matrix from $k-1$ to k, B_{k-1} represents the system control matrix at time $k-1$, U_{k-1} represents the noise that influences the current motion law of the targets, such as acceleration and deceleration of the moving targets, Q is the self-covariance of U_{k-1}, and V_k represents the observation noise, which in this experiment represents the deviation during the tracking process. Therefore, the prediction equation of the next coordinate $X_{k+1,k}$ and its covariance matrix $P_{k+1,k}$ can be obtained as:

$$\begin{cases} X_{k+1,k} = A_{k+1,k} X_k + B_k U_k \\ P_{k+1,k} = A_{k+1,k} P_k A_{k+1,k}^T + B_k Q_k B_k^T \end{cases} \tag{6}$$

2.4   UAV Path Planning

This paper focuses on the optimization of ant colony algorithm which is slow in convergence and easy to fall into the local optimal solution. The probability of stagnation of the algorithm is reduced by limiting the increment of pheromone update and the dynamic change of volatile factors, also by using blending inheritance algorithm; thus, the UAV dynamic path planning can be realized. The specific steps are as follows:

Step 1: Information initialization:

$$\tau_{ij}(0) = m/d_{\min} \tag{7}$$

where $\tau_{ij}(0)$ represents the pheromone from target i to target j, m represents the number of ants, and $d_{\min}$ represents the distance between the closest two targets.

Step 2: Path construction by all ant state transition.

Calculate the probability of transition of each ant k from target i to target j by the following formula at time t:

$$p_{ij}^k(t) = \begin{cases} \dfrac{\tau_{ij}^{\alpha}(ij)\eta_{ij}^{\beta}}{\sum_{s \in \text{allowed}_k} \tau_{ij}^{\alpha}(ij)\eta_{ij}^{\beta}}, & j \in \text{allowed}_k \\ 0, & \text{other} \end{cases} \tag{8}$$

where $\eta_{ij}(t) = 1/d_{ij}$ represents the degree of expectation of ants from target i to target j, α represents the information heuristic factor, which indicates the influence of pheromone on the transition probability, β represents the expected heuristic factor, which indicates the influence of visibility on the transition probability of ants, allowed $= \{1, 2, \ldots, n\} - \text{tabu}_k$ represents the targets which ant k has not visited, and tabu_k represents the targets that the ant k has passed.

Step 3: Pheromone update according to the steps which ants construct.

Ants will release pheromone on the path and update the pheromone locally according to the following formula. In order to increase the difference between the best and the worst paths, this paper adopts the maximum and minimum ant system to strengthen the optimal solution and weaken the worst solution:

$$\tau_{ij}(t+1) = \begin{cases} (1 - \rho)\tau_{ij}(t) + \varepsilon(L_{\text{worst}} - L_{\text{best}}), & \text{if } ij \text{ is the best path.} \\ (1 - \rho)\tau_{ij}(t) - \varepsilon(L_{\text{worst}} - L_{\text{best}}), & \text{if } ij \text{ is the worst path.} \\ (1 - \rho)\tau_{ij}(t), & \text{other} \end{cases} \quad (9)$$

where ρ represents the pheromone volatility coefficient, ε represents the pheromone enhancement coefficient, and $L_{\text{best}}, L_{\text{worst}}$ represent the length of the optimal path and the worst path, respectively, for all ants in this time.

Step 4: Update the number of iterations.

If the number of iterations is less than the set value N_{max}, repeat step 2.

Finally, the ant colony will gradually converge on the best path through the constant updating of pheromone. The specific process is shown in Fig. 3.

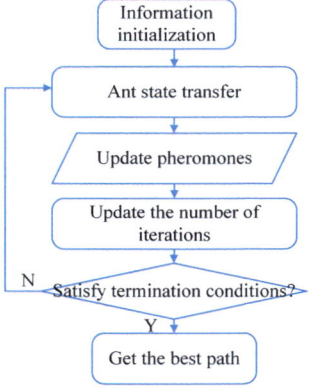

Fig. 3. Flowchart of ant colony algorithm

3   Simulation Experiment

3.1   Single-Target Tracking Experiment

In order to demonstrate the effectiveness and feasibility of the algorithm, there are two simulation experiments for the single-target tracking and multi-target tracking of UAV.

The first experiment is the simulation of UAV's approaching a single moving target. At first, a moving target is marked for tracking. Then, the marked target is tracked by the KCF tracking algorithm to obtain the target motion and position information, and the Kalman filter method is used to predict the moving target position. There are two UAVs for comparison in the experiment, one of which is an experimental UAV and the other is a comparison UAV. The experimental UAV is the UAV that tracks pre-targeted positions, and the comparison UAV is the UAV that tracks the current position of the target. In the experiment, the UAVs' movement speed was set to 2 pixels per frame, and the target movement speed was about 1 pixel per frame. The experimental results are shown in Fig. 4.

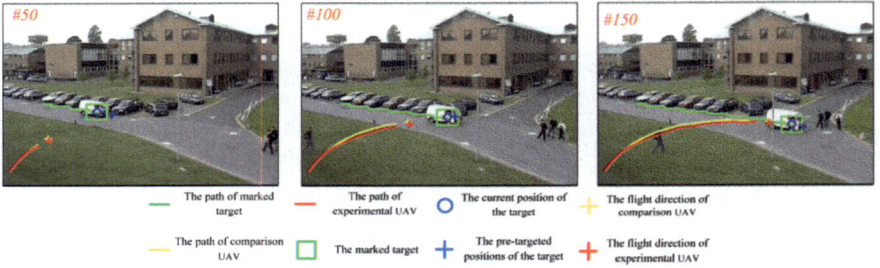

Fig. 4. Experimental results of the simulation of a single moving target

The result shows that the flight path of the experimental UAV is shifted in advance to the moving direction, and the required path is relatively shorter. It means the algorithm can improve tracking efficiency and optimize the UAV path.

3.2   Multi-target Tracking Experiment

The second experiment is the simulation of the path planning of multiple moving targets for UAV. Based on the first experiment, this experiment expands the single target to multi-targets, then by using the improved ant colony algorithm to plan the UAV's paths. The experimental results are as shown in Fig. 5.

The experimental result shows that the algorithm can effectively adapt to track the targets' movement and get the tracking sequence of the moving target to plan the UAV's paths. It means that the UAV path planning algorithm is feasibility. And at the 50th frame, two UAVs have the same sequence of tracking the targets. But at 70 frames, the experimental UAV changed the tracking experimental, but the tracking sequence of the comparison UAV did not change. At 90 frames, the tracking sequence

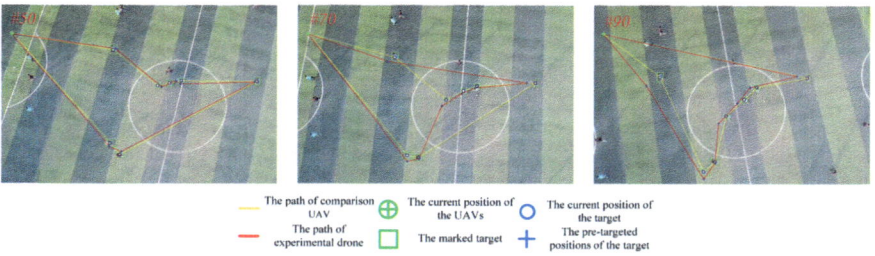

Fig. 5. Experimental results of the simulation of the path planning of multiple targets

of the comparison UAV changes to the same as that of the experimental UAV. It means that the UAV path planning algorithm can effectively predict the moving target positions, and the algorithm can update the tracking sequence of the moving targets in advance to optimize the UAV tracking path.

4   Conclusions

In this paper, aiming at the problem of UAV path planning for multiple moving targets, firstly, the KCF algorithm is used to track the moving targets to obtain the positions of the targets. Then, Kalman filter is used to predict the positions of the moving targets, and finally, the improved ant colony algorithm is used to complete the dynamic path planning to achieve the UAV path planning for multiple moving targets. The experimental results show that the proposed scheme can effectively complete the dynamic optimization of the UAV path, thus improving the intelligence of the drone in the scene of multiple moving target tracking.

Acknowledgments. This work was supported in part by the National Natural Science Foundation of China under Project No. 41701531. It was also supported in part by the Natural Science Foundation of Jiangsu Province under Project No. BK20170782. And this work was supported by the Open Research Fund of State Key Laboratory of Tianjin Key Laboratory of Intelligent Information Processing in Remote Sensing under grant No. 2016-ZW-KFJJ-01.

References

1. Wang X. Vision-based detection and tracking of a mobile ground target using a fixed-wing UAV. Int J Adv Robot Syst. 2014;11(156):1–11. https://doi.org/10.5772/58989.
2. Han P, Chen M, Chen SD, et al. Path planning for UAVs based on improved ant colony algorithm. J Jilin Univ. 2013;31(1):66–72. https://doi.org/10.3969/j.issn.1671-5896.2013.01.011.
3. Henriques JF, Rui C, Martins P, et al. High-speed tracking with kernelized correlation filters. IEEE Trans Pattern Anal Mach Intell. 2015;37(3):583–96. https://doi.org/10.1109/tpami.2014.2345390.

4. Kalman RE. A new approach to linear filtering and prediction problems. Trans ASME J Basic Eng. 1960;82(1):35–45. https://doi.org/10.1115/1.3662552.
5. Dantzig G, Johnson S. Solution of a large-scale traveling-salesman problem. Oper Res. 2010;2(4):393–410. https://doi.org/10.2307/166695.
6. Dorigo M, Maniezzo V, Colorni A. Ant system: optimization by a colony of cooperating agents. IEEE Trans Syst Man Cybern B. 1996;26(1):29. https://doi.org/10.1109/3477.484436.
7. Stutzle T, Hoos H. MAX-MIN ant system and local search for the traveling salesman problem. In: IEEE international conferences on evolutionary computation; 2002. p. 309–14. https://doi.org/10.1109/icec.1997.592327.
8. Li S, Zhang Y, Gong Y. The research on the optimal path of intelligent transportation based on ant colony algorithm. J Changchun Univ Sci Technol. 2015;4:122–6. https://doi.org/10.3969/j.issn.1672-9870.2015.04.027.

A Least Square Dynamic Localization Algorithm Based on Statistical Filtering Optimal Strategy

Xiaozhen Yan[1,2(✉)], Zhihao Han[1], Yipeng Yang[1], Qinghua Luo[1,2], and Cong Hu[2]

[1] School of Information Science and Engineering,
Harbin Institute of Technology at Weihai, Weihai, China
{yxz_heu, luoqinghua081519}@163.com
[2] Guangxi Key Laboratory of Automatic Detecting Technology and Instruments,
Guilin University of Electronic Technology, Guilin, China
yiqi@guet.edu.cn

Abstract. In wireless sensor network localization, many anchor nodes and target node exchange information at specified time intervals to obtain the distance information between each anchor node and the target node. With this information, the coordinates of the target node can be achieved through the calculation of the positioning algorithm. However, as there are numerous negative factors like non-line-of-sight measurement, complex multipath fading, which leads to high-level localization error. To improve localization accuracy, an improved least square localization algorithm is proposed, which combines the least square localization method with the statistical filtering optimization strategy. The simulation results show that this algorithm can effectively reduce localization error and achieve more accurate localization.

Keywords: Wireless sensor networks · Localization · Least square method · Extended Kalman filter · Particle filter

1 Introduction

With the development of science and technology, people's demand for localization is increasing day by day, and the demand for localization accuracy is also greatly increased. As a mature positioning scheme, GPS has good user experience and positioning accuracy in an open and unobstructed environment. However, after encountering obstacles, the satellite signal strength decreases greatly, which leads to a significant decline in GPS positioning accuracy. And it is difficult to meet the requirements of high-precision positioning results. Wireless sensor network localization could solve the problem.

The use of wireless sensor network in a small range of high-precision positioning has become the current trend. Wireless sensor network is a self-organizing network composed of a large number of randomly distributed small nodes, which integrated with sensors, data processing units, and communication modules. There are two import steps in range-based localization method. Firstly, the distances between anchor nodes

© Springer Nature Singapore Pte Ltd. 2020
Q. Liang et al. (eds.), *Communications, Signal Processing,
and Systems*, Lecture Notes in Electrical Engineering 516,
https://doi.org/10.1007/978-981-13-6504-1_165

and unknown node (to be localized) should be estimated via a specific method. And then localization result can be gained based on distances estimation results through specific localization computation method. When we choose a suitable algorithm, we can get high-precision positioning results.

The remainder of this paper is organized as follows: In Sect. 2, we review the related research works. Section 3 introduces the least square method and its implementation principle. Section 4 describes the proposed improved localization algorithm based on least square and statistical filtering strategy. Section 5 simulates the proposed algorithm and compares it with other related works. Finally, we conclude this paper.

2   Related Works

There are numerous negative factors during distance estimation and localization computation, which lead to different degrees of error in distance estimation and localization error, or even more serious consequence. Therefore, in recent years, scholars of various countries have proposed many optimization algorithms.

Literature [1] proposed a localization algorithm based on mobile anchor nodes. The author used mobile anchor node localization to avoid the accumulation of localization errors caused by multi-hop and long-distance transmission in wireless sensor networks. Literature [2] proposed a moving anchor node localization estimation algorithm based on weighted least square method. The author moves the anchor nodes along the linear model [3, 4]; at the same time, using the weighted least square method reduces the distance estimation error. This algorithm has good performance on localization estimation [5]. In the literature [6], the author designed an indoor wireless localization platform based on NanoLoc and used an improved particle filter algorithm. Compared with the ordinary least square method, it had a certain improvement in accuracy. In the literature [7], the problem of limited precision of the traditional least square localization algorithm based on the received signal strength indicator (RSSI) is addressed. Starting from the lower signal-to-noise influence, the author proposed a modified least square-BFGS localization algorithm based on dynamic t-test [8]. In reference [9], the author put forward a Monte Carlo mobile node localization algorithm based on least square method. Based on the continuity of the movement, this algorithm used the method of least square curve fitting and calculated where the unknown nodes may be in the next moment.

3   The Least Square Localization Algorithm

The principle of least square localization algorithm is shown in Fig. 1. Take n reference nodes as P_1, P_2, P_3, ..., P_n. We suppose their coordinates are (x_1, y_2), (x_2, y_2), $(x_3, y_3), \ldots, (x_n, y_n)$, respectively. The distance between them and the unknown node D is d_1, d_2, $d_3, \ldots, d_n$. And suppose the coordinate of point D is (x, y), which to be determined. According to geometric constraint relation among anchor nodes and unknown node, we can get the equations as shown in (1).

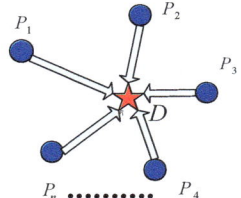

Fig. 1. Principle of least square localization algorithm

$$\begin{cases} (x_1 - x)^2 + (y_1 - y)^2 = d_1^2 \\ \quad\vdots \\ (x_n - x)^2 + (y_n - y)^2 = d_n^2 \end{cases} \tag{1}$$

According to the idea of least square method, the target node coordinate (x, y) should get the smallest error sum of squares between all measured distances $d_i(i = 1, 2, \dots, n)$ and its corresponding actual distances $\sqrt{(x_i - x)^2 + (y_i - y)^2}$. The equation can be formed as followed.

$$\min_{x,y} \sum_{i=1}^{n} (\sqrt{(x_i - x)^2 + (y_i - y)^2} - x)^2 \underset{m}{\Delta} e(x, y) \tag{2}$$

Taking the partial derivative to $e(x, y)$ and let it be zero, we can get the equation as followed.

$$\frac{\delta e}{\delta x} = \sum_{i=1}^{n} 2(\sqrt{(x_i - x)^2 + (y_i - y)^2} - d_i) \frac{2(x_i - x) \cdot (-1)}{2\sqrt{(x_i - x)^2 + (y_i - y)^2}} = 0 \tag{3}$$

$$\frac{\delta e}{\delta y} = \sum_{i=1}^{n} 2(\sqrt{(x_i - x)^2 + (y_i - y)^2} - d_i) \frac{2(y_i - y) \cdot (-1)}{2\sqrt{(x_i - x)^2 + (y_i - y)^2}} = 0 \tag{4}$$

Then, we can simplify them and get the equations as followed.

$$\begin{cases} \sum_{i=1}^{n} \frac{\mu_i - d_i}{\mu_i} (x_i - x) = 0 \\ \sum_{i=1}^{n} \frac{\mu_i - d_i}{\mu_i} (x_i - x) = 0 \end{cases} \tag{5}$$

The coordinates of target nodes can be obtained by solving the equations.

4   The Improved Least Square Localization Algorithm

This section mainly introduces the optimized least square localization algorithm, which uses extended Kalman filter or particle filter to optimize localization results to improve localization accuracy. The following is mainly discussed for the nonlinear model.

4.1   The Least Square Localization Algorithm Based on Extended Kalman Filtering

For nonlinear filtering system, transforming it into an approximate linear filtering problem by linearization technique is a commonly used treatment method. And one of the most widely used is extended Kalman filter method. Extended Kalman filter based on linear Kalman filter, the core idea is that below. For a nonlinear system, we can expand the nonlinear function f (*) and h (*) into Taylor series around the filter value $X(k)$, and omit the second order and above terms. Then, we will get an approximate linear model, and apply linear Kalman filter to estimate filter result and other processes.

The flowchart of extended Kalman filter localization algorithm is shown in Fig. 2.

First of all, an approximate linear model is obtained by local linearization of the nonlinear motion model. For a nonlinear system, its dynamic equations can be expressed as shown in (6) and (7).

$$X(k+1) = f[k, X(k)] + G(k)W(k) \tag{6}$$

$$Z(k) = h[k, X(k)] + V(k) \tag{7}$$

Based on the local linearization characteristics of nonlinear functions, we can expand the nonlinear function (6) and (7) into Taylor series in first order around the filter value $\hat{X}(k)$. The equations can be formed as followed.

The state equation is presented as following:

$$X(k+1) = \phi(k+1|k)X(k) + G(k)W(k) + \phi(k) \tag{8}$$

The observational equation is noted as follows:

$$Z(k) = H(k)X(k) + y(k) + V(k) \tag{9}$$

where the matrix is defined as follows, respectively:

$$\frac{\partial f}{\partial \hat{X}(k)} = \frac{\partial f\left[\hat{X}(k), k\right]}{\partial \hat{X}(k)}\Bigg|_{\hat{X}(k)=X(k)} = \phi(k+1|k) \tag{10}$$

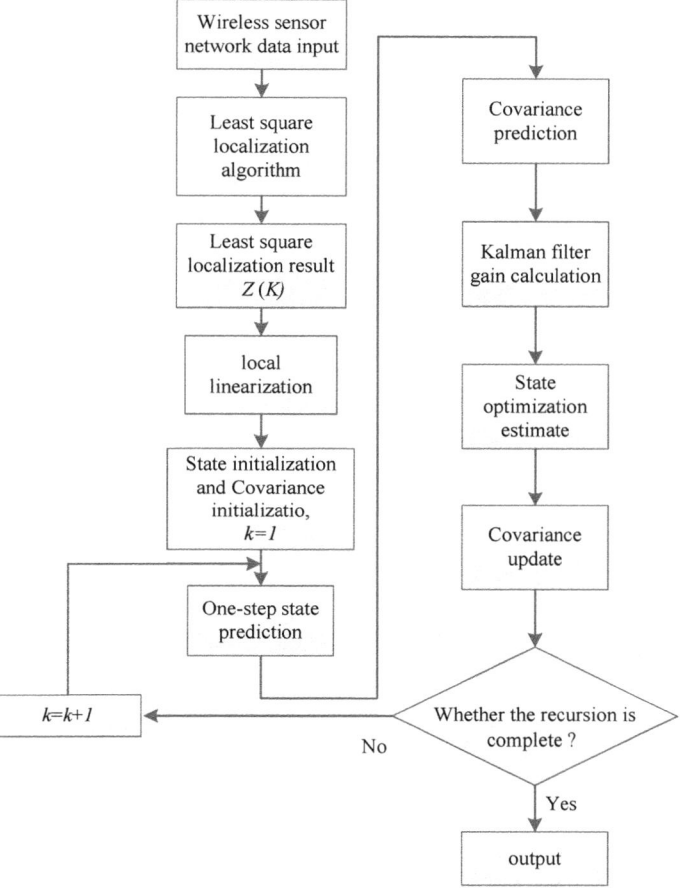

Fig. 2. Framework of extended Kalman filter localization algorithm based on least square method

$$f\left[\hat{X}(k),k\right] - \left.\frac{\partial f}{\partial X(k)}\right|_{\hat{X}(k)=X(k)} \hat{X}(k) = \phi(k) \tag{11}$$

$$\left.\frac{\partial h}{\partial \hat{X}(k)}\right|_{X(k)=\hat{X}(k)} = H(k) \tag{12}$$

$$y(k) = h\left[\hat{X}(k|k-1),k\right] - \left.\frac{\partial h}{\partial \hat{X}(k)}\right|_{X(k)=\hat{X}(k)} \hat{X}(k|k-1) \tag{13}$$

After local linearization, we can use the basic equations of linear Kalman filter to get the recurrence equations of extended Kalman filter.

One-step state prediction equation and covariance prediction equation are formed as followed.

$$\hat{X}(k|k+1) = f(\hat{X}(k|k)) \tag{14}$$

$$P(k+1|k) = \phi(k+1|k)P(k|k)\phi^T(k+1|k) + Q(k+1) \tag{15}$$

Then, we can use the formula (16) to get the Kalman filter gain.

$$K(k+1) = P(k+1|k)H^T(k+1)\left[H(k+1)P(k+1|k)H^T(k+1) + R(k+1|k)\right]^{-1} \tag{16}$$

Next, we can have the following equation for state optimization estimate:

$$\hat{X}(k+1|k+1) = \hat{X}(k+1|k) + K(k+1)\left[Z(k+1) - h(\hat{X}(k+1|k))\right] \tag{17}$$

where $Z(k+1)$ is the localization result obtained by the least square localization algorithm.

Finally, update the covariance with the following equation so that it can be used for the next recursive calculation.

$$P(k+1) = [I - K(k+1)H(k+1)]P(k+1|k) \tag{18}$$

We can recursively compute the above equations to obtain the filtering estimate of the least square localization result.

4.2   The Least Square Localization Algorithm Based on Particle Filtering

Particle filter is an approximate Bayesian filtering algorithm based on Monte Carlo simulation. Its core idea is to use some discrete random sampling points to approximate the probability density function of system random variables. Based on that, we can use the sample mean instead of the integral operation to obtain the minimum variance estimate of the state.

The flowchart of particle filter localization algorithm is presented in Fig. 3.

For N particles, we need to make predictions based on the state transfer function during sampling. The relationship is as follows.

$$X_k^{(i)} \sim q\left(X_k | X_{0:k-1}^{(i)}, Z_{1:k}\right) \tag{19}$$

When calculating the weight, substitute $\Delta(t) = Z(t) - Zpre(t)$ as the error value into the weight calculation equation, where $Zpre(t)$ is the observed value of the sampling point and $Z(t)$ is the localization result obtained by the least square method. The choice of $Z(t)$ affects the accuracy of the filter estimation.

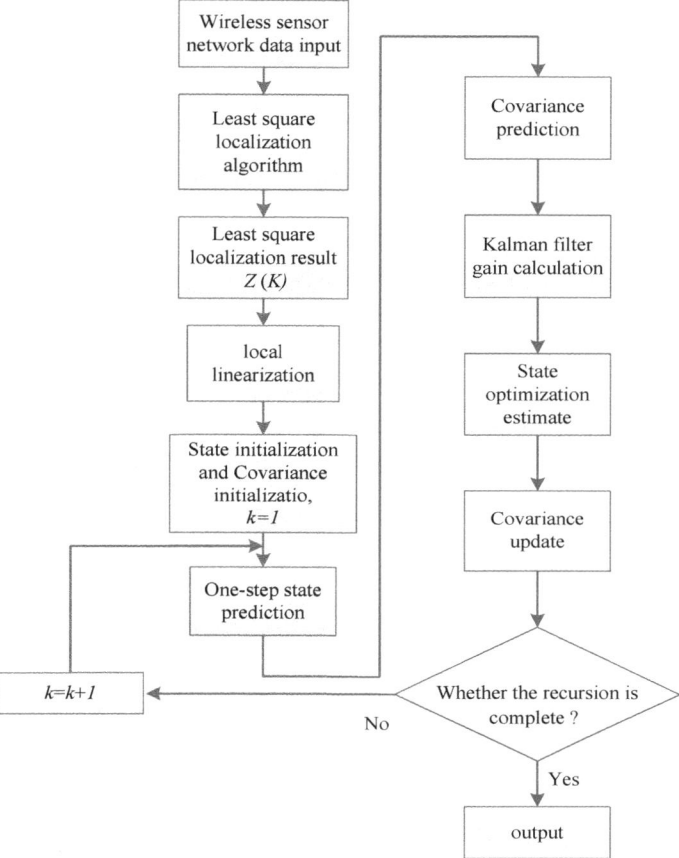

Fig. 3. Framework of particle filter localization algorithm based on least square method

Based on the position of the particle, we can get the corresponding weight. This estimation method will determine the effect of each particle, and by re-sampling, it will remove some particles that have little effect on the estimation of the filter. Then, we can estimate a result with the highest probability, which is our optimized localization result, by the weighted average method.

5   Evaluation

In this section, we will validate and evaluate the performance of our optimized least square localization algorithm. The performance of the localization is evaluated in terms of localization accuracy and computation time. We adopt absolute localization error (e_{ER}) to indicate localization accuracy, which is calculated according to (20), where e_{ER} is the absolute localization error, ($\hat{x}_n, \hat{y}_n$) is the localization result of the unknown node,

(x_n, y_n) is the exact coordinate of unknown node, n is the serial number of the localization point, and N is the total number of localization experiments. The smaller the e_{ER} is, the more accurate the localization result becomes. Computational time can reflect the cost of the algorithm.

$$e_{ER} = \sqrt{(x_n - \hat{x}_n)^2 + (y_n - \hat{y}_n)^2} \qquad (20)$$

During the simulation, five anchor nodes were randomly distributed in the 100 m * 100 m motion region. The simulation results are shown in Figs. 4, 5, and 6.

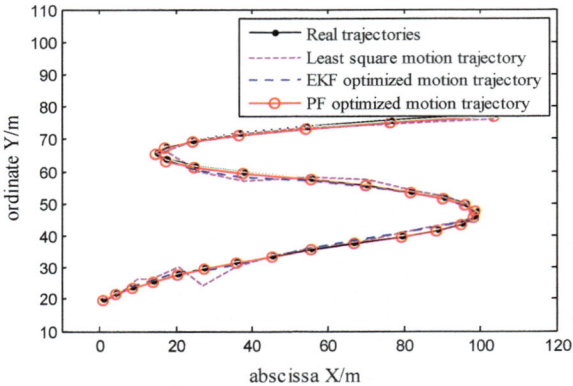

Fig. 4. Trajectory of different dynamic localization methods

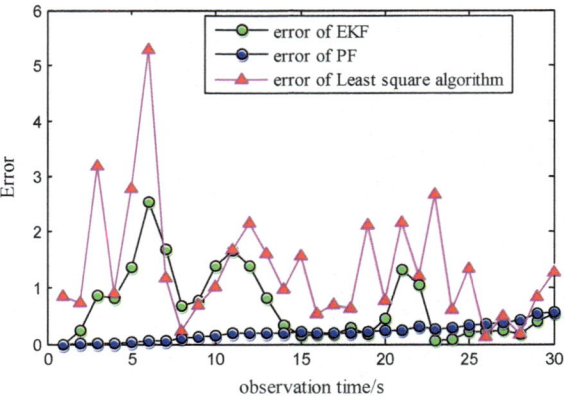

Fig. 5. Localization error of different methods

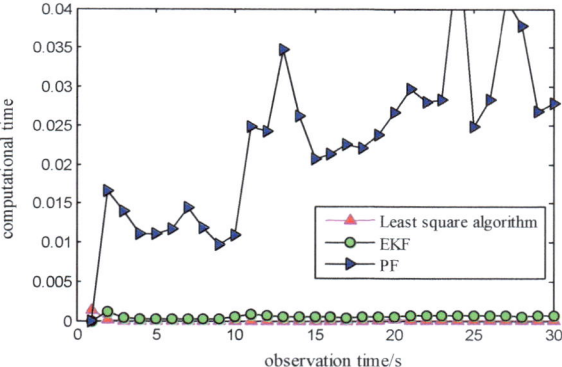

Fig. 6. Computational time of different methods

Under the nonlinear model, the localization trajectory obtained by different localization methods is illustrated in Fig. 4. It can be seen from Fig. 4 that all three localization methods can basically complete the task of localization.

Figure 5 is a comparison of localization errors of the three localization methods. Based on the distance between nodes measured by the wireless sensor, the least square localization algorithm can get the location result by solving the equation. However, due to noise and error, this algorithm has a relatively large localization error. As for extended Kalman filter algorithm based on the least square method, the localization error is reduced compared with the result before optimization. At the same time, localization errors fluctuate less. Least square localization algorithm based on particle filter has the best localization accuracy most of the time. However, because of accumulative error, the localization error will gradually increase.

Comparison of calculation time of three localization methods is shown in Fig. 6. The calculation time can reflect the overhead problem of the algorithm. In areas where real-time requirements are relatively high, this indicator is extremely concerned. As can be seen from Fig. 6, the least square method and the extended Kalman filter localization algorithm based on the least square method have shorter calculation time and better real-time performance. As for the particle filter localization algorithm based on the least square method, it has longer calculation time and more overhead than the previous two algorithms. Therefore, it is suitable for applications where the demand for localization accuracy is high, but the demand for real-time performance is not superior.

6   Conclusion

To improve the localization accuracy with a nonlinear dynamic model, we proposed an optimized least square localization algorithm. It can also be divided into two types of algorithms, which are based on the extended Kalman filter algorithm and the particle filter algorithm, respectively. In these two methods, we take advantage of filtering estimates to minimize the impact of negative factor during the localization procedure.

Relative to least square localization method, EKF algorithm based on the least square method has higher accuracy with approximate efficiency. As for PF algorithm based on the least square method, it has higher accuracy but less computational efficiency relative to the EKF.

Acknowledgments. The research presented in this paper is supported by the National Natural Science Foundation of China (61671174, 61601142), the Natural Science Foundation of Shandong Province of China (ZR2015FM027), WeiHai Research program of Science and Technology (16), the Laboratory of Satellite Navigation System and Equipment Technology (EX166840037, EX166840044), the Guangxi Key Laboratory of Automatic Detecting Technology and Instruments (YQ18206,YQ15203), the Natural Scientific Research Innovation Foundation of the Harbin Institute of Technology (HIT.NSRIF.2015122), the State Key Laboratory of Geoinformation Engineering (SKLGIE2014-M-2-4), and Discipline Construction Guiding Foundation in Harbin Institute of Technology (Weihai) (WH20150211).

References

1. Peng Y, Wang D. A review: wireless sensor network localization. J Electron Meas Instrum. 2011;25(5):389–99.
2. Kim E, Kim K. Distance estimation with weighted least squares for mobile beacon-based localization in wireless sensor networks. IEEE Signal Process Lett. 2010;17(6):559–62.
3. Cheng J, Yang L, Li Y, et al. Seamless outdoor/indoor navigation with WIFI/GPS aided low cost Inertial Navigation System. Phys Commun. 2014;13(PA):31–43.
4. Ades M, Van Leeuwen PJ. An exploration of the equivalent weights particle filter. Q J R Meteorol Soc. 2013;139(672):820–40.
5. Huang CH, Lee LH, Ho CC, et al. Real-time RFID indoor positioning system based on Kalman-filter drift removal and Heron-Bilateration location estimation. IEEE Trans Instrum Meas. 2015;64(3):728–39.
6. Jiaojiao W. Research on wireless localization technology of indoor moving target. TianJing University master thesis; 2013. p. 19–31.
7. Juan M, Li H, Yanan L, et al. Modified least squares-BFGS positioning algorithm based on dynamic T-test. Comput Appl Softw. 2016;22(6):126–9.
8. Mirzaei HR, Akbari A, Gockenbach E, et al. A novel method for ultra-high-frequency partial discharge localization in power transformers using the particle swarm optimization algorithm. IEEE Electr Insul Mag. 2013;29(2):26–39.
9. Yao Y. Polaronic quantum diffusion in dynamic localization regime. New J Phys. 2017;19(4): 043015.

Design and Implementation of an UWB-Based Anti-lose System

Yue Wang$^{(\boxtimes)}$ and Yunxin Yuan

Tianjin Key Laboratory of Wireless Mobile Communications
and Power Transmission, Tianjin Normal University, Tianjin 300387, China
ywang_tjnu@163.com

Abstract. Accurate and reliable distance information is essential to a variety of wireless applications, and ultra-wideband (UWB) signal can theoretically achieve centimeter-level ranging accuracy. In this paper, we design and implement an UWB-based anti-lose system, using commercial ScenSor DWM1000 module. The centimeter-level ranging accuracy of the system is shown under both line-of-sight (LoS) and non-line-of-sight (NLoS) experimental conditions.

1 Introduction

Accurate and reliable distance information is essential to a variety of wireless applications, such as anti-lose devices, indoor/outdoor positioning and navigation, and wireless sensor networks [1]. High-accuracy ranging can be achieved by using high-resolution time measurement, which is reversal to the bandwidth of the ranging signal [2]. Ultra-wideband (UWB) signal has huge bandwidth, i.e., more than 500 MHz, thus it can theoretically achieve centimeter-level ranging accuracy [3].

In this paper, we design and implement an UWB-based anti-lose system using ScenSor DWM1000 module, which is the first commercial module conformed to the IEEE 802.15.4a standard with low cost [4]. In the experiments, ranging performance is measured under two perpendicular antenna radiation angles and two signal prorogation conditions, i.e., line-of-sight (LoS) and non-line-of-sight (NLoS).

The rest of the paper is organized as follows. Section 2 introduces the design and implementation of the anti-loss system. Section 3 gives the experimental ranging results. A conclusion in Sect. 4 wraps up this paper.

© Springer Nature Singapore Pte Ltd. 2020
Q. Liang et al. (eds.), *Communications, Signal Processing,
and Systems*, Lecture Notes in Electrical Engineering 516,
https://doi.org/10.1007/978-981-13-6504-1_166

2   System Design and Implementation

The anti-loss system contains a pair of nodes; the distance between them is measured using two-way time-of-arrival (TW-ToA) ranging protocol [5], as shown in Fig. 1. The distance between node A and node B is calculated by

$$\hat{d}_{AB} = c \cdot \hat{t}_p = c \cdot \left(\frac{t^A_{RTT} - t^B_{TAT}}{2} \right),\tag{1}$$

where c is the propagation speed of UWB signal, $\hat{t}_p$ is the calculated propagation time, t^A_{RTT} is the round-trip-time (RTT) measured by node A, and t^B_{TAT} is the turn-around-time (TAT) measured by node B.

Each node in Fig. 1 contains four subsystems, as shown in Fig. 2; they are power subsystem, UWB ranging subsystem, micro-controller subsystem, and alarm subsystem.

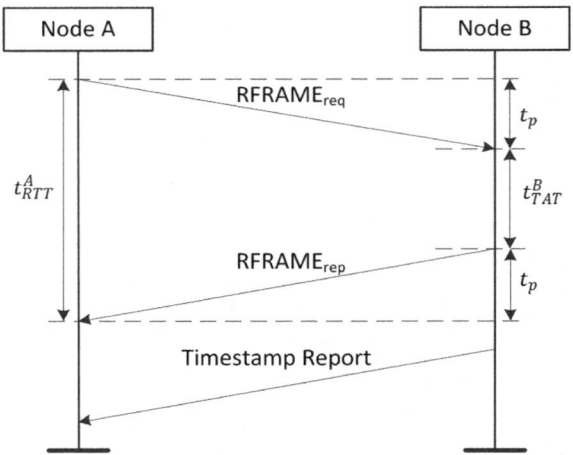

Fig. 1. TW-ToA ranging protocol.

The key element of the UWB ranging subsystem is ScenSor DWM1000 module and its peripheral circuit, which integrates DW1000 chip, onboard antenna, power management, clock management, and serial peripheral interface (SPI). By using DWM1000 module, the size of ranging node can be reduced and the RF circuit design is omitted. The micro-controller subsystem can configure the registers of DWM1000 and control its send/receive status via SPI.

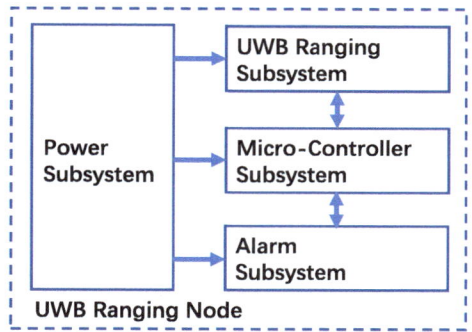

Fig. 2. Hardware structure of the ranging node.

The micro-controller subsystem adopts low-power and ARM-embedded STM32F411 chip. By designing the embedded program, the micro-controller can read the ranging information from the UWB ranging subsystem and transmit this information to the computer display via SPI to conduct ranging experiments.

The power subsystem uses single lithium battery TP054 to provide 5V power source, which can be charged by USB port.

The alarm subsystem contains a buzzer and a 2-bit dial switch. By using the 2-bit dial switch, there are four kinds of alarm distance settings; they are 0.5, 1, 5, and 10 m. If the ranging result is larger than the setting distance, the buzzer will produce the alarm sound.

The implemented ranging node is shown in Fig. 3.

Fig. 3. UWB ranging node prototype.

3   Experimental Ranging Results

In our ranging experiments, one ranging node is fixed on a tripod, the other ranging node is mounted on a moving cart. The distance between two nodes is preset, and the increasing distance step is 1 m. For each distance, 50 ranging measurements are carried out and recorded. We define the ranging error between node A and node B as

$$e_{AB} = \left| \hat{d}_{AB} - d_{AB} \right|, \tag{2}$$

where $\hat{d}_{AB}$ is the measured distance between node A and node B, and d_{AB} is the real distance between them.

In the first experiment, we measure the ranging accuracy for two perpendicular directions, i.e., 0° and 90°, between the antenna on each ranging node in the indoor environments, since the antenna gain is different for different directions. Two antennas on two nodes are parallel for 0°, and two antennas on two nodes are perpendicular for 90°.

The cumulative distribution function (CDF) for two directions is shown in Figs. 4 and 5. Under both 0° and 90° directions, the ranging error is less than 0.07 m with about 50% probability, and less than 0.15 m with about 90% probability.

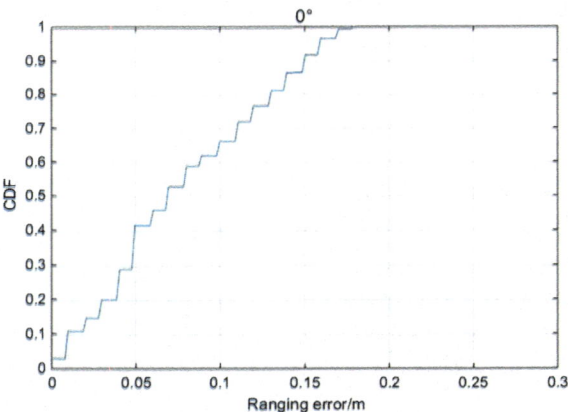

Fig. 4. CDF of 0° direction.

The box plot for two directions is shown in Figs. 6 and 7. Although the median for both directions is small, the outliers for 90° are larger than that for 0°.

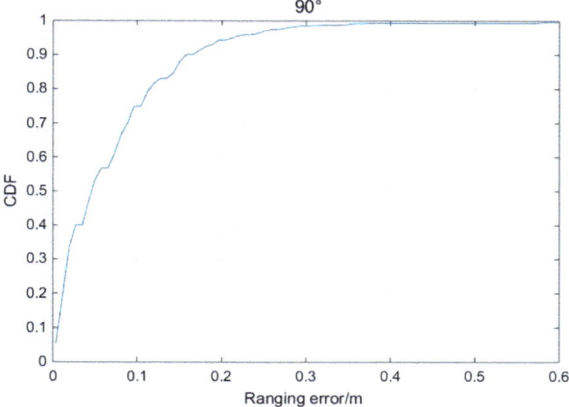

Fig. 5. CDF of 90° direction.

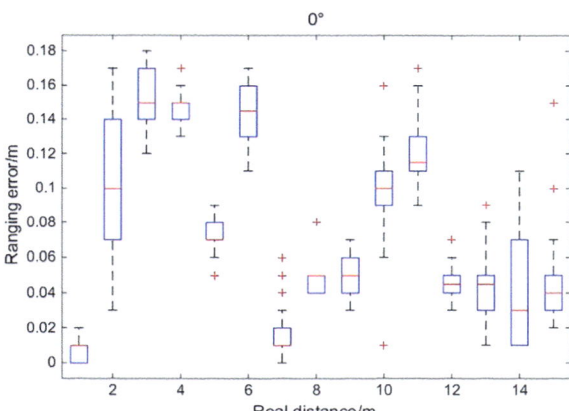

Fig. 6. Box plot of 0° direction.

In the second experiment, we compare the ranging accuracy for LoS and NLoS conditions. The experimental setting is shown in Fig. 8, where the NLoS condition is caused by a wooden door with 10 cm thickness.

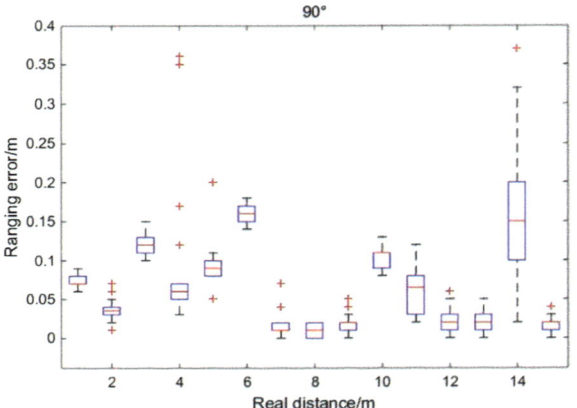

Fig. 7. Box plot of 90° direction.

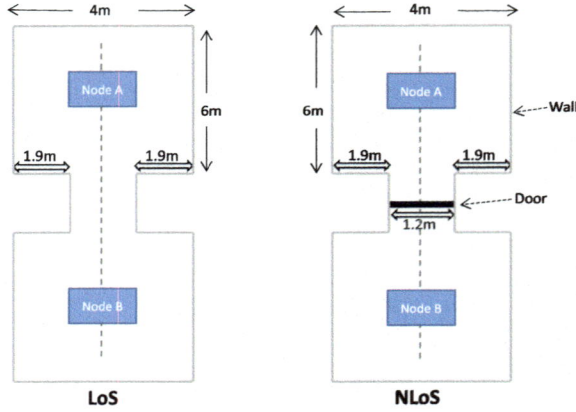

Fig. 8. LoS and NLoS ranging experimental settings.

The CDF for ranging error in LoS and NLoS conditions is shown in Figs. 9 and 10, respectively. Under LoS condition, the ranging error is less than 0.1 m with about 50% probability, and less than 0.13 m with about 90% probability. Under NLoS condition, the ranging error is less than 0.17 m with about 50% probability, and less than 0.25 m with about 90% probability.

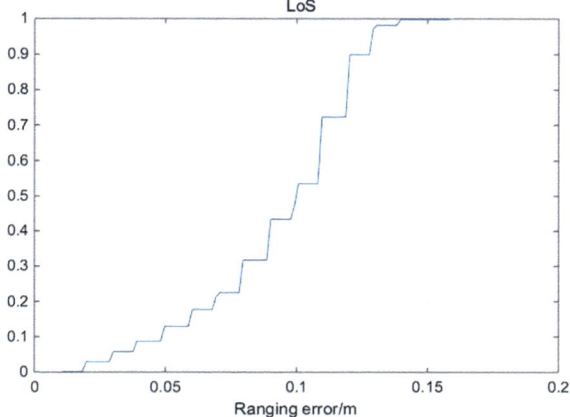

Fig. 9. CDF of LoS condition.

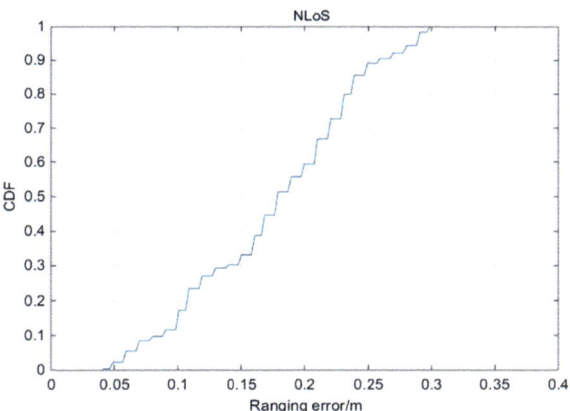

Fig. 10. CDF of NLoS condition.

The box plot for ranging error in LoS and NLoS conditions is shown in Figs. 11 and 12, respectively. The ranging error becomes larger in NLoS condition due to the signal block by the door. Even in the NLoS condition, the ranging error is relatively low thanks to the good penetrating property of UWB signal.

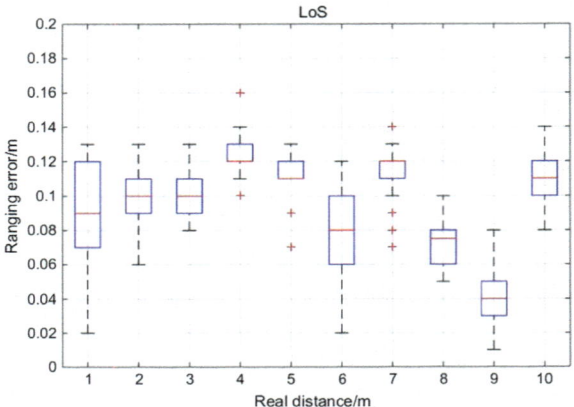

Fig. 11. Box plot of LoS condition.

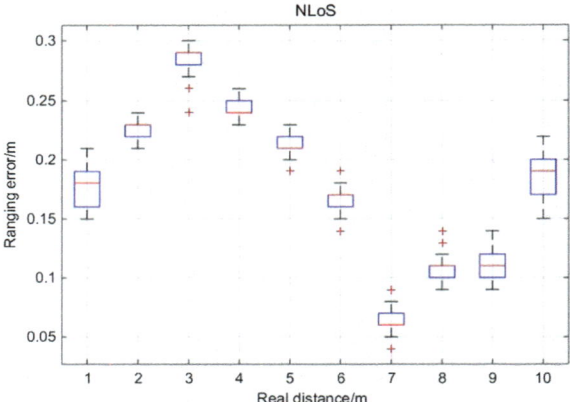

Fig. 12. Box plot of NLoS condition.

4   Conclusions

In this paper, an UWB-based anti-loss system is design and implemented. The centimeter-level accuracy of the system is shown under both LoS and NLoS experimental environments.

Acknowledgments. This work was supported by the Scientific Research Plan Project of the Committee of Education in Tianjin (No. JW1708).

References

1. Dardari D, Luise M, Falletti E. Satellite and terrestrial radio positioning techniques: a signal processing perspective. Oxford: Elsevier; 2012.
2. Gezici S, Poor HV. Position estimation via ultra-wide-band signals. Proc IEEE. 2009;97(2):386–403.
3. Dardari D, Conti A, Ferner U, Giorgetti A, Win MZ. Ranging with ultrawide bandwidth signals in multipath environments. Proc IEEE. 2009;97(2):404–26.
4. www.decawave.com
5. Wang C, Wang Y, et al. Ultra-wideband ranging system prototype design and implementation. In: Proceedings of the international conference in communications, signal processing and systems (CSPS), Harbin; July 14–16; 2017. p. 1848–55.

Indoor and Outdoor Seamless Localization Method Based on GNSS and WLAN

Yongliang Sun, Jing Shang$^{(\boxtimes)}$, and Yang Yang

School of Computer Science and Technology,
Nanjing Tech University, Nanjing 211816, Jiangsu, China
1254938049@qq.com

Abstract. Localization technology has been widely applied in various fields such as military investigation, natural disaster prevention, address search, and travel route planning. In order to guarantee the coverage range and localization performance of localization technology in both indoor and outdoor environments, research on indoor and outdoor seamless localization using global navigation satellite system (GNSS) and wireless local area network (WLAN) has attracted lots of attention. In this paper, a seamless localization method based on GNSS and WLAN is proposed. The method is able to switch smoothly from GNSS to WLAN localization in indoor and outdoor environments and outperforms either the GNSS localization or WLAN trilateration localization.

Keywords: Global Navigation Satellite System · Wireless Local Area Network · Seamless Localization · Trilateration

1 Introduction

In recent years, with the emergence of smart terminal devices and the rapid development of Internet of Things, people have more demands on location-based services (LBS) [1–3]. In outdoor environments, people can get satisfactory LBS with global navigation satellite system (GNSS) that consists of Beidou, GLONASS, global positioning system (GPS), and so on [4,5]. However, in indoor environments, satellite signals that are blocked by buildings are usually very weak [6]. The terminal devices can occasionally receive the satellite signals indoors. So indoor localization has been one of the research hot spots. Currently, wireless local area network (WLAN) is almost ubiquitous in people's daily life. WLAN-based indoor localization systems are able to meet the people's requirements for high-accuracy and low-cost LBS [7,8]. Therefore, in order to achieve seamless localization in indoor and outdoor environments, we propose an indoor and outdoor seamless localization method based on GNSS and WLAN.

© Springer Nature Singapore Pte Ltd. 2020
Q. Liang et al. (eds.), *Communications, Signal Processing, and Systems*, Lecture Notes in Electrical Engineering 516,
https://doi.org/10.1007/978-981-13-6504-1_167

The proposed localization method has strong localization reliability and is able to switch smoothly from GNSS to WLAN localization in indoor and outdoor environments.

2 Related Localization Methods

2.1 Satellite-Based Localization

Satellite-based localization technology usually locates a terminal device with four satellites as shown in Fig. 1. The ground terminal device measures the satellite signals and computes the location coordinates and clock bias. Let the device location coordinates be (x, y, z), ith satellite location and satellite time be (x_i, y_i, z_i, s_i), $i = 1, 2, 3, 4$, the signal reception time denoted by the terminal device clock be t_i', and the true signal reception time from ith satellite be $t_i = t_i' - b$, where b is the clock bias of the terminal device. For ith satellite, we can have the distance between the terminal device and ith satellite as follows:

$$d_i = (t_i' - b - s_i) \times c \tag{1}$$

where c is the speed of light and $d_i = \sqrt{(x_i - x)^2 + (y_i - y)^2 + (z_i - z)^2}$. When four satellites are available, we have four equations to determine the four unknown parameters (x, y, z, b).

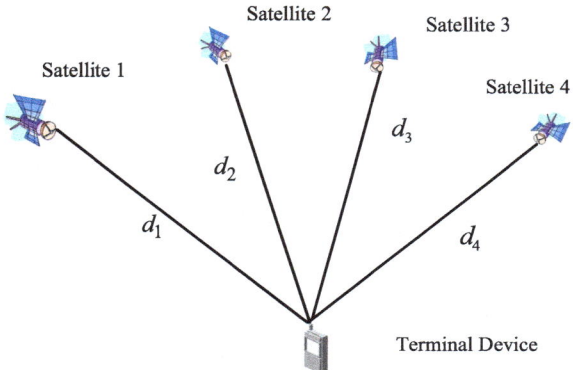

Fig. 1. Satellite-Based Localization Method

2.2 WLAN Trilateration Localization

When WLAN signals are transmitted, there will be power loss that is closely related to the propagation distance. WLAN trilateration localization is actually based on this relationship to obtain the propagation distances between a terminal device and access points (APs). In our experiment, we use the propagation model denoted by (2) to estimate the distances.

$$r = -(10N \log_{10} d + A) \tag{2}$$

where, r is the RSS value measured by a WLAN terminal device, d is the distance between the terminal device and an AP, N and A are propagation model parameters. With the propagation model, three distances between the terminal device and APs are estimated, then WLAN trilateration localization can be performed [5].

3   Experimental Setup and Results of GNSS and WLAN Localization

3.1   Experimental Setup

The experimental area is a part of the top floor of the School of Computer Science and Technology in Nanjing Tech University. As shown in Fig. 2, there is an outdoor terrace with dimensions of $20.4\,\mathrm{m} \times 9.2\,\mathrm{m}$, which makes the floor a proper experimental area for indoor and outdoor seamless localization. We totally select 46 testing points (TPs) with $0.6\,\mathrm{m}$ gaps in the outdoor terrace and indoor office area denoted by the green dots. We collect satellite signals and RSS samples using an OPPO smart phone. The smart phone is placed on a tripod with a height of $1.2\,\mathrm{m}$.

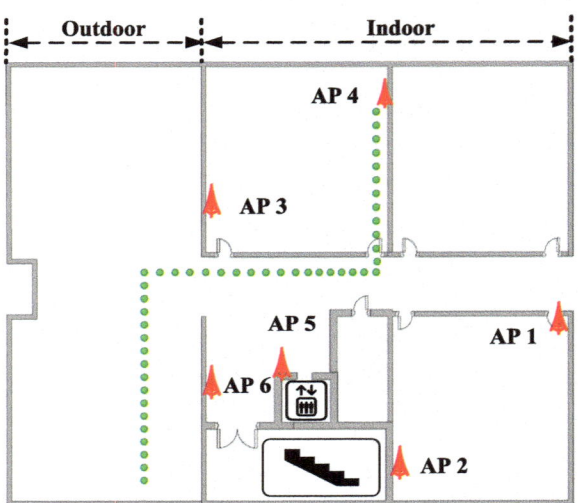

Fig. 2. Experimental Area Plan.

3.2 GNSS Localization Experiment

As shown in Fig. 2, 20 TPs are in the outdoor terrace and the other 26 TPs are in the indoor environment. We find that when we enter into the building from the terrace about 4 meters, satellite signals are difficult to obtain. Therefore, besides the 20 outdoor TPs, we collect satellite signals at extra six indoor TPs and calculate the localization results of the 26 TPs using GNSS. Then, we transform the localization results into a coordinate system we establish based on the experimental area plan. Within the signal collection duration, some latitude and longitude outliers appear, which may be caused by the ionosphere or troposphere effects, satellite clock errors, or human interference factors. In order to improve the localization accuracy, we filter the outliers and then sample the data. The means of the latitude and longitude are used to compute the localization coordinates. As shown in Fig. 3, the localization errors of the 26 TPs are calculated and the mean error is 5.13 m.

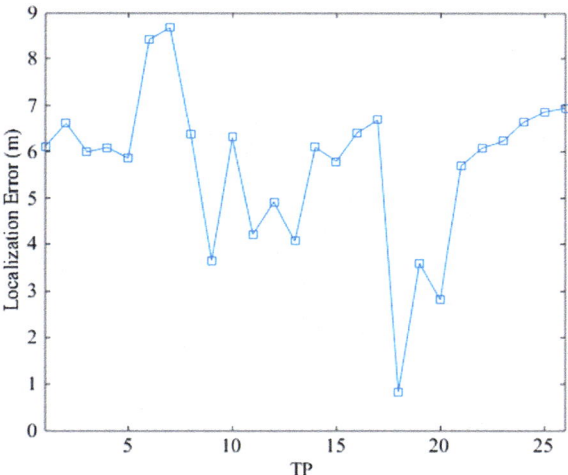

Fig. 3. GNSS Localization Errors.

3.3 WLAN Localization Experiment

The WLAN localization experiment is also carried out under the experimental area shown in Fig. 2. A total of six APs of WLAN are mounted in the experimental area with a height of 2.2 m and 60 RSS samples are collected at each TP. After filtering the RSS outliers from the six APs, the means of RSS samples collected at each TP are shown in Fig. 4.

According to empirical data, we set the propagation model parameters N and A be 1.73 and 21.31, respectively. Using the RSS data measured at each

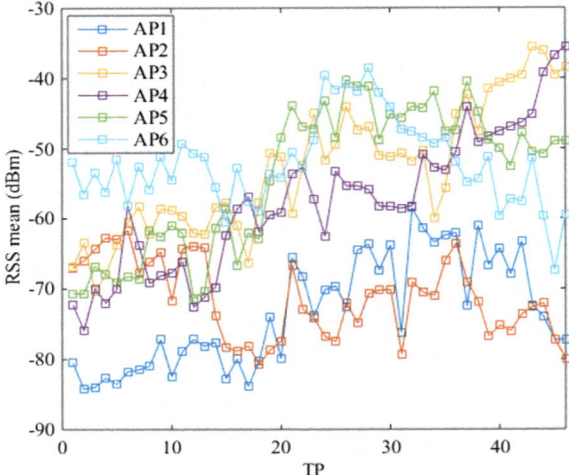

Fig. 4. Means of RSS Samples Collected at TPs.

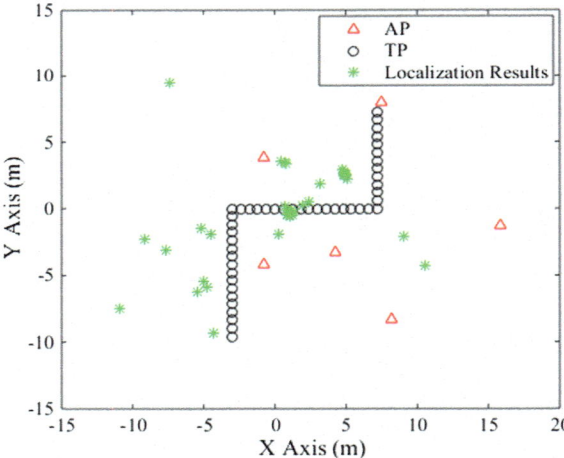

Fig. 5. Trilateration Localization Results.

TP, the distances between each TP and three corresponding APs are computed, so trilateration localization can be performed. The localization results as well as actual locations of TPs and APs are shown in Fig. 5. The mean error of the WLAN trilateration localization is 6.91 m.

4   Experiment of Indoor and Outdoor Seamless Localization

The experimental area can be divided into three subareas: GNSS localization subarea, handover subarea, and WLAN localization subarea. Satellite signals are used in the GNSS localization subarea and RSS data are used in the WLAN localization subarea. In the handover subarea, RSS may be strong and satellite signals are also available, so localization method can be selected according to a handover mechanism. We select the nearest three APs to the handover subarea that are AP3, AP5, and AP6, and set an RSS threshold T_{rss} be -65 dBm for localization method selection. When the measured RSS values from the nearest three APs are smaller than -65 dBm, then GNSS localization is selected in the handover subarea. If these RSS values are equal or greater than the threshold T_{rss}, then the counting parameter α increases by 1, otherwise the counting parameter α will multiply with a reduction rate β, which can be denoted by:

$$\begin{cases} \alpha\left(i\right) = \alpha\left(i-1\right) + 1, rss\left(i\right) \geq T_{rss} \\ \alpha\left(i\right) = \alpha\left(i-1\right) \times \beta, rss\left(i\right) < T_{rss} \end{cases} \tag{3}$$

After the counting parameter α is greater than a counting threshold, which means the RSS values are equal or greater than the threshold T_{rss} for a while, then WLAN-based trilateration localization is applied. The general flowchart of the handover mechanism is shown in Fig. 6.

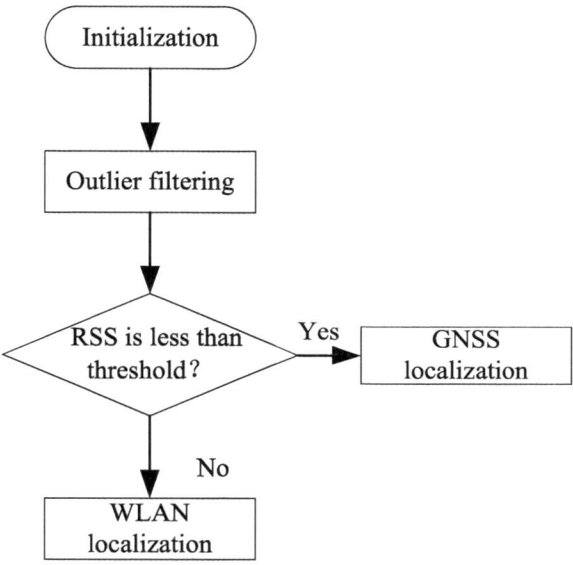

Fig. 6. Means of RSS Samples Collected at TPs.

From the experimental results, we find that the localization coordinates of outdoor TP1, TP2, ..., TP18 are computed using GNSS localization and the localization coordinates of the other TPs are computed using WLAN trilateration localization. The localization errors of all the TPs are shown in Fig. 7. After performing the indoor and outdoor seamless localization, the mean error of the localization results is reduced to 4.60 m. Compared with the mean errors of 5.13 m using GNSS localization only and 6.91 m using WLAN trilateration localization only, the proposed indoor and outdoor seamless localization method outperforms any of the two localization methods.

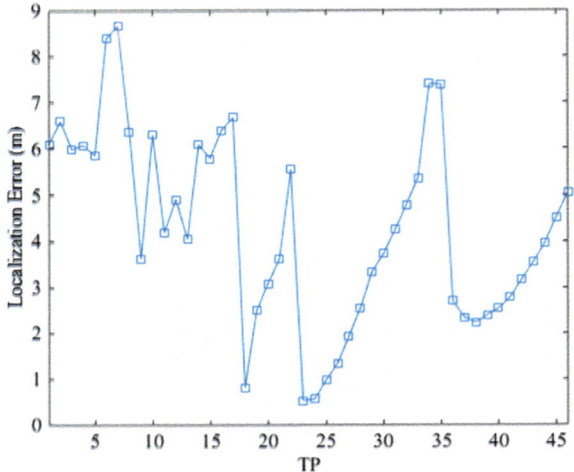

Fig. 7. Seamless Localization Errors.

5   Conclusions

This paper proposes an indoor and outdoor seamless localization method based on GNSS and WLAN. The method utilizes a handover mechanism to achieve seamless localization using GNSS and WLAN. The handover mechanism is used to select the GNSS localization or WLAN trilateration localization in the handover subarea. The experiment is performed in a real environment with an outdoor terrace and indoor office area. The experimental results demonstrate that the proposed indoor and outdoor seamless localization method using GNSS and WLAN outperforms either the GNSS localization or WLAN trilateration localization.

Acknowledgments. This work was supported by the Natural Science Foundation of the Jiangsu Higher Education Institutions of China under Grant No. 16KJB510014, the Natural Science Foundation of Jiangsu Province under Grant No. BK20171023, and the National Natural Science Foundation of China under Grant No. 61701223.

References

1. Gu YY, Lo A, Niemegeers I. A survey of indoor positioning systems for wireless personal networks. IEEE Commun Surv Tutorials. 2009;11(1):13–32.
2. Han GJ, Jiang JF, Zhang CY, et al. A survey on mobile anchor node assisted localization in wireless sensor networks. IEEE Commun Surv Tutorials. 2016;18(3):2220–43.
3. Zhou, M., Tang, Y.X., Tian, Z.S., Xie, L.B, Nie, W. Robust neighborhood graphing for semi-supervised indoor localization with light-loaded location fingerprinting. IEEE Internet Things J. 2017;(99): 1–1.
4. Kim J, Cheng J, Guivant J, Nieto J. Compressed fusion of GNSS and inertial navigation with simultaneous localization and mapping. IEEE Aerosp Electron Syst Mag. 2017;32(8):22–36.
5. Zou DY, Meng WX, Han S, He K, Zhang ZZ. Toward ubiquitous lbs: multi-radio localization and seamless positioning. IEEE Wirel Commun. 2016;23(6):107–13.
6. Sun YL, Xu YB. Error estimation method for matrix correlation-based Wi-Fi indoor localization. KSII Trans Internet and Inf Syst. 2013;7(11):2657–75.
7. Zhou M, Tang YX, Tian ZS, Xie LB, Geng XL. Semi-supervised learning for indoor hybrid fingerprint database calibration with low effort. IEEE Access. 2017;5(1):4388–400.
8. Sun YL, Meng WX, Li C, Zhao N, Zhao KL, Zhang NT. Human localization using multi-source heterogeneous data in indoor environments. IEEE Access. 2017;5:812–22.

Land Subsidence Monitoring System Based on BeiDou High-Precision Positioning

Yuan Chen[1], Xiaorong Li[2], Yue Yue[1], and Zhijian Zhang[1(✉)]

[1] Nanjing University, Nanjing, China
{281508958,765727692}@qq.com,njuzzj@nju.edu.cn
[2] Traffic Business Department of CIECC, Beijing, China
13911537198@163.com

Abstract. Land subsidence is a geological disaster caused by natural or human activities. The rate of change in early settlements is often extremely small and presents a challenge to monitoring. This experiment includes BeiDou positioning, multiple antenna, and high-precision baseline solution. It developed the BeiDou deformation monitoring system and used static relative positioning for high-precision land subsidence monitoring. We have adopted integrated hardware design, equipped with a variety of communication modules, satellite receivers, and embedded module in one. In addition, we have developed the corresponding communication protocol for data transmission. Finally, a corresponding monitoring interface software was designed on the client to intuitively reflect the settlement process in a graphical manner.

Keywords: Land subsidence · Baseline solution · High-precision positioning

1 Introduction

Since entering the twenty-first century, China has made remarkable achievements in the process of modernization. However, in the process of rapid development, many potential hidden dangers are often ignored by people. Land subsidence is one of them, which is a kind of local subsidence movement caused by the loosening of underground structures and stratum crushing due to human engineering activities, resulting in the reduction of the elevation of the upper crust.

The secondary consequences of land subsidence have caused heavy casualties and loss of social wealth. In order to avoid or reduce the disasters caused by land subsidence, it is necessary to carry out effective monitoring in the early stage of land subsidence.

Current monitoring methods are mainly divided into traditional manual monitoring and automatic monitoring systems. In the last century, land subsidence monitoring has been dominated by traditional methods which is often difficult

© Springer Nature Singapore Pte Ltd. 2020
Q. Liang et al. (eds.), *Communications, Signal Processing,*
and Systems, Lecture Notes in Electrical Engineering 516,
https://doi.org/10.1007/978-981-13-6504-1_168

to meet the monitoring needs due to the poor accuracy. With the development of information, communication, and automation technology, automatic monitoring system has been rapidly developed in recent years. Compared to traditional methods, the automatic monitoring system has advantages such as real time, accuracy, and stability.

2 Measurement Technology

There are many ways in traditional geodesy, such as leveling, GNSS, InSAR, and layerwise mark [1]. Among them, GNSS measurement technology has been developed rapidly in the past decades. It may provide high-precision three-dimensional coordinate which can be widely used in land subsidence monitoring. The relative baseline precision of GNSS can achieve 110^{-7}.

Current GNSS measurement technology mainly includes precise point positioning (PPP), differential positioning, and network real-time kinematic (RTK).

2.1 Precise Point Positioning

Precise point positioning, also called as absolute positioning, uses single receiver for positioning. GNSS performances are optimal in an open sky when many satellites are in view and the signals are uncorrupted [2]. After years of development, PPP has been widely used in high-precision measurement, satellite orbit determination, aeronautical measurements, and surface deformation monitoring. The system architecture of PPP is relatively simple, while precise satellite orbit and clock products required by it always suffer a latency [3].

Traditional precise point positioning using undifferenced and ionosphere-free pseudo-range and phase combination measurements can obtain positioning results of centimeter level [4].

2.2 Differential Positioning

Differential positioning, also known as relative positioning, places some GPS receivers as reference station for observation. Based on the known precise coordinates of the reference station, the distance correction to satellite is calculated, and the reference station sends data continuously. After years of improvement, real-time differential positioning technology has been well developed which can significantly improve the positioning accuracy and reliability [5].

2.3 Network RTK

Network RTK is a new technology based on continuous operation reference station (CORS) network and conventional RTK technology which have been widely used in real-time high-precision navigation, survey, and mapping [6].

3   Land Subsidence Monitoring System

For a real-time precise positioning service, at least three components including precise orbit determination (POD), precise clock estimation (PCE), and precise point positioning (PPP) are necessary.

Multi-global navigation satellite system (GNSS) combined positioning [7] has become an inevitable trend in GNSS-based navigation. In addition to BeiDou as the main reference, the system also introduced GPS, GLONASS, and GALILEO auxiliary navigation systems.

3.1   System Schematic

The automatic land subsidence monitoring system mainly includes data receiving system, communication network, and monitoring center as Fig. 1 shows.

The receiving system consists of several GNSS stations, and satellite raw data will be transmitted to communication modules and then be forwarded to communication network to the cloud server.

Data of all the station will be transmitted to the solution terminal. When completed, results will be uploaded to the cloud server and broadcast to the monitoring center.

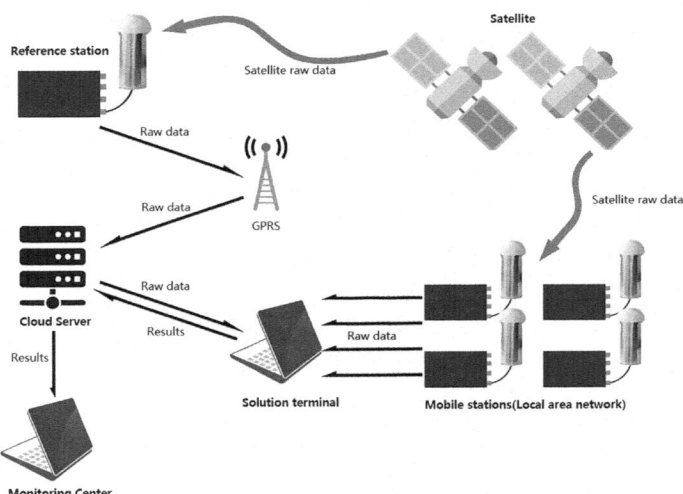

Fig. 1. Land subsidence monitoring system

3.2   Receiving System

In order to obtain data in a specific format and verify it, commands need inputting to the serial port of GNSS receiver. As for stability and safety, and relatively low real-time requirements, ARM embedded modules are recommended to be applied in the subsystem.

The GNSS receiver and ARM embedded module together form a receiving system. Orbit and ephemeris parameters will be transmitted to the GNSS receiver through high-frequency (HF) antenna. ARM embedded module is responsible for messaging with the receiver and as the controlling core.

3.3 Communication Network and Monitoring Center

In actual situations, the reference point of geological stability often exceeds the distance of the monitoring point beyond the scope of the general local area network, so that the cloud server will play the role of forwarding data.

4 Data Processing

The processing of GPS measurements adopts GAMIT/GLOBK, which is a comprehensive GPS analysis package developed by MIT, Scripps Institution of Oceanography, and Harvard University [8]. IRTF2000 is adopted as reference frame, and the model is set to the baseline solution (detailed in Fig. 2) with an interval of 24 h.

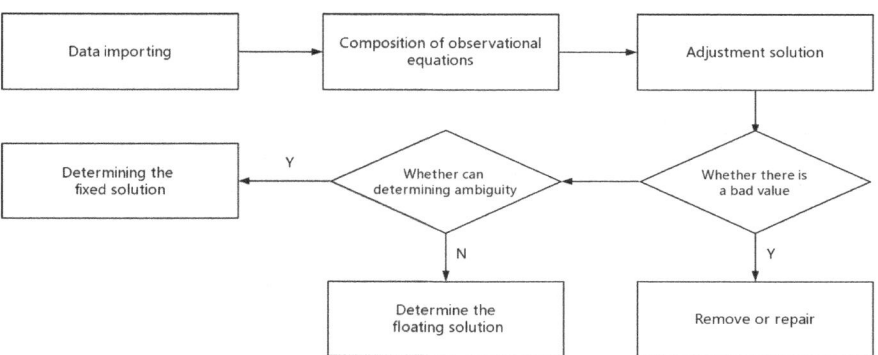

Fig. 2. Process of baseline solution

4.1 Basline Solution

The principle of the baseline solution can be summarized in the following steps:

(1) Adjustment initialization: Uncertain parameters can be solved based on the double-differenced observables, and the error equation is as follows.

Undetermined parameters:

$$\widehat{X} = \begin{bmatrix} \widehat{X} \ C \\ \widehat{X} \ N \end{bmatrix} \tag{1}$$

Cofactor matrix of the undetermined parameters:

$$Q = \begin{bmatrix} Q_{\widehat{X_C}\widehat{X_C}} & Q_{\widehat{X_C}\widehat{X_N}} \\ Q_{\widehat{X_N}\widehat{X_C}} & Q_{\widehat{X_N}\widehat{X_N}} \end{bmatrix} \tag{2}$$

After initialization, the variable of integer ambiguity is obtained.

(2) Determination of ambiguity: There are many ways to confirm the ambiguity. Now, a more reliable and reliable way is based on the search method which takes each ambiguity as the origin and then uses the error as a radius to confirm the integer solution of all the ambiguities.

(3) Determination of baseline fix solution: When confirming the integer solution of ambiguity, baseline integer solution [9] that the integer ambiguity located can be obtained. The integer solution can be used as a reference, which is helpful to the subsequent baseline solution, and to evaluate the quality of baseline.

4.2 Kalman Filtering

The essence of GLOBK is a Kalman filter, which makes the best estimate of the system state by analyzing the input and output observation data [10].

The Kalman filter estimates the process through a form of feedback control: The filter estimates the process state at a certain time and then obtains the feedback as a measurement. Therefore, the Kalman filter equations are divided into two groups: time-updated and measured equations, which can be expressed as follows:

$$\widehat{x_k}^- = A\widehat{x_{k-1}} + B\widehat{u_{k-1}} \tag{3}$$

$$P_k^- = AP_{k-1}A^T + Q \tag{4}$$

$$K_k = P_k^- H^T (HP_k^- H^T + R)^{-1} \tag{5}$$

$$\widehat{x_k} = \widehat{x_k}^- + K(z_k - H\widehat{x_k}^-) \tag{6}$$

$$P_k = (I - K_k H)P_k^- \tag{7}$$

The schematic diagram of Kalman filter is detailed in Fig. 3.

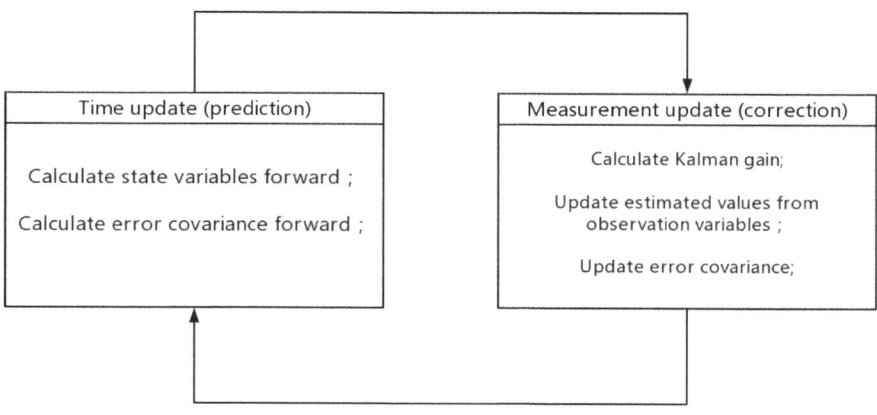

Fig. 3. Process of Kalman filtering

5   Results and Analysis

To simulate the actual settlement, observation antenna was placed on a precision mobile platform with an precision of 0.1 mm. The platform was moved in the vertical direction of 1.0 cm on March 24.

5.1   Results

Table 1 and Fig. 4 show the daily average observations from 23 to 28, which reflect the average change trend of the elevation. Compared with the actual adjustment, the absolute measurement accuracy has reached mm level.

Table 1. Daily average settlement data

Station	Date	Altitude (m)	Settlement (mm)
NJ-01	March 23	154.0375	0
NJ-01	March 24	154.0327	4.8
NJ-01	March 25	154.0275	10
NJ-01	March 26	154.0269	10.6
NJ-01	March 27	154.0273	10.2
NJ-01	March 28	154.0273	10.2

5.2   Error Analysis

The data of March 23 is taken as a benchmark to analyze the accuracy of the system.

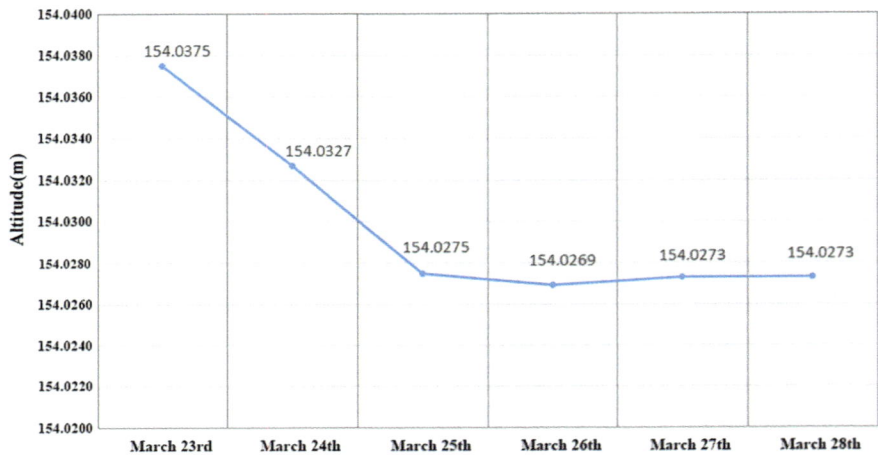

Fig. 4. Settlement curve

In order to quantitatively reflect the accuracy, it is necessary to introduce absolute difference and standard deviation which can be expressed as follows:

$$\sigma = \sqrt{\frac{\sum_{i=1}^{N}(S_i - S_{avg})^2}{N}} \tag{8}$$

According to the data in Table 1, substituting them into formula (8) respectively, the average absolute error and standard deviation can be calculated as 1.2 and 2.2 mm, having reached the mm level.

6 Conclusions

The chapter first introduced the current situation, background, and hazards of land subsidence in China. In order to take necessary measures in time for the hidden safety hazards of land subsidence, high-precision areas need to be monitored in real time. However, traditional monitoring methods have been difficult to meet the monitoring needs. Therefore, automated subsidence monitoring systems using modern communications, computer science, and satellite navigation technologies have emerged. With the completion of the BeiDou navigation system, more and more GNSS receivers are now beginning to be compatible with BeiDou Navigation Satellite System (BDS) information. Compared with other navigation systems, BeiDou provides unique short message communication means on the basis of ensuring safety, reliability, and high performance and can be used as an emergency communication method when a disaster occurs. Therefore, this experiment uses BDS as the main body and uses GPS, GLONASS, etc., as auxiliary references to establish a multi-mode fusion ground subsidence monitoring system to ensure the positioning accuracy in the case of real time and stability.

Acknowledgments. This work was supported by State Key Laboratory of Smart Grid Protection and Control of NARI Group Corporation.

References

1. Wang A, Sun Z. Multi-geodesy techniques data fusing and analyzing for land subsidence monitoring. In: International workshop on earth observation and remote sensing applications, 2014. p. 345–8.
2. Angrisano A, Gaglione S, Gioia C. Performance assessment of GPS/GLONASS single point positioning in an urban environment. Acta Geod Geophys. 2013;48(2):149–61.
3. Shi J, Yuan X, Cai Y, Wang G. GPS real-time precise point positioning for aerial triangulation. GPS Solutions, 2017. p. 1–10.
4. Qu L, Zhao Q, Guo J, Wang G, Guo X, Zhang Q, Jiang K, Luo L. BDS/GNSS real-time kinematic precise point positioning with un-differenced ambiguity resolution. Lect Notes Electr Eng. 2015;342:13–29.
5. Wang L, Li Z, Yuan H, Zhao J, Zhou K, Yuan C. Influence of the time-delay of correction for BDS and GPS combined real-time differential positioning. Electron Lett. 2016;52(12):1063–5.
6. Yang C, Wu D, Lu Y, Yu Y. Research on network RTK positioning algorithm aided by quantum ranging. Sci China Inf Sci. 2010;53:248–57.
7. Li X, Ge M, Dai X, Ren X, Fritsche M, Wickert J, Schuh H. Accuracy and reliability of multi-GNSS real-time precise positioning: GPS, GLONASS, BeiDou, and Galileo. J Geodesy. 2015;89(6):607–35.
8. Li X, Xu L, Fang Y, Zhang Y, Ding J, Liu H, Deng X. Estimation of the precipitable water vapor from ground-based GPS with GAMIT/GLOBK. IEEE. 2010;1:210–4.
9. Schwarz KP, Lachapelle G. Kinematic systems in geodesy, surveying, and remote sensing. New York: Springer; 1991.
10. King RW. Documentation for the GAMIT GPS analysis software. Massachusetts Institute of Technology, 1995.

Multi-layer Location Verification System in MANETs

Jingyi Dong[✉]

School of Electrical Engineering and Telecommunications,
The University of New South Wales, Sydney, Australia
jennied_jiyue@163.com

Abstract. The mobility and feasibility of mobile ad hoc networks (MANETs) have to deeply rely on the accurate location information to support multiple applications. A wrong announced location of a node may cause some serious consequences. Thus, the localization and the location security should be considered as important parts of the whole design of the MANETs. In the traditional way, the location verification schemes need complex calculation and multi-step communication with base stations and other vehicles, which are strengthen the burden of the MANETs routings and increased the complexity of protocol. In this paper, an improved multi-layer location verification system (MLVS) based on the optimal common neighbor's knowledge between claimer and verifier in MANETs is been proposed and discussed. In this system, each node in MANETs could have a trust value, and a mutually shared token scheme is provided to make the decision of the MLVS. Furthermore, the MLVS shows a reliable performance on the ability of attacker defense and accuracy in high node density networks.

Keywords: Mobile ad hoc networks · Location verification system · MLVS · Location security

1 Introduction

As several innovative technologies of multi-hop ad networks, the mobile ad hoc networks (MANETs), it has been one of the most popular and interesting topics for researchers. The major characteristic of MANETs is that many mobile terminals and low-cost sensors are consisted, and the packets are delivered between wireless interfaces following the geographic routing. To achieve a better performance and support possible applications, some localization schemes should be deployed into sensors and terminals, such as Global Positioning Systems (GPS). However, the openness and public ability of the wireless communication makes that the localization of MANETs could be easily attacked by malicious nodes [1]. Hence, the accuracy of the location information plays a significant role to protect the sensors' privacy, and the way to authenticate location information has been concerned [2–6]. Figure 1 illustrates the basic concept of location verification.

© Springer Nature Singapore Pte Ltd. 2020
Q. Liang et al. (eds.), *Communications, Signal Processing, and Systems*, Lecture Notes in Electrical Engineering 516,
https://doi.org/10.1007/978-981-13-6504-1_169

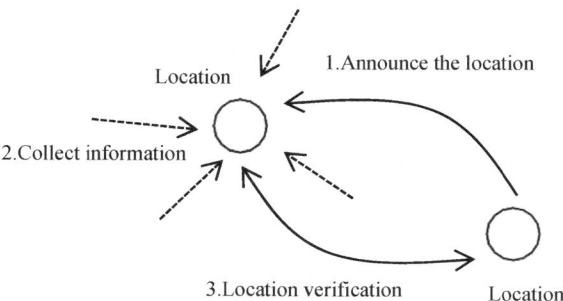

Fig. 1. Basic concept of location verification

A basic binary rule is proposed for location verification system between claimer and verifier in [7]. However, some measurements are difficult to acquire without line-of-sight (LOS). Therefore, the cooperative nodes' location communication scheme is proposed with the protocol and routing techniques' help [8, 9]. Compared to the other routing techniques, geographic routing has got foremost attention for information transmission in vehicular communication [10–12], and the information of neighbors can be obtained [13].

In this paper, a two-layer location verification system with neighbors' knowledge in mobile ad hoc networks is proposed; it follows the basic principle of mutually shared region-based location verification (MSRLV) in WSNs [14]. In layer one, LVS would roughly eliminate some malicious nodes and create a trust value table for nodes. Then, in layer two, the trust weight of common neighbors would be signed as the preparation to create a mutually shared model; the selected common neighbors in shared region between verifier and claimant provide the common knowledge to both verifier and claimant. With the information collected from neighbors, the system shows a good performance with higher node density. However, the location error would be increased in sparse MANETs.

2 System Model

In this section, some assumptions are considered for the network as follows: The mobile terminals and sensors have the same wireless transmission range r; assume that the network contains N nodes. Some of them are malicious nodes n_m, which may provide a fake location to others. The nodes' location is assumed to be obtained by GPS which is considered as the true location neglecting the localization error; we assume that the node (source) who sends the data first is not a malicious node, and following the protocol, the hops between source and destination must do the location verification before the data transmission.

The transmitted data of nodes is $n_i\{\text{data}_i, L_i, \text{RV}_i\}$, where data_i is the data transmitted by the node n_i, L_i is the announced location, and RV_i is a random value which would be discussed in next section.

2.1   Layer 1: Trust Value Table (TV)

The trust value table (TV) is constructed by several factors which are RSSI, moving direction and relative speed.

The RSSI is capable of roughly eliminating some malicious nodes who provide fake locations. The node n could obtain the received signal power $P^r_{n_j}$ from its one-hop neighbors n_j to measure the distance d_{n_j} using RSSI. At the same time, the $P^r_{n_j^*}$ could be calculated with announced location from n_j with $d_{n_j^*}$. In ideal state, the malicious node could pass the estimation only if $d_{n_j} = d_{n_j^*}$. However, considering the mobility of the MANETs, some error may be caused by complex wireless environment. In this term, a filter could be proposed to eliminate some malicious points with the threshold α; the neighbors who pass the distance filter can consist into a set S_{neighbor} with their different level.

$$\beta_i = 1 - \left| P^r_{n_j} - P^r_{n_j^*} \right| / P^r_{n_j^*} \tag{1}$$

where β_i is the i-th neighbor's distance different level of n_i. The β_i would be stored in $S_{\text{neighbor}}\left\{ n, \beta_n; \left| P^r_{n_j} - P^r_{n_j^*} \right| < \alpha \right\}$. However, the RSSI could only eliminate some malicious nodes which fake their location with a large distance difference; in some cases, the system may be suffered from the similar distance-based malicious node (SDM attack). Hence, a mutually shared region token scheme (MSR) is considered to deploy as layer two of the LVS [14]. To achieve a better performance, a trust value is regarded as the standard about common neighbors' selection in layer two. The probability that the neighbors passed the distance filtering, the relative speed and the relative distance between verifier and claimant are the factors contributing to the trust value. The model of the network is shown in Fig. 2, the node n could be seen as the verifier n_v, and the next-hop node could be seen as the claimant n_c.

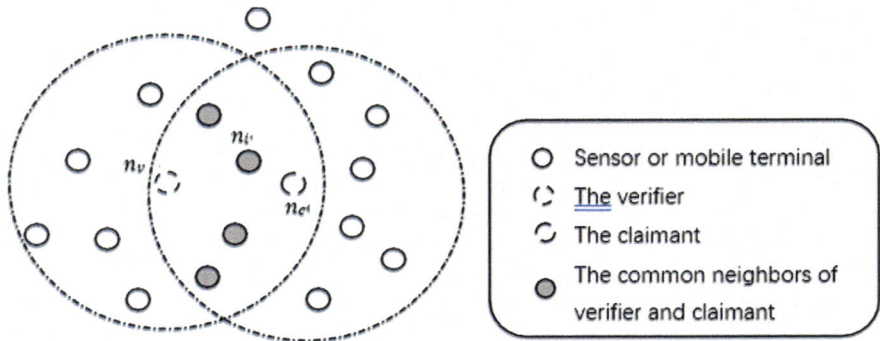

Fig. 2. Sensors and mobile terminals in network

We assume the speed of n_v, n_c, and n_i are s_v, s_v and s_i, respectively, which the unit is m/s. The speed difference level γ is calculated by the following equation:

$$\gamma_i = 1 - \frac{|s_v - s_i| + |s_v - s_i|}{2s_i} \tag{2}$$

The distance difference between the verifier and claimant level μ is measured with:

$$\mu_i = \frac{r}{d_{v-i} + d_{v-c}} \tag{3}$$

where d_{v-i} is the distance between the verifier and the i-th neighbor, and d_{v-c} is the distance between the claimant and i-th neighbor; the location information is acquired from the GPS. r is the communication range. Hence, the trust value TV_{n_i} of the i-th neighbor is represented by Eq. (4):

$$\text{TV}_{n_i} = \gamma_i + \mu_i + \beta_i \tag{4}$$

2.2 Layer 2: Mutually Shared Token Scheme

As Fig. 2 shows, when the nodes finished the distance filtering, it could collect $\{\text{RV}_i\}$ into a packet $\text{data}_{\text{mst}}^{n_c}$ from the common neighbors, the mutually shared token follows the bellowing calculation:

$$\text{MST}_{n_c} = \{n_i \in \text{SCN}: \quad \text{TV}_{n_i} > \varphi, \quad \text{data}_{\text{mst}}^{n_c} = \sum \text{RV}_i\} \tag{5}$$

where φ is a threshold of trust value depending on the wireless environment, and SCN is the set of common neighbors. And then $\text{data}_{\text{mst}}^{n_c}$ would be sent to n_v. As the packet received by the verifier n_v, the n_v would do mutually shared token in the same way:

$$\text{MST}_{n_v} = \{n_j \in \text{SCN}^*: \quad \text{TV}_{n_j} > \varphi, \quad \text{data}_{\text{mst}}^{n_c} = \sum \text{RV}_j\} \tag{6}$$

and if the MST_{n_c} matched with MST_{n_v} ($\text{MST}_{n_c} = \text{MST}_{n_v}$), the location information of n_c could be accepted and regarded the n_c as the next-hop node.

3 Performance Analysis

We assume that the malicious nodes n_{m_k} fake its location and broadcast a wrong location to the transmission environment $n_{m_k}^* \{L_{m_k}^*\}$, $L_{m_k}^* \neq L_{m_k}^{\text{true}}$. With the suitable selection of common neighbors, the verifier and the claimant can receive the same MST. However, if a malicious node pretends its position at a wrong place, such as the situation shows in Fig. 3, even it is an SDM attack, it is hard to collect all necessary packets from signed nodes. Hence, the multi-layer location verification system can defense several attacks discussed above.

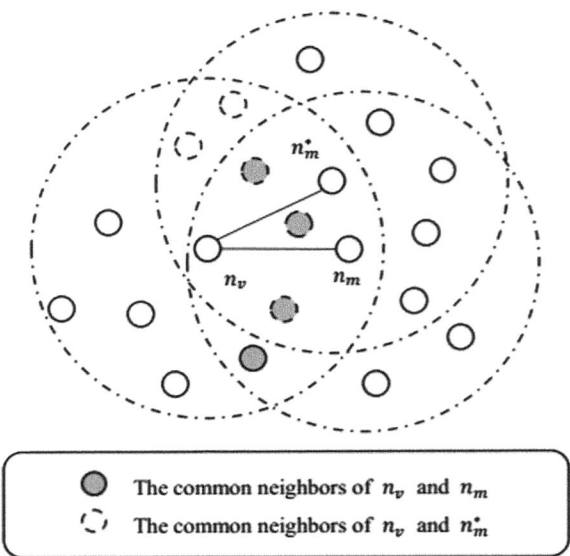

Fig. 3. Mutually shared region model with malicious nodes

The performance is evaluated in terms of the ability of the malicious nodes rejection and the location error in different situation. The simulations are built on the MATLAB. The parameter of the simulation is shown in Table 1 which is similar to the [15].

Table 1. Parameters of the simulation

Parameters	Values	Parameters	Values
Simulation area	1000×1000 m^2	Propagation	Shadowing
Nodes' speed	0–10 m/s	MAC data rate	10 Mps
Simulation time	20	MAC protocol	IEEE 802.11
Number of nodes	0–500 nodes	Frequency	5.9 GHz
Sources and destinations	20 nodes	Packet type	UDP
Transmission range	150 m	Packet size	512 bytes

The ability of the malicious nodes' defense means that the probability of location error has been discovered under the MLVS. It is assumed that the total number of the malicious nodes is N_m, and the times that packets sent to a malicious node in i-th simulation can be regarded as T_m^i. Then the location error e_L happened in the MANETs with multi-layer location verification is following the equation:

$$e_L(\%) = \left\{ \frac{\sum_{i=1}^{20} \frac{T_m^i}{N_m}}{20} \right\} \times 100 \tag{7}$$

In addition, we assume that a single-fake location error in this simulation. Single-fake location error is considered that only one malicious node fakes its location around honest nodes. And Fig. 4a illustrates the performance of MLVS in MANETs with different maximum nodes which contains 10%, 20%, and 30% malicious nodes respectively. The MLVS is a location verification scheme which verifies nodes relying on the neighbors' information; thus, the density of the MANETs deeply impacts the performance of the system. As Fig. 4a shows, the system has a lower location error as the maximum number of nodes is more than 400.

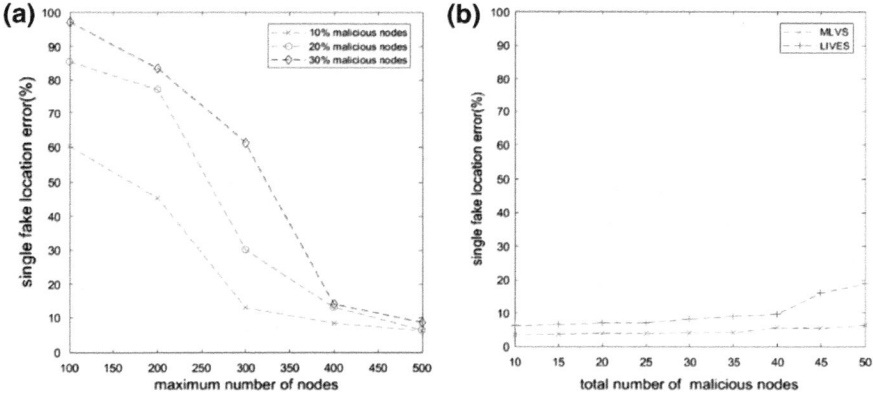

Fig. 4. **a** Performance of MLVS with difference number of malicious nodes and **b** the performance of MLVS compares with LIVES

In Fig. 4b, it shows the performance of MLVS with different number of malicious node and the maximum number of nodes is 500 compared with LIVES illustrated in [15]. In LIVES, the number of malicious nodes impacts the probability of fake position rejection. However, the MLVS has a stable performance with higher node density.

4 Conclusion

In this paper, a multi-layer location verification technique is discussed based on geographic routing in MANETs. The first layer performs a distance filtering for verifiers and claimants and then calculates the trust value for their neighbors as the preparation of common neighbor selection. The second layer evaluates the information collected from common based on the principle of mutually shared token. The trust value calculated in layer one considers the mobility of the network to ensure the verifier, and the claimant can collect information from the same group of neighbors. The MLVS shows a good performance with high-density network; however, it cannot be deployed in sparse MANETs or higher mobility ad hoc networks, such as VANETs. In future, an improved location verification system for VANETs will be considered with the concept of cooperative position.

References

1. Tippenhauer NO, Rasmussen KB, Pöpper C, Capkun S. iPhone and iPod location spoofing: attacks on public WLAN-based positioning systems. SysSec technical Report. Swiss Federal Institute of Technology; 2012 Apr. 2008.
2. Malaney RA. A location enabled wireless security system. In Global telecommunications conference, 2004. GLOBECOM'04. IEEE; 2004. pp. 2196–200.
3. Faria DB, Cheriton DR. Detecting identity-based attacks in wireless networks using signalprints. In Proceedings of the 5th ACM workshop on wireless security; 2006. pp. 43–52.
4. Papadimitratos P, Gligor V, Hubaux JP. Securing vehicular communications-assumptions, requirements and principles. Proc ESCAR. 2006:5–14.
5. Papadimitratos P, Buttyan L, Holczer T, Schoch E, Freudiger J, Raya M, Hubaux JP. Secure vehicular communication systems: design and architecture. IEEE Commun Mag. 2009;46 (11):100–9.
6. Bauer K, McCoy D, Anderson E, Breitenbach M, Grudic G, Grunwald D, Sicker D. The directional attack on wireless localization: how to spoof your location with a tin can. In: Proceedings of the 28th IEEE conference on global telecommunications; 2009. pp. 4125–30.
7. Yan S, Malaney R. Location verification systems in emerging wireless networks. arXiv preprint; 2013 arXiv:1307.3348.
8. Abumansoor O, Boukerche A. A secure cooperative approach for nonline-of-sight location verification in VANET. IEEE Trans Veh Technol. 2012;61(1):275–85.
9. Vora A, Nesterenko M. Secure location verification using radio broadcast. IEEE Trans Dependable Secure Comput. 2006;3(4):2006.
10. Kaiwartya O, Kumar S. Cache agent-based geocasting in VANETs. Int J Inf Commun Technol. 2015;7(6):562–84.
11. Suthaputchakun C, Sun Z. Routing protocol in intervehicle communication systems: a survey. IEEE Commun Mag. 2011;49(12):150–156.
12. Ansari K, Feng Y, Singh J. Study of a geo-multicast framework for efficient message dissemination at unmanned level crossings. IET Intell Transp Syst. 2013;8(4):425–34.
13. Cao Y, Sun Z, Wang N, Riaz M, Cruickshank H, Liu X. Geographic-based spray-and-relay (GSaR): an efficient routing scheme for DTNs. IEEE Trans Veh Technol. 2015;64(4): 1548–64.
14. Kim IH, Kim BS, Song J. An efficient location verification scheme for static wireless sensor networks. Sensors. 2017;17(2):225.
15. Kargl F, Klenk A, Schlott S. Weber M. Advanced detection of selfish or malicious nodes in ad hoc networks. In: European workshop on security in Ad-hoc and sensor networks; 2014. pp. 152–165.

Design of Multi-antenna BeiDou High-Precision Positioning System

Kunzhao Xie[1(✉)], Zhicong Chen[1], Rongwu Tang[1], Xisheng An[1], and Xiaorong Li[2]

[1] Laibin Power Supply Bureau of Guangxi, Power Grid Co., Ltd., Laibin, China
275856140@qq.com,
{chen_zc.lbg,tang_rw.lbg,an_xs.lbg}@gx.csg.cn
[2] Traffic Business Department of CIECC, Beijing, China

Abstract. With the progress of society and economic development and the expanding deployment scale of substations, the automatic monitoring of the foundation settlement in construction station becomes an important issue of the operation and maintenance in the power grid. BeiDou satellite navigation system is a global navigation and positioning system, which is independently developed to provide navigation, positioning, and timing services in China and its surrounding areas. This article describes the principles of BeiDou high-precision positioning. According to the land subsidence monitoring and warning requirements for substations, multi-antenna technology is introduced in order to solve the problem of the cost of large-scale land subsidence monitoring system. The monitoring interface software is designed to visually reflect the settlement situation. The monitoring system can effectively improve the monitoring efficiency and early warning capability of the ground subsidence in the power company's substation. The system will also provide a basis for decision-making management and improve the reliability, safety, and stability of power grid operation based on high-precision field accuracy and trend.

Keywords: Land subsidence · Multi-antenna ·
BeiDou satellite positioning

1 Introduction

With the development of social economy, land resources are decreasing more and more land and low-lying areas are being exploited and utilized. Grid transmission and transformation should be synchronized with the development of social economy, when choosing the location of the substation, and some of them have to be established in some areas that are not suitable for them. Meanwhile, substation foundation sedimentation problem is increasingly prominent. Hidden dangers that may cause substation equipment accidents happen sometimes, which bring great influence to the safety of substations [1, 2]. Substation foundation sedimentation problems are divided into two parts: uniform sedimentation and non-uniform sedimentation. On the occasion of uniform sedimentation, there is little internal stress on electrical equipment on the ground, doing no harm to the substation. However, on the other occasion, the internal stress is much larger. It is easy to cause the electric equipment to break, tilt, or even collapse, resulting in accidents. In order to avoid above-mentioned problems and keep substations work under safe conditions, monitoring the non-uniform sedimentation of the substation foundation is of great significance. Nowadays, the automatic level of land subsidence and deformation monitoring systems is limited [3, 4].

According to the ground sedimentation monitoring and warning requirements for substations, this paper studies and develops a BeiDou high-precision network deformation monitoring system, focusing on the development of multi-antenna technology to achieve time-sharing reception of multi-channel antenna signals by satellite receiver boards. That is, a receiver board can receive antenna signals at multiple points, which greatly reduce the cost of large-scale deployment of satellite positioning monitoring networks.

The system provides the power company with an advanced, practical, and reliable online monitoring and analysis system for ground sedimentation in substation areas. It is able to remotely monitor and analyze the geological deformation of the site in the substation area and provide timely warning of various dangerous situations that may occur. Based on BeiDou's high-precision field accuracy and trend, it will also provide a basis for decision management and improve the reliability, safety, and stability of power grid operation [5].

2   Problems

With the improvement of satellite receiver hardware performance and software processing technology, satellite positioning technology has been applied and popularized in many fields such as slope deformation monitoring, geodetic survey, crustal deformation monitoring, and precision engineering measurement. The application of satellites to slope deformation monitoring is not only highly accurate, unaffected by climatic conditions, but also enables unattended automated real-time monitoring modes.

BeiDou satellite navigation system is a global satellite navigation system independently developed by China. At present, the system is actively promoting and building in accordance with the principles of openness, autonomy, compatibility, and gradual progress. Now, it has the ability to cover the positioning, navigation, and communication services in the Asia-Pacific region. Around 2020, the system will include 5 GEO satellites, 3 IGSO satellites, and 27 MEO satellites, eventually achieving global coverage [6]. In Asia-Pacific areas, it is even better than GPS, which is verified by researchers from different countries. The basic method of ground sedimentation monitoring is to obtain ground elevation changes through high-precision positioning technology and to analyze vertical displacement components based on complex algorithms to achieve monitoring of ground subsidence. The BeiDou second-generation satellites' high-precision positioning technology plays an active and important role in it. The application of satellite technology to a wide range of geological monitoring is mainly limited by the following factors:

- The precision measurement satellite receiving equipment are expensive. In a conventional way, one satellite antenna is equipped with one receiver. If it is utilized in a large-scale deformation monitoring, the cost could be high.
- Generally, the landslide that needs to be monitored has a large monitoring range and many monitoring points. The amount of original satellite monitoring data collected at the monitoring site is large, and how to use effective means to ensure the massive and reliable transmission of massive data to the control center is worth focusing.

3   Receiver with Multi-antenna Design

One receiver is connected to multiple antennas. At each monitoring point, satellite antennas are installed with no receivers. Multiple monitoring points share one satellite receiver. Compared to the conventional idea that one satellite antenna uses one satellite receiver, the cost of such a monitoring system can be significantly reduced by the reduction in the number of satellite receivers [7,8]. The satellite multi-antenna controller is shown in Fig. 1.

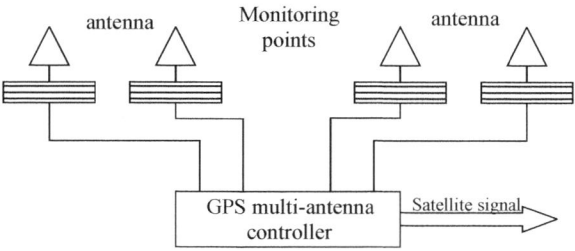

Fig. 1. Multi-antenna technology

The satellite multi-antenna controller is a key device in the landslide monitoring system. It is an organic integrated body of multiple satellite antennas and microwave switches with multiple channels. Each satellite antenna is connected to a corresponding channel of the microwave switch, and the on/off state of the plurality of signal channels of the microwave switch is controlled by the switch control circuit in real time through computer programming. The rotation of the monitoring point data acquisition is realized by controlling the on and off states of the channel.

Figure 2 provides a multichotomous antenna signal controller, including a power supply module, a control module, and a switch module. The power supply module provides the voltage for the system. The control system consists of a microprocessor unit and a relay drive unit. The switch system includes a relay group, receiving antenna bases, and a sending antenna base. Through the selector in the module, signals are sent to the sending antenna base and transmitted to the receiver board.

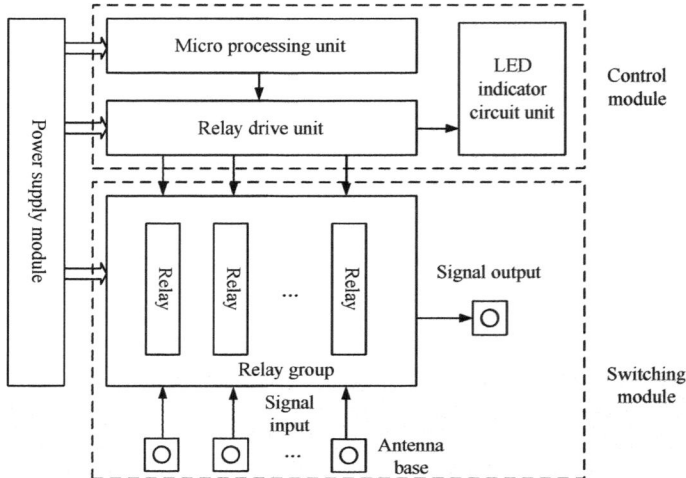

Fig. 2. Multi-antenna controller

4   Multi-antenna BeiDou High-Precision Positioning System

According to the requirements of data acquisition and power system security considerations, the ground sedimentation monitoring system consists of monitoring terminals, a reference terminal, a substation local server, a cloud server, and a client, which is illustrated in Fig. 3. To monitor one substation, there are always more than one monitoring terminals that are distributed around the substation. The monitoring terminal is deployed at where is geological hard and about 3 kms away from the substation.

4.1   System Structure

The receiving terminal is mainly responsible for receiving the radio-frequency signal transmitted by the satellite through the antenna and performing preliminary processing of the satellite signal to obtain the original data and transmitting it to the local server terminal.

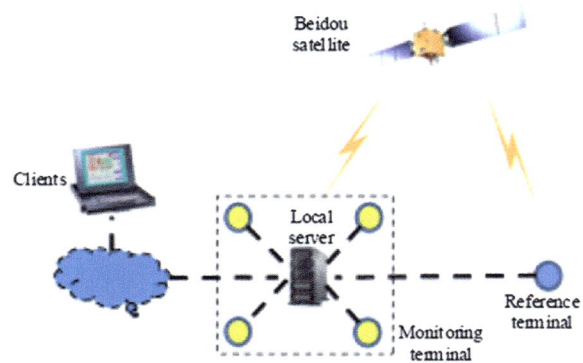

Fig. 3. Overall design of land subsidence monitoring system

4.2   Reference Terminal

The reference station terminal consists of a power supply module, a GNSS board, a control module, and a communication module. The power supply module mainly contains DC-DC voltage conversion circuits. The GNSS board is responsible for receiving satellite radio-frequency signals through the satellite antenna and transmitting the coordinate data to the control module. The control module is made up of a microprocessor unit, a storage unit, an LED indicator unit, and an associated external circuit. It is mainly responsible for transmitting the satellite positioning coordinate data of the desired format to the receiver board module. Meanwhile, it transmits the coordinate data of the desired format to the communication module. The coordinate data is stored locally. The communication module includes a wired network unit and a 4G mobile network unit.

4.3   Monitoring Terminal with Multi-antenna

This part includes power supply modules, a module that has one receiver with multiple antennas, a GNSS board, a control module, and a communication module. Different from the reference station terminal, the monitoring station terminal has a module that has one receiver with multiple antennas. Other parts are consistent with the reference station terminal. The multichotomous switch module is composed of a relay driving unit, an RF relay group, and a receiving antenna base. The microprocessor unit outputs a control signal at a preset time interval, and the control signal can cause the corresponding pin of the relay driving unit to output a current which is high enough to drive the corresponding relay to realize the function of switching the control circuit. This module enables the GNSS board to receive signals from multiple antennas in a time-division manner.

4.4 Substation Local Server

The substation local server deploys the data solving subsystem and receives the terminal data. The main functions implemented include: (1) Receive raw data sent by the satellite positioning information receiving subsystem, and store it locally in real time. (2) Set the solution interval, sampling interval, filter settings according to the instructions sent by the center, and start solving. (3) Keep a long connection with the central station, send the solution result (site coordinates) to the center through the cloud center, and receive instructions from the center.

4.5 Client

In order to better realize the management and maintenance of the settlement monitoring system, the central station software is designed. The main functions realized are: (1) Download the solution data forwarded from the cloud center, and store it to a local database in real time. (2) Display the solution data by graphics. (3) Control the receiving terminal subsystem partially.

5 Experiments and Results

5.1 Experiments

The ground subsidence monitoring system designed in this paper was tested at a substation in Laibin, Guangxi province. The offset of XYZ was adjusted by the coordinate adjustment table, compared with the system data measurement. Figure 4 is the measurement data in the X-direction; Fig. 5 is measurement data in the Y-direction; Fig. 6 is measurement data in the Z (elevation)-direction.

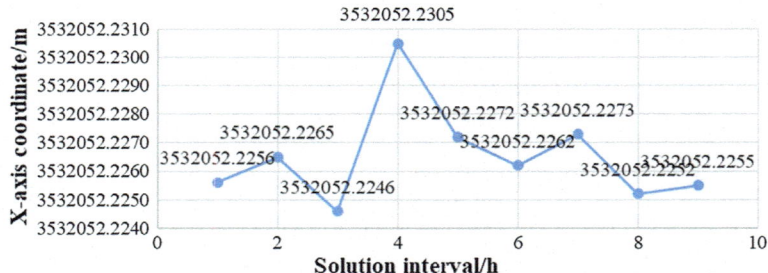

Fig. 4. Measurement data of X-direction

Figure 6 shows the monitoring data for the experiment of simulation settling. For measurement data of the elevation test, we reduced the height of the monitoring station by 16 mm during the sixth data solution calculation. We find that the offset between 1–5 sets of data and 7–9 sets of data is about 15 mm.

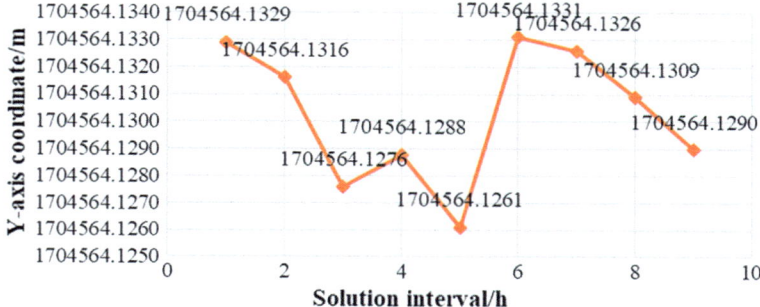

Fig. 5. Measurement data of Y-direction

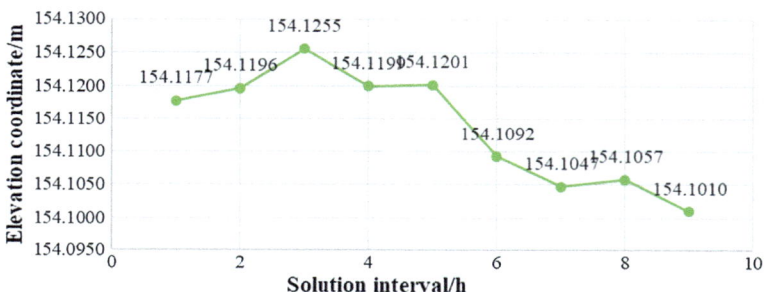

Fig. 6. Measurement data of Z-direction

5.2   Results

We use the range and standard deviation as the evaluation criteria to analyze the above test data. Table 1 shows the test data. The standard deviation is within 3 mm, and the error meets the design accuracy requirements.

Table 1. The data analysis

	Range(R) (mm)	Standard deviation(σ) (mm)
X	6	1.7
Y	7	2.5
Z(1–5)	7.8	2.9
Z(7–9)	4.6	2.5

The range is the difference between the maximum and minimum values of a set of measurements, which is also known as range error or full distance, expressed as R:

$$Range : R = X_{max} - X_{min}, \tag{1}$$

$$Standard deviation : \sigma = \sqrt{\frac{1 \sum_{i=1}^{N}(x_i - \mu)^2}{N}}. \tag{2}$$

6 Conclusion

According to the ground subsidence monitoring and early warning requirements of the substation, this paper proposed a BeiDou high-precision network deformation monitoring system, especially developed a multi-antenna technology and module. The system reduced the system cost and effectively realized the automatic collection, monitoring, early warning, and management of ground settlement monitoring data, which greatly improved the automation level of grid operation and maintenance. The results of experiment show that the system function is stable, and the positioning accuracy meets the design requirements.

References

1. Gang Liu. On the safety management of substation operation and accident prevention. Value Eng. 2011;1656(1):113–23.
2. Mi Chen, Roberto Toms, Zhenhong Li, Mahdi Motagh, Tao Li, Leyin Hu, Huili Gong, Xiaojuan Li, Jun Yun, Xulong Gong. Imaging land subsidence induced by groundwater extraction in beijing (china) using satellite radar interferometry. Remote Sens. 2016;8(6).
3. Nasipuri A, Cox R, Conrad J, Van Der Zel L. Design considerations for a large-scale wireless sensor network for substation monitoring. In: Local computer networks; 2010. p. 866–73.
4. Guoqing Yao, Jingqin Mu. D-insar technique for land subsidence monitoring. Earth Sci Front. 2008;15(4):239–43.
5. Yue Yang, Qi Kang. Operation situation analysis of substation equipment online monitoring system in inner mongolia power grid. Inner Mongolia Electric Power. 2014.
6. Xuefeng Lu, Yongfeng Liao, Bo Li, Lan Deng. Beidou integrated disaster reduction application platform. China Telecom. 2015;12(8):169–82.
7. Juang JC, Tsai CT, Chen YH. Development of a multi-antenna gps/beidou receiver for troposphere/ionosphere monitoring. In: Proceedings of international technical meeting of the satellite division of the institute of navigation; 2012. p. 947–952.
8. Yongqi Chen, Xiaoli Ding, Dingfa Huang, Jianjun Zhu. A multi-antenna gps system for local area deformation monitoring. Earth Planets Space. 2000;52(10):873–6.

Research on Sound Source Localization Algorithm of Spatial Distributed Microphone Array Based on PHAT Model

Yong Liu, Jia qi Zhen[✉], Yan chao Li, and Zhi qiang Hu

School of Electronics Engineering, Heilongjiang University,
Harbin 150080, China
zhenjiaqi2011@163.com

Abstract. With the development of artificial intelligence voice technology, array signal processing technology has been widely used in intelligent human–computer interaction. In this field, the sound source localization technology of the microphone array system is also one of the key technologies in the composition of intelligent systems, and it is also a research hotspot technology, which is of great significance for improving human–computer interaction ability. In this paper, the microphone array technology is used to demonstrate and improve the sound source localization of indoor speech signals. A spatial distributed microphone array localization algorithm is proposed to improve the accuracy of the indoor voice source signal localization in the presence of indoor noise and reverberation.

Keywords: Spatial distributed microphone array · Sound source localization · Room impulse response · Phase transformation weighted generalized cross-correlation algorithm

1 Introduction

The microphone array technology uses a focused method to generate a beam and align it with the estimated position of the sound source, thereby reducing the interference signals of other unwanted signals, environmental noise, indoor reverberation, etc., to a certain extent, so that the collected elements are acquired [1, 2].

The time delay of arrival (TDOA) sound source localization method is the most commonly used in all positioning fields [3]. It first estimates the time delay of the sound source reaching each element of the receiver and obtains the distance difference. The mathematical method of space geometry is used to solve the position of the target sound source [4].

However, the sound source localization technology of steerable beamforming is difficult to apply in practice [5]. Even though some iterative methods can be used to reduce the computation time and enhance the real-time performance, this may result in

© Springer Nature Singapore Pte Ltd. 2020
Q. Liang et al. (eds.), *Communications, Signal Processing, and Systems*, Lecture Notes in Electrical Engineering 516,
https://doi.org/10.1007/978-981-13-6504-1_171

the inability to search for effective global peaks and make the positioning effect less than ideal [6]. In addition, the algorithm relies on the spectral characteristics of the sound source. In practical applications, a series of interference signals in the sound source environment make the spectral characteristics of the sound source difficult to obtain, which also limits the application range of the algorithm [7].

2   Sound Source Localization Algorithm Based on PHAT for Spatial Distributed Microphone Array

Establishing a spatial distributed microphone array model, the indoor space is a cube, as shown in Fig. 1. The origin of the coordinate is also the coordinate position of the microphone element $M_1(0, 0, 0)$, and the positions of the remaining seven elements are $M_2(L, 0, 0)$, $M_3(L, L, 0)$, $M_4(0, L, 0)$, $M_5(0, L, L)$, $M_6(0, 0, L)$, $M_7(L, 0, L)$, $M_8(L, L, L)$, and L represents the length of the room model. The coordinates of the speaker are $S(x, y, z)$. The azimuth is $\varphi(0° \leq \varphi \leq 360°)$. The pitch angle is $\theta(0° \leq \theta \leq 90°)$. The distance from the first microphone receiver element M_1 is r. $d_i(2 \leq i \leq 8)$ is the distance difference between the sound source to the position 2 of the first microphone element and the position of the sound source to the third element position 3. $d_i(2 \leq i \leq 8)$ is the distance difference between the position M_1 of the sound source to the first microphone element and the position M_i of the source to the ith element.

According to the geometric relationship between each receiver element and the position of the sound source:

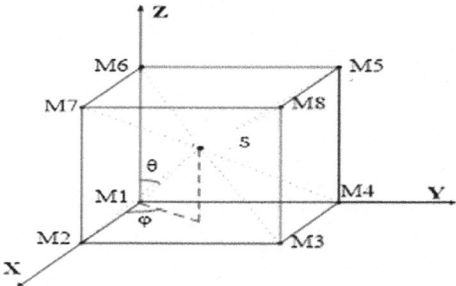

Fig. 1. Spatial distributed microphone array

$$\begin{cases} S(O)M_1 : x^2 + y^2 + z^2 = r^2 \\ SM_2 : (L-x)^2 + y^2 + z^2 = (r+d_2)^2 \\ SM_3 : (L-x)^2 + (L-y)^2 + z^2 = (r+d_3)^2 \\ SM_4 : x^2 + (L-y)^2 + z^2 = (r+d_4)^2 \\ SM_5 : x^2 + (L-y)^2 + (L-z)^2 = (r+d_5)^2 \\ SM_6 : x^2 + y^2 + (L-z)^2 = (r+d_6)^2 \\ SM_7 : (L-x)^2 + y^2 + (L-z)^2 = (r+d_7)^2 \\ SM_8 : (L-x)^2 + (L-y)^2 + (L-z)^2 = (r+d_8)^2 \end{cases} \tag{2.1}$$

From formula (2.1):

$$r = \frac{c \cdot \left[\sum_{i=2}^{8} \hat{\tau}_i^2 - 4(\hat{\tau}_2 + \hat{\tau}_4 + \hat{\tau}_6) \right]}{8(\hat{\tau}_2 + \hat{\tau}_4 + \hat{\tau}_6) - 2 \sum_{i=2}^{8} \hat{\tau}_i} \tag{2.2}$$

From formula (2.1) can obtain the elevation angle θ and the azimuth angle φ:

$$\begin{cases} \tan \theta = \dfrac{L^2 - 2rd_4 - d_4^2}{L^2 - 2rd_2 - d_2^2} \\ \tan \varphi = \dfrac{\sqrt{(L^2 - 2rd_2 - d_2^2)^2 + (L^2 - 2rd_4 - d_4^2)^2}}{L^2 - 2rd_6 - d_6^2} \end{cases} \tag{2.3}$$

From formula (2.3), $d_i = c\hat{\tau}_i$, c is the speed of sound. $\hat{\tau}_i = (2 \le i \le 6)$ represents an estimated value of the signal delay received by the first microphone element M_1 and the ith microphone element M_i.

Using the above method, the azimuth of the sound source and the tangent of the pitch angle are obtained. At the same time, since the arctangent function belongs to the transcendental function, it is calculated by the coordinate rotation numerical calculation method Coordinate Rotation Digital Computer (CORDIC). This method is a simple displacement and addition and subtraction operation, which will improve the accuracy of the iterative calculation.

The azimuth angle r obtained by formula (2) and the azimuth angle φ obtained by formula (3), and the elevation angle θ, the spherical coordinates can be obtained as (r, φ, θ), which can be expressed as (4):

$$\begin{cases} x = r \sin \varphi \cos \theta \\ y = r \sin \varphi \sin \theta \\ z = r \cos \varphi \end{cases} \tag{2.4}$$

Formula (2.4) can determine the position coordinates of the sound source.

3   Conclusion

It can be seen from above that in the case of indoor background noise and reverberation in the room, there is a certain deviation between the actual result and the simulation result. Compared with the traditional quaternary cross-indoor positioning effect, the indoor positioning effect of the spatial distributed microphone array has the characteristics of improved positioning accuracy.

Acknowledgments. This paper is supported by the Natural Science Foundation of China 61501176 and Natural Science Foundation of Heilongjiang province F2018025.

References

1. Dmochowski JP, Goubran RA. Decoupled beamforming and noise cancellation. IEEE Trans. Instrumentation Meas. 2007;56(1):80–8.
2. Talagala Dumidu S, Zhang W, Abhayapala TD. Broadband DOA estimation using sensor arrays on complex-shaped rigid bodies. IEEE Trans Audio, Speech Lang Processing. 2013;21 (8):1573–85.
3. Minotto VP, Jung CR, Lee B. Simultaneous-speaker activity detection and localization using mid-fusion of SVM and HMMs. IEEE Trans Multimed. 2014;16(4):1032–44.
4. Yong-guang C, Xiu-he L. Indoor positioning technology based on signal strength. Electron. J. 2010;7(9):1457–8.
5. Zuo-liang Y. Experimental research on MUSIC sound source location based on microphone array. Ha er bin: Harbin Institute of Technology, 2011.
6. Huan Y, Meng-rao Z, Xiao-qiang Z. Microphone array consistency analysis based on time delay estimation. Fudan J (Nat Sci Ed.), 2017;56(02):175–81.
7. Yi-bo Z, Wei J, Li-fu W. Speech signal endpoint detection method based on microphone array adaptive nonlinear filtering. Sci Bull. 2017;33(04):199–203.

A Research on the Improvement of Resource Allocation Algorithm for D2D Users

Yan-Jun Liang[1] and Hui Li[1,2(✉)]

[1] College of Information Science and Technology,
Hainan University, Haikou, China
1307431141@qq.com, hitlihui1112@163.com
[2] Marine Communication and Network Engineering Technology Research
Center of Hainan Province, Hainan University, Haikou, China

Abstract. In the 5G communication network, D2D technology is considered to be one of the important components of the future [1]. D2D users bring multiplexing gain to the channel resources of multiplexed cellular users, improve system communication capacity, and reduce communication delay and terminal performance [2]. D2D also brings considerable interference to the reuse of cellular user channel resources [3]. The research on the resource allocation algorithm in D2D communication system mainly aims to reduce the interference between D2D users and cellular users. The improved bilateral rejection algorithm based on the best algorithm improves the stability and system performance of the whole communication system to a large extent [4].

Keywords: Wireless communication · Resource allocation · D2D · Interference suppression · Bilateral rejection algorithm

1 Introduction

In the past research on network communication systems, D2D technology has become an important research direction in the 5G network framework due to its low latency, higher transmission rate, and higher spectrum utilization [5]. The D2D user brings multiplexing gain to the channel resources of the multiplexed cellular users, improves the system communication capacity, and reduces the communication delay and the power consumption of the terminal [6]. However, while D2D users access traditional cellular communication networks with the reuse of communication resources, the problem of inter-user interference is caused [7]. An effective resource management strategy will solve the interference problem, and it's an important way to make D2D users and traditional cellular users gain higher spectrum utilization and increase the capacity of communacation system [8]. This paper proposes a new scheme to satisfy the communication between D2D users and cellular devices sharing spectrum resources, and studies the optimization of D2D communication resource allocation algorithm under this model.

The main work of this paper includes: (1) Propose and build a new D2D into the existing cellular network communication system model. (2) Try an existing resource allocation algorithm for the model to satisfy the frequency reuse of D2D users and cellular users. (3) Improve the existing resource allocation algorithm, propose a new

© Springer Nature Singapore Pte Ltd. 2020
Q. Liang et al. (eds.), *Communications, Signal Processing, and Systems*, Lecture Notes in Electrical Engineering 516,
https://doi.org/10.1007/978-981-13-6504-1_172

bilateral elimination algorithm, and compare the simulation. (4) Simulation shows that the improved bilateral culling algorithm can effectively improve the stability of the system communication link and increase the system communication capacity (Fig 1).

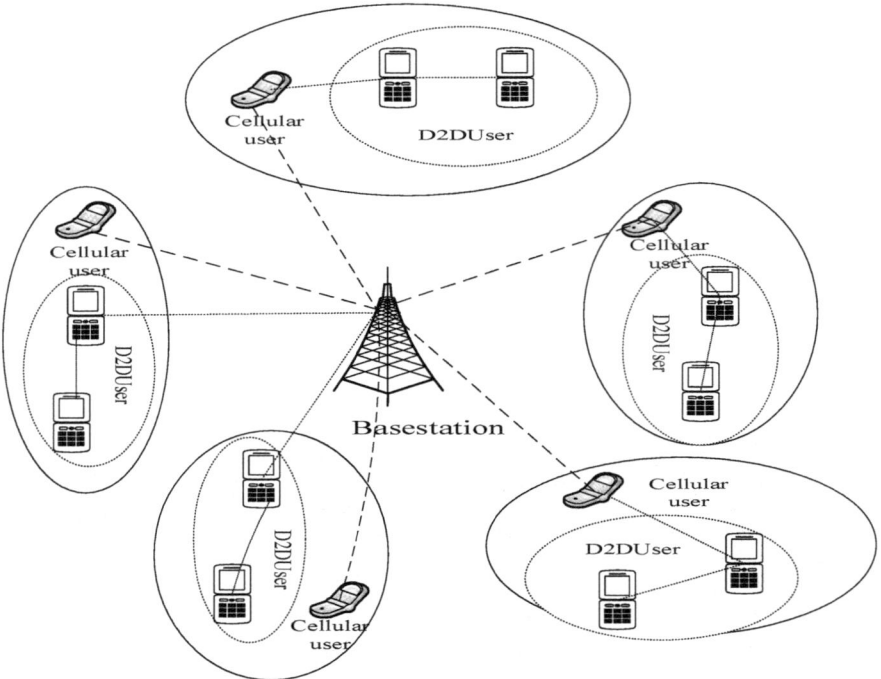

Fig. 1. Cell model

2 System Model

There are M cellular users and N D2D users in one cell, and each pair of D2D users includes one transmitting user D_n and receiving user $D_{n'}$, and when the nth pair of D2D users multiplex the channel resources of the mth cellular user, the D2D user interference occurs for the mth cellular user, and the base station in the system also interferes with the D2D receiving user [9]. The number of channel resources in the cell is K, because some D2D users in the cell multiplex cellular user channel resources, and some D2D users occupy dedicated channel resources, so the total channel resource quantity of the system is greater than the number of cellular users, and is smaller than the sum of the number of D2D users and the number of cellular users [10], that is, $N < M < K < N + M$.

The SINR of receiving part in the D2D pairing-user could be calculated by formula $\gamma_{n,m} = \frac{\alpha_{n,m} P_{n,m} G_{n,n'}}{I_{BS,n'} + N_0}$, $\alpha_{n,m}$ is the nth pair of D2D users multiplex the channel resources of the mth cellular user, $P_{n,m}$ represents the transmit power of the D2D pair of users transmitting the user D_n of the mth cellular user communication resource, $I_{BS,n'}$ is the transmission interference from the base station received by the D2D to the receiving user in the user, and N_0 is the received noise power of D_n, $G_{n,n'}$ is the link gain between the nth D2D user pair [11].

The signal-to-noise ratio (SINR) of cellular users is:

$$\gamma_m = \frac{P_{BS,m} G_{BS,m}}{I_{n,m} + N_0} \tag{1}$$

where $P_{BS,m}$ is the base station transmit power received by the mth cellular user, $G_{BS,m}$ is the link gain between the mth cellular users of the base station, and $I_{n,m}$ is the interference of the sender D_n from the nth pair D2D users to the cellular user. $I_{n,m} = P_{n,m}, G_{n,m}, G_{n,m}$ represents the link gain for the sender for the cellular user and the D2D user.

According to the Shannon formula, the communication capacity of the system is as follows:

$$C = \sum_{n=1}^{N} \sum_{m=1}^{M} \sum_{k=1}^{K} \log_2(1 + \text{SINR}_{n,m}^k) + \sum_{m=1}^{M} \sum_{k=1}^{K} \log_2(1 + \text{SINR}_m^k) \tag{2}$$

3   Algorithm Analysis

Using the appropriate resource allocation algorithm to improve the communication capacity of the communication system is the focus of this paper. Based on the above-mentioned system model, the related algorithms are proposed to realize the rational allocation of resources.

3.1   Game Playing Algorithm

Establish a game model and define the communication network power control model for cellular users and D2D users to coexist with users as follows:

$$G = (L, P, C) \tag{3}$$

L   Link set between communication users.
P   The set of user transmit powers participating in the game, values $[0, P_D^{\max}]$, $P_D^{\max}$ is the maximum transmit power of the D2D user.
C   Cost function set.

$$C_n(P_n, P_{-n}) = \left(\frac{G_{n,m} P_n}{I_{m,n} + N_0} - \text{SINR} \right)^2 + \beta_n P_n \tag{4}$$

P_n The transmit power policy value of the nth D2D user pairs.
P_{-n} Other users (except the nth pair D2D user pairs) interfere with the link's power policy value.

Under this model, each D2D user can choose its own optimal cost function to determine its own transmit power, and establish a minimum cost function for the communication link:

$$\min_{p_n \in [0, P_D^{\max}]} C_n(P_n, P_{-n}) \tag{5}$$

Then, the distributed iterative algorithm is used to find the Nash equilibrium point power that satisfies the objective function, and the partial derivative of the cost function is obtained.

$$\frac{\partial C_n}{\partial P_n} = \frac{2G_{n,m}}{I_{m,n} + N_0} \left(\frac{G_{n,m} P_n}{I_{m,n} + N_0} - \text{SINR} \right) + \beta_n \tag{6}$$

Let the formula be equal to 0 and we can get the follows:

$$P_n = \frac{I_{m,n} + N_0}{G_{n,m}} \left(\text{SINR}_{n,m} - \frac{\beta_n (I_{m,n} + N_0)}{2G_{n,m}} \right) \tag{7}$$

The best response function of the link l_n can be obtained by using the above formula.

$$P^* = f(P_n) \tag{8}$$

According to the given optimal power response function, the optimal power can be achieved by the distributed power allocation algorithm. We get the simplification of the above formula:

$$I_{m,n} + N_0 = I_n \tag{9}$$

$$P_n^{q+1} = \frac{I_{-n}^q}{G_{n,m}} \left(\text{SINR}_{n,m} - \frac{\beta_n I_{-n}^q}{2} \right) \tag{10}$$

In one cycle, all transmitting users should adjust the transmission power of the user by accepting the interference value and the best response function of the user feedback until the transmission power satisfies $\left| P_n^{q+1} - P_n^q \right| < \Delta_P$, and the task link converges to the Nash equilibrium state.

3.2   Optimal Algorithm

The process of multiplexing the mth cellular user for the nth D2D user is to be analyzed. During the D2D link establishment process, it is interfered by the cellular user and the signal-to-noise ratio that the D2D communication establishment condition needs to satisfy is:

$$\gamma_{n,m} = \frac{P_{n,m}H_n}{I_{m,n} + N_0} \geq \gamma_{th}^D \qquad (11)$$

$P_{n,m}$ is the transmit power of the D2D pair user transmitting user D_n multiplexed with the mth cellular user communication resource, H_n is the link gain between D_n and D_n, $I_{m,n}$ is the D2D for the user receiving user D_n The transmission interference from the cellular user, N_0 is the noise received by D_n. As we have set the threshold of SINR for the communication system,only when the receiving SINR of the receiving part is greater than the threshold that the quality of the D2D user can be guaranteed.

Then, the minimum transmission power of the sender of the D2D user pair can be calculated as:

$$P_{n,m} = \frac{\gamma_{th}^D(I_{m,n} + N_0)}{H_{n,n'}} \qquad (12)$$

In this communication system, cellular users have higher priority than D2D users. So base stations can determine whether D2D users can access cellular communication systems. To facilitate the algorithm calculation, a control factor θ is set to indicate whether the D2D user is allowed to multiplex the channel resources of the cellular user, and the value is 0/1. I_0 is set as the interference threshold of the base station. If $I_{n,m} \leq I_0$, the system allows the nth D2D users to multiplex the channel resources of the mth cellular user, $\theta = 1$, and vice versa,$\theta = 0$.

If N_1 D2D user pairs are unable to multiplex cellular user channel resources, when $N_1 \leq K + M$, these multiplexed D2D users can still access the communication system through dedicated communication channel resources. When $N_1 > K + M$, insufficient dedicated communication channel resources result in $N_1 - K + M$ D2D user pairs being completely excluded from the communication system.

The threshold for setting the interference of the D2D user to the base station is C_1, and the objective function for establishing the D2D communication link is:

$$\min_{P_{n,m}} \sum_{n=1}^{N_2} \sum_{m=1}^{K} p_{n,m}I_{n,m} \qquad (13)$$

If $\min_{p_{n,m}} \sum_{n=1}^{N_2} \sum_{m=1}^{K} p_{n,m}I_{n,m} > C_1$, then the final channel allocation matrix is calculated.

3.3 Bilateral Rejection Algorithm

Set the link gain between cellular users and D2D users as a fixed value. The user acceptance of noise power is also a fixed value. The transmission power of the user was determined by the interference value. Then the objective function of the maximum system capacity can be converted to the maximum interference value that the cellular user can withstand.

$$\min_{\beta_{n,m}^k} \sum_{n=1}^{N} \sum_{m=1}^{M} \sum_{k=1}^{K} \beta_{n,m}^k I_{m,n}^k + \sum_{n=1}^{N} \sum_{m=1}^{M} \sum_{k=1}^{K} I_{n,m}^k \qquad (14)$$

In the communication system, some D2D users multiplex cellular user communication resources, and another part of D2D users use dedicated communication resources. D2D users do not interfere with other cellular users when using dedicated communication resources, and are not interfered by other cellular users. The above formula can be turned into:

$$\min_{\beta_{n,m}^k} \sum_{n=1}^{N'} \sum_{m=1}^{M} \sum_{k=1}^{K} \beta_{n,m}^k I_{m,n}^k + \sum_{n=1}^{N'} \sum_{m=1}^{M} \sum_{k=1}^{K} I_{n,m}^k \qquad (15)$$

$$\begin{cases} \beta_{n,m}^k = \{0,1\}, \forall n, m, k; \\ \sum_{n=1}^{N'} \sum_{m=1}^{M} \beta_{n,m}^k = 1, \forall k; \\ \sum_{n=1}^{N'} \sum_{m=1}^{M} \beta_{n,m}^k = 1, \forall m; \\ \sum_{m=1}^{M} \sum_{k=1}^{K} \beta_{n,m}^k = 1, \forall n; \end{cases} \qquad (16)$$

Algorithm steps:

1. Setting parameters(number of communication resource blocks K, number of cellular users M, number of D2D users N, interference threshold of D2D users I_{th}^D, and interference threshold of cellular users I_{th}^c)
2. Calculating the interference matrix of the D2D user multiplexed cellular user communication resources $I_{N \times M}$ and the interference matrix of the D2D users of the multiplexed communication resources by the cellular users $I_{M \times N}$.
3. Calculating $I_{N \times M}$, we can get the minimum value $I_{n,m}^{\min}$, and if $I_{n,m}^{\min} C_{th}$, we find $I_{m,n}$ in $I_{M \times N}$, when $I_{m,n} D_{th}$, the BS will allocate resource block k to the cellular user m and the D2D user n. And then Excluding resource block k from K to indicate that the D2D user n and the cellular user m are no longer involved in the resource allocation progress. If it is not satisfied, the same operation is performed to the second smallest $I_{n,m}$ in the sorting, and so on to the end.
4. If there is a cellular user failing to allocate communication resources, then the resource blocks in K will be preferentially assigned to these cellular users, and after the cellular users are fully satisfied, the allocation of dedicated communication resources will be performed.

4   Simulation Analysis

We simulate and analyze the above optimal algorithm, game algorithm, and bilateral culling algorithm. To reduce the complexity of the algorithm, a base station is set in a cell, so that the D2D user pair and the cellular user are evenly distributed in the cell, and each cellular user is preset to occupy only one channel resource, and only one D2D user pair can be restored. Using a cellular user channel, the average is taken after multiple simulations (Fig. 2).

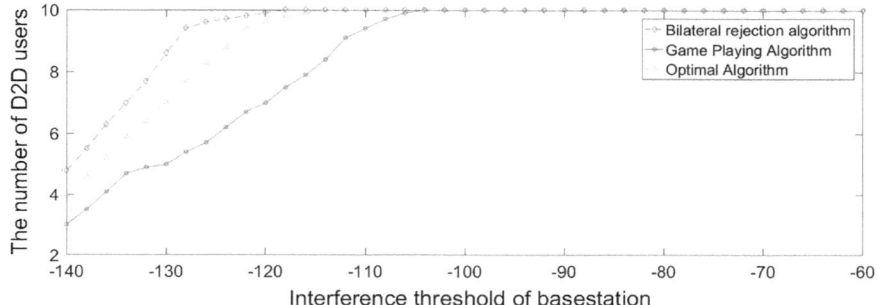

Fig. 2. Relationship between the interference threshold of a base station and the number of D2D users allowed to access

The figure above shows the relationship between the interference threshold of the base station and the number of D2D users allowed to communicate. It can be seen that as the interference threshold of the base station becomes larger, the number of D2D users that can be allowed to communicate is gradually increasing, but eventually an equilibrium value will be reached, i.e., the system will allow up to 10 pairs of D2D users to communicate. The simulation results can clearly see the comparison of the three algorithms. We name the improved algorithm proposed in this paper as the bilateral culling algorithm.It could reach the maximum allowable communication D2D user pair when the interference threshold is −120 dBm. The signal is reached at −115 dBm, and the game algorithm reaches the D2D user-to-communication upper limit at −108 dBm. From −140 dBm, the bilateral culling algorithm can allow five pairs of D2D users to communicate with each other, and the best algorithm and game. The algorithm can only do four pairs and three pairs.

As shown in Fig. 3, it can be seen that as the distance between users increases, the number of D2D users that can access the system will decrease. This is because the D2D user's interference value to the system will also vary with the D2D users. The maximum distance increases and increases, but the stability of the resource allocation strategy based on the double-quenching algorithm has obvious advantages. When the distance keep under 160m, the maximum number of accessers could be maintained at the number 10, and is exceeded. In the case of more than 160 m, the system still maintains considerable stability. In the case of 160 m, it still maintains considerable stability. In contrast, the communication link based on the best algorithm and the game algorithm shows considerable instability.

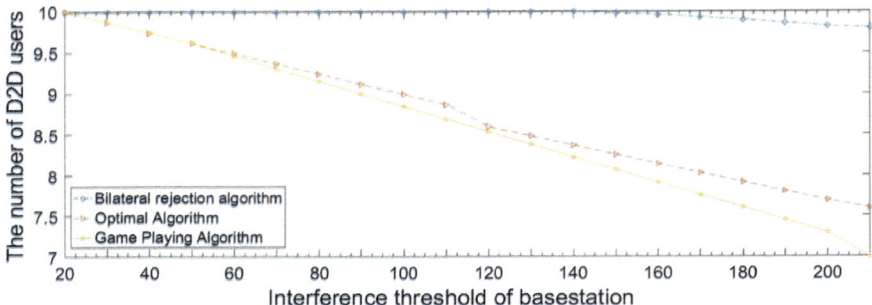

Fig. 3. Relationship between the maximum distance between D2D users and the number of D2D users allowed to access

5   Conclusions

In this paper, the D2D communication system based on cellular users is established firstly, and the optimal algorithm and the resource allocation method of the game algorithm in D2D communication are analyzed in detail. Finally, an improved method is proposed based on the optimal algorithm. The simulation results of the bilateral culling algorithm shows that the proposed bilateral culling algorithm could effectively reduce the complexity of algorithm, and it keeps the considerable advantage of the system communication link. The bilateral culling algorithmgreatly greatly increases the accessibility of D2D users to the cellular communication systems.

Acknowledgments. This work was supported by Hainan Provincial Key R. & D. Projects of China (ZDYF2016010 and ZDYF2018012) and the National Natural Science Foundation of China (No. 61661018).

References

1. Fodor G. Design aspects of network assisted device-to-device communications. IEEE Commun Mag. 2012;50(3):170–7.
2. Yu CH, Tirkkonen O, Doppler K, Ribeiro C. On the performance of device-to-device underlay communication with simple power control. In: Proceeding of IEEE Vehicular Technology Conference; 2009. p. 1–5.
3. Moltchanov D. Distance distributions in random networks. Ad Hoc Netw. 2012;10(6):1146–66.
4. Lin X, Andrews J. Optimal spectrum partition and mode selection in device-to-device overlaid cellular networks. In: Proceedings of IEEEGLOBECOM; 2013. p. 1837–42.
5. Lei L, Zhong Z, Lin C, Shen X. Operator controlled device to-device communications in LTE-advanced networks. IEEE Wireless Commun. 2012;19(3):96–104.
6. Xiao X, Tao X, Lu J. A QoS-aware power optimization scheme in OFDMA systems with integrated device-to-device (D2D) communications. In: Proceedings of IEEE Vehicular Technology Conference; 2011. p. 1–5.

7. Jindal N, Weber S, Andrews JG. Fractional power control for decentralized wireless networks. IEEE Trans Wireless Commun. 2008;7(12):5482–92.
8. Zhang X, Haenggi M. Random power control in Poisson networks. IEEE Trans Commun. 2012;60(9):2602–11.
9. Gu J, Bae SJ, Choi BG, Chung MY. Dynamic power control mechanism for interference coordination of device-to-device communication in cellular networks. In: Proceeding of 3rd International Conference on Ubiquitous and Future Networks; 2011. p. 71–5.
10. Corson M. Toward proximity-aware internetworking. IEEE. Wireless Commun. 2010;17 (6):26–33.
11. Janis P. Device-to-device communication underlaying cellular communications systems. Int J Commun Netw Syst Sci. 2009;2(3):169–78.

Author Index

© Springer Nature Singapore Pte Ltd. 2020
Q. Liang et al. (eds.), *Communications, Signal Processing,
and Systems*, Lecture Notes in Electrical Engineering 516,
https://doi.org/10.1007/978-981-13-6504-1